The Princeton Review®

AP® CALCULUS AB PREMIUM PREP

12th Edition

David S. Kahn

PrincetonReview.com

Penguin Random House

The Princeton Review
110 East 42nd Street, 7th Floor
New York, NY 10017
princetonreview.com
penguinrandomhouse.com

Published in the United States by Penguin Random House LLC, New York.

ISBN: 978-0-593-51821-2
ISSN: 2690-5299

The material in this book is up-to-date at the time of publication. However, changes may have been instituted by the testing body in the test after this book was published.

If there are any important late-breaking developments, changes, or corrections to the materials in this book, we will post that information online in the Student Tools. Register your book and check your Student Tools to see if there are any updates posted there.

Editor: Chris Chimera
Production Editors: Emma Parker and Kathy Carter
Production Artist: Deborah Weber

Manufactured in the United States of America

10 9 8 7 6 5 4 3 2 1

12th Edition

EU Contact:
Penguin Random House Ireland
32 Nassau Street
Dublin D02 YH68
https://eu-contact.penguin.ie

The Princeton Review Publishing Team
Rob Franek, Editor-in-Chief
David Soto, Senior Director, Data Operations
Stephen Koch, Senior Manager, Data Operations
Deborah Weber, Director of Production
Jason Ullmeyer, Production Design Manager
Jennifer Chapman, Senior Production Artist
Selena Coppock, Director of Editorial
Aaron Riccio, Director, Editorial Admissions Content
Orion McBean, Senior Editor
Meave Shelton, Senior Editor
Chris Chimera, Editor
Patricia Murphy, Editor
Laura Rose, Editor
Isabelle Appleton, Editorial Assistant

Penguin Random House Publishing Team
Tom Russell, VP, Publisher
Alison Stoltzfus, Senior Director, Publishing
Emily Hoffman, Managing Editor
Mary Ellen Owens, Assistant Director of Production
Suzanne Lee, Senior Designer
Eugenia Lo, Publishing Assistant

For customer service, please contact **editorialsupport@review.com**, and be sure to include:

- full title of the book
- ISBN
- page number

Acknowledgments

First of all, I would like to thank Arnold Feingold and Peter B. Kahn for once again doing every problem, reading every word, and otherwise lending their invaluable assistance. I also want to thank Chris Chimera for being a terrific editor on this edition of the book, Gary King for his first-rate reading, analysis, and contributions, and the production team of Liz Dacey, Kathy Carter, and Deborah Weber. Thanks to Frank, without whose advice I never would have taken this path. Thanks to Jeffrey, Miriam, and Vicki for moral support. Thanks Mom.

Finally, I would like to thank the people who really made all of this effort worthwhile—my students. I hope that I haven't omitted anyone, but if I have, the fault is entirely mine.

Aaron and Sasha, Aaron, Aaron T., Abby B., Abby F., Abby H., Abby L., Abbye, Abigail H., Addie, Aidan, Alan M., Albert G., Alec G, Alec M., Alec R., Alex and Claire, Alex A., Alex B., Alex and Daniela B., Alex F., Alex D., Alex G., Alex H., Alex I., Alex S., Alex and Gabe, Alexa, Alexandra, Natalie and Jason, Alexes, Alexis and Brittany, Ali and Jon, Ali and Amy Z., Ali H., Alice, Alice C., Alicia, Alisha, Allegra, Allie and Lauren, Allison and Andrew, Allison H., Allison and Matt S., Allison R., Ally, Ally M., Ally T., Ally W., Aly and Lauren, Alyssa C., Alyssa TDC, Alyssa and Courtney, Alvin, Amanda, Brittany, and Nick A., Amanda and Pamela B., Amanda C., Amanda E., Amanda H., Amanda M., Amanda R., Amanda S., Amanda Y., Amber and Teal, Amparo, Andrea T., Andrea V., Andrew A., Andrei, Andrew B., Andrew C., Andrew D., Andrew E., Andrew H., Andrew M., Andrew S., Andy and Sarah R., Andy and Allison, Angela, Angela F., Angela L., Anisha, Ann, Anna C-S., Anna D., Anna and Jon, Anna and Max, Anna L., Anna M., Anna S., Anna W., Annabelle M., Annerose, Annie, Annie B., Annie W., Antigone, Antonio, Anu, April, Ares, Ariadne, Ariane, Ariel, Arielle and Gabrielle, Arthur and Annie, Arya, Asheley and Freddy, Ashley A., Ashley K., Ashley and Sarah, Ashley and Lauren, Aubrey, Avra, Becca A., Becky B., Becky S., Becky H., Ben S., Ben and Andrew Y., Ben D., Ben S., Benjamin D., Benjamin H., Benji, Beth and Sarah, Bethany and Lesley, Betsy and Jon, Blythe, Bianca and Isabella, Biz, Bonnie, Bonnie C., Boris, Braedan, Brendan, Brett, Brett A., Brette and Josh, Brian, Brian C., Brian N., Brian W., Brian Z., Brigid, Brin, Brittany E., Brittany S., Brooke, Devon, and Megan, Brooke and Lindsay E., Butch, Cailyn H., Caitlin, Caitlin F., Caitlin M., Caitlin and Anna S., Camilla and Eloise, Camryn, Candace, Cara, Cara G., Carinna, Caroline A., Caroline C., Caroline H., Caroline S., Caroline and Peter W., Carrie M., Catherine W., Cecilia, Celia, Chad, Channing, Charlie, Charlotte, Charlotte B., Charlotte M., Chase, Chelsea, Chloe, Chloe K., Chris B., Chris C., Chris K., Chris P., Chrissie, Christa S., Christen C., Christian, Christian C., Christina F., Christine, Christine W., CJ, Claire H., Clare, Claudia, Clio, Colette, Conor G., Corey, Corinne, Coryn, Courtney and Keith, Courtney B., Courtney F., Courtney S., Courtney W., Craig, Dan M., Dana J., Dani and Adam, Daniel, Daniel and Jen, Daniella, Daniella C., Danielle and Andreas, Danielle and Nikki D., Danielle G., Danielle H., Danny K., Dara and Stacey, Dara M., Darcy, Darya, David B., David C., David R., David S., Deborah and Matthew, Deniz and Destine, Deshaun, Deval, Devin, Devon and Jenna, Deyshaun, Diana H., Dilly, Disha, Dong Yi, Dora, Dorrien, Eairinn, Eddy, Elana, Eleanor K., Elexa and Nicky, Elisa, Eliza, Elizabeth, Elizabeth and Mary C., Elizabeth F., Ella, Ellie, Elly B., Elyssa T., Emily A., Emily B., Emily and Allison, Emily and Catie A., Emily C., Emily G., Emily H., Emily K., Emily L., Emily and Pete M., Emily R., Emily R-H., Emily S., Emily T., Emma, Emma and Sophie, Emma D., Emma S., Eric N., Erica F., Erica H., Erica R., Erica S., Eric and Lauren, Erica and Annie, Erika, Erin, Erin I., Ernesto, Esther, Ethan, Eugenie, Eva, Evan, Eve H., Eve M., Fifi, Francesca, Frank, Franki, Gabby, Geoffrey, George and Julian., George M., Gloria, Grace M., Grace P., Gracie, Graham and Will, Greg, Greg F., Griffin, Gussie, Gylianne, Haley, Halle R., Hallie, Hannah C., Hannah J., Hannah R., Hannah and Paige, Harrison, Annabel and Gillian, Harry, Hayley and Tim D., Hazel, Heather D., Heather E., Heather F., Heather and Gillian, Hernando and Vicki, Hilary and Lindsay, Hilary F., Holli, Holly G., Holly K., Honor, Ian, Ian P., India, Ingrid, Ira, Isa, Isabel, Isabella and Simone, Isabelle T., Ismini, Isobel, Ivy, Jabari, Jacob and Kara, Jack, Jack B.,

Jackie and Vicky, Jackie, Jackie A., Jackie S., Jackson, Jaclyn and Adam, Jacob D., Jada, James S., Jamie, Jason P., Jason and Andrew, Jay, Jay K., Jae and Gideon, Jake T., Jarrod, Jason L., Jayne and Johnny, Jed D., Jed F., Jeffrey M., Jenna, Jen, Jen M., Jenn N., Jennifer B., Jennifer W., Jenny K, Jenny and Missy, Jeremy C., Jeremy and Caleb, Jess, Jess P., Jesse, Jessica, Jessica and Eric H., Jessica L., Jessica T., Jessie, Jessie C., Jessie and Perry N., Jill, Jillian, Daniel and Olivia, Jillian S., Jimin, Jimmy C., Jimmy P., Joanna, Joanna C., Joanna G., Joanna M., Joanna and Julia M., Joanna W., Jocelyn, Jody and Kim, Joe B., Johanna, John, John and Dan, Jonah and Zoe, Jonathan G., Jonathan P., Jonathan W., Jongwoo, Jordan C., Jordan and Blair F., Jordan G., Jordan P., Jordana, Jordyn, Jordyn K., Josh and Jesse, Josh and Noah, Judie and Rob, Julia and Caroline, Julia, Julia F., Julia G., Julia H., Julia and Charlotte P., Julia T., Julie, Julie H., Julie J., Julie and Dana, Julie P., Juliet, Juliette R., Justin L., Kara B., Kara O., Kasey, Kasia and Amy, Kat C., Kat R., Kate D., Kate F., Kate G., Kate L., Kate P., Kate S., Kate W., Katie C., Katie F., Katherine C., Kathryn, Leslie, and Travis, Katie, Katie M., Katrina, Keith M., Kelly C., Ken D., Kenzie, Keri, Kerri, Kim H., Kimberly, Kimberly H., Kirsten, Kitty and Alex, Kori, Kripali, Krista, Erika, and Karoline, Kristen, Kristen, Kristen Y., Magan, and Alexa, Laila and Olivia R., Laura B., Laura F., Laura G., Laura R., Laura S., Laura T., Laura Z., Lauren and Eric, Lauren R., Lauren S., Lauren T., Lauren and Allie, Lea K., Leah, Lee R., Leigh and Ruthie, Leigh, Lexi S., Lila, Lila M., Lilaj, Lili C., Lilla, Lily, Lily M-R., LilyHayes, Lisbeth and Charlotte, Lina, Lindsay F., Lindsay K., Lindsay L., Lindsay N., Lindsay R, Lindsay and Jessie S., Lindsey and Kari, Lisa, Liz D., Liz H., Liz M., Lizzi B., Lizzie A., Lizzie M., Lizzie W., Lizzy, Lizzy C., Lizzy R., Lizzy T., Louis, Lucas, Luke C., Lucinda, Lucy, Lucy D., Mackenzie, Maddie and JD, Maddie H., Maddie P., Maddy N., Maddy T., Maddy W., Madeline, Magnolia, Maggie L., Maggie T., Manuela, Mara and Steffie, Marcia, Margaret S., Maria B., Mariel C., Mariel L., Mariel S., Marielle K., Marielle S., Marietta, Marisa B., Marisa C., Marissa, Marlee, Marnie and Sam, Mary M., Matt, Matthew B., Matthew K., Matt F., Matt G., Matt and Caroline, Matt V., Matthew G., Mavis H., Max B., Max and Chloe K., Max M., Maxx, Ben, and Sam, Maya and Rohit, Maya N., Mckyla, Meesh, Megan G., Meliha, Melissa, Melissa and Ashley I., Meredith, Meredith and Gordie B., Meredith and Katie D., Meredith R., Meredith S-K., Michael and Dan, Michael H., Michael R., Michaela, Michal, Michele S., Mike C., Mike R., Miles, Milton, Miranda, Moira, Molly L., Molly R., Molly and Annie, Morgan, Morgan C., Morgan, Zoe, and Lila, Morgan K., Nadia and Alexis, NaEun, Nancy, Nanette, Naomi, Natalie and Andrew B., Natasha and Mikaela B., Natasha L., Nathania, Nico, Nick, Nick A., Nicky S., Nicole B., Nicole J., Nicole N., Nicole P., Nicole S., Nicolette, Nidhi, Nikki G., Nina R., Nina V., Noah, Nora, Nushien, Odean, Ogechi, Oli and Arni, Oliver, Oliver W., Olivia and Daphne, Olivia P., Omar, Oren, Pam, Paige, Pasan, Paul A., Peter, Philip and Peter, Phoebe, Pierce, Pooja, Priscilla, Quanquan, Quinn and Chris, Rachel A, Rachel B., Rachel and Adam, Rachel H., Rachel I., Rachel and Jake, Rachel K., Rachel L., Rachel and Mitchell M., Rachel S., Rachel and Eli, Rachel and Steven, Rachel T., Rachel W., Rae, Ramit, Randi and Samantha, Rasheeda, Ray B., Rayna, Rebecca A., Rebecca B., Rebecca G., Rebecca and Gaby K., Rebecca R., Rebecca W., Renee F., Resala, Richard, Tina, and Alice, Richie D., Ricky, Riley C., Rob L., Robin P., Romanah, Rose, Ryan, Ryan B., Ryan S., Saahil, Sabrina and Arianna, Sabrina O., Sally, Sam B., Sam C., Sam G., Sam L., Sam R., Sam S., Sam W., Sam Z., Samantha and Savannah, Samantha M., Samuel C., Samar, Sara G., Sara H., Sara L., Sara R., Sara S., Sara W., Sarah, Patty, and Kat, Sarah and Beth, Sarah A., Sarah C., Sarah D., Sarah F., Sarah G., Sarah and Michelle K., Sarah L., Sarah M-D., Sarah and Michael, Sarah P., Sarah S., Sascha, Samara, Saya, Sean, Selena G., Serena, Seung Woo, Shadae, Shannon, Shayne, Shayne, Shorty, Siegfried, Simon W., Simone, Sinead, Siobhan, Skye F., Skye L., Sofia G., Sofia M., Sonja, Sonja and Talya, Sophia A., Sophia B., Sophia J., Sophia S., Sophie and Tess, Sophie D., Sophie S., Sophie W., Stacey, Stacy, Steffi, Steph, Stephanie, Stephanie K., Stephanie L., Stephen C., Sumair, Sunaina, Suzie, Syd, Sydney, Sydney S., Tammy and Hayley, Tara and Max, Taryn, Tasnia, Tayler, Taylor, Tenley and Galen, Terrence, Tess, Theodora, Thomas B., Timothy H., Tina, Tita, Tom, Tori, Torri, Tracy, Tracy K., Tripp H., Tripp W., Tyler, Tyrik, Uma, Vana, Vanessa and John, Veda, Veronica M., Vicky B., Victor Z., Victoria, Victoria H., Victoria J., Victoria M., Vinny and Eric, Vivek, Vladimir, Waleed, Will A., Will, Teddy, and Ellie, William M., William Z., Wyatt and Ryan, Wyna, Xianyuan, Yakir, Yesha, Yujin, Yuou, Zach, Zach B., Zach S., Zach Y., Zachary N., Zaria, Zoë, Zoe L., and Zoey and Rachel.

Contents

at **PrincetonReview.com/prep**

As easy as *1•2•3*

1 Go to PrincetonReview.com/prep or scan the **QR code** and enter the following ISBN to register your book:

9780593518212

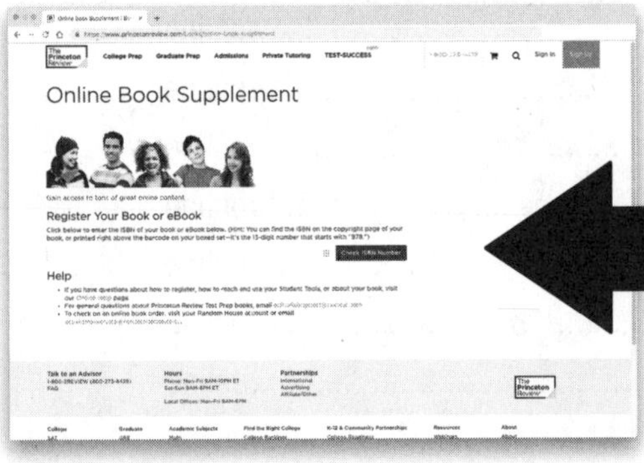

2 Answer a few simple questions to set up an exclusive Princeton Review account. *(If you already have one, you can just log in.)*

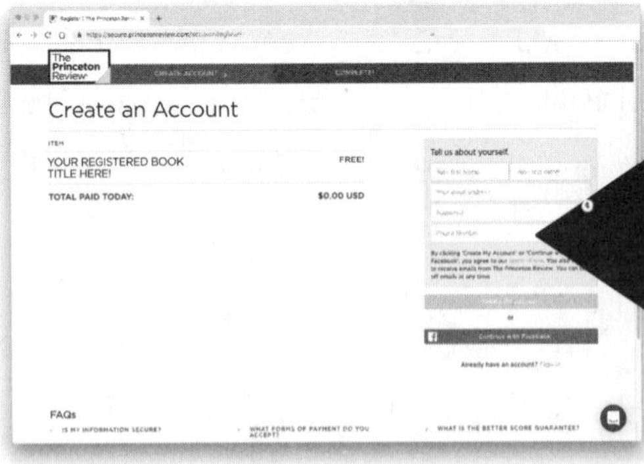

3 Enjoy access to your **FREE** content!

Access Your Online Test Practice

Your AP Prep book comes with new interactive practice exams to help you prepare for digital test-taking! Find these tests in your online Student Tools, provided in at least one of the following formats:

- Digital versions with a timer option to simulate the hybrid exam experience
- Downloadable interactive PDFs with digital features like clicking your answer in Section I

Check back often as we continue to update the included AP Student Tools.

PLUS, IN YOUR ACCOUNT YOU CAN:

- Get valuable advice about the college application process, tips for essay writing, and financial aid info
- Use our searchable rankings to learn more about your dream school
- Access additional full-length practice AP Calculus AB tests, comprehensive study guides, bonus questions, a calculator appendix, answer bubble sheets, and important formulas
- Check whether there have been any updates or corrections to this edition

Need to report a potential **content** issue?

Contact **EditorialSupport@review.com** and include:

- full title of the book
- ISBN
- page number

Need to report a **technical** issue?

Contact **TPRStudentTech@review.com** and provide:

- your full name
- email address used to register the book
- full book title and ISBN
- Operating system (Mac/PC) and browser (Chrome, Firefox, Safari, etc.)

Look For These Icons Throughout The Book

- ONLINE ARTICLES
- PROVEN TECHNIQUES
- APPLIED STRATEGIES
- MORE GREAT BOOKS
- STUDY BREAK
- GOING DEEPER

Part I
Using This Book to Improve Your AP Score

- Preview: Your Knowledge, Your Expectations
- Your Guide to Using This Book
- How to Begin

PREVIEW: YOUR KNOWLEDGE, YOUR EXPECTATIONS

Your route to a high score on the AP Calculus AB Exam depends a lot on how you plan to use this book. Start thinking about your plan by responding to the following questions.

1. Rate your level of confidence about your knowledge of the content tested by the AP Calculus AB Exam:

   A. Very confident—I know it all
   B. I'm pretty confident, but there are topics for which I could use help
   C. Not confident—I need quite a bit of support
   D. I'm not sure

2. If you have a goal score in mind, circle your goal score for the AP Calculus AB Exam:

   5 4 3 2 1 I'm not sure yet

3. What do you expect to learn from this book? Circle all that apply to you.

   A. A general overview of the test and what to expect
   B. Strategies for how to approach the test
   C. The content tested by this exam
   D. I'm not sure yet

YOUR GUIDE TO USING THIS BOOK

This book is organized to provide as much—or as little—support as you need, so you can use this book in whatever way will be most helpful for improving your score on the AP Calculus AB Exam.

- The remainder of **Part I** will provide guidance on how to use this book and help you determine your strengths and weaknesses.

- **Part II** of this book contains Practice Test 1, the Diagnostic Answer Key, answers and explanations for each question, and a scoring guide. (Bubble sheets can be found in the back of this book and can also be printed from your free online Student Tools for easy use.) We strongly recommend that you take this test before going any further, in order to realistically determine:
  - your starting point right now
  - which question types you're ready for and which you might need to practice
  - which content topics you are familiar with and which you should carefully review

  Once you have nailed down your strengths and weaknesses with regard to this exam, you can focus your test preparation, build a study plan, and be efficient with your time. Our Diagnostic Answer Key will assist you with this process.

- **Part III** of this book will:
  - provide information about the structure, scoring, and content of the AP Calculus AB Exam
  - help you to make a study plan
  - point you toward additional resources

- **Part IV** of this book will explore the following strategies:
  - how to attack multiple-choice questions
  - how to write a high-scoring free-response answer
  - how to manage your time to maximize the number of points available to you

- **Part V** of this book covers the content you need for your exam.

- **Part VI** of this book contains two additional practice tests, along with their answers, explanations, and scoring guides. If you skipped Practice Test 1, we recommend that you take it in conjunction with the remaining tests (with at least a day or two between them) to chart your progress and identify lingering issues: if you get a certain type of question wrong every time, you probably need to review it. If you got it wrong only once, you may have run out of time or been distracted by something. In either case, this will allow you to focus on the factors that caused the discrepancy and to be as prepared as possible on the day of the test.

- **Online Resources** contain four additional practice tests. Follow the study guide found there based on the amount of time you have to study for the exam.

You may choose to use some parts of this book over others, or you may work through the entire book. This will depend on your needs and how much time you have. Let's now look at how to make this determination.

Once you register your book online using the QR code below, you can download and print the bubble sheets for all eight practice tests. You can also find a handy Appendix that contains all the formulas you should know.

HOW TO BEGIN

1. **Take a Test**

   Before you can decide how to use this book, you need to take a practice test. Doing so will give you insight into your strengths and weaknesses, and the test will also help you make an effective study plan. If you're feeling test-phobic, remind yourself that a practice test is a tool for diagnosing yourself—it's not how well you do that matters but how you use information gleaned from your performance to guide your preparation.

   So, before you read further, take the AP Calculus AB Practice Test 1 starting on page 7 of this book. Be sure to do so in one sitting, following the instructions that appear before the test.

Scoring Worksheets
We've included a scoring worksheet for each of the Practice Tests. Remember that these worksheets are meant to serve as a rough guideline only. AP exam scores are weighted according to a statistical process that varies slightly every year based on how students perform on the exam. But you can use the worksheets to approximate your score!

2. Check Your Answers

Using the Diagnostic Answer Key on page 42, follow our three-step process to identify your strengths and weaknesses with regard to the tested topics. This will help you determine which content review chapters to prioritize when studying this book. Don't worry about the explanations for now, and don't worry about missed questions. We'll get to that soon.

3. Reflect on the Test

After you take your first test, respond to the following questions:

- How much time did you spend on the multiple-choice questions?
- How much time did you spend on each free-response question?
- How many multiple-choice questions did you miss?
- Do you feel you had the knowledge to address the subject matter of the free-response questions?
- Do you feel your free responses were well organized and thoughtful?
- Circle the content areas that were most challenging for you and draw a line through the ones in which you felt confident/did well.
    - Limits and Continuity
    - Basic Derivative Rules
    - Composite, Implicit, and Inverse Functions
    - Contextual Applications of Differentiation
    - Analytical Applications of Differentiation
    - Integration and Accumulation of Change
    - Differential Equations
    - Applications of Integration

4. Read Part III and Complete the Self-Evaluation

Part III will provide information on how the test is structured and scored. It will also set out areas of content that are tested.

As you read Part III, reevaluate your answers to the questions above. At the end of Part III, you will revisit the questions above and refine your answers to them. You will then be able to make a study plan, based on your needs and time available, that will allow you to use this book most effectively.

5. **Engage with Parts IV and V as Needed**

Notice the word *engage*. You'll get more out of this book if you use it intentionally than if you read it passively, hoping for an improved score through osmosis.

Strategy chapters will help you think about your approach to the question types on this exam. Part IV will open with a reminder to think about how you approach questions now and then close with a reflection section asking you to think about how/whether you will change your approach in the future.

Content chapters in Part V are designed to provide a review of the content tested on the AP Calculus AB Exam, including the level of detail you need to know and how the content is tested. You will have the opportunity to assess your grasp of the content of each chapter through test-appropriate questions and a reflection section.

6. **Take Another Test and Assess Your Performance**

Once you feel you have developed the strategies you need and have gained the knowledge you lacked, you should take Test 2, which starts on page TBD of this book. You should do so in one sitting, following the instructions at the beginning of the test.

When you complete the test, check your answers to the multiple-choice section with the correct responses on page TBD. Then, see whether a teacher will read your essays and provide feedback for your improvement.

Once you have taken the test, reflect on what areas you still need to work on, and revisit the chapters in this book that address those topics. Part VI of this book includes practice tests 2 and 3, and five more tests can be found online in your Student Tools. Finally, don't forget to consult your class materials (e.g., textbook, handouts) as well.

7. **Keep Working**

As discussed in Part III, there are other resources available to you, including a wealth of information on the official AP Students website. Here, you can continue to explore areas to improve upon and engage with right up until the day of the test. If you've followed our recommended approach, you'll have a total of eight practice tests (three in the book, five online), and you should use a mix of Web resources and book review to solidify your understanding of any questions or subjects that you keep getting wrong.

Use the QR code below to access four more online practice tests in your Student Tools so you can be extra prepared!

Another Course? Of Course!

If you can't get enough AP Calculus AB and want to review this material with an expert, we also offer an online Cram Course that you can sign up for here: www.princetonreview.com/ap-cram-course or by using the QR code below.

Part II
Practice Test 1

- Practice Test 1
- Practice Test 1: Diagnostic Answer Key and Explanations
- Practice Test 1: Free-Response Questions Grading Rubrics
- How to Score Practice Test 1

Practice Test 1

AP® Calculus AB Exam

SECTION I, PART A: Multiple Choice

At a Glance

Total Time
1 hour
Number of Questions
30
Percent of Total Grade
$33\,^1/_3$%

DISCLAIMER: The official AP Calculus AB will be administered in a hybrid digital format. Instructions for the digital exam may differ from this practice test.

Instructions

Section I, Part A, of this examination contains 30 multiple-choice questions. Fill in only the ovals for numbers 1 through 30 on your answer sheet.

No calculator is allowed for this part of the exam.

Indicate all of your answers to the multiple-choice questions on the answer sheet. Give only one answer to each question.

Use your time effectively, working as quickly as you can without losing accuracy. Do not spend too much time on any one question. Go on to other questions and come back to the ones you have not answered if you have time. It is not expected that everyone will know the answers to all the multiple-choice questions.

Your total score on the multiple-choice section is based only on the number of questions answered correctly. Points are not deducted for incorrect answers or unanswered questions.

SECTION I, PART B: Multiple Choice

At a Glance

Total Time
45 minutes
Number of Questions
15
Percent of Total Grade
$16\,^2/_3$%

DISCLAIMER: The official AP Calculus AB will be administered in a hybrid digital format. Instructions for the digital exam may differ from this practice test.

Instructions

Section I, Part B, of this examination contains 15 multiple-choice questions. Fill in only the ovals for numbers 31 through 45 on your answer sheet.

A graphing calculator is required for some questions on this part of the exam. You may use a handheld calculator or a Desmos graphing calculator application. **Make sure your calculator is in radian mode.**

Indicate all of your answers to the multiple-choice questions on the answer sheet. Give only one answer to each question.

Use your time effectively, working as quickly as you can without losing accuracy. Do not spend too much time on any one question. Go on to other questions and come back to the ones you have not answered if you have time. It is not expected that everyone will know the answers to all the multiple-choice questions.

Your total score on the multiple-choice section is based only on the number of questions answered correctly. Points are not deducted for incorrect answers or unanswered questions.

AP CALCULUS AB

SECTION I, Part A

Time—60 Minutes

Number of questions—30

No calculator is allowed for this part of the exam.

Directions: Solve each of the following problems, using the available paper for scratchwork. After examining the choices, select the best of the choices given.

In this test: Unless otherwise specified, the domain of a function f is assumed to be the set of all real numbers x for which $f(x)$ is a real number. The inverse of a trigonometric function f may be indicated using the inverse function notation f^{-1} or with the prefix "arc" (e.g., $\sin^{-1} x = \arcsin x$).

1 Mark for Review

On what interval(s) is the graph of $y = x - x^3$ concave up?

(A) $(-\infty, 0)$

(B) $(0, \infty)$

(C) $\left(-\infty, -\sqrt{\frac{1}{3}}\right) \cup \left(\sqrt{\frac{1}{3}}, \infty\right)$

(D) $\left(-\sqrt{\frac{1}{3}}, \sqrt{\frac{1}{3}}\right)$

GO ON TO THE NEXT PAGE.

2 Mark for Review

The Intermediate Value Theorem guarantees a value c, such that $f(c) = 0$ for $f(x) = x^3 + x - 3$ on which of the following intervals?

Ⓐ (–1, 0)

Ⓑ (0, 1)

Ⓒ (1, 2)

Ⓓ (2, 3)

---

3 Mark for Review

$\lim\limits_{x \to 2^-} \dfrac{9}{x-2} =$

Ⓐ $-\infty$

Ⓑ 0

Ⓒ 9

Ⓓ The limit does not exist.

---

4 Mark for Review

Evaluate $\int \sin^3 x \cos x \, dx$.

Ⓐ $\dfrac{\sin^2 x}{2} + C$

Ⓑ $\dfrac{\sin^3 x}{3} + C$

Ⓒ $\dfrac{\sin^4 x \cos^2 x}{8} + C$

Ⓓ $\dfrac{\sin^4 x}{4} + C$

GO ON TO THE NEXT PAGE.

5 Mark for Review

Find the average value of $f(x) = 2e^{2x}$ on the interval [1, 5].

Ⓐ $\frac{1}{2}\left(e^{10} - e^{2}\right)$

Ⓑ $\frac{1}{4}\left(e^{10} - e^{2}\right)$

Ⓒ $\frac{1}{4}\left(e^{8}\right)$

Ⓓ $\frac{1}{2}\left(e^{8}\right)$

---

6 Mark for Review

A box with a square base, rectangular sides, and no top is to have a volume of 108 feet. What are the dimensions of the rectangular sides that minimize the box's surface area?

Ⓐ 6 feet by 12 feet

Ⓑ 12 feet by 12 feet

Ⓒ 12 feet by 3 feet

Ⓓ 6 feet by 3 feet

GO ON TO THE NEXT PAGE.

7 Mark for Review

Given $f(x)$ below, for what values of a and b is $f(x)$ differentiable for all values of x?

$$f(x)=\begin{cases} x^{\frac{5}{3}}+b; & x<1 \\ ax^{\frac{4}{3}}; & x\geq 1 \end{cases}$$

Ⓐ $a=\frac{5}{4}$ and $b=\frac{1}{4}$

Ⓑ $a=\frac{5}{4}$ and $b=\frac{9}{4}$

Ⓒ $a=\frac{4}{5}$ and $b=-\frac{1}{5}$

Ⓓ $a=\frac{4}{5}$ and $b=\frac{9}{5}$

---

8 Mark for Review

Find $f'(x)$ if $f(x)=\frac{\sec x}{\sqrt{x}}$.

Ⓐ $\frac{\sec x\tan x}{\left(\frac{1}{2\sqrt{x}}\right)}$

Ⓑ $\frac{\csc x}{\left(\frac{1}{2\sqrt{x}}\right)}$

Ⓒ $\frac{\sec x(2x\tan x+1)}{2x\sqrt{x}}$

Ⓓ $\frac{\sec x(2x\tan x-1)}{2x\sqrt{x}}$

GO ON TO THE NEXT PAGE.

9 Mark for Review

Evaluate $\int x\left(\sqrt[3]{x+8}\right)dx$.

Ⓐ $\frac{3}{4}(x+8)^{\frac{4}{3}}+C$

Ⓑ $\frac{4}{3}(x+8)^{\frac{4}{3}}+C$

Ⓒ $\frac{3}{7}(x+8)^{\frac{7}{3}}-6(x+8)^{\frac{4}{3}}+C$

Ⓓ $\frac{3}{7}(x+8)^{\frac{7}{3}}+6(x+8)^{\frac{4}{3}}+C$

---

10 Mark for Review

Evaluate the limit $\lim_{x\to 0}\frac{\sin^2(3x)}{x}$.

Ⓐ 0

Ⓑ 1

Ⓒ 3

Ⓓ ∞

---

11 Mark for Review

A shoe manufacturer's profit can be found by the function $P(x) = -x^3 + 6x^2 + 125$, $(x > 0)$, where x is the number of shoes sold (in thousands). The maximum profit for the manufacturer will be when it sells how many shoes?

Ⓐ 0

Ⓑ 1000

Ⓒ 4000

Ⓓ 12,000

GO ON TO THE NEXT PAGE.

12 Mark for Review

Find $f'(2)$ if $f(x) = x^2 e^{x^2}$.

Ⓐ $4e^4$

Ⓑ $8e^4$

Ⓒ $20e^4$

Ⓓ $32e^4$

---

13 Mark for Review

The velocity of a particle at certain times is given in the table below. Approximate the total distance traveled on the interval [0, 3].

Time (hours)	Velocity (mph)
0	12
1	18
2	22
3	24

Ⓐ 19

Ⓑ 20

Ⓒ 29

Ⓓ 58

GO ON TO THE NEXT PAGE.

14 Mark for Review

What value of k makes $f(x)$ continuous for all values of x?

$$f(x)=\begin{cases}-x^2+2kx+66; & x>3\\ 4kx+x^3; & x\leq 3\end{cases}$$

(A) 0

(B) 3

(C) 5

(D) There is no value of k for which $f(x)$ is continuous for all values of x.

---

15 Mark for Review

$$\lim_{x\to 8}\frac{x^2-6x-16}{x^2-2x-48}=$$

(A) 0

(B) $\frac{1}{2}$

(C) $\frac{5}{7}$

(D) The limit does not exist.

GO ON TO THE NEXT PAGE.

16 Mark for Review

Find $\frac{d}{dx}\left[f^{-1}(x)\right]$ at $x = 2$ if $f(x)=\frac{x^3-17}{5}$.

(A) $\frac{5}{27}$

(B) 1

(C) $\frac{12}{5}$

(D) $\frac{27}{2}$

17 Mark for Review

Find the absolute maximum of $y = 6x^4 - 3x^2 - 5$ on the interval $[-1, 2]$.

(A) –5

(B) $\frac{43}{8}$

(C) 2

(D) 79

18 Mark for Review

Find $\frac{dy}{dx}$ if $y = (\sec(3x))(\ln(5x))$.

(A) $(\sec(3x))\left(\frac{5}{x}\right)+(\sec(3x)\tan(3x))(\ln(5x))$

(B) $(\sec(3x))\left(\frac{1}{5x}\right)+3(\sec(3x)\tan(3x))(\ln(5x))$

(C) $(\sec(3x))\left(\frac{5}{x}\right)+3(\sec(3x)\tan(3x))(\ln(5x))$

(D) $(\sec(3x))\left(\frac{1}{x}\right)+3(\sec(3x)\tan(3x))(\ln(5x))$

GO ON TO THE NEXT PAGE.

19 Mark for Review

$\lim_{\theta \to 0} \frac{\tan 7\theta}{\theta} =$

Ⓐ 0

Ⓑ 1

Ⓒ 7

Ⓓ The limit does not exist.

---

20 Mark for Review

Find $\frac{dy}{dx}$ if $y = \cos^5(1 - x^2)$.

Ⓐ $10x\cos^4(1 - x^2)\sin(1 - x^2)$

Ⓑ $-10x\cos^4(1 - x^2)\sin(1 - x^2)$

Ⓒ $5x\cos^4(1 - x^2)$

Ⓓ $10x\cos^4(1 - x^2)$

---

21 Mark for Review

Find $\frac{dy}{dx}$ if $y = \frac{5x^4 - 3x^3 + x^2}{x^3}$.

Ⓐ $\frac{20x^3 - 9x^2 + 2x}{x^6}$

Ⓑ $5 + \frac{1}{x^2}$

Ⓒ $\frac{20x^3 - 9x^2 + 2x}{3x^2}$

Ⓓ $5 - \frac{1}{x^2}$

GO ON TO THE NEXT PAGE.

22 Mark for Review

Find $\frac{dy}{dx}$ if $y = \arcsin\left(4\sqrt{x}\right)$.

(A) $\frac{2}{x - 16x^2}$

(B) $\frac{1}{4\sqrt{x - 16x^2}}$

(C) $\frac{4}{x - x^2}$

(D) $\frac{2}{\sqrt{x - 16x^2}}$

---

23 Mark for Review

Evaluate $\lim_{x \to 0} \frac{x2^x}{2^x - 1}$.

(A) 0

(B) $\frac{1}{\ln 2}$

(C) $\frac{1}{2}$

(D) The limit does not exist.

GO ON TO THE NEXT PAGE.

24 Mark for Review

Find $\frac{dy}{dx}$ at the point (2, 1) if $3x^2 + 2xy^2 - 3y^2 = 13$.

Ⓐ $\frac{dy}{dx} = -14$

Ⓑ $\frac{dy}{dx} = -7$

Ⓒ $\frac{dy}{dx} = 7$

Ⓓ $\frac{dy}{dx} = 14$

---

25 Mark for Review

Find $y(-2)$ if $\frac{dy}{dt} = -4y$ and $y(0) = 12$.

Ⓐ 12

Ⓑ $12e^{-8}$

Ⓒ $12e^{8}$

Ⓓ $12e^{16}$

---

26 Mark for Review

Find $\frac{dy}{dx}$ if $y = \frac{\sin 2x}{\sin x} - \frac{\cos 2x}{\cos x}$.

Ⓐ $\sec x \tan x$

Ⓑ $2\sec x \tan x$

Ⓒ $4\sec^2 x$

Ⓓ $4\sec^2 x \tan x$

GO ON TO THE NEXT PAGE.

27 Mark for Review

Evaluate $\lim_{x \to 0} \frac{2\tan(3x)}{x}$.

Ⓐ 0

Ⓑ 2

Ⓒ 6

Ⓓ The limit does not exist.

---

28 Mark for Review

The average value of $f(x) = \frac{1}{x}$ from $x = 1$ to $x = e$ is

Ⓐ $\frac{1}{e+1}$

Ⓑ $\frac{1}{1-e}$

Ⓒ $e - 1$

Ⓓ $\frac{1}{e-1}$

---

29 Mark for Review

Which of the following is a solution to the differential equation $\frac{d^2y}{dx^2} = 4y$?

Ⓐ $y = e^{2x}$

Ⓑ $y = \sin 2x$

Ⓒ $y = \cos 2x$

Ⓓ $y = x^4$

GO ON TO THE NEXT PAGE.

30 Mark for Review

If the position of a particle is given by the function $s(t) = 2t^3 - 21t^2 + 60t - 42$, $t \geq 0$, for what value(s) of t is the particle changing direction?

(A) $t = 2$

(B) $t = 5$

(C) $t = 2, 5$

(D) The particle does not change direction.

END OF PART A, SECTION I

IF YOU FINISH BEFORE TIME IS CALLED, YOU MAY CHECK YOUR WORK ON PART A ONLY.

DO NOT GO ON TO PART B UNTIL YOU ARE TOLD TO DO SO.

AP CALCULUS AB

SECTION I, Part B

Time—45 Minutes

Number of questions—15

A graphing calculator is required for some questions on this part of the exam. You may use a handheld calculator or a Desmos graphing calculator application. **Make sure your calculator is in radian mode.**

Directions: Solve each of the following problems, using the available paper for scratchwork. After examining the choices, select the best of the choices given.

In this test:

1. The **exact** numerical value of the correct answer does not always appear among the choices given. When this happens, select from among the choices the number that best approximates the exact numerical value.

2. Unless otherwise specified, the domain of a function f is assumed to be the set of all real numbers x for which $f(x)$ is a real number. The inverse of a trigonometric function f may be indicated using the inverse function notation f^{-1} or with the prefix "arc" (e.g., $\sin^{-1} x = \arcsin x$).

---

31 Mark for Review

Find $\frac{d}{dx}\int_{5}^{\sqrt{x}} \cos\left(t^2\right) dt$.

(A) $\frac{\cos x}{2\sqrt{x}}$

(B) $\cos x$

(C) $-\sin x$

(D) $-\frac{\sin x}{2\sqrt{x}}$

GO ON TO THE NEXT PAGE.

32 Mark for Review

If $g(x)=\int_{0}^{x}(t^3-t^2+1)\,dt$, find $g(2)$.

(A) 4

(B) 5

(C) $\frac{10}{3}$

(D) $\frac{13}{3}$

---

33 Mark for Review

Evaluate $\int_{0}^{1}\frac{12dx}{1+36x^2}$.

(A) arctan 6

(B) 2arctan 6

(C) 6arctan 6

(D) 12arctan 6

---

34 Mark for Review

If $\int_{-2}^{2}f(x)dx=8$, $\int_{2}^{5}f(x)dx=-16$, and $\int_{10}^{5}f(x)dx=18$, what is $\int_{-2}^{10}f(x)dx$?

(A) –26

(B) 6

(C) 10

(D) 42

GO ON TO THE NEXT PAGE.

35 Mark for Review

On what interval(s) is $f = 2x\cos x$ increasing on $[0, \pi]$?

(A) $(1.1571, \pi)$

(B) $(0, 1.571)$

(C) $(0, 0.860)$

(D) $(0.860, \pi)$

---

36 Mark for Review

Evaluate $\lim_{h \to 0} \frac{(3+h)^4 - 81}{h}$.

(A) 0

(B) 81

(C) 108

(D) The limit does not exist.

---

37 Mark for Review

If a particle's acceleration is given by $a(t) = t - 2$, $t \geq 0$, with its initial velocity $v(0) = 12$ and initial position $s(0) = 40$, find the position equation $s(t)$.

(A) $s(t) = t^2 - 2t + 12$

(B) $s(t) = \frac{t^2}{2} - 2t + 12$

(C) $s(t) = \frac{t^3}{6} - t^2 + 12t + 40$

(D) $s(t) = \frac{t^3}{2} - t^2 + 12t + 40$

GO ON TO THE NEXT PAGE.

38 Mark for Review

Which of the following is the graph of $f'(x)$ if the graph of $f(x)$ is

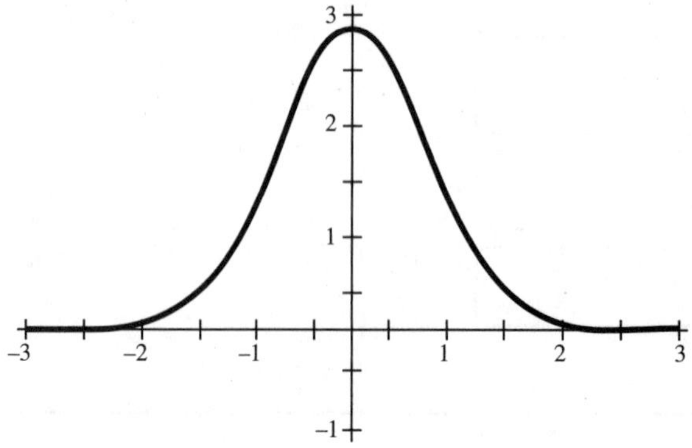

(A)

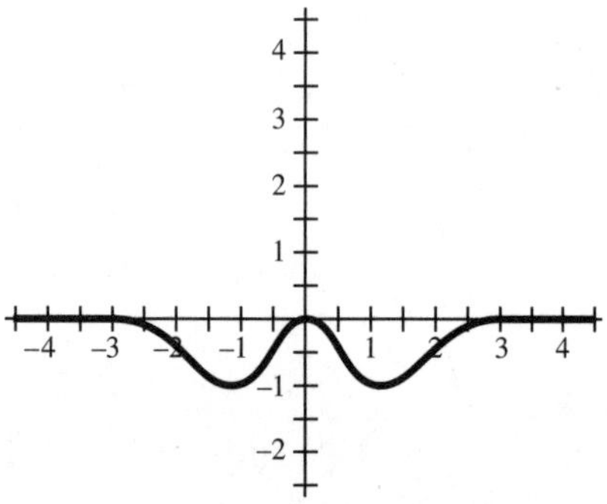

(C)

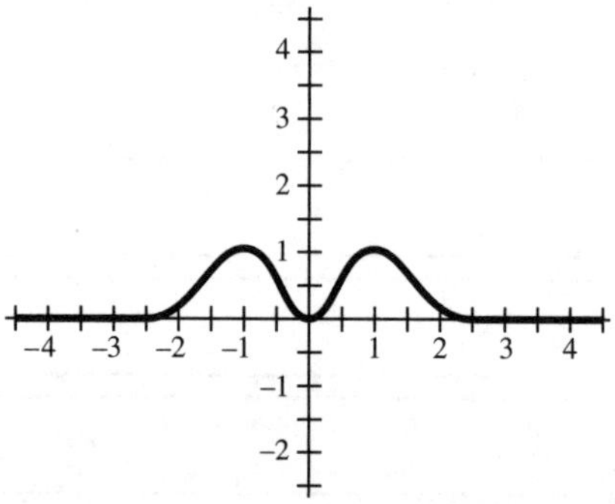

(B)

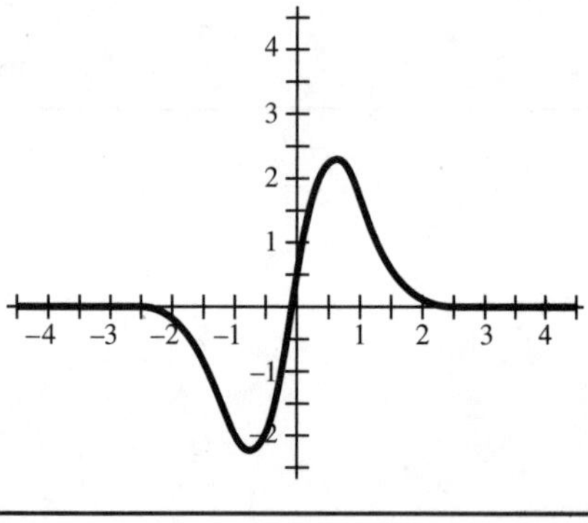

(D)

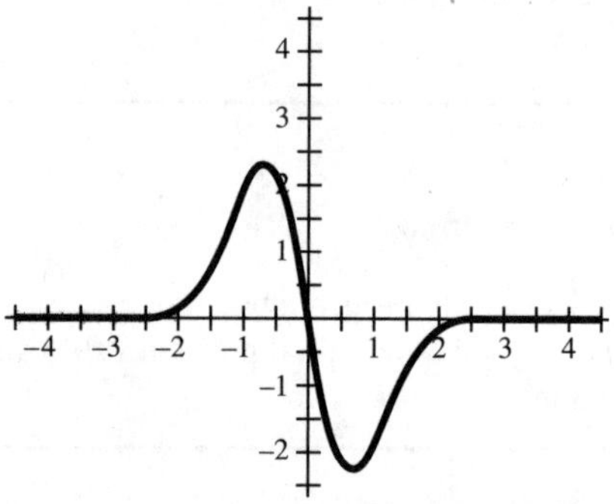

GO ON TO THE NEXT PAGE.

39 Mark for Review

Evaluate $\int \frac{\sin x}{3-\cos x}dx$.

(A) $-\ln|3-\cos x|+C$

(B) $\ln|3-\cos x|+C$

(C) $\frac{(3-\cos x)^2}{2}+C$

(D) $-\frac{(3-\cos x)^2}{2}+C$

---

40 Mark for Review

The table below gives values for $f(x)$ and $g(x)$, and their derivatives, for certain values of x. If $h(x) = f(x^2)g(x)$, find $h'(2)$.

x	$f(x)$	$g(x)$	$f'(x)$	$g'(x)$
–2	8	3	0	1
–1	6	1	3	12
2	5	6	–2	24
4	7	–1	5	4
16	0	2	1	–24

(A) 120

(B) 198

(C) 288

(D) 328

GO ON TO THE NEXT PAGE.

41 Mark for Review

Find the area between $y = \sin x$ and $y=\frac{1}{2}x$ on the interval $[0, \pi]$.

Ⓐ –1.309

Ⓑ –0.467

Ⓒ 1.309

Ⓓ 1.574

---

42 Mark for Review

If the velocity of a particle in meters per second is given by $v(t) = t^2 - 7t + 10$, $t \geq 0$, find the distance that the particle travels in the time interval [0, 3].

Ⓐ 7

Ⓑ 9.833

Ⓒ 10.666

Ⓓ 16

GO ON TO THE NEXT PAGE.

43 Mark for Review

Find y if $\frac{dy}{dx}=\frac{1}{(x-1)^2}$ and $y(0) = 10$.

(A) $y=\frac{1}{(x-1)^3}+10$

(B) $y=\frac{1}{(x-1)^3}+9$

(C) $y=-\frac{1}{x-1}+10$

(D) $y=-\frac{1}{x-1}+9$

---

44 Mark for Review

Find the value of c that is guaranteed by the Mean Value Theorem for $f(x)=x+\frac{1}{x}$ on the interval [1, 3].

(A) 1.155

(B) 1.414

(C) 1.732

(D) There is no value of c.

GO ON TO THE NEXT PAGE.

45 Mark for Review

Approximate the area between the parabola $y = 6x - x^2$ and the x-axis using four right-hand rectangles on the interval [0, 6].

(A) 9

(B) 23.625

(C) 33.75

(D) 36

STOP

END OF PART B, SECTION I

IF YOU FINISH BEFORE TIME IS CALLED, YOU MAY CHECK YOUR WORK ON PART B ONLY.

DO NOT GO ON TO SECTION II UNTIL YOU ARE TOLD TO DO SO.

AP CALCULUS AB

SECTION II

GENERAL INSTRUCTIONS

You may wish to look over the problems before starting to work on them, since it is not expected that everyone will be able to complete all parts of all problems. All problems are given equal weight, but the parts of a particular problem are not necessarily given equal weight.

A GRAPHING CALCULATOR IS REQUIRED FOR SOME PROBLEMS OR PARTS OF PROBLEMS ON THIS SECTION OF THE EXAMINATION.

- You may use the available paper for scratch work, but you must write your answers in the free-response booklet. In the free-response booklet, write your solution for each part of each question in the space provided for that part. For questions that have sub-parts, be sure to label those clearly in your solution. Use a pencil or pen with black or dark blue ink.
- You are permitted to use your calculator to solve an equation, find the derivative of a function at a point, or calculate the value of a definite integral. However, you must clearly indicate the setup of your problem, namely the equation, function, or integral you are using. If you use other built-in features or programs, you must show the mathematical steps necessary to produce your results.
- Show all of your work, even though a question may not explicitly remind you to do so. Clearly label any functions, graphs, tables, or other objects that you use. Justifications require you to give mathematical reasons and that you verify the needed conditions under which relevant theorems, properties, definitions, or tests applied. Your work will be scored on the correctness and completeness of your methods as well as your answers. Answers without supporting work will usually not receive any credit.
- Your work must be expressed in standard mathematical notation rather than calculator syntax. For example, $\int_1^5 x^2\,dx$ may not be written as fnInt (X^2, X, 1, 5).
- Unless otherwise specified, answers (numeric or algebraic) need not be simplified. If you use decimal approximations in your calculations, your work will be scored on accuracy. Unless otherwise specified, your final answers should be accurate to three places after the decimal point.
- Unless otherwise specified, the domain of a function f is assumed to be the set of all real numbers x for which $f(x)$ is a real number.

GO ON TO THE NEXT PAGE.

AP CALCULUS AB

SECTION II, PART A

Time—30 minutes

Number of problems—2

A graphing calculator is required for some questions on this part of the exam. You may use a handheld calculator or a Desmos graphing calculator application. **Make sure your calculator is in radian mode.**

During the timed portion for Part A, you may work only on the problems in Part A.

On Part A, you are permitted to use your calculator to solve an equation, find the derivative of a function at a point, or calculate the value of a definite integral. However, you must clearly indicate the setup of your problem, namely the equation, function, or integral you are using. If you use other built-in features or programs, you must show the mathematical steps necessary to produce your results.

1 Mark for Review

Oil is being pumped into a cavern for storage. The cavern is 400 meters deep. The area of the horizontal cross-section of the chamber at a depth y is given by the function A, where $A(y)$ is measured in square meters. The function A is continuous and decreases as depth increases. Selected values for $A(y)$ are given in Table 1.

Table 1

y (meters)	0	100	150	250	400
$A(y)$ (square meters)	65	38.4	30.6	20.2	11.3

A. Use a right Riemann sum with the four subintervals indicated by the data in Table 1 to approximate the volume of the chamber. Indicate the units of measure.

B. Does the approximation in part A overestimate or underestimate the volume of the cavern? Explain your reasoning.

C. The area in square meters of the horizontal cross-section at depth y is modeled by the function f given by

$f(y) = \dfrac{65}{e^{0.004y} + .002y}$. Based on this model, find the volume of the tank. Indicate the units of measure.

D. Oil is pumped into the cavern. When the depth of the oil is 200 meters, the depth is increasing at a rate of 0.45 meter per minute. Using the model from part C, find the rate at which the volume of oil is changing with respect to time when the depth of the oil is 200 meters. Indicate the units of measure.

GO ON TO THE NEXT PAGE.

2 Mark for Review

When a subway leaves its initial station, there are 25 passengers on board. Passengers leave a subway at a rate modeled by the function $f(t)$ given by

$$f(t)=12+(0.5t)\sin\left(\frac{t^2}{40}\right);\ 0\leq t\leq 60$$

where $f(t)$ is measured in passengers per minute and t is measured in minutes since the subway leaves its initial station. After the subway has been traveling for 20 minutes, passengers board the subway at a rate modeled by

$$g(t) = 20 + 3.2\ln(t^2 + 4t);\ 20 \leq t \leq 60$$

where the function $g(t)$ is measured in passengers per minute and t is the number of minutes since the subway leaves its initial station.

A. How many passengers leave the subway during the time interval $0 \leq t \leq 20$? Round to the nearest passenger.

B. Find $f'(20)$. Using the correct units, explain the meaning of $f'(20)$ in the context of this problem.

C. Is the number of passengers on the subway increasing or decreasing at time $t = 20$? Justify your reason for your answer.

D. How many passengers are on the subway at time $t = 30$? Round to the nearest passenger.

END OF PART A, SECTION II

IF YOU FINISH BEFORE TIME IS CALLED, YOU MAY CHECK YOUR WORK ON PART A ONLY.

DO NOT GO ON TO PART B UNTIL YOU ARE TOLD TO DO SO.

AP CALCULUS AB

SECTION II, PART B

Time—1 hour

Number of problems—4

A calculator is not allowed for this part of the exam.

During the timed portion for Part B, you may continue to work on the problems in Part A without the use of any calculator.

3 Mark for Review

Figure 1

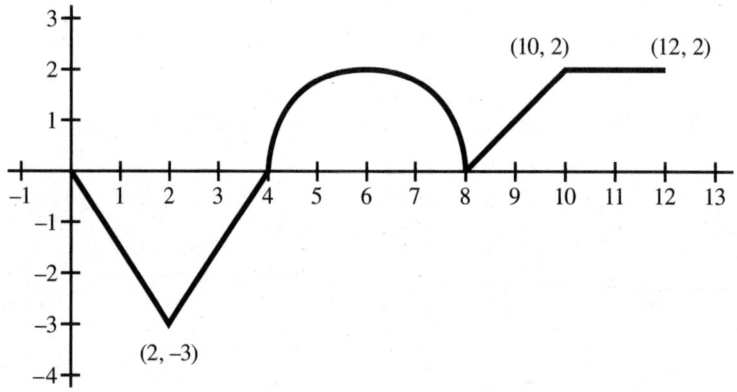

Graph of $f'(x)$

The function f is differentiable on the closed interval [0, 12] and $f(0) = 7$. The graph of $f'(x)$ in Figure 1 consists of a semicircle and four line segments.

A. Find the values of $f(4)$ and $f(12)$.

B. On which intervals is f decreasing? Justify your answer.

C. Find the absolute minimum value of f on the closed interval [0, 12]. Justify your answer.

D. Find $f''(1)$.

GO ON TO THE NEXT PAGE.

4 Mark for Review

At time $t = 0$, an iron ingot is taken from a furnace and placed on an anvil to cool. The internal temperature of the ingot is 286 degrees Celsius at time $t = 0$ and the internal temperature is greater than room temperature, 30 degrees Celsius. The internal temperature of the ingot at time t minutes can be modeled by the function I that satisfies the differential equation $\frac{dI}{dt} = -\frac{1}{32}(I - 30)$, where $I(t)$ is measured in degrees Celsius and $I(0) = 286$.

A. Write an equation for the line tangent to the graph of I at $t = 0$. Use this equation to approximate the internal temperature of the ingot at time $t = 8$.

B. Use $\frac{d^2I}{dt^2}$ to determine whether your answer in part A is an overestimate or an underestimate of the internal temperature of the ingot at time $t = 8$.

C. If the ingot is immersed in a cooling bath, another model of the internal temperature of the ingot at time t minutes is the function B that satisfies the differential equation $\frac{dB}{dt} = -(B - 30)^{\frac{3}{4}}$, where $B(t)$ is measured in degrees Celsius and $B(0) = 286$. Using this model, what is the internal temperature of the ingot at time $t = 8$?

GO ON TO THE NEXT PAGE.

5 Mark for Review

Consider the curve given by $2xy - y^2 = 3$.

A. Write an equation for the tangent line to the curve at the point (2, 1).

B. Find the coordinates of all points on the curve at which the line tangent to the curve at that point is vertical.

C. Evaluate $\frac{d^2y}{dx^2}$ at the points on the curve where $x = 2$ and $y = 1$.

GO ON TO THE NEXT PAGE.

6 Mark for Review

Two particles move along the x-axis. For $0 \leq t \leq 10$, the velocity of particle A at time t is given by $v_A(t) = t^2 - 5t + 4$ and the position of particle B at time t is given by $x_B(t) = \frac{t^3}{3} - 4t^2 + 12t$. Particle A is at position $x = 3$ at time $t = 0$.

A. For $0 \leq t \leq 10$, when is particle B moving to the left?

B. For $0 \leq t \leq 10$, find all times t when the two particles are traveling in the same direction.

C. Find the acceleration of particle A at time $t = 3$. Is its speed increasing, decreasing, or neither at time $t = 3$? Justify your answer.

D. Find the position of particle A the first time that it changes direction.

STOP

END OF EXAM

Practice Test 1: Diagnostic Answer Key and Explanations

PRACTICE TEST 1: DIAGNOSTIC ANSWER KEY

Let's take a look at how you did on Practice Test 1. Follow the three-step process in the diagnostic answer key below and read the explanations for any questions you got wrong or you struggled with but got correct. Once you finish working through the answer key and the explanations, go to the next chapter to make your study plan.

STEP 1 » Check your answers and mark any correct answers with a ✔ in the appropriate column.

Section I—Multiple Choice							
Q #	Ans.	✔	Chapter #, Section Title	Q #	Ans.	✔	Chapter #, Section Title
1	A		**7,** Determining Concavity of Functions over Their Domains	24	B		**5,** Implicit Differentiation
2	C		**3,** Working with the Intermediate Value Theorem (IVT)	25	C		**9,** Finding Particular Solutions Using Initial Conditions and Separation of Variables
3	A		**3,** Connecting Infinite Limits and Vertical Asymptotes	26	A		**5,** Selecting Procedures for Calculating Derivatives
4	D		**8,** Integrating Using Substitution	27	C		**3,** Determining Limits Using the Squeeze Theorem
5	B		**10,** Finding the Average Value of a Function on an Interval	28	D		**10,** Finding the Average Value of a Function on an Interval
6	D		**7,** Solving Optimization Problems	29	A		**9,** Verifying Solutions for Differential Equations
7	A		**4,** Connecting Differentiability and Continuity: Determining When Derivatives Do and Do Not Exist	30	C		**6,** Straight-Line Motion: Connecting Position, Velocity, and Acceleration
8	D		**4,** The Quotient Rule	31	A		**8,** The Fundamental Theorem of Calculus and Definite Integrals
9	C		**8,** Integrating Using Substitution	32	C		**8,** The Fundamental Theorem of Calculus and Accumulation Functions
10	A		**3,** Determining Limits Using the Squeeze Theorem	33	B		**8,** Integrating Using Substitution
11	C		**7,** Solving Optimization Problems	34	A		**8,** Applying Properties of Definite Integrals
12	C		**4,** The Product Rule	35	C		**7,** Determining Intervals on Which a Function Is Increasing or Decreasing
13	D		**8,** Approximating Areas with Riemann Sums	36	C		**4,** Defining the Derivative of a Function and Using Derivative Notation
14	C		**3,** Defining Continuity at a Point	37	C		**6,** Straight-Line Motion: Connecting Position, Velocity, and Acceleration
15	C		**6,** Using L'Hospital's Rule for Determining Limits of Indeterminate Forms	38	D		**7,** Sketching Graphs of Functions and Their Derivatives
16	A		**5,** Differentiating Inverse Functions	39	B		**8,** Integrating Using Substitution
17	D		**7,** Using the Candidates Test to Determine Absolute (Global) Extrema	40	C		**4,** The Product Rule
18	D		**4,** The Product Rule	41	C		**10,** Finding the Area Between Curves Expressed as Functions of *x*
19	C		**6,** Using L'Hospital's Rule for Determining Limits of Indeterminate Forms	42	B		**6,** Straight-Line Motion: Connecting Position, Velocity, and Acceleration
20	A		**5,** The Chain Rule	43	D		**9,** Finding Particular Solutions Using Initial Conditions and Separation of Variables
21	D		**5,** Selecting Procedures for Calculating Derivatives	44	C		**7,** Using the Mean Value Theorem
22	D		**5,** Differentiating Inverse Trigonometric Functions	45	C		**8,** Approximating Areas with Riemann Sums
23	B		**6,** Using L'Hospital's Rule for Determining Limits of Indeterminate Forms				

Section II—Free Response			
Q #	Ans.	✔	Chapter #, Section Title
1A	See Explanation		**8,** Approximating Areas with Riemann Sums
1B	See Explanation		**8,** Approximating Areas with Riemann Sums
1C	See Explanation		**10,** Using Accumulation Functions and Definite Integrals in Applied Contexts
1D	See Explanation		**6,** Introduction to Related Rates
2A	See Explanation		**10,** Using Accumulation Functions and Definite Integrals in Applied Contexts
2B	See Explanation		**6,** Rates of Change in Applied Contexts Other Than Motion
2C	See Explanation		**6,** Interpreting the Meaning of the Derivative in Context
2D	See Explanation		**10,** Using Accumulation Functions and Definite Integrals in Applied Contexts
3A	See Explanation		**8,** Interpreting the Behavior of Accumulation Functions Involving Area
3B	See Explanation		**7,** Determining Intervals on Which a Function Is Increasing or Decreasing
3C	See Explanation		**7,** Using the Candidates Test to Determine Absolute (Global) Extrema
3D	See Explanation		**5,** Calculating Higher-Order Derivatives
4A	See Explanation		**6,** Approximating Values of a Function Using Local Linearity and Linearization
4B	See Explanation		**6,** Approximating Values of a Function Using Local Linearity and Linearization
4C	See Explanation		**9,** Finding Particular Solutions Using Initial Conditions and Separation of Variables
5A	See Explanation		**5,** Implicit Differentiation
5B	See Explanation		**5,** Implicit Differentiation
5C	See Explanation		**5,** Implicit Differentiation
6A	See Explanation		**6,** Straight-Line Motion: Connecting Position, Velocity, and Acceleration
6B	See Explanation		**6,** Straight-Line Motion: Connecting Position, Velocity, and Acceleration
6C	See Explanation		**6,** Straight-Line Motion: Connecting Position, Velocity, and Acceleration
6D	See Explanation		**6,** Straight-Line Motion: Connecting Position, Velocity, and Acceleration

STEP 2 » **Tally your correct answers from Step 1 by chapter. For each chapter, write the number of correct answers in the appropriate box. Then, divide your correct answers by the number of total questions (which we've provided) to get your percent correct.**

CHAPTER 3 TEST SELF-EVALUATION

# CORRECT ANSWERS	# TOTAL QUESTIONS	PERCENT CORRECT
	5	%

CHAPTER 4 TEST SELF-EVALUATION

# CORRECT ANSWERS	# TOTAL QUESTIONS	PERCENT CORRECT
	6	%

CHAPTER 5 TEST SELF-EVALUATION

# CORRECT ANSWERS	# TOTAL QUESTIONS	PERCENT CORRECT
	10	%

CHAPTER 6 TEST SELF-EVALUATION

# CORRECT ANSWERS	# TOTAL QUESTIONS	PERCENT CORRECT
	15	%

CHAPTER 7 TEST SELF-EVALUATION

# CORRECT ANSWERS	# TOTAL QUESTIONS	PERCENT CORRECT
	9	%

CHAPTER 8 TEST SELF-EVALUATION

# CORRECT ANSWERS	# TOTAL QUESTIONS	PERCENT CORRECT
	12	%

CHAPTER 9 TEST SELF-EVALUATION

# CORRECT ANSWERS	# TOTAL QUESTIONS	PERCENT CORRECT
	4	%

CHAPTER 10 TEST SELF-EVALUATION

# CORRECT ANSWERS	# TOTAL QUESTIONS	PERCENT CORRECT
	6	%

STEP 3 » **Use the results above to customize your study plan. You may want to start with, or give more attention to, the chapters with the lowest percents correct.**

ANSWERS AND EXPLANATIONS TO SECTION I

1. **A** First, take the first and second derivatives.

$$\frac{dy}{dx} = 1 - 3x^2$$

$$\frac{d^2y}{dx^2} = -6x$$

In order to find where the graph of y is concave up, you need to know where the second derivative is positive. You just need to know where $-6x$ is positive, which is where $x < 0$, so the interval is $(-\infty, 0)$. The answer is (A).

2. **C** The Intermediate Value Theorem guarantees that if a function $f(x)$ is continuous on the interval $[a, b]$ and if $f(a)$ and $f(b)$ have opposite signs, then there exists at least one value c in the interval (a, b) such that $f(c) = 0$. What this means is that you are looking for an interval in which $f(x)$ is positive on one end of the interval and negative on the other end. Find the values of $f(x)$ at each of the endpoints of the various intervals:

$$f(-1) = (-1)^3 + (-1) - 3 = -5$$

$$f(0) = (0)^3 + (0) - 3 = -3$$

$$f(1) = (1)^3 + (1) - 3 = -1$$

$$f(2) = (2)^3 + (2) - 3 = 7$$

$$f(3) = (3)^3 + (3) - 3 = 27$$

On the interval $(1, 2)$, you have $f(1) < 0$ and $f(2) > 0$. The answer is (C).

3. **A** If you plug in 2 for x, you get zero in the denominator. Remember, however, that when you take the limit, you aren't actually plugging in 2 but a number *very* close to 2 and, because of the negative sign, slightly less than 2. This means that the denominator is a number *very* close to zero and negative. Therefore, $\lim_{x \to 2^-} \frac{9}{x-2} = -\infty$. The answer is (A).

4. **D** Use u-substitution. Let $u = \sin x$. Then $du = \cos x\, dx$. Substitute into the integrand to get $\int \sin^3 x \cos x\, dx = \int u^3 du$. Integrate: $\int u^3 du = \frac{u^4}{4} + C$. And substitute back: $\frac{u^4}{5} + C = \frac{\sin^4 x}{4} + C$. The answer is (D).

5. **B** Find the average value of $f(x)$ on the interval $[a, b]$ by evaluating the integral $\frac{1}{b-a}\int_a^b f(x)dx$.

Here you need to evaluate: $\frac{1}{5-1}\int_1^5 2e^{2x}dx = \frac{1}{2}\int_1^5 e^{2x}dx$. Integrate: $\frac{1}{2}\int_1^5 e^{2x}dx = \frac{1}{2}\frac{e^{2x}}{2}\Big|_1^5 = \frac{1}{4}\left(e^{10} - e^2\right)$.

The answer is (B).

6. **D** Call the sides of the base X and the height of the box Y. The volume of the box is then $V = X^2Y = 108$. The surface area of the box is $S = X^2 + 4XY$. Solve the volume equation for Y: $Y = \frac{108}{X^2}$. Now plug this in for Y in the surface area equation to get $S = X^2 + 4X\left(\frac{108}{X^2}\right) = X^2 + \frac{432}{X}$. Next, take the derivative and set it equal to zero:

$$S = 2X - \frac{432}{X^2}$$

$$2X - \frac{432}{X^2} = 0$$

$$2X^3 = 432$$

$$X = 6$$

Finally, plug in $X = 6$ to solve for Y: $Y = \frac{108}{6^2} = 3$. The answer is (D).

7. **A** First, show that $f(x)$ is continuous. Plug 1 into the top and bottom expressions and set them equal to each other: $f(x) = \begin{cases} (1)^{\frac{5}{3}} + b; & x < 1 \\ a(1)^{\frac{4}{3}}; & x \geq 1 \end{cases}$. You get $1 + b = a$. Next, take the derivative of the top and bottom expressions: $f'(x) = \begin{cases} \frac{5}{3}x^{\frac{2}{3}}; & x < 1 \\ a\left(\frac{4}{3}x^{\frac{1}{3}}\right); & x > 1 \end{cases}$. Now plug 1 into the top and bottom expressions and set them equal to each other: $f'(x) = \begin{cases} \frac{5}{3}(1)^{\frac{2}{3}}; & x < 1 \\ a\left(\frac{4}{3}(1)^{\frac{1}{3}}\right); & x > 1 \end{cases}$. You get $\frac{4}{3}a = \frac{5}{3}$, which gives you $a = \frac{5}{4}$. Now $1 + b = a$, and you get $b = \frac{1}{4}$. The answer is (A).

8. **D** Use the Quotient Rule:

$$f'(x) = \frac{\left(\sqrt{x}\right)(\sec x \tan x) - (\sec x)\left(\frac{1}{2\sqrt{x}}\right)}{\left(\sqrt{x}\right)^2}$$

This simplifies to $f'(x) = \frac{\left(\sqrt{x}\right)(\sec x \tan x) - \left(\frac{\sec x}{2\sqrt{x}}\right)}{x}$.

Next, combine the terms in the numerator with a common denominator of $2\sqrt{x}$:

$$f'(x)=\frac{\frac{(\sqrt{x})(\sec x\tan x)(2\sqrt{x})}{2\sqrt{x}}-\left(\frac{\sec x}{2\sqrt{x}}\right)}{x}=\frac{\frac{(2x)(\sec x\tan x)-\sec x}{2\sqrt{x}}}{x}$$

This can then be reduced to $f'(x)=\frac{\frac{(2x)(\sec x\tan x)-\sec x}{2\sqrt{x}}}{x}=\frac{\sec x(2x\tan x-1)}{2x\sqrt{x}}$.

The answer is (D).

9. **C** Use u-substitution. Let $u = x + 8$. Then $x = u - 8$ and $du = dx$. Substituting into the integrand, you get: $\int x\left(\sqrt[3]{x+8}\right)dx=\int(u-8)u^{\frac{1}{3}}du=\int\left(u^{\frac{4}{3}}-8u^{\frac{1}{3}}\right)du$. Integrate: $\int\left(u^{\frac{4}{3}}-8u^{\frac{1}{3}}\right)du=\frac{u^{\frac{7}{3}}}{\frac{7}{3}}-8\frac{u^{\frac{4}{3}}}{\frac{4}{3}}+C=\frac{3}{7}u^{\frac{7}{3}}-6u^{\frac{4}{3}}+C$. And substitute back: $\frac{3}{7}(x+8)^{\frac{7}{3}}-6(x+8)^{\frac{4}{3}}+C$.

The answer is (C).

10. **A** First, break up the limit into the product of two limits: $\lim_{x\to0}\frac{\sin^2(3x)}{x}=\left(\lim_{x\to0}\frac{\sin(3x)}{x}\right)\left(\lim_{x\to0}\sin(3x)\right)$.

The limit on the left is $\lim_{x\to0}\frac{\sin(3x)}{x}=3$ and the limit on the right is $\lim_{x\to0}\sin(3x)=0$. Therefore, the limit is $\lim_{x\to0}\frac{\sin^2(3x)}{x}=(3)(0)=0$. The answer is (A).

11. **C** Take the derivative: $P'(x) = -3x^2 + 12x$. Set it equal to zero and solve for x:

$$-3x^2 + 12x = 0$$

$$-3x(x - 4) = 0$$

$$x = 0 \text{ or } x = 4$$

Since the question stipulates that $x > 0$, reject $x = 0$. Now, use the Second Derivative Test. Remember that if c is a critical value and $f''(c) < 0$, then c is at a maximum and if $f''(c) > 0$, then c is at a minimum. First, find the second derivative: $P''(x) = -6x + 12$. At $x = 4$, $P''(4) = -12$. Thus, the maximum occurs at $x = 4$, or 4,000 shoes. The answer is (C).

12. **C** Use the Product Rule to find $f'(x)$: $f'(x)=x^2\left(2xe^{x^2}\right)+(2x)e^{x^2}=\left(1+x^2\right)2xe^{x^2}$. Now plug in $x = 2$ to get $f'(2)=\left(1+(2)^2\right)2(2)e^{(2)^2}=20e^4$. The answer is (C).

13. **D** The approximate total distance is the sum of the width of each time interval times the velocity during that interval. The width of each time interval is 1. Use the left end of each interval to get that the total distance traveled is (1)[12 + 18 + 22] = 52. Use the right end of each interval to get that the total distance traveled is (1)[18 + 22 + 24] = 64. Average the two values to get that the total distance traveled is 58. The answer is (D).

14. **C** Simply plug 3 into the top and bottom expressions and set them equal to each other to get

$$f(x)=\begin{cases}-(3)^2+2k(3)+66=57+6k;\ x>3\\ 4k(3)+(3)^3=12k+27;\ x\le 3\end{cases}$$

Solve $57 + 6k = 12k + 27$ to get $k = 5$. The answer is (C).

15. **C** Plug in 8 for x to get $\lim\limits_{x\to 8}\dfrac{x^2-6x-16}{x^2-2x-48}=\dfrac{(8)^2-6(8)-16}{(8)^2-2(8)-48}=\dfrac{0}{0}$. Before you decide that the limit does not exist, factor the numerator and the denominator: $\lim\limits_{x\to 8}\dfrac{x^2-6x-16}{x^2-2x-48}=\lim\limits_{x\to 8}\dfrac{(x-8)(x+2)}{(x-8)(x+6)}$. Now cancel $(x - 8)$ from the numerator and denominator: $\lim\limits_{x\to 8}\dfrac{(x-8)(x+2)}{(x-8)(x+6)}=\lim\limits_{x\to 8}\dfrac{(x+2)}{(x+6)}$. Now take the limit to get $\lim\limits_{x\to 8}\dfrac{(x+2)}{(x+6)}=\dfrac{5}{7}$. The answer is (C).

16. **A** First, find the x-value when $f(x) = 2$. Solve $2=\dfrac{x^3-17}{5}$ to get $x = 3$. Next, find $f'(x)$: $f'(x)=\dfrac{1}{5}(3x^2)=\dfrac{3x^2}{5}$. Finally, find $\dfrac{1}{f'(3)}$: $\dfrac{1}{\frac{3(3)^2}{5}}=\dfrac{5}{27}$. The answer is (A).

17. **D** First, take the derivative: $\dfrac{dy}{dx}=24x^3-6x$. Next, set the derivative equal to zero and solve.

$$24x^3 - 6x = 0$$

$$6x(4x^2 - 1) = 0$$

$$x = 0, \pm\frac{1}{2}$$

Now, plug these values and the values of the endpoints into the original function. The largest value of y is the absolute maximum.

$$y(-1) = 6(-1)^4 - 3(-1)^2 - 5 = -2$$

$$y\left(-\frac{1}{2}\right)=6\left(-\frac{1}{2}\right)^4-3\left(-\frac{1}{2}\right)^2-5=-\frac{43}{8}$$

$$y(0) = 6(0)^4 - 3(0)^2 - 5 = -5$$

$$y\left(\frac{1}{2}\right)=6\left(\frac{1}{2}\right)^4-3\left(\frac{1}{2}\right)^2-5=-\frac{43}{8}$$

$$y(2) = 6(2)^4 - 3(2)^2 - 5 = 79$$

Therefore, the absolute maximum is 79. The answer is (D).

18. **D** Use the Product Rule: $\dfrac{dy}{dx}=(\sec(3x))\left(\dfrac{5}{5x}\right)+3(\sec(3x)\tan(3x))(\ln(5x))$, which simplifies to $\dfrac{dy}{dx}=(\sec(3x))\left(\dfrac{1}{x}\right)+3(\sec(3x)\tan(3x))(\ln(5x))$. The answer is (D).

19. **C** Multiply the numerator and denominator by 7: $\lim_{\theta \to 0} \frac{7\tan 7\theta}{7\theta}$. Now evaluate the limit:

$$\lim_{\theta \to 0} \frac{7\tan 7\theta}{7\theta} = 7\lim_{\theta \to 0} \frac{\tan 7\theta}{7\theta} = 7\lim_{\theta \to 0} \frac{\frac{\sin 7\theta}{\cos 7\theta}}{7\theta} = 7\lim_{\theta \to 0} \frac{\sin 7\theta}{7\theta}\frac{1}{\cos 7\theta} = 7(1)(1) = 7.$$

Here's a shortcut: any limit of the form $\lim_{\theta \to 0} \frac{\tan a\theta}{\theta} = a$. The answer is (C).

20. **A** Use the Chain Rule: $\frac{dy}{dx} = 5\cos^4\left(1-x^2\right)\left(-\sin\left(1-x^2\right)\right)(-2x) = 10x\cos^4\left(1-x^2\right)\sin\left(1-x^2\right)$. The answer is (A).

21. **D** Before you reflexively use the Quotient Rule, notice that you can factor x^3 out of every term in the numerator and then cancel the like terms. You get $y = \frac{\left(x^3\right)\left(5x-3+x^{-1}\right)}{x^3} = 5x-3+\frac{1}{x}$. Isn't that easier to differentiate? Now use the Power Rule to find the derivative: $\frac{dy}{dx} = 5-\frac{1}{x^2}$. The answer is (D).

22. **D** Remember, the derivative of $\arcsin(u) = \frac{1}{\sqrt{1-u^2}}\frac{du}{dx}$. Here you get

$$\frac{dy}{dx} = \frac{1}{\sqrt{1-\left(4\sqrt{x}\right)^2}}\frac{4}{2\sqrt{x}} = \frac{1}{\sqrt{1-16x}}\frac{2}{\sqrt{x}} = \frac{2}{\sqrt{x-16x^2}}$$

The answer is (D).

23. **B** First, check that you can use L'Hospital's Rule. Plug in $x = 0$: $\frac{(0)2^0}{2^0-1} = \frac{0}{0}$. So yes, you can use L'Hospital's Rule. Take the derivative of the numerator and the denominator: $\lim_{x \to 0} \frac{x\left(2^x \ln 2\right)+2^x}{2^x \ln 2}$, which simplifies to $\lim_{x \to 0} \frac{2^x(x\ln 2+1)}{2^x \ln 2} = \lim_{x \to 0} \frac{x\ln 2+1}{\ln 2}$. Now plug in $x = 0$ to evaluate the limit: $\lim_{x \to 0} \frac{x\ln 2+1}{\ln 2} = \frac{1}{\ln 2}$. The answer is (B).

24. **B** Use Implicit Differentiation to find $\frac{dy}{dx}$. Take the derivative of every term: $6x+\left(2x(2y)\frac{dy}{dx}+2y^2\right)-6y\frac{dy}{dx} = 0$. Next, plug in (2, 1): $6(2)+2(2)(2(1))\frac{dy}{dx}+2(1)^2-6(1)\frac{dy}{dx} = 0$. Now solve for $\frac{dy}{dx}$:

$$12+8\frac{dy}{dx}+2-6\frac{dy}{dx} = 0$$

$$14+2\frac{dy}{dx} = 0$$

$$\frac{dy}{dx} = -7$$

The answer is (B).

25. **C** Use Separation of Variables to solve the differential equation: $\frac{dy}{y} = -4dt$. Integrate both sides:

$$\int \frac{dy}{y} = -\int 4dt$$

$$\ln|y| = -4t + C$$

$$y = e^{-4t+C} = Ae^{-4t}$$

Now substitute the initial condition to solve for the constant to get $12 = Ae^{-4(0)} = A$, so $y = 12e^{-4t}$. Therefore, $y(-2) = 12e^{-4(-2)} = 12e^{8}$. The answer is (C).

26. **A** Before you use the Quotient Rule, simplify the expression. Using the trigonometric identities $\sin 2x = 2\sin x \cos x$ and $\cos 2x = 2\cos^2 x - 1$, you can rewrite the numerators as $y = \frac{2\sin x \cos x}{\sin x} - \frac{2\cos^2 x - 1}{\cos x}$. This simplifies to $y = 2\cos x - 2\cos x + \frac{1}{\cos x} = \frac{1}{\cos x}$. And, because $\sec x = \frac{1}{\cos x}$, you can simplify this further to $\frac{1}{\cos x} = \sec x$. The derivative then is $\sec x \tan x$. The answer is (A).

27. **C** Remember that $\lim_{x\to 0} \frac{\tan(x)}{x} = 1$. Multiply the numerator and denominator of the limit by 3 to get $\lim_{x\to 0} \frac{2\tan(3x)}{x}\frac{3}{3} = \lim_{x\to 0} \frac{6\tan(3x)}{3x}$. Next, factor out the 6: $6\lim_{x\to 0} \frac{\tan(3x)}{3x}$. And take the limit: $6\lim_{x\to 0} \frac{\tan(3x)}{3x} = 6(1) = 6$. The answer is (C).

28. **D** In order to find the average value, you use the Mean Value Theorem for Integrals, which says that the average value of $f(x)$ on the interval $[a, b]$ is $\frac{1}{b-a}\int_a^b f(x)\,dx$.

Here you have $\frac{1}{e-1}\int_1^e \frac{1}{x}\,dx$. The answer is (D).

29. **A** The simplest thing to do is to take the second derivative of each of the answer choices and see which one satisfies the differential equation.

Choice (A): $\frac{dy}{dx} = 2e^{2x}$; $\frac{d^2y}{dx^2} = 4e^{2x}$, which works. Just in case, check the others.

Choice (B): $\frac{dy}{dx} = 2\cos 2x$; $\frac{d^2y}{dx^2} = -4\sin 2x$, which is not $4\sin 2x$.

Choice (C): $\frac{dy}{dx} = -2\sin 2x$; $\frac{d^2y}{dx^2} = -4\cos 2x$, which is not 4cos 2*x*.

Choice (D): $\frac{dy}{dx} = 4x^3$; $\frac{d^2y}{dx^2} = 12x^2$, which is not $4x^4$.

The answer is (A).

30. **C** The particle is changing direction at value(s) of t where the velocity changes sign. First, find the velocity by taking the derivative: $v(t) = 6t^2 - 42t + 60$. Next, set the velocity equal to zero and solve for t:

$$6t^2 - 42t + 60 = 0$$

$$6(t^2 - 7t + 10) = 0$$

$$6(t - 2)(t - 5) = 0$$

$$t = 2, 5$$

Make sure that the velocity is changing sign at each value of t. Make a number line and place $t = 2, 5$ on the line. Choose a value of t less than 2 (say, 0), between 2 and 5 (say, 3), and greater than 5 (say, 6) and plug each value into the derivative to check the sign of the velocity:

At $t = 0$, $v(t) = 60$.

At $t = 3$, $v(t) = -12$.

At $t = 6$, $v(t) = 24$.

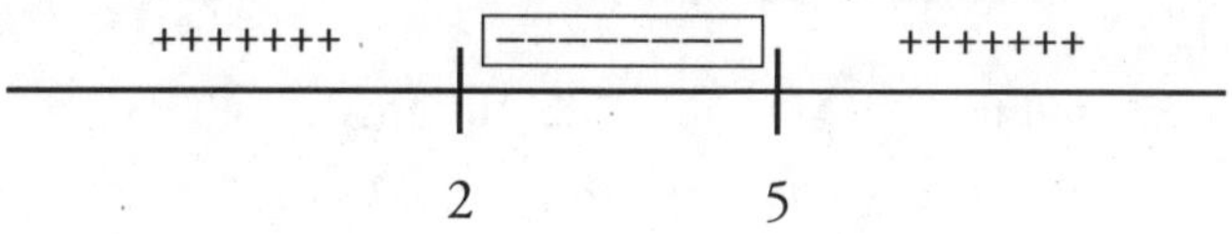

The velocity is thus changing sign at $t = 2, 5$, so those are the values of t where the particle is changing direction. The answer is (C).

31. **A** According to the Fundamental Theorem of Calculus, $\frac{d}{dx}\int_c^{f(x)} g(t)\,dt = g(f(x))f'(x)$. Here, you get $\frac{d}{dx}\int_5^{\sqrt{x}} \cos(t^2)\,dt = \cos\left(\sqrt{x}^2\right)\left(\frac{1}{2\sqrt{x}}\right) = \frac{\cos x}{2\sqrt{x}}$. The answer is (A).

32. **C** To find $g(2)$, evaluate the integral $g(2) = \int_0^2 (t^3 - t^2 + 1)\,dt$ to get

$$g(2) = \int_0^2 (t^3 - t^2 + 1)\,dt = \frac{t^4}{4} - \frac{t^3}{3} + t\Big|_0^2 = \frac{10}{3}$$

The answer is (C).

33. **B** First, rewrite the denominator of the integrand as $\int_0^1 \frac{12dx}{1+(6x)^2}$. Next, use u-substitution. Let $u = 6x$. Then $du = 6dx$. Substitute into the integrand to get $\int \frac{12dx}{1+(6x)^2} = \int \frac{2du}{1+u^2} = 2\arctan u + C$. Substitute back to get $\int_0^1 \frac{12dx}{1+(6x)^2} = 2\arctan(6x)\Big|_0^1 = 2\arctan 6$. The answer is (B).

34. **A** According to the First Fundamental Theorem of Calculus, $\int_a^b f(x)dx + \int_b^c f(x)dx = \int_a^c f(x)dx$. This means that $\int_{-2}^2 f(x)dx + \int_2^5 f(x)dx + \int_5^{10} f(x)dx = \int_{-2}^{10} f(x)dx$. Also, note that if $\int_{10}^5 f(x)dx = 18$, then $\int_5^{10} f(x)dx = -18$. Put all of these together to get $\int_{-2}^{10} f(x)dx = 8 - 16 - 18 = -26$. The answer is (A).

35. **C** First, take the derivative using the Product Rule: $f'(x) = 2\cos x - 2x\sin x$. Next, set the derivative equal to zero and solve. Use your calculator to get $x = 0.860$. Now, put the value of x on the number line and sign test it. Pick a number less than 0.860 and plug it into the derivative. For example, at $x = 0.5$, the derivative is positive. Then pick a number greater than 0.860 and sign test it. For example, at $x = 1$, the derivative is negative. Thus, the function is increasing on the interval (0, 0.860). The answer is (C).

36. **C** This limit is in the form of the definition of the derivative where $f(x) = x^4$ and you are evaluating the derivative at $x = 3$. In other words, $\lim_{h\to 0} \frac{f(3+h)-f(3)}{h} = \lim_{h\to 0} \frac{(3+h)^4 - 81}{h}$. The derivative of $f(x) = x^4$ is $f'(x) = 4x^3$, so find $f'(3) = 4(3)^3 = 108$. The answer is (C).

37. **C** First, find the particle's velocity by integrating the acceleration equation with respect to t: $v(t) = \int (t-2)\,dt = \frac{t^2}{2} - 2t + C$. Next, use the initial condition to solve for the constant: $12 = \frac{0^2}{2} - 2(0) + C$, so $C = 12$. The velocity equation is thus $v(t) = \frac{t^2}{2} - 2t + 12$. Next, find the particle's position by integrating the velocity equation with respect to t: $s(t) = \int \left(\frac{t^2}{2} - 2t + 12\right)dt = \frac{t^3}{6} - t^2 + 12t + C_1$. Next, use the initial condition to solve for the constant: $40 = \frac{(0)^3}{6} - (0)^2 + 12(0) + C_1$, so $C_1 = 40$. Therefore, the position equation is $s(t) = \frac{t^3}{6} - t^2 + 12t + 40$. The answer is (C).

38. **D** Starting from the left end of the graph, the slope starts at zero, then becomes positive, and then gets back to zero at $x = 0$. That eliminates (A) and (B). Then the slope is negative and gets back to zero. That eliminates (C). Therefore, the answer is (D).

39. **B** Use u-substitution. Let $u = 3 - \cos x$. Then $du = \sin x\, dx$. Substitute into the integrand to get $\int \frac{\sin x}{3 - \cos x} dx = \int \frac{du}{u}$. Integrate: $\int \frac{du}{u} = \ln|u| + C$. And substitute back: $\ln|3 - \cos x| + C$. The answer is (B).

40. **C** Find $h'(x)$ using the Product Rule to get $h'(x) = f(x^2)g'(x) + f'(x^2)(2x)g(x)$. Then $h'(2) = f(4)g'(2) + f'(4)(4)g(2)$. Use the table to get $h'(2) = (7)(24) + (5)(4)(6) = 288$. The answer is (C).

41. **C** Find the area by finding the integral of the difference between the two curves over the interval.

First, draw the graph:

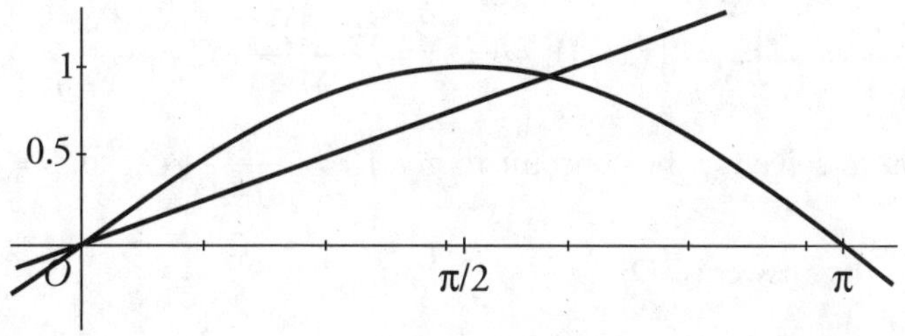

Use a graphing calculator to find that the curves intersect at approximately (1.895, 0.945). Note that between $x = 0$ and $x = 1.895$, the curve $y = \sin x$ is above the curve $y = \frac{1}{2}x$ and between $x = 1.895$ and $x = \pi$, the curve $y = \frac{1}{2}x$ is above the curve $y = \sin x$. This means that you will need to evaluate the sum of two integrals: $\int_0^{1.895}\left(\sin x - \frac{1}{2}x\right)dx$ and $\int_{1.895}^{\pi}\left(\frac{1}{2}x - \sin x\right)dx$. Use your calculator to get $\int_0^{1.895}\left(\sin x - \frac{1}{2}x\right)dx \approx 0.4208$ and $\int_{1.895}^{\pi}\left(\frac{1}{2}x - \sin x\right)dx \approx 0.8882$. The sum of the two integrals is approximately 1.309. The answer is (C).

42. **B** First, check the sign of the velocity over the interval. Factor the equation for the velocity to get $v(t) = (t - 2)(t - 5)$. Next, use a number line and sign test the velocity.

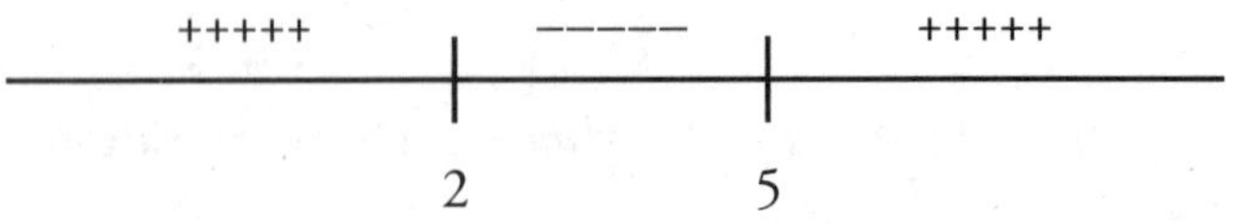

Note that the interval provided by the question is [0, 3], so you only need to be concerned up to $t = 3$ rather than $t = 5$. Because the velocity is negative between $t = 2$ and $t = 3$, find the distance traveled by evaluating $\int_0^2 \left(t^2 - 7t + 10\right) dt - \int_2^3 \left(t^2 - 7t + 10\right) dt$. Use your calculator to get $8.6667 - (-1.1667) \approx 9.833$. The answer is (B).

43. **D** Use Separation of Variables to solve the differential equation: $dy = \dfrac{dx}{(x-1)^2} = (x-1)^{-2}\,dx$. Integrate both sides: $\int dy = \int (x-1)^{-2}\,dx$; $y = \dfrac{(x-1)^{-1}}{-1} + C = -\dfrac{1}{x-1} + C$. Now substitute the initial condition to solve for the constant to get $10 = -\dfrac{1}{0-1} + C$, so $C = 9$. Thus, the solution is $y = -\dfrac{1}{x-1} + 9$. The answer is (D).

44. **C** The Mean Value Theorem says that if $f(x)$ is continuous on the interval $[a, b]$ and differentiable on the interval (a, b), then there exists at least one value c on the interval (a, b) such that $f'(c) = \dfrac{f(b) - f(a)}{b - a}$. Here, find c such that $f'(c) = \dfrac{f(3) - f(1)}{3 - 1} = \dfrac{\frac{10}{3} - 2}{2} = \dfrac{2}{3}$. The derivative is $f'(x) = 1 - \dfrac{1}{x^2}$, so you need $1 - \dfrac{1}{c^2} = \dfrac{2}{3}$, so $c = \pm\sqrt{3} \approx \pm 1.732$. Throw out the negative answer because it isn't in the interval. The answer is (C).

45. **C** First, draw the graph:

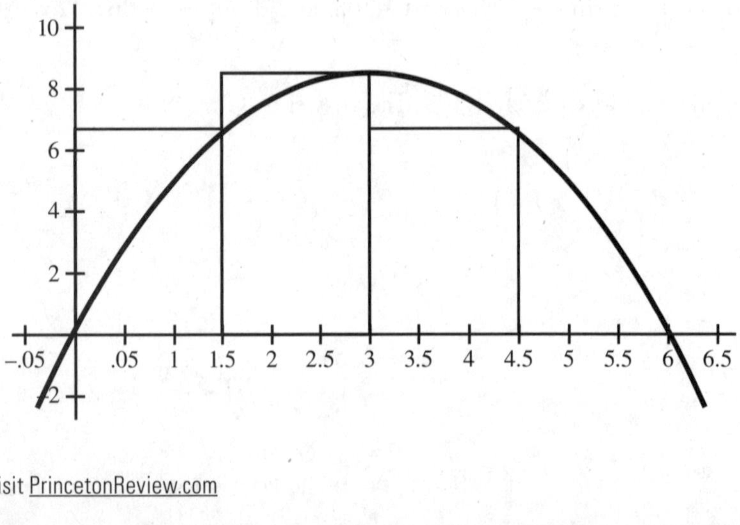

Divide the interval from $x = 0$ to $x = 6$ into four equal intervals. The width of each rectangle is thus $\frac{6-0}{4} = 1.5$. Find the height of each rectangle by plugging the value of x on the right end of each interval. Then the area of each rectangle is the width times the height. You get

$$y(1.5) = 6(1.5) - (1.5)^2 = 6.75$$

$$y(3) = 6(3) - (3)^2 = 9$$

$$y(4.5) = 6(4.5) - (4.5)^2 = 6.75$$

$$y(6) = 6(6) - (6)^2 = 0$$

Thus the areas of the rectangles are as follows:

$$(1.5)(6.75) = 10.125$$

$$(1.5)(9) = 13.5$$

$$(1.5)(6.75) = 10.125$$

$$(1.5)(0) = 0$$

The area under the parabola is approximately $10.125 + 13.5 + 10.125 + 0 = 33.75$. The answer is (C).

ANSWERS AND EXPLANATIONS TO SECTION II

1. Oil is being pumped into a cavern for storage. The cavern is 400 meters deep. The area of the horizontal cross-section of the chamber at a depth y is given by the function A, where $A(y)$ is measured in square meters. The function A is continuous and decreases as depth increases. Selected values for $A(y)$ are given in the table below.

y (meters)	0	100	150	250	400
$A(y)$ (square meters)	65	38.4	30.6	20.2	11.3

A. Use a right Riemann sum with the four subintervals indicated by the data in the table to approximate the volume of the chamber. Indicate the units of measure.

To evaluate the Riemann sum, add up the areas of each subinterval, using the width of the subinterval and $A(y)$ as the height to get $(100 - 0)A(100) + (150 - 100)A(150) + (250 - 150)A(250) + (400 - 250)A(400) = (100)(38.4) + (50)(30.6) + (100)(20.2) + (150)(11.3) = 9{,}085$ cubic meters.

B. Does the approximation in part A overestimate or underestimate the volume of the cavern? Explain your reasoning.

The function is decreasing and a right Riemann sum is used, so the approximation is an underestimate.

C. The area in square meters of the horizontal cross-section at depth y is modeled by the function f given by $f(y)=\dfrac{65}{e^{0.004y}+.002y}$. Based on this model, find the volume of the tank. Indicate the units of measure.

Find the volume by evaluating the integral $V(y)=\int_0^{400} f(y)\,dy$ to get $\int_0^{400}\dfrac{65}{e^{0.004y}+.002y}dy=11{,}507.7723$. Therefore, the volume is 11,507.7723 cubic meters.

D. Oil is pumped into the cavern. When the depth of the oil is 200 meters, the depth is increasing at a rate of 0.45 meter per minute. Using the model from part C, find the rate at which the volume of oil is changing with respect to time when the depth of the oil is 200 meters. Indicate the units of measure.

Use the model $V(y)=\int_0^h f(y)\,dy$. If you differentiate the volume with respect to time, you get $\dfrac{dV}{dt}=\dfrac{dV}{dh}\dfrac{dh}{dt}=f(h)\dfrac{dh}{dt}$. At a depth of 200 meters, you get $f(200)(0.45) = 11.141$. Thus, at $y = 200$, the volume of the oil is changing at a rate of 11.141 cubic meters per minute.

2. When a subway leaves its initial station, there are 25 passengers on board. Passengers leave a subway at a rate modeled by the function $f(t)$ given by

$$f(t)=12+(0.5t)\sin\left(\frac{t^2}{40}\right);0\le t\le 60$$

where $f(t)$ is measured in passengers per minute and t is measured in minutes since the subway leaves its initial station. After the subway has been traveling for 20 minutes, passengers board the subway at a rate modeled by

$$g(t) = 20 + 3.2\ln(t^2 + 4t);\ 20 \le t \le 60$$

where the function $g(t)$ is measured in passengers per minute and t is the number of minutes since the subway leaves its initial station.

A. How many passengers leave the subway during the time interval $0 \le t \le 20$? Round to the nearest passenger.

Find the number of passengers who leave in the time interval by integrating $f(t)$ over $0 \le t \le 20$ to get $\int_0^{20} 12+(0.5t)\sin\left(\frac{t^2}{40}\right)dt=258.391$. Therefore, 258 passengers leave in the time interval $0 \le t \le 20$.

B. Find $f'(20)$. Using the correct units, explain the meaning of $f'(20)$ in the context of this problem.

Use the calculator to find that $f'(20) = -8.663$. The rate at which passengers are leaving the subway is changing at a rate of 8.663 passengers per minute per minute (or passengers per minute squared).

C. Is the number of passengers on the subway increasing or decreasing at time $t = 20$? Justify your reason for your answer.

Find the number of passengers on the subway at $t = 20$ by looking at the difference between the number of passengers boarding and the number of passengers leaving: $g(20) - f(20) = 33.196$. Because the difference is positive, the number of passengers on the subway is increasing at $t = 20$.

D. How many passengers are on the subway at time $t = 30$? Round to the nearest passenger.

Find the total number of passengers on the subway at $t = 30$ by taking the initial number of passengers, adding the number of passengers who boarded and subtracting the number of passengers who left to get $25 + \int_{20}^{30} g(t)\,dt - \int_{0}^{30} f(t)\,dt = 56.648$. There are 57 passengers on the train at the time $t = 30$.

3.

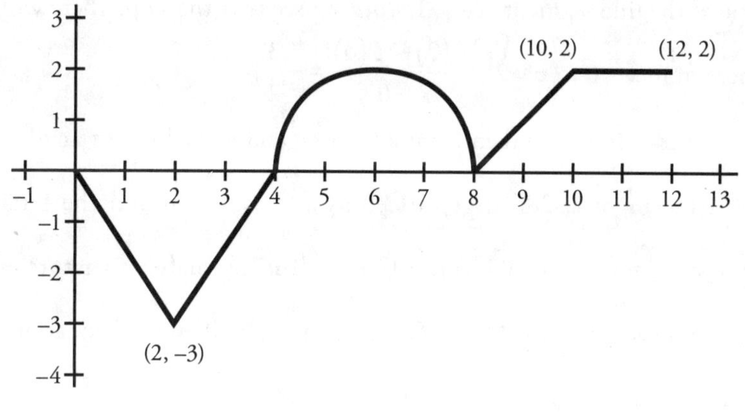

Graph of $f'(x)$

The function f is differentiable on the closed interval [0, 12] and $f(0) = 7$. The graph of $f'(x)$ above consists of a semicircle and four line segments.

A. Find the values of $f(4)$ and $f(12)$.

Find $f(4)$ by the sum of the value of f and the integral of $f'(x)$ from $x = 0$ to $x = 4$ to get $7 + \int_0^4 f'(x)\,dx = 7 - \frac{1}{2}(4)(3) = 1$.

Find $f(12)$ by the sum of the value of f and the integral of $f'(x)$ from $x = 0$ to $x = 12$ to get

$$7+\int_0^{12} f'(x)\,dx = 7-\frac{1}{2}(4)(3)+\frac{1}{2}\pi\left(2^2\right)+\frac{1}{2}(2)(2)+(2)(2) = 7+2\pi.$$

B. On which intervals is f decreasing? Justify your answer.

f is decreasing where $f'(x)$ is negative. This occurs on the interval (0, 4).

C. Find the absolute minimum value of f on the closed interval [0, 12]. Justify your answer.

The absolute minimum will be at a critical point where $f'(x) = 0$ or at an endpoint. $f'(x) = 0$ at $x = 0$, $x = 4$, and $x = 8$. Evaluate f at each of those values.

x	$f(x)$
0	7
4	1
8	$1 + 2\pi$
12	$7 + 2\pi$

The absolute minimum occurs at $f(4) = 1$.

D. Find $f''(1)$.

Find the slope of the line segment at $x = 1$. You can see that the slope there will be the same as the slope anywhere from $x = 0$ to $x = 2$: $\dfrac{f(2)-f(0)}{2-0} = -\dfrac{3}{2}$.

4. At time $t = 0$, an iron ingot is taken from a furnace and placed on an anvil to cool. The internal temperature of the ingot is 286 degrees Celsius at time $t = 0$ and the internal temperature is greater than room temperature, 30 degrees Celsius. The internal temperature of the ingot at time t minutes can be modeled by the function I that satisfies the differential equation $\dfrac{dI}{dt} = -\dfrac{1}{32}(I-30)$, where $I(t)$ is measured in degrees Celsius and $I(0) = 286$.

A. Write an equation for the line tangent to the graph of I at $t = 0$. Use this equation to approximate the internal temperature of the ingot at time $t = 8$.

First, you need the slope of the tangent line. $I'(0) = -\dfrac{1}{32}(286-30) = -8$. You also know that $I(0) = 286$, so the equation of the tangent line is $I - 286 = -8(t - 0)$ or $I = -8t + 286$.

Use the equation to find that the approximate internal temperature at time $t = 8$ is $I = -8(8) + 286 = 222$ degrees Celsius.

B. Use $\frac{d^2I}{dt^2}$ to determine whether your answer in part A is an overestimate or an underestimate of the internal temperature of the ingot at time $t = 8$.

$\frac{d^2I}{dt^2} = -\frac{1}{32}\frac{dI}{dt} = -\frac{1}{32}\left(-\frac{1}{32}(I-30)\right) = \frac{1}{1{,}024}(I-30)$. This is positive so the graph of I is concave up for all $t > 0$. This means that the answer in part A is an underestimate because the tangent line will be below the curve.

C. If the ingot is immersed in a cooling bath, another model of the internal temperature of the ingot at time t minutes is the function B that satisfies the differential equation $\frac{dB}{dt} = -(B-30)^{\frac{3}{4}}$, where $B(t)$ is measured in degrees Celsius and $B(0) = 286$. Using this model, what is the internal temperature of the ingot at time $t = 8$?

Solve this differential equation using separation of variables. First, separate the variables:

$$\frac{dB}{(B-30)^{\frac{3}{4}}} = -dt$$

Integrate both sides:

$$\int (B-30)^{-\frac{3}{4}}\,dB = -\int dt$$

$$4(B-30)^{\frac{1}{4}} = -t + C$$

Use the initial condition to solve for C:

$$4(286-30)^{\frac{1}{4}} = 0 + C$$

$$C = 16$$

This gives you $4(B-30)^{\frac{1}{4}} = -t + 16$. Finally, isolate B:

$$(B-30)^{\frac{1}{4}} = -\frac{1}{4}t + 4$$

$$B = \left(-\frac{1}{4}t + 4\right)^4 + 30$$

Plug in $t = 8$ to get that the internal temperature of the ingot is $B = \left(-\frac{1}{4}t + 4\right)^4 + 30 = 46$ degrees Celsius.

5. Consider the curve given by $2xy - y^2 = 3$.

A. Write an equation for the tangent line to the curve at the point (2, 1).

Use implicit differentiation to find $\frac{dy}{dx}$.

You get $2x\frac{dy}{dx} + 2y - 2y\frac{dy}{dx} = 0$. Next, group the terms with $\frac{dy}{dx}$ on one side of the equals sign and the terms without $\frac{dy}{dx}$ on the other side: $2y = 2y\frac{dy}{dx} - 2x\frac{dy}{dx}$. Factor out $\frac{dy}{dx}$: $2y = (2y - 2x)\frac{dy}{dx}$.

And isolate $\frac{dy}{dx}$: $\frac{dy}{dx} = \frac{2y}{2y - 2x} = \frac{y}{y - x}$.

Now plug in (2, 1) to get the slope of the tangent line: $\frac{dy}{dx} = \frac{1}{1-2} = -1$. Therefore, the tangent line is $y - 1 = -1(x - 2)$.

B. Find the coordinates of all points on the curve at which the line tangent to the curve at that point is vertical.

The tangent line will be vertical at all points where the denominator of the derivative is zero, providing the numerator is not also zero there. Here you get that the denominator of the derivative is zero where $y = x$. Now go back to the original equation and substitute x for y to get $2x(x) - (x)^2 = 3$. Solve this for x: $2x(x) - (x)^2 = 3$.

$$x^2 = 3$$

$$x = \pm\sqrt{3}$$

Because $y = x$, the points are $(\sqrt{3}, \sqrt{3})$ and $(-\sqrt{3}, -\sqrt{3})$.

C. Evaluate $\frac{d^2y}{dx^2}$ at the points on the curve where $x = 2$ and $y = 1$.

Take the derivative of $\frac{dy}{dx}$: $\frac{d^2y}{dx^2} = \frac{(y-x)\frac{dy}{dx} - y\left(\frac{dy}{dx} - 1\right)}{(y-x)^2}$. You know from part A above that at (2, 1), $\frac{dy}{dx} = -1$, so substitute into the second derivative to get $\frac{d^2y}{dx^2} = \frac{(-1)(-1) - (-1-1)}{(1-2)^2} = 3$.

6. Two particles move along the x-axis. For $0 \le t \le 10$, the velocity of particle A at time t is given by $v_A(t) = t^2 - 5t + 4$ and the position of particle B at time t is given by $x_B(t) = \frac{t^3}{3} - 4t^2 + 12t$. Particle A is at position $x = 3$ at time $t = 0$.

A. For $0 \le t \le 10$, when is particle B moving to the left?

Particle B will be moving to the left where its velocity (the derivative of position) is negative. You get $x_B'(t) = v_B(t) = t^2 - 8t + 12$. You can find where this is negative by setting the derivative equal to zero and then sign testing the intervals between critical values.

$v_B(t) = t^2 - 8t + 12 = (t-2)(t-6) = 0$ gives you critical values of $t = 2$ and $t = 6$. Next do the sign test:

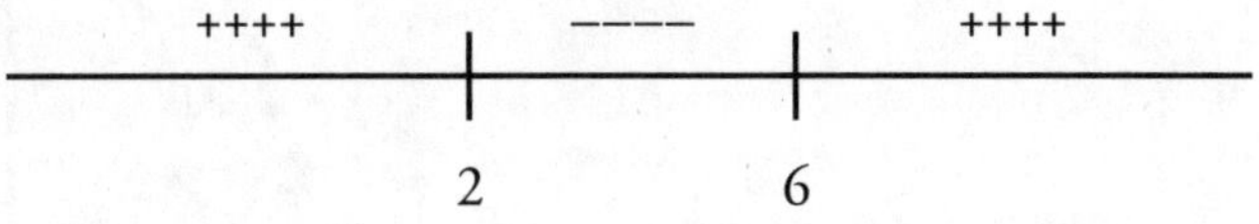

Thus, the particle is moving to the left on the interval $2 < t < 6$.

B. For $0 \le t \le 10$, find all times t when the two particles are traveling in the same direction.

Now perform the same analysis for particle A. You already have the velocity, so factor the derivative to find the critical values: $v_A(t) = t^2 - 5t + 4 = (t-1)(t-4)$. Thus, it has critical values at $t = 1$ and $t = 4$. Again, do a sign test:

Both particles move in the same direction for $0 < t < 1$, $2 < t < 4$, and $6 < t < 10$ because v_A and v_B have the same sign on those intervals.

C. Find the acceleration of particle A at time $t = 3$. Is its speed increasing, decreasing, or neither at time $t = 3$? Justify your answer.

Find the acceleration by taking the derivative of the velocity to get $a_A(t) = v_A'(t) = 2t - 5$. If you want to know whether the speed is increasing or decreasing at $t = 3$, find the signs of the acceleration and velocity. If the signs are the same, the particle's speed is increasing, and if the opposite, the particle's speed is decreasing. At $t = 3$, $a_A(3) = 2(3) - 5 = 1$ and $v_A(3) = (3)^2 - 5(3) + 4 = -2$. The velocity and acceleration have opposite signs, so the speed of the particle is decreasing.

D. Find the position of particle A the first time that it changes direction.

You know from part B above that particle A changes direction for the first time at time $t = 1$. You can find its position by evaluating $x_A = 3 + \int_0^1 (t^2 - 5t + 4)\,dt$ to get

$$x_A = 3 + \int_0^1 (t^2 - 5t + 4)\,dt = 3 + \left(\frac{t^3}{3} - \frac{5t^2}{2} + 4t\right)\Bigg|_0^1 = \frac{29}{6}$$

PRACTICE TEST 1: FREE-RESPONSE QUESTIONS GRADING RUBRICS

Scoring Guidelines for Question 1

A.	Sets up correct sum	**1 point**
	Answer	**1 point**
B.	Answer	**1 point**
	Explanation	**1 point**
C.	Correct integral	**1 point**
	Answer	**1 point**
	Correct units	**1 point**
D.	Correct derivative	**1 point**
	Answer	**1 point**
Total		**9 points**

Scoring Guidelines for Question 2

A.	Correct integral	**1 point**
	Answer	**1 point**
B.	Answer	**1 point**
	Explanation	**1 point**
	Correct units	**1 point**
C.	Answer	**1 point**
	Justification	**1 point**
D.	Correct integral	**1 point**
	Answer	**1 point**
Total		**9 points**

Scoring Guidelines for Question 3

A.	Correct integral	**1 point**
	Correct evaluation of $f(4)$	**1 point**
	Correct evaluation of $f(12)$	**1 point**
B.	Correct justification	**1 point**
	Answer	**1 point**
C.	Answer	**1 point**
	Correct justification	**1 point**
D.	Correct slope	**1 point**
	Answer	**1 point**
Total		**9 points**

Scoring Guidelines for Question 4

A.	Correct slope of tangent line	**1 point**
	Correct equation of tangent line	**1 point**
	Answer	**1 point**
B.	Correct second derivative	**1 point**
	Correct justification	**1 point**
	Answer (underestimate)	**1 point**
C.	Correct integration	**1 point**
	Correct constant of integration	**1 point**
	Answer	**1 point**
Total		**9 points**

Scoring Guidelines for Question 5

A.	Correct derivative	**1 point**
	Correct slope	**1 point**
	Correct equation of tangent line	**1 point**
B.	Finds where denominator of derivative is zero	**1 point**
	Solutions	**2 points**
C.	Correct derivative	**2 points**
	Answer	**1 point**
Total		**9 points**

Scoring Guidelines for Question 6

A.	Correct critical values	**1 point**
	Answer	**1 point**
B.	Correct critical values	**1 point**
	Answer	**1 point**
C.	Finding acceleration	**1 point**
	Correctly finds "decreasing"	**1 point**
	Justification	**1 point**
D.	Correct integral	**1 point**
	Answer	**1 point**
Total		**9 points**

HOW TO SCORE PRACTICE TEST 1

Section I: Multiple-Choice

__________ × 1.6667 = __________

Number of Correct (out of 45) | Weighted Section I Score (Do not round)

Exact scoring can vary from administration to administration. Therefore, this scoring should only be used as an estimate.

Section II: Free-Response

(See if you can find a teacher or classmate to score your Free-Response questions.)

Question 1 __________ (out of 9) × 1.3889 = __________ (Do not round)

Question 2 __________ (out of 9) × 1.3889 = __________ (Do not round)

Question 3 __________ (out of 9) × 1.3889 = __________ (Do not round)

Question 4 __________ (out of 9) × 1.3889 = __________ (Do not round)

Question 5 __________ (out of 9) × 1.3889 = __________ (Do not round)

Question 6 __________ (out of 9) × 1.3889 = __________ (Do not round)

Sum = __________

Weighted Section II Score (Do not round)

AP Score Conversion Chart Calculus AB

Composite Score Range	AP Score
112–150	5
98–111	4
80–97	3
55–79	2
0–54	1

Composite Score

__________ + __________ = __________

Weighted Section I Score | Weighted Section II Score | Composite Score (Round to nearest whole number)

Part III
About the AP Calculus AB Exam

- Calculus AB vs. Calculus BC
- The Structure of the AP Calculus AB Exam
- The AP Calculus AB Exam Is Hybrid Digital
- How the AP Calculus AB Exam Is Scored
- Past AP Calculus AB Score Distributions
- Overview of Content Topics
- General Overview of This Book
- How AP Exams Are Used
- Other Resources
- Designing Your Study Plan

Please note that this book will focus only on the topics that will appear on the AP Calculus AB Exam. If you are looking to prepare for the AP Calculus BC Exam, please purchase *Princeton Review AP Calculus BC Prep,* in stores now!

CALCULUS AB VS. CALCULUS BC

AP Calculus is divided into two types: AB and BC. The former is supposed to be the equivalent of a semester of college calculus; the latter, a year. In truth, Calculus AB covers closer to three-quarters of a year of college calculus. In fact, the main difference between the two is that Calculus BC tests some more theoretical aspects of calculus and covers a few additional topics. In addition, Calculus BC is harder than Calculus AB. The AB exam usually tests straightforward problems in each topic. They're not too tricky, and they don't vary too much. The BC exam asks harder questions. But neither exam asks trick questions, nor do they test esoteric aspects of calculus. Rather, both tests tend to focus on testing whether you've learned the basics of differential and integral calculus. The tests are difficult because of the breadth of topics that they cover, not the depth. You will probably find that many of the problems in this book seem easier than the problems you've had in school. This is because your teacher is giving you problems that are harder than those on the AP Exam.

Be Calculating

Students may use a four-function, scientific, or graphing calculator. However, be aware that some models have unapproved features (like keyboards, styluses, or wireless capability) and are not permitted, so please check the College Board website. If you're still unsure, you can take two calculators so that if one is rejected (or not functioning), you do not have to rely on a school-provided backup that you may be unfamiliar with.

THE STRUCTURE OF THE AP CALCULUS AB EXAM

Section/Type	Time	Score	More Score Info
Section I, Part A: 30 multiple-choice questions	60 minutes	Along with I-B, 50% of complete AP Exam score	No guessing penalty, calculator NOT permitted
Section I, Part B: 15 multiple-choice questions	45 minutes	Along with I-A, 50% of complete AP Exam score	No guessing penalty, calculator allowed
Section II, Part A: 2 open-ended questions	30 minutes	Along with II-B, 50% of complete AP Exam score	Calculator allowed
Section II, Part B: 4 open-ended questions	60 minutes	Along with II-A, 50% of complete AP Exam score	Calculator NOT permitted

Sign in to your online Student Tools to download and print a handy appendix of tips aimed at getting the most out of the Desmos calculator.

There is no guessing penalty on the test, so don't leave anything blank in Section I. Section II, however, requires you to not only write out your solutions, but also show your steps. You can get partial credit, so make sure you put everything down on paper. You'll note, too, that some parts of the test allow you to use a graphing calculator and its programs. But here's the truth about calculus: most of the time, you don't need a calculator, so don't get stuck trying to find the right program. Go with what you know.

THE AP CALCULUS AB EXAM IS HYBRID DIGITAL

While the multiple-choice section of the AP Calculus AB Exam is administered digitally via the College Board's Bluebook testing app, the free-response questions are viewed digitally, but your answers must be written in the paper exam booklet given to you and returned for scoring.

For the digital portion of the AP Calculus AB, access to the Bluebook app is available via Windows, Mac laptops/desktops, iPads, tablets, and Chromebooks. If your computer or tablet is owned or managed by your school, your school official will likely already have this installed. The test cannot be taken on a smartphone.

The Bluebook app allows you to annotate and highlight texts, eliminate answers, and flag questions for review later. You will also have access to scratch paper to plan responses to essay questions. Unlike other digital exams like the SAT, the digital AP Exam will NOT be adaptive: this means that the difficulty of questions will not change depending on how well you do on earlier sections of the test.

It is important to note that the digital AP Calculus AB Exam includes the same number of sections, types of questions, and timing as the previous paper exam. This means the strategies used for success on paper versions of the exam are just as applicable to the digital version. So don't worry, the practice tests in this book will still prepare you for test day and digital practice tests are available via your student tools.

AP Exams Go Digital in 2025

College Board announced that a number of AP Exams are now fully digital or hybrid digital! For the latest information regarding the specific AP subjects going digital and testing accommodations, visit the College Board website at apcentral.collegeboard.org/exam-administration-ordering-scores/digital-ap-exams.

HOW THE AP CALCULUS AB EXAM IS SCORED

A numeral score of 1 to 5 is going to be assigned to your test, based on the number of questions you've answered correctly.

5 = Extremely Well Qualified
4 = Well Qualified
3 = Qualified
2 = Possibly Qualified
1 = No Recommendation

Colleges decide for themselves the minimum score they will accept for college credit and/or advanced placement. The American Council on Education recommends the acceptance of grades 3 and above, and many colleges adhere to these standards. About 60 percent of students who take the AP Calculus AB Exam receive a score of 3 or higher. Check the website for each college you plan to apply to so that you know its policy on granting credit or advanced placement.

PAST AP CALCULUS AB SCORE DISTRIBUTIONS

Score	Percentage 2024	Credit Recommendation	College Grade Equivalent
5	21.4%	Extremely Well Qualified	A
4	27.8%	Well Qualified	A–, B+, B
3	15.3%	Qualified	B–, C+, C
2	22.7%	Possibly Qualified	–
1	12.9%	No Recommendation	–

Scores taken from May 2024 test administration. Data taken from the College Board website.

OVERVIEW OF CONTENT TOPICS

The content in this book has been organized to align with the College Board's recommended syllabus, as that is likely what your AP instructor has been using. If your course has been taught in a different order, you should still be able to find the appropriate subjects by using the Table of Contents.

Chapter 3: Limits and Continuity (10–12)%

- You should be able to calculate limits algebraically or to estimate them from a graph or from a table of data.
- You do **not** need to find limits using the Delta-Epsilon definition of a limit.
- You should understand asymptotes in terms of limits involving infinity.
- You should be able to connect limits at infinity with horizontal asymptotes.
- You should be able to determine limits using the Squeeze Theorem.
- You should be able to test the continuity of a function in terms of limits.
- You should understand continuous functions graphically.
- You should understand the Intermediate Value Theorem (IVT).

Chapter 4: Differentiation: Definition and Basic Derivative Rules (10–12%)

- You should be able to find the instantaneous rate of change of a function using the derivative or the limit of the average rate of change of a function.
- You should be able to approximate the rate of change of a function from a graph or from a table of values.
- You should be able to find a derivative by finding the limit of the difference quotient.

- You should also know the relationship between differentiability and continuity. That is, if a function is differentiable at a point, it's continuous there. But if a function is continuous at a point, it's not necessarily differentiable there.
- You should know the Power Rule, the Product Rule, and the Quotient Rule.
- You should be able to find the derivatives of trigonometric functions, e^x, and $\ln x$.
- You should be able to find the slope of a curve at a point and the tangent and normal lines to a curve at a point.

Chapter 5: Differentiation: Composite, Implicit, and Inverse Functions (9–13%)

- You should be able to differentiate using the Chain Rule.
- You should be able to find Higher-Order Derivatives and to use Implicit Differentiation.
- You should be able to find the derivative of the inverse of a function, including Inverse Trigonometric Functions.

Chapter 6: Contextual Applications of Differentiation (10–15%)

- You should be able to solve Rectilinear Motion (a.k.a. Position, Velocity, and Acceleration) problems using derivatives.
- You should be able to solve Related Rates problems.
- You should also be able to use local linear approximation and differentials to estimate the tangent line to a curve at a point.
- You should know L'Hospital's Rule for determining limits of Indeterminate Forms.

Chapter 7: Analytical Applications of Differentiation (15–18%)

- You should know the Mean Value Theorem for derivatives and Rolle's Theorem.
- You should know the Extreme Value Theorem.
- You should be able to relate the graph of a function to the graph of its derivative, and vice versa. This is tricky.
- You should know the relationship between the sign of a derivative and whether the function is increasing or decreasing (positive derivative means increasing; negative means decreasing).
- You should know how to find relative and absolute maxima and minima.
- You should know the relationship between concavity and the sign of the second derivative (positive means concave up; negative means concave down).
- You should know how to find points of inflection.
- You should be able to sketch a curve using first and second derivatives and be able to analyze the critical points.
- You should be able to solve optimization problems (max/min problems).

There's More Online
You can find a full list of all the formulas in this book in your online Student Tools, which also contain a wealth of other resources, like calculator shortcuts.

Chapter 8: Integration and Accumulation of Change (17–20%)

- You should be able to use integrals to evaluate accumulation functions.
- You should be able to approximate areas using Riemann sums, specifically using left, right, and midpoint evaluations and the Trapezoid Rule.
- You should know the Fundamental Theorem of Calculus and be able to use it to find the derivative of an integral and for the analytical and graphical analysis of functions.
- You should be able to find specific antiderivatives using initial conditions.
- You should be able to integrate using the Power Rule and *u*-substitution.
- You should be able to integrate functions using long division and completing the square.

Chapter 9: Differential Equations (6–12%)

- You should be able to interpret differential equations via slope fields. Don't be intimidated. These look harder than they are.
- You should be able to solve differential equations using Separation of Variables.

Chapter 10: Applications of Integration (10–15%)

- You should be able to find the Average Value of a Function.
- You should be able to solve Rectilinear Motion (a.k.a. Position, Velocity, and Acceleration) problems using integrals.
- You should be able to find the area of a region.
- You should be able to find the volume of a solid of known cross-section.
- You should be able to find the volume of a solid of revolution.

GENERAL OVERVIEW OF THIS BOOK

The key to doing well on the exam is to memorize a variety of techniques for solving calculus problems and to recognize when to use them. There's so much to learn in AP Calculus that it's difficult to remember everything. Instead, you should be able to derive or figure out how to do certain things based on your understanding of a few essential techniques. In addition, you'll be expected to remember a lot of the math that you did before calculus—particularly trigonometry. You should be able to graph functions and find zeros, derivatives, and integrals with the calculator.

Furthermore, if you can't derive certain formulas, you should memorize them! A lot of students don't bother to memorize the trigonometry special angles and formulas because they can do them on their calculators. This is a big mistake. You'll be expected to be very good with these in calculus, and if you can't recall them easily, you'll be slowed down and the problems will seem much harder. Make sure that you're also comfortable with analytic geometry. If you rely on your calculator to graph for you, you'll get a lot of questions wrong because you won't recognize the curves when you see them.

This advice is going to seem backward compared with what your teachers are telling you. In school, you're often told not to simply memorize things. Teachers tell you to understand the concepts, not just memorize the answers. Well, things are different here. The understanding will come later, after you're comfortable with the mechanics. In the meantime, you should learn techniques and practice them, and, through repetition, you will ingrain them in your memory.

No, Really, There's Even More Online
Once you've registered your copy of the book, you can access drills on fundamental math skills and also work on challenging drills that put everything to the test.

HOW AP EXAMS ARE USED

Different colleges use AP Exams in different ways, so it is important that you go to a particular college's website to determine how it uses AP Exams. The three items below represent the main ways in which AP Exam scores can be used.

- **College Credit.** Some colleges will give you college credit if you score well on an AP Exam. These credits count toward your graduation requirements, meaning that you can take fewer courses while in college. Given the cost of college, this could be quite a benefit, indeed.
- **Satisfy Requirements.** Some colleges will allow you to "place out" of certain requirements if you do well on an AP Exam, even if they do not give you actual college credits. For example, you might not need to take an introductory-level course, or perhaps you might not need to take a class in a certain discipline at all.
- **Admissions Plus.** Even if your AP Exam will not result in college credit or even allow you to place out of certain courses, most colleges will respect your decision to push yourself by taking an AP course or even an AP Exam outside of a course. A high score on an AP Exam shows proficiency in more difficult content than is taught in many high school courses, and colleges may take that into account during the admissions process.

More Great Books
Check out The Princeton Review's college guide books, including *The Best 390 Colleges, The Complete Book of Colleges, Paying for College,* and many more!

OTHER RESOURCES

Break up your review into manageable portions. Download our helpful study guide for this book, once you register online.

There are many resources available to help you improve your score on the AP Calculus AB Exam, not the least of which are your **teachers**. If you are taking an AP class, you may be able to get extra attention from your teacher, such as obtaining feedback on your free-response questions. If you are not in an AP course, reach out to a calculus teacher and ask them to review your free-response questions or otherwise help you with content.

Another wonderful resource is **AP Students**, the official site of the AP Exams. The scope of the information at this site is quite broad and includes:

- a course description, which includes details on what content is covered
- sample test questions
- free-response prompts from previous years

The AP Students home page address is apstudent.collegeboard.org.

The AP Calculus AB Exam Course home page address is apcentral.collegeboard.org/courses/ap-calculus-ab.

Finally, **The Princeton Review** offers tutoring and small group instruction. Our expert instructors can help you refine your strategic approach and add to your content knowledge. For more information, call 1-800-2REVIEW.

DESIGNING YOUR STUDY PLAN

Looking to Guarantee a 4 or 5?
We now offer one-on-one tutoring for a guaranteed 5 or an online course for a guaranteed 4 on the AP Calculus AB Exam. For information on rates, availability, and to learn more about the guarantee, visit PrincetonReview.com/college/ap-test-prep.

As part of the Introduction, you identified some areas of potential improvement. Let's now delve further into your performance on Practice Test 1, with the goal of developing a study plan appropriate to your needs and time commitment.

Read the answers and explanations associated with the multiple-choice questions (starting on page 45). After you have done so, respond to the following questions:

- Review the Overview of Content Topics on pages 68–70. Next to each topic, indicate your rank of the topic as follows: "1" means "I need a lot of work on this," "2" means "I need to beef up my knowledge," and "3" means "I know this topic well."
- How many days/weeks/months away is your exam?
- What time of day is your best, most focused study time?
- How much time per day/week/month will you devote to preparing for your exam?
- When will you do this preparation? (Be as specific as possible: Mondays and Wednesdays from 3 to 4 P.M., for example.)
- Based on the answers above, will you focus on strategy (Part IV) or content (Part V) or both?
- What are your overall goals in using this book?

Part IV
Test-Taking Strategies for the AP Calculus AB Exam

PREVIEW ACTIVITY

Review your responses to the first three questions on page 4 of Part I, and then respond to the following questions:

- How many multiple-choice questions did you miss even though you knew the answer?
- On how many multiple-choice questions did you guess randomly?
- How many multiple-choice questions did you miss after eliminating some answers and guessing based on the remaining answers?
- Did you find any of the free-response questions easier or harder than the others, and if so, why?

HOW TO USE THE CHAPTERS IN THIS PART

Before you read the following Strategy chapters, take a moment to think about what you are doing now. As you read and engage in the directed practice, be sure to appreciate the ways you can change your approach. At the end of each chapter in Part IV, you will have the opportunity to reflect on how you will change your approach.

Chapter 1
How to Approach Multiple-Choice Questions

CRACKING THE MULTIPLE-CHOICE QUESTIONS

Section I of the AP Calculus Exam consists of 45 multiple-choice questions, which you're given 105 minutes to complete. This section is worth 50% of your score.

All the multiple-choice questions will have a similar format: each will be followed by four answer choices. At times, it may seem that there could be more than one possible correct answer. There is only one! Remember that the committee members who write these questions are calculus teachers. So, when it comes to calculus, they know how students think and what kind of mistakes they make. Answers resulting from common mistakes are often included in the four answer choices to trap you.

Let's Get Digital

Remember that the multiple-choice section of the AP Calculus AB exam will be administered in a digital format, so you will be taking it on the Bluebook app. It's not taken online, but it is taken digitally (Section I, that is). The test will be administered at your school on a desktop or laptop computer that has the Bluebook app installed. To prepare for this, it's best if you practice as much as you can in a digital environment so that you will be used to the look and feel of the test and know how to use the in-app Desmos calculator (if you wish to use that instead of your own calculator). Head to your online Student Tools (found at princetonreview.com/prep) and register your book to access a digital practice test.

Should You Guess?

Use Process of Elimination (POE) to rule out answer choices you know are wrong and increase your chances of guessing the right answer. Read all the answer choices carefully. Use the built-in elimination tool to cross out the ones that you know are wrong. If you have only one answer choice left, *choose it,* even if you're not completely sure why it's correct. Remember that questions in the multiple-choice section are graded by a computer, so it doesn't care *how* you arrived at the correct answer.

Proven Techniques
Use POE and the Two-Pass System to help boost your score.

Even if you can't eliminate answer choices, go ahead and guess: there is no penalty for doing so. You will be assessed only on the total number of correct answers, so be sure to fill in all the bubbles even if you have no idea what the correct answers are. When you get to questions that are too time-consuming, or that you don't know the answer to (and can't eliminate any options), don't just fill in any answer. Use what we call your "letter of the day" (LOTD). Selecting the same answer choice each time you guess will increase your odds of getting a few of those skipped questions right.

The Two-Pass System

The AP Calculus AB Exam covers a broad range of topics. There's no way, even with our extensive review, that you will know everything about every topic in calculus. So, what should you do?

Adopt what we call the two-pass system. The two-pass system entails going through the test and answering the easy questions first. Save the more time-consuming questions for later. (Don't worry—you'll have time to do them later!) First, read the question and decide if it is a "Now" or "Later" question. If you decide this is a "Now" question, solve it in your scratch paper and click on the answer in the Bluebook app. If it is a "Later" question, use the Mark for Review tool to add a little flag next to the question. Then keep on moving.

Once you've finished all the "Now" questions, go back to the questions that you Marked for Review and start in on those. Start with the easier questions first. These are the ones that require calculations or that require you to eliminate the answer choices (in essence, the correct answer does not jump out at you immediately).

REFLECT

Respond to the following questions:

- How long will you spend on each multiple-choice question?
- How will you change your approach to multiple-choice questions?
- What is your multiple-choice guessing strategy?

Bonus Tips and Tricks...
Check us out on YouTube for additional test-taking tips and must-know strategies at www.youtube.com/ThePrincetonReview.

Chapter 2
How to Approach Free-Response Questions

CRACKING THE FREE-RESPONSE QUESTIONS

Section II is worth 50% of your score on the AP Calculus Exam. This section is composed of two parts. Part A contains two free-response questions (you may use a calculator on this part); Part B contains four free-response questions for which calculators are not allowed. You're given a total of 90 minutes for this section.

Clearly Explain and Justify Your Answers

Remember that your answers to the free-response questions are graded by *readers* and not by computers. Communication is a very important part of AP Calculus. Compose your answers in precise sentences. Just getting the correct numerical answer is not enough. You should be able to *explain* your reasoning behind the technique that you selected and *communicate* your answer in the context of the problem. Even if the question does not explicitly say so, always explain and *justify* every step of your answer, including the final answer. Do not expect the graders to read between the lines. Explain everything as though somebody with no knowledge of calculus is going to read it. Be sure to present your solution in a systematic manner using solid logic and appropriate language. And remember, although you won't earn points for neatness, the graders can't give you a grade if they can't read and understand your solution!

Room to Write

On exam day, you will view the free-response questions in the Bluebook testing app and handwrite your responses in a paper exam booklet. You can download a sample AP Calculus AB booklet at apcentral.collegeboard.org/media/pdf/ap-calculus-ab-hybrid-sample-free-response-booklet.pdf.

You can use scrap paper for the FRQs in this book. For the practice tests, we also provide printable free-response booklets, which are available in your Student Tools via PrincetonReview.com/prep.

Use Only the Space You Need

Do not try to fill up the space provided for each question. The space given is usually more than enough. The people who design the tests realize that some students write in big letters or make mistakes and need extra space for corrections. So if you have a complete solution, don't worry about the extra space. Writing more will not earn you extra credit. In fact, many students tend to go overboard and shoot themselves in the foot by making a mistake after they've already written the right answer.

Read the Whole Question!

Some questions might have several subparts. Try to answer them all, but don't give up on the question if one part is giving you trouble. For example, if the answer to part B depends on the answer to part A, but you think you got the answer to part A wrong, you should still go ahead and do part B using your answer to part A as required. Chances are that the grader will not mark you wrong twice, unless it is obvious from your answer that you should have discovered your mistake.

REFLECT

Respond to the following questions:

- How much time will you spend on each free-response question?
- How will you change your approach to the free-response questions?
- Will you seek further help, outside of this book (such as a teacher, tutor, or AP Central), on how to approach the calculus exam?

Go Online!
Check us out on YouTube for test-taking tips and techniques to help you ace your next exam at www.youtube.com/ThePrincetonReview.

Part V
Content Review for the AP Calculus AB Exam

HOW TO USE THE CHAPTERS IN THIS PART

You may need to review the following content chapters more than once. Your goal is to obtain command of the content you are missing, and a single read of a chapter may not be sufficient. At the end of each chapter, you will have an opportunity to reflect on whether you truly have mastered the content of that chapter.

Chapter 3 Limits and Continuity

INTRODUCING CALCULUS: CAN CHANGE OCCUR AT AN INSTANT?

Suppose we are told that a car is traveling from point A to point B, which are separated by 3 miles. It leaves A and arrives at B 6 minutes later. We know that the average velocity of the car is $\frac{3}{6}=\frac{1}{2}$ miles per minute (mpm). We find that by taking the total distance, 3 miles, and dividing it by the total time, 6 minutes. But suppose that we want to know the car's velocity at exactly the 3-minute mark. We know that it averaged $\frac{1}{2}$ miles per minute, but that doesn't mean that its velocity was constant. It could have gone faster and slower throughout the trip. So how can we find the car's speed at that exact time? This is called the instantaneous velocity.

If at 2.9 minutes, the car had traveled 1.4 miles and at 3.1 minutes, the car had traveled 1.47 miles, the car's average velocity for that period is $\frac{1.47-1.4}{3.1-2.9}=\frac{0.07}{0.2}=0.35$ mpm. If at 2.99 minutes, the car had traveled 1.428 miles, and at 3.01 minutes, the car had traveled 1.436 miles, the car's average velocity for that period is $\frac{1.436-1.428}{3.01-2.99}=\frac{0.08}{0.02}=0.40$ mpm. This is a very short time interval, so we are probably close to the instantaneous velocity. The difficulty is that, in order to know the instantaneous velocity, we will have a time interval of zero in the denominator.

This was a problem that baffled mathematicians and scientists for a very long time. The solution uses limits, which we will learn about in this chapter. Then, in Chapter 4, we will learn how to find instantaneous velocity with calculus.

DEFINING LIMITS AND USING LIMIT NOTATION

In order to understand calculus, you need to know what a "limit" is. A limit is the value a function (which usually is written as "$f(x)$" on the AP Exam) approaches as the variable within that function (usually "x") gets nearer and nearer to a particular value. In other words, when x is very close to a certain number, what is $f(x)$ very close to?

Let's look at an example of a limit: What is the limit of the function $f(x) = x^2$ as x approaches 2? In limit notation, the expression "the limit of $f(x)$ as x approaches 2" is written like this: $\lim_{x \to 2} f(x)$. In order to evaluate the limit, let's check out some values of $\lim_{x \to 2^-} f(x)$ as x increases and gets closer to 2 (without ever exactly getting there).

When $x = 1.9$, $f(x) = 3.61$.
When $x = 1.99$, $f(x) = 3.9601$.
When $x = 1.999$, $f(x) = 3.996001$.
When $x = 1.9999$, $f(x) = 3.99960001$.

As x increases and approaches 2, $f(x)$ gets closer and closer to 4. This is called the **left-hand limit** and is written like this: $\lim_{x \to 2^-} f(x)$. Notice the little minus sign!

What about when x is bigger than 2?

When $x = 2.1$, $f(x) = 4.41$.
When $x = 2.01$, $f(x) = 4.0401$.
When $x = 2.001$, $f(x) = 4.004001$.
When $x = 2.0001$, $f(x) = 4.00040001$.

As x decreases and approaches 2, $f(x)$ still approaches 4. This is called the **right-hand limit** and is written like this: $\lim_{x \to 2^+} f(x)$. Notice the little plus sign!

We got the same answer when evaluating both the left- and right-hand limits, because when x is 2, $f(x)$ is 4. You should always check both sides of the independent variable because, as you'll see shortly, sometimes you don't get the same answer. Therefore, we write that $\lim_{x \to 2} x^2 = 4$.

We didn't really need to look at all of these decimal values to know what was going to happen when x got really close to 2. But it's important to go through the exercise because, typically, the problems get a lot more complicated. Let's do a few examples.

Example 1: Find $\lim_{x \to 5} x^2$.

The approach is simple: plug in 5 for x, and you get 25.

Example 2: Find $\lim_{x \to 3} x^3$.

Here the answer is 27.

Example 3: Find $\lim_{x \to 0} \left(x^2 + 5x\right)$.

Plug in 0, and you get 0.

So far, so good. All you do to find the limit of a simple polynomial is plug in the number that the variable is approaching and see what the answer is. Naturally, the process can get messier—especially if x approaches zero.

PRACTICE PROBLEM SET 1

Try these 6 problems to test your skills. The answers are at the end of the chapter, starting on page 116.

1. $\lim_{x\to 8}\left(x^2-5x-11\right)=$

2. $\lim_{x\to 5}\left(\frac{x+3}{x^2-15}\right)=$

3. Let $f(x)=\begin{cases} x^2-5, & x\le 3 \\ x+2, & x>3 \end{cases}$

Find: (a) $\lim_{x\to 3^-} f(x)$; (b) $\lim_{x\to 3^+} f(x)$; and (c) $\lim_{x\to 3} f(x)$.

4. Let $f(x)=\begin{cases} x^2-5, & x\le 3 \\ x+1, & x>3 \end{cases}$

Find: (a) $\lim_{x\to 3^-} f(x)$; (b) $\lim_{x\to 3^+} f(x)$; and (c) $\lim_{x\to 3} f(x)$.

5. Find $\lim_{x\to \frac{\pi}{4}} 3\cos x$.

6. Find $\lim_{x\to 0} 3\frac{x}{\cos x}$.

ESTIMATING LIMIT VALUES FROM GRAPHS

One way to find limits is to look at a graph of a function. This will help us see the value that a function approaches as we take the limit.

Example 4: Find $\lim_{x\to 0}\frac{1}{x^2}$.

If you plug in some very small values for *x*, you'll see that this function approaches ∞. And it doesn't matter whether *x* is positive or negative, you still get ∞. Look at the graph of $y=\frac{1}{x^2}$:

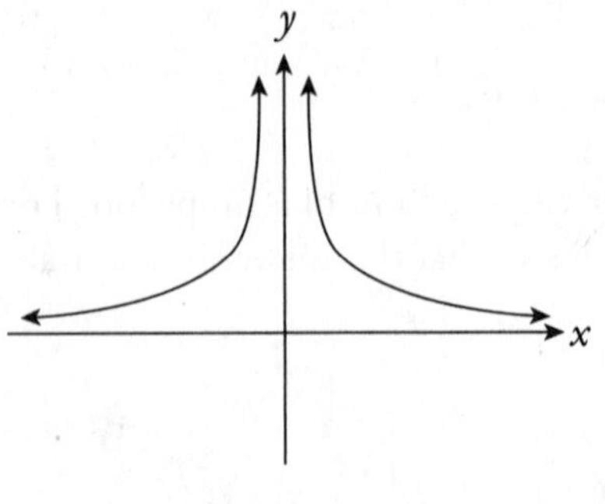

On either side of $x = 0$ (the y-axis), the curve approaches ∞.

Example 5: Find $\lim_{x \to 0} \frac{1}{x}$.

Here you have a problem. If you plug in some very small positive values for x (0.1, 0.01, 0.001, and so on), you approach ∞. In other words, $\lim_{x \to 0^+} \frac{1}{x} = \infty$. But, if you plug in some very small negative values for x (–0.1, –0.01, –0.001, and so on), you approach $-\infty$. That is, $\lim_{x \to 0^-} \frac{1}{x} = -\infty$. Because the right-hand limit is not equal to the left-hand limit, the limit does not exist.

Look at the graph of $\frac{1}{x}$.

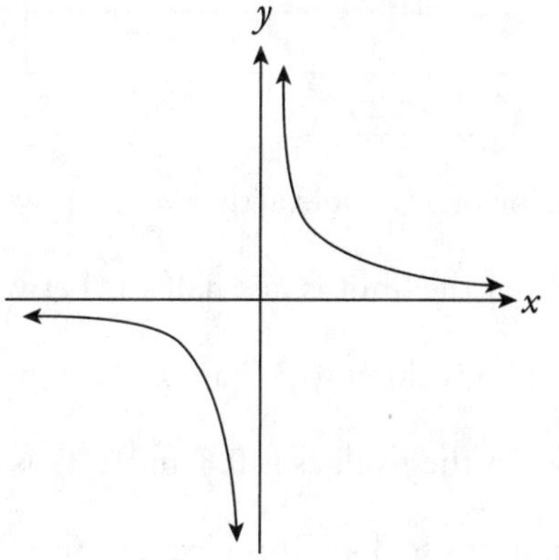

You can see that on the left side of $x = 0$, the curve approaches $-\infty$, and on the right side of $x = 0$, the curve approaches ∞. There are some very important points that we need to emphasize from the last two examples.

Why do we state the limit in Example 4 but not for Example 5? Because when we have $\frac{k}{x^2}$, in which k is positive constant, the function is always positive no matter what the sign of x is and thus the function has the same limit from the left and the right. But when we have $\frac{k}{x}$, the function's sign depends on the sign of x, and you get a different limit from each side.

(1) If the left-hand limit of a function is not equal to the right-hand limit of the function at a point, then the limit does not exist.

(2) A limit equal to infinity is different from a limit that does not exist, but sometimes you will see the expression "no limit," which serves both purposes. If $\lim_{x \to a} f(x) = \infty$, the limit, technically, does not exist.

(3) If k is a positive constant, then $\lim_{x \to 0^+} \frac{k}{x} = \infty$, $\lim_{x \to 0^-} \frac{k}{x} = -\infty$, and $\lim_{x \to 0} \frac{k}{x}$ does not exist.

(4) If k is a positive constant, then $\lim_{x \to 0^+} \frac{k}{x^2} = \infty$, $\lim_{x \to 0^-} \frac{k}{x^2} = \infty$, and $\lim_{x \to 0} \frac{k}{x^2} = \infty$.

ESTIMATING LIMIT VALUES FROM TABLES

Sometimes we will be asked to estimate a limit without knowing the actual function or graph of the function, but instead will simply be given a table of values. The usual procedure is to interpolate, using the two closest x-values and their corresponding y-values to estimate the limit. Let's do an example.

Example 6: Estimate $\lim_{x\to 3} f(x)$ given the following table of values:

x	$f(x)$
0.5	1.8
1.25	2.1
2	2.55
2.75	3.4
3.5	4.2
4.25	5.2

Note that we don't have the value of $f(x)$ at exactly $x = 3$, but we do have values at $x = 2.75$ and at $x = 3.5$. We could estimate that the limit is just halfway between the two f-values. This gives us $\lim_{x\to 3} f(x) = 3.8$. However, 2.75 is closer to 3 than 3.5 is, so maybe we should factor that into our estimate. The spread between the f-values is 0.8 and 2.75 is twice as close to 3 as 3.5 is, so we should take $\frac{1}{3}$ of 0.8 and add this to 3.4. This gives us $\lim_{x\to 3} f(x) = 3.667$. This isn't an exact answer, but it's close enough such that when you have answer choices to choose between, you'll be able to pick the best one.

Let's do another example.

Example 7: Estimate $\lim_{x\to 0} f(x)$ given the following table of values:

x	$f(x)$
–0.1	4.2
–0.01	4.249
–0.001	4.2499
0.001	4.2501
0.01	4.251
0.1	4.3

Again, note that we don't have the value of $f(x)$ at exactly $x = 0$, but we do have values at $x = 0.001$ and at $x = -0.001$, both of which are pretty close to zero. We could estimate that the limit is just halfway between the two f-values. This gives us $\lim_{x\to 0} f(x) = 4.25$. Notice also that the values on both sides of $x = 0$ seem to be "closing in" on 4.25, so that's very likely the limit.

DETERMINING LIMITS USING ALGEBRAIC PROPERTIES OF LIMITS

Just as we can do algebra with variables, we can do algebra with limits.

There are some simple algebraic rules of limits that you should know.

$$\lim_{x\to a} kf(x) = k\lim_{x\to a} f(x)$$

Sample: $\lim_{x\to 5} 3x^2 = 3\lim_{x\to 5} x^2 = 3(25) = 75$

If $\lim_{x\to a} f(x) = L_1$ and $\lim_{x\to a} g(x) = L_2$, then $\lim_{x\to a}\left[f(x) + g(x)\right] = L_1 + L_2$

Sample: $\lim_{x\to 5}\left[x^2 + x^3\right] = \lim_{x\to 5} x^2 + \lim_{x\to 5} x^3 = 25 + 125 = 150$

If $\lim_{x\to a} f(x) = L_1$ and $\lim_{x\to a} g(x) = L_2$, then $\lim_{x\to a}\left[f(x)\cdot g(x)\right] = L_1 \cdot L_2$

Sample: $\lim_{x\to 5}\left[\left(x^2+1\right)\sqrt{x-1}\right] = \lim_{x\to 5}\left(x^2+1\right)\lim_{x\to 5}\sqrt{x-1} = (26)(2) = 52$

DETERMINING LIMITS USING ALGEBRAIC MANIPULATION

Sometimes we can use algebra to simplify a limit, making it easier to solve. Suppose we wanted to evaluate $\lim_{x \to 5} \frac{x^2 - 8x + 15}{x^2 - 3x - 10}$. Notice that if we plug in 5 for x, we get $\frac{0}{0}$. Before we decide that the limit does not exist (abbreviated *DNE*), let's factor the numerator and the denominator. We get $\lim_{x \to 5} \frac{x^2 - 8x + 15}{x^2 - 3x - 10} = \lim_{x \to 5} \frac{(x-3)(x-5)}{(x+2)(x-5)}$. We can cancel the $(x - 5)$ terms and get $\lim_{x \to 5} \frac{(x-3)}{(x+2)}$.

Now we plug in $x = 5$ and we get $\lim_{x \to 5} \frac{(x-3)}{(x+2)} = \frac{2}{7}$.

How did we know we can do this? When we are given a limit as x approaches a number (not infinity) of a polynomial divided by another polynomial, and when we plug in that number we get $\frac{0}{0}$, then we can factor the numerator and denominator and take the limit of the reduced function. That is, if we have $\lim_{x \to a} \frac{f(x)}{g(x)} = \frac{0}{0}$, then we will be able to factor $x - a$ out of the numerator and the denominator.

Example 8: Find $\lim_{x \to 10} \frac{x^2 - 7x - 30}{x^2 - 12x + 20}$.

Notice that if we plug in 10 for x, we get $\frac{0}{0}$. Let's factor the numerator and the denominator. We get $\lim_{x \to 10} \frac{x^2 - 7x - 30}{x^2 - 12x + 20} = \lim_{x \to 10} \frac{(x+3)(x-10)}{(x-2)(x-10)}$. We can cancel the $x - 10$ terms and get $\lim_{x \to 10} \frac{(x+3)}{(x-2)}$. Now we plug in $x = 10$ and we get $\lim_{x \to 10} \frac{(x+3)}{(x-2)} = \frac{13}{8}$.

Other places where we can use algebra will be with limits involving infinity.

Let's look at a few examples in which the independent variable approaches infinity.

Example 9: Find $\lim_{x \to \infty} \frac{1}{x}$.

As x gets bigger and bigger, the value of the function gets smaller and smaller. Therefore, $\lim_{x \to \infty} \frac{1}{x} = 0$.

Example 10: Find $\lim_{x \to -\infty} \frac{1}{x}$.

It's the same situation as the one in Example 9; as x decreases (approaches negative infinity), the value of the function increases (approaches zero). We write the following:

$$\lim_{x \to -\infty} \frac{1}{x} = 0$$

We don't have the same problem here that we did when x approached zero because "positive zero" is the same thing as "negative zero," whereas positive infinity is different from negative infinity.

Here's another rule.

> If k and n are constants with $n > 0$, then $\lim_{x \to \infty} \frac{k}{x^n} = \lim_{x \to -\infty} \frac{k}{x^n} = 0$.

Example 11: Find $\lim_{x \to \infty} \frac{3x+5}{7x-2}$.

When you have a rational expression (that is, a fraction with a polynomial in both the numerator and the denominator), you can't just plug ∞ into the expression. You'll get $\frac{\infty}{\infty}$. Solve these by using the following technique:

> Limits of the form $\frac{\infty}{\infty}$ are called "Indeterminate Forms," and we will study these in Chapter 6 when we learn L'Hospital's Rule.

> When an expression consists of a polynomial divided by another polynomial, divide each term of the numerator and the denominator by the highest power of x that appears in the expression.

The highest power of x in this case is x^1, so we divide every term in the expression (both top and bottom) by x, like so:

$$\lim_{x \to \infty} \frac{3x+5}{7x-2} = \lim_{x \to \infty} \frac{\frac{3x}{x} + \frac{5}{x}}{\frac{7x}{x} - \frac{2}{x}} = \lim_{x \to \infty} \frac{3 + \frac{5}{x}}{7 - \frac{2}{x}}$$

Now when we take the limit, the two terms containing x approach zero. We're left with $\frac{3}{7}$.

Example 12: Find $\lim_{x\to\infty}\dfrac{8x^2-4x+1}{16x^2+7x-2}$.

Divide each term by x^2. You get

$$\lim_{x\to\infty}\frac{8-\frac{4}{x}+\frac{1}{x^2}}{16+\frac{7}{x}-\frac{2}{x^2}}=\frac{8}{16}=\frac{1}{2}$$

Remember to focus your attention on the highest power of *x*.

Example 13: Find $\lim_{x\to\infty}\dfrac{-3x^{10}-70x^5+x^3}{33x^{10}+200x^8-1000x^4}$.

Divide each term by x^{10}.

$$\lim_{x\to\infty}\frac{-3x^{10}-70x^5+x^3}{33x^{10}+200x^8-1,000x^4}=\lim_{x\to\infty}\frac{-3-\frac{70}{x^5}+\frac{1}{x^7}}{33+\frac{200}{x^2}-\frac{1,000}{x^6}}=-\frac{3}{33}=-\frac{1}{11}$$

The other powers don't matter, because they're all going to disappear. Now we have two new rules for evaluating the limit of a rational expression as *x* approaches infinity.

PRACTICE PROBLEM SET 2

Now try these problems. The answers are at the end of the chapter, starting on page 117.

1. $\lim_{x\to 3}\left(\dfrac{x^2-2x-3}{x-3}\right)=$

2. $\lim_{x\to 0^+}\left(\dfrac{x}{|x|}\right)=$

3. Find $\lim_{h\to 0}\dfrac{(3+h)^2-9}{h}$.

4. Find $\lim_{h\to 0}\dfrac{\frac{1}{x+h}-\frac{1}{x}}{h}$.

5. $\lim_{x\to 6^+}\left(\dfrac{x+2}{x^2-4x-12}\right)=$

6. $\lim_{x\to 6^-}\left(\dfrac{x+2}{x^2-4x-12}\right)=$

SELECTING PROCEDURES FOR DETERMINING LIMITS

How do we know which technique to use when determining a limit? First, just try plugging in the number that the limit is approaching. If we get a number, we are done. Of course, this will almost never happen! Second, if we get $\frac{0}{0}$, try factoring the numerator and the denominator and canceling a common factor. Third, if we get a number over zero, then we need to look at the limits from the left side to see if they agree or not. If not, then the limit does not exist. Finally, we can try evaluating a limit at infinity by the techniques we have learned.

Let's do some examples.

Example 14: Find $\lim\limits_{x \to 9} \frac{x^2 - 4x - 45}{x^2 - 11x + 18}$.

Notice that if we plug in $x = 9$, we get $\frac{0}{0}$. Factor the numerator and denominator: $\lim\limits_{x \to 9} \frac{x^2 - 4x - 45}{x^2 - 11x + 18} = \lim\limits_{x \to 9} \frac{(x-9)(x+5)}{(x-9)(x-2)}$. Cancel the factor $x - 9$: $\lim\limits_{x \to 9} \frac{(x+5)}{(x-2)}$. Now we can plug in $x = 9$ and we get $\frac{(9+5)}{(9-2)} = 2$.

Example 15: Find $\lim\limits_{x \to \infty} \frac{5x^2 - 3x + 2}{3x^2 - 7x + 100}$.

Divide each term in the numerator and denominator by the highest power of x, namely x^2. We get

$\lim\limits_{x \to \infty} \frac{\frac{5x^2}{x^2} - \frac{3x}{x^2} + \frac{2}{x^2}}{\frac{3x^2}{x^2} - \frac{7x}{x^2} + \frac{100}{x^2}} = \lim\limits_{x \to \infty} \frac{5 - \frac{3}{x} + \frac{2}{x^2}}{3 - \frac{7}{x} + \frac{100}{x^2}}$. Now take the limit: $\lim\limits_{x \to \infty} \frac{5 - \frac{3}{x} + \frac{2}{x^2}}{3 - \frac{7}{x} + \frac{100}{x^2}} = \frac{5 - 0 + 0}{3 - 0 + 0} = \frac{5}{3}$.

Example 16: Find $\lim\limits_{x \to 2} \frac{8}{x-2}$.

Notice that if we plug in $x = 2$, we get $\frac{8}{0}$. This could be a problem. In order to figure out whether the limit exists, we need to find the left- and right-hand limits. The limit from the left is $\lim\limits_{x \to 2^-} \frac{8}{(x-2)} = -\infty$ and the limit from the right is $\lim\limits_{x \to 2^+} \frac{8}{(x-2)} = +\infty$. Because the two limits don't agree, we get $\lim\limits_{x \to 2} \frac{8}{x-2} = DNE$.

Example 17: Find $\lim_{x\to 2}\frac{8}{(x-2)^2}$.

Notice that if we plug in $x = 2$, we again get $\frac{8}{0}$. We need to find the left- and right-hand limits.

The limit from the left is $\lim_{x\to 2^-}\frac{8}{(x-2)^2} = +\infty$ and the limit from the right is $\lim_{x\to 2^+}\frac{8}{(x-2)^2} = +\infty$.

Because the two limits agree, we get $\lim_{x\to 2}\frac{8}{(x-2)^2} = \infty$.

DETERMINING LIMITS USING THE SQUEEZE THEOREM

Sometimes we can figure out a limit even if we are unable to evaluate the function at the limit. One way to do this is with the **Squeeze Theorem**. The Squeeze Theorem says that given $g(x) \le f(x) \le h(x)$ for all values of x in the interval $[a, b]$, then if $\lim_{x\to c} g(x) = L$ and $\lim_{x\to c} h(x) = L$ where c is in $[a, b]$, then $\lim_{x\to c} f(x) = L$. It also holds for if $\lim_{x\to\infty} g(x) = L$ and $\lim_{x\to\infty} h(x) = L$ where c is in $[a, b]$, then $\lim_{x\to\infty} f(x) = L$.

Let's do an example.

Example 18: Use the Squeeze Theorem to evaluate $\lim_{x\to 0} x^2 \sin^2 \frac{1}{x}$.

Hint: You will usually see the Squeeze Theorem when you are asked to evaluate the limit of a rational expression with sines or cosines in the numerator and a power of x in the denominator.

First, we cannot evaluate $\lim_{x\to 0} \sin\frac{1}{x}$ because the value of sine oscillates between -1 and 1. However, if we multiply through by x^2, we get $0 \le x^2 \sin^2 \frac{1}{x} \le x^2$. We know that $\lim_{x\to 0} 0 = 0$ and $\lim_{x\to 0} x^2 = 0$. Therefore, according to the Squeeze Theorem, $\lim_{x\to 0} x^2 \sin^2 \frac{1}{x} = 0$.

Remember that the $\lim_{x\to 0} \sin x = 0$.

At some point during the exam, you'll have to find the limit of certain trig expressions, usually as x approaches either zero or infinity. There are four standard limits that you should memorize—with those, you can evaluate all of the trigonometric limits that appear on the test. As you'll see throughout this book, calculus requires that you remember all of your trig from previous years.

Rule No. 1: $\lim\limits_{x \to 0} \dfrac{\sin x}{x} = 1$ (x is in radians, *not* degrees)

This may seem strange, but if you look at the graphs of $f(x) = \sin x$ and $f(x) = x$, they have approximately the same slope near the origin (as x gets closer to zero). Because x and the sine of x are about the same as x approaches zero, their quotient will be very close to one. Furthermore, because $\lim\limits_{x \to 0} \cos x = 1$ (review cosine values if you don't get this!), we know that $\lim\limits_{x \to 0} \tan x = \lim\limits_{x \to 0} \dfrac{\sin x}{\cos x} = 0$.

Now we will find a second rule. Let's evaluate the limit $\lim\limits_{x \to 0} \dfrac{\cos x - 1}{x}$. First, multiply the top and bottom by $\cos x + 1$. We get $\lim\limits_{x \to 0} \left(\dfrac{\cos x - 1}{x}\right)\left(\dfrac{\cos x + 1}{\cos x + 1}\right)$. Now, simplify the limit to $\lim\limits_{x \to 0} \dfrac{\cos^2 x - 1}{x(\cos x + 1)}$. Next, we can use the trigonometric identity $\sin^2 x = 1 - \cos^2 x$ and rewrite the limit as $\lim\limits_{x \to 0} \dfrac{-\sin^2 x}{x(\cos x + 1)}$. Now, break this into two limits: $\lim\limits_{x \to 0} \dfrac{-\sin x}{x} \dfrac{\sin x}{(\cos x + 1)}$. The first limit is –1 (see Rule No. 1) and the second is 0, so the limit is 0.

Rule No. 2: $\lim\limits_{x \to 0} \dfrac{\cos x - 1}{x} = 0$

Example 19: Find $\lim\limits_{x \to 0} \dfrac{\sin 3x}{x}$.

Use a simple trick: multiply the top and bottom of the expression by 3. This gives us $\lim\limits_{x \to 0} \dfrac{3 \sin 3x}{3x}$.

Next, substitute a letter for $3x$; for example, a. Now, we get the following:

$$\lim_{a \to 0} \frac{3 \sin a}{a} = 3 \lim_{a \to 0} \frac{\sin a}{a} = 3(1) = 3$$

Example 20: Find $\lim_{x \to 0} \frac{\sin 5x}{\sin 4x}$.

Now we get a bit more sophisticated. First, divide both the numerator and the denominator by x, like so:

$$\lim_{x \to 0} \frac{\frac{\sin 5x}{x}}{\frac{\sin 4x}{x}}$$

Next, multiply the top and bottom of the numerator by 5, and the top and bottom of the denominator by 4, which gives us

$$\lim_{x \to 0} \frac{\frac{5\sin 5x}{5x}}{\frac{4\sin 4x}{4x}}$$

From the work we did in Example 19, we can see that this limit is $\frac{5}{4}$.

Two more rules:

Rule No. 3: $\lim_{x \to 0} \frac{\sin ax}{x} = a$

Rule No. 4: $\lim_{x \to 0} \frac{\sin ax}{\sin bx} = \frac{a}{b}$

Example 21: Find $\lim_{x \to 0} \frac{x^2}{1 - \cos^2 x}$.

Using trigonometric identities, you can replace $(1 - \cos^2 x)$ with $\sin^2 x$.

$$\lim_{x \to 0} \frac{x^2}{1 - \cos^2 x} = \lim_{x \to 0} \frac{x^2}{\sin^2 x} = \lim_{x \to 0} \left(\frac{x}{\sin x} \cdot \frac{x}{\sin x} \right) = 1 \cdot 1 = 1$$

Notice that in the above example, we used the idea that $\lim_{x \to 0} \frac{x}{\sin x} = 1$, even though previously we had only established that $\lim_{x \to 0} \frac{\sin x}{x} = 1$. Because the number 1 is its own reciprocal, if $\frac{\sin x}{x}$ approaches 1, then $\frac{x}{\sin x}$ will as well.

PRACTICE PROBLEM SET 3

Now try these problems. The answers are at the end of the chapter, starting on page 118.

1. Find $\lim_{x \to 0} 3\frac{x}{\sin x}$.

2. Find $\lim_{x \to 0} \frac{\tan 7x}{\sin 5x}$.

3. Find $\lim_{x \to \infty} \sin x$.

4. Find $\lim_{x \to \infty} \sin \frac{1}{x}$.

5. Find $\lim_{x \to 0} \frac{\sin^2 7x}{\sin^2 11x}$.

CONNECTING MULTIPLE REPRESENTATIONS OF LIMITS

Now that we have seen the many different ways that limits can be represented, let's do some practice problems.

Example 22: Find $\lim_{x \to 3} \frac{x-3}{x+2}$.

If you plug in 3 for *x*, you get $\lim_{x \to 3} \frac{3-3}{3+2} = \frac{0}{5} = 0$.

Example 23: Find $\lim_{x \to 3} \frac{x+2}{x-3}$.

If plugging in the value of *x* results in the denominator equaling zero and you cannot factor the quotient anymore, then check the left- and right-hand limits to find the limit of the expression.

The left-hand limit is $\lim_{x \to 3^-} \frac{x+2}{x-3} = -\infty$.

The right-hand limit is $\lim_{x \to 3^+} \frac{x+2}{x-3} = \infty$.

These two limits are not the same. Therefore, the limit does not exist.

Example 24: Find $\lim\limits_{x \to 3} \dfrac{x+2}{(x-3)^2}$.

The left-hand limit is $\lim\limits_{x \to 3^-} \dfrac{x+2}{(x-3)^2} = \infty$.

The right-hand limit is $\lim\limits_{x \to 3^+} \dfrac{x+2}{(x-3)^2} = \infty$.

These two limits are the same, so the limit is ∞.

Example 25: Find $\lim\limits_{x \to -4} \dfrac{x^2+6x+8}{x+4}$.

If $\lim\limits_{x \to a} f(x) = \dfrac{0}{0}$, then $x - a$ is a factor of the top and the bottom.

If you plug –4 into the top and bottom, you get $\dfrac{0}{0}$. You have to factor the top into $(x + 2)(x + 4)$ to get this: $\lim\limits_{x \to -4} \dfrac{(x+2)(x+4)}{(x+4)}$.

Now it's time to cancel like terms: $\lim\limits_{x \to -4} \dfrac{(x+2)(x+4)}{(x+4)} = \lim\limits_{x \to -4} (x+2) = -2$.

Example 26: Find $\lim\limits_{x \to \infty} \dfrac{15x^2-11x}{22x^2+4x}$.

Divide each term by x^2.

$$\lim_{x \to \infty} \frac{15x^2-11x}{22x^2+4x} = \lim_{x \to \infty} \frac{15-\frac{11}{x}}{22+\frac{4}{x}} = \frac{15}{22}$$

Example 27: Find $\lim\limits_{x \to 0} \dfrac{4x}{\tan x}$.

Replace $\tan x$ with $\dfrac{\sin x}{\cos x}$, which changes the expression into

$$\lim_{x \to 0} \frac{4x}{\tan x} = \lim_{x \to 0} \frac{4x}{\frac{\sin x}{\cos x}} = \lim_{x \to 0} \frac{4x \cos x}{\sin x}$$

Because $\lim\limits_{x \to 0} \dfrac{\sin x}{x} = 1$, the $\lim\limits_{x \to 0} \dfrac{x}{\sin x} = 1$ as well. Thus, because $\lim\limits_{x \to 0} \dfrac{x}{\sin x} = 1$ and $\lim\limits_{x \to 0} \cos x = 1$, the answer is 4.

Example 28: Find $\lim_{h\to 0}\frac{(5+h)^2-25}{h}$.

First, expand and simplify the numerator.

$$\lim_{h\to 0}\frac{(5+h)^2-25}{h}=\lim_{h\to 0}\frac{25+10h+h^2-25}{h}=\lim_{h\to 0}\frac{10h+h^2}{h}$$

Note: Pay careful attention to this solved problem. It will be very important when you work on problems in Chapter 5.

Next, factor h out of the numerator and the denominator.

$$\lim_{h\to 0}\frac{10h+h^2}{h}=\lim_{h\to 0}\frac{h(10+h)}{h}=\lim_{h\to 0}(10+h)$$

Taking the limit, you get $\lim_{h\to 0}(10+h)=10$.

Example 29: Find $\lim_{x\to\infty}\frac{\cos 2x}{x^2}$.

The difficulty here is that cos $2x$ oscillates between –1 and 1, so we can't find a value for cos $2x$ at infinity. Fortunately, we can use the Squeeze Theorem. We know that $-1 \le \cos 2x \le 1$, so $-\frac{1}{x^2}\le\frac{\cos 2x}{x^2}\le\frac{1}{x^2}$. Find the limits of the expressions on the ends: $\lim_{x\to\infty}\frac{-1}{x^2}=0$ and $\lim_{x\to\infty}\frac{1}{x^2}=0$. Therefore, $\lim_{x\to\infty}\frac{\cos 2x}{x^2}=0$.

EXPLORING TYPES OF DISCONTINUITIES

Every AP Exam has a few questions on continuity, so it's important to understand the basic idea of what it means for a function to be continuous. The concept is very simple: if the graph of the function doesn't have any breaks or holes in it within a certain interval, the function is continuous over that interval.

Simple polynomials are continuous everywhere; it's the other ones—trigonometric, rational, piecewise—that might have continuity problems. Most of the test questions concern these last types of functions. In order to learn how to test whether a function is continuous, you'll need some more mathematical terminology.

Intermediate Learning
The Intermediate Value Theorem, which we will cover later in this chapter, says that if a function is continuous on the interval $[a, b]$, then for any value c in the interval (a, b), there exists a value $f(c)$ between $f(a)$ and $f(b)$, inclusive. In other words, if the function is continuous in some interval, then all of the possible values of f between and including the two end values of f exist in that interval.

Types of Discontinuities

There are three types of discontinuities you have to know: jump, essential, and removable.

A jump discontinuity occurs when the curve "breaks" at a particular place and starts somewhere else. The limits from the left and the right will both exist, but they will not match.

An example of jump discontinuity looks like this.

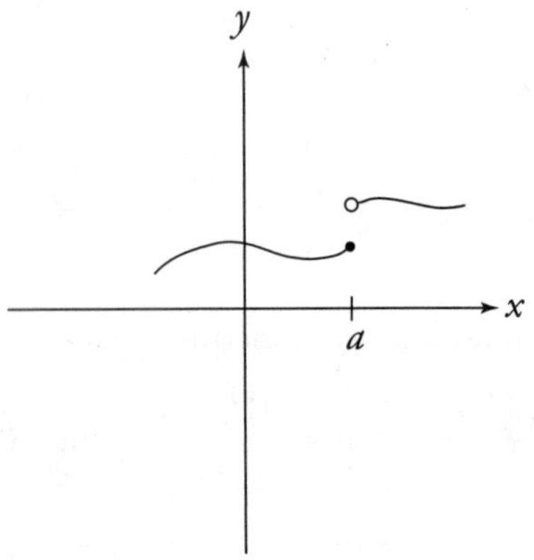

An essential discontinuity (also known as an "infinite discontinuity") occurs when the curve has a vertical asymptote.

This is an example of an essential discontinuity.

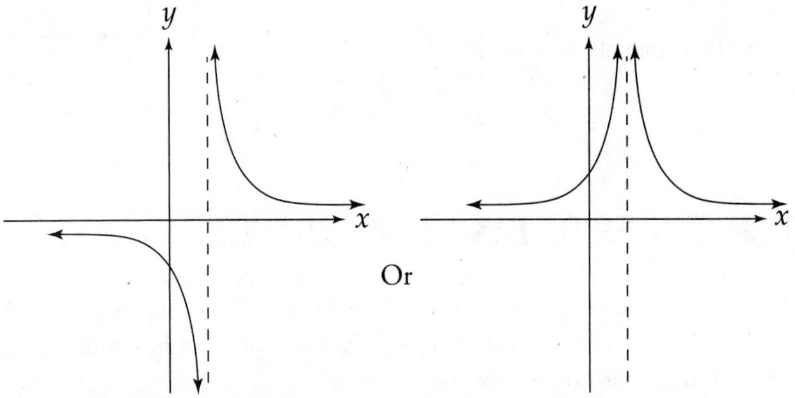

A removable discontinuity, also known as a point discontinuity, occurs when an otherwise continuous curve has a "hole" in it. This type of discontinuity is called "removable" because one can remove the discontinuity simply by filling the hole.

There are two types of removable discontinuity. As the following graphs illustrate, both curves are continuous everywhere except at the "hole."

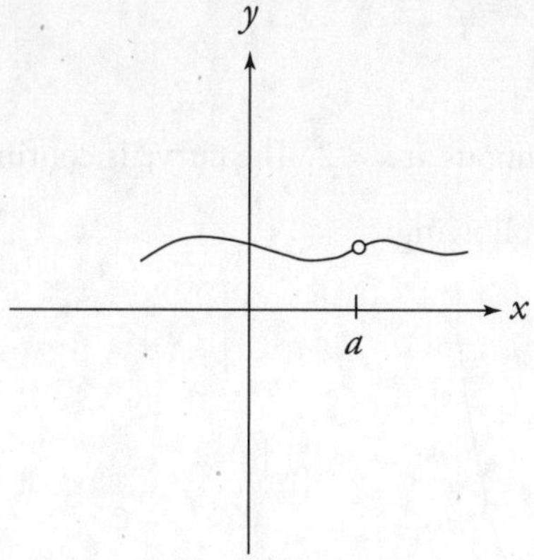

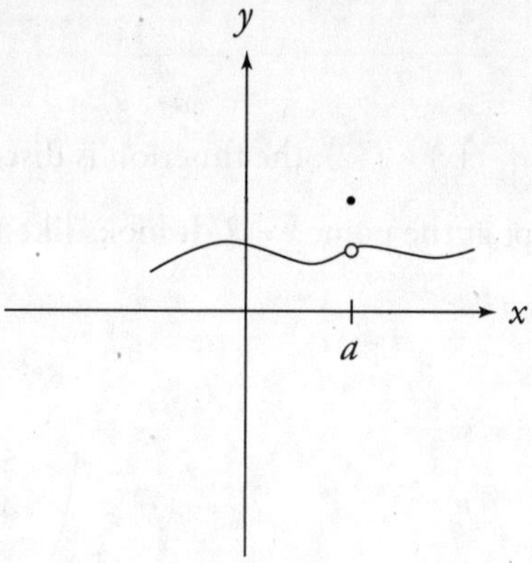

In the left graph, the function is not defined at all at the x-value of the hole, while in the right graph, the function is defined at the x-value of the hole, but the function value does not match the y-coordinate of the hole. In both cases, $\lim_{x\to a^-} f(x) = \lim_{x\to a^+} f(x)$, which means that $\lim_{x\to a} f(x)$ exists, and it is equal to the y-coordinate of the hole.

Now that you know what these three types of discontinuities look like, let's see what types of functions are not continuous everywhere.

Example 30: Consider the following function:

$$f(x) = \begin{cases} x+3, & x \le 2 \\ x^2, & x > 2 \end{cases}$$

The left-hand limit is 5 as x approaches 2, and the right-hand limit is 4 as x approaches 2. Because the curve has different values on each side of 2, the curve is discontinuous at $x = 2$. We say that the curve "jumps" at $x = 2$ from the left-hand curve to the right-hand curve because the left- and right-hand limits differ. It looks like the following:

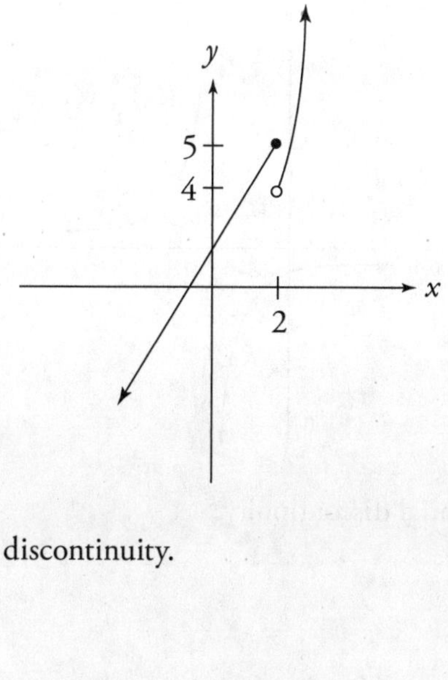

This is an example of a jump discontinuity.

Example 31: Consider the following function:

$$f(x) = \begin{cases} x^2, \; x \neq 2 \\ 5, \; x = 2 \end{cases}$$

Because $\lim_{x \to 2} f(x) \neq f(2)$; the function is discontinuous at $x = 2$. The curve is continuous everywhere except at the point $x = 2$. It looks like the following:

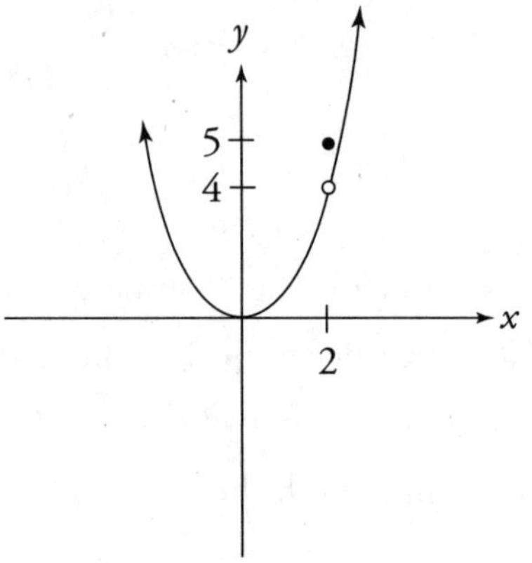

This is an example of a point discontinuity.

Example 32: Consider the following function: $f(x) = \dfrac{5}{x-2}$.

The function is discontinuous because it's possible for the denominator to equal zero (at $x = 2$). This means that $f(2)$ doesn't exist, and the function has an asymptote at $x = 2$. In addition, $\lim_{x \to 2^-} f(x) = -\infty$ and $\lim_{x \to 2^+} f(x) = \infty$.

The graph looks like the following:

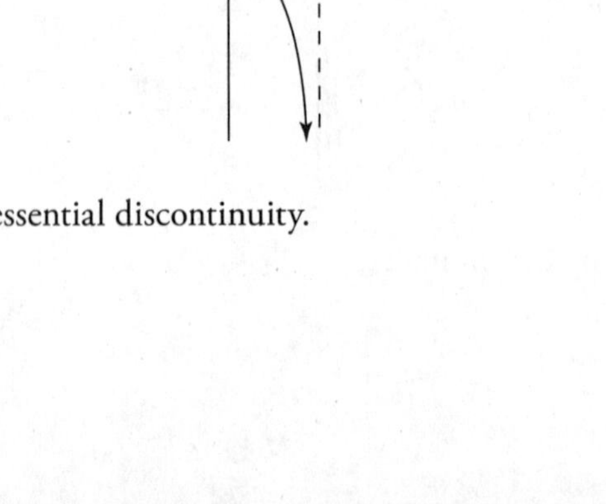

This is an example of an essential discontinuity.

Example 33: Consider the following function:

$$f(x) = \frac{x^2 - 8x + 15}{x^2 - 6x + 5}$$

If you factor the top and bottom, you can see where the discontinuities are.

$$f(x) = \frac{x^2 - 8x + 15}{x^2 - 6x + 5} = \frac{(x-3)(x-5)}{(x-1)(x-5)}$$

The function has a zero in the denominator when $x = 1$ or $x = 5$, so the function is discontinuous at those two points. But you can cancel the term $(x - 5)$ from both the numerator and the denominator, leaving you with

$$f(x) = \frac{x-3}{x-1}$$

Now the reduced function *is* continuous at $x = 5$. Thus, the original function has a removable discontinuity at $x = 5$. Furthermore, if you now plug $x = 5$ into the reduced function, you get

$$f(5) = \frac{2}{4} = \frac{1}{2}$$

The discontinuity is at $x = 5$, and there's a hole at $\left(5, \frac{1}{2}\right)$. In other words, if the original function were continuous at $x = 5$, it would have the value $\frac{1}{2}$. Notice that this is the same as $\lim_{x \to 5} f(x)$.

These are the types of discontinuities that you can expect to encounter on the AP Exam.

DEFINING CONTINUITY AT A POINT

In order for a function $f(x)$ to be continuous at a point $x = c$, it must fulfill *all three* of the following conditions:

Condition 1: $f(c)$ exists.

Condition 2: $\lim_{x \to c} f(x)$ exists.

Condition 3: $\lim_{x \to c} f(x) = f(c)$.

Let's look at a simple example of a continuous function.

Example 34: Is the function $f(x)=\begin{cases} x+1, & x<2 \\ 2x-1, & x\geq 2 \end{cases}$ continuous at the point $x = 2$?

Condition 1: Does $f(2)$ exist?

Yes. It's equal to $2(2) - 1 = 3$.

Condition 2: Does $\lim_{x\to 2} f(x)$ exist?

You need to look at the limit from both sides of 2. The left-hand limit is $\lim_{x\to 2^-} f(x) = 2 + 1 = 3$. The right-hand limit is $\lim_{x\to 2^+} f(x) = 2(2) - 1 = 3$.

Because the two limits are the same, the limit exists.

Condition 3: Does $\lim_{x\to 2} f(x) = f(2)$?

The two equal each other, so yes; the function is continuous at $x = 2$.

A simple and important way to check whether a function is continuous is to sketch the function. If you can't sketch the function without lifting your pencil from the paper at some point, then the function is not continuous.

Now let's look at some examples of functions that are not continuous.

Example 35: Is the function $f(x) = \begin{cases} x+1, & x<2 \\ 2x-1, & x>2 \end{cases}$ continuous at $x = 2$?

Condition 1: Does $f(2)$ exist?

The function of x is defined if x is greater than or less than 2, but not if x is equal to 2. Therefore, the function is not continuous at $x = 2$. Notice that we don't have to bother with the other two conditions. Once you find a problem, the function is automatically not continuous, and you can stop.

Example 36: Is the function $f(x) = \begin{cases} x+1, & x<2 \\ 2x+1, & x\geq 2 \end{cases}$ continuous at $x = 2$?

Condition 1: Does $f(x)$ exist?

Yes. It is equal to $2(2) + 1 = 5$.

Condition 2: Does $\lim_{x\to 2} f(x)$ exist?

The left-hand limit is $\lim_{x\to 2^-} f(x) = 2 + 1 = 3$.

The right-hand limit is $\lim_{x \to 2^+} f(x) = 2(2) + 1 = 5$.

The two limits don't match, so the limit doesn't exist, and the function is not continuous at $x = 2$.

Example 37: Is the function $f(x) = \begin{cases} x+1, & x<2 \\ x^2, & x=2 \\ 2x-1, & x>2 \end{cases}$ continuous at $x = 2$?

Condition 1: Does $f(2)$ exist?

Yes. It's equal to $2^2 = 4$.

Condition 2: Does $\lim_{x \to 2} f(x)$ exist?

The left-hand limit is $\lim_{x \to 2^-} f(x) = 2 + 1 = 3$.

The right-hand limit is $\lim_{x \to 2^+} f(x) = 2(2) - 1 = 3$.

Because the two limits are the same, the limit exists.

Condition 3: Does $\lim_{x \to 2} f(x) = f(2)$?

The $\lim_{x \to 2} f(x) = 3$, but $f(2) = 4$. Because these aren't equal, the answer is "no" and the function is not continuous at $x = 2$.

CONFIRMING CONTINUITY OVER AN INTERVAL

Continuity exists not just at a point but at an interval. A function is continuous on an interval if it is continuous at every point on that interval. For example, polynomials are continuous at all points on their domains. So too are exponential, logarithmic, trigonometric, rational, and power functions. Notice the phrase *at all points on their domains.* For example, tan x is not defined at $x = \pm\frac{\pi}{2}$ (among other points). So tan x would be continuous on the interval $\left(-\frac{\pi}{2}, \frac{\pi}{2}\right)$ but not $\left[-\frac{\pi}{2}, \frac{\pi}{2}\right]$.

Example 38: On what intervals is the function $f(x)=\dfrac{x^2-8x+12}{x-2}$ continuous?

Note that the function is undefined at $x = 2$. If we factor *f*, we get $f(x)=\dfrac{(x-2)(x-6)}{x-2}$.

Simplify: $f(x)=\dfrac{(x-2)(x-6)}{x-2}=x-6$. Now we have a line that is continuous everywhere.

Therefore, *f* is continuous on the intervals $(-\infty, 2)$ and $(2, \infty)$.

REMOVING DISCONTINUITIES

If we have a point discontinuity, we can "remove" the discontinuity by redefining the function without that point in the domain.

For example, suppose we have the function from Example 30 on page 103. Note that at $x = 2$, $f(x) = 5$, which is what makes the function discontinuous. If we defined the function instead as just $f(x) = x^2$, we have "removed" the discontinuity.

Let's do an example.

Example 39: Given the function $f(x)=\dfrac{x^2-4x-45}{x+5}$, redefine the function so that *f* is continuous for all real values of *x*.

We can see that the function *f* is not continuous at $x = -5$. If we factor *f*, we get

$f(x)=\dfrac{x^2-4x-45}{x+5}=\dfrac{(x-9)(x+5)}{x+5}$. If we cancel the factor $x + 5$, we get $f(x) = x - 9$.

This function is continuous for all real values of *x*.

PRACTICE PROBLEM SET 4

Now try these problems. The answers are at the end of the chapter, starting on page 120.

1. Is the function $f(x)=\begin{cases} x+7, & x<2 \\ 9, & x=2 \\ 3x+3, & x>2 \end{cases}$ continuous at $x = 2$?

2. Is the function $f(x)=\begin{cases} 4x^2-2x, & x<3 \\ 10x-1, & x=3 \\ 30, & x>3 \end{cases}$ continuous at $x = 3$?

3. Is the function $f(x) = \sec x$ continuous everywhere?

4. Is the function $f(x) = \sec x$ continuous on the interval $\left[-\frac{\pi}{2}, \frac{\pi}{2}\right]$?

5. Is the function $f(x) = \sec x$ continuous on the interval $\left(-\frac{\pi}{2}, \frac{\pi}{2}\right)$?

6. For what value(s) of k is the function $f(x) = \begin{cases} 3x^2-11x-4, & x \le 4 \\ kx^2-2x-1, & x>4 \end{cases}$ continuous at $x = 4$?

7. At what point is the removable discontinuity for the function $f(x) = \dfrac{x^2+5x-24}{x^2-x-6}$?

8. Given the graph of $f(x)$ on the right, find

(a) $\lim\limits_{x \to -\infty} f(x)$

(b) $\lim\limits_{x \to \infty} f(x)$

(c) $\lim\limits_{x \to 3^-} f(x)$

(d) $\lim\limits_{x \to 3^+} f(x)$

(e) $f(3)$

(f) Any discontinuities

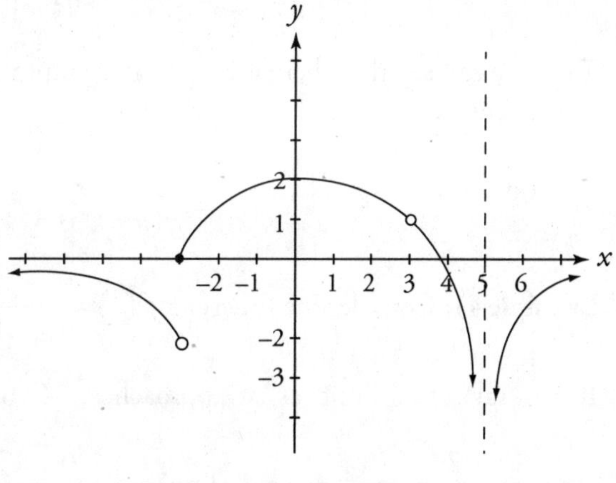

CONNECTING INFINITE LIMITS AND VERTICAL ASYMPTOTES

The limit of some functions will go toward infinity or negative infinity as x approaches a number. This is called a vertical asymptote and is an essential discontinuity. Let's do an example.

Example 40: Consider the function $f(x)=\frac{9}{x-3}$. Evaluate $\lim_{x\to 3}\frac{9}{x-3}$.

As x approaches 3 from the left, the values will get increasingly smaller (more negative). That is, $\lim_{x\to 3^-}\frac{9}{x-3}=-\infty$. As x approaches 3 from the right, the values will get increasingly larger. That is, $\lim_{x\to 3^+}\frac{9}{x-3}=+\infty$. If we were to graph the function, it would look like this:

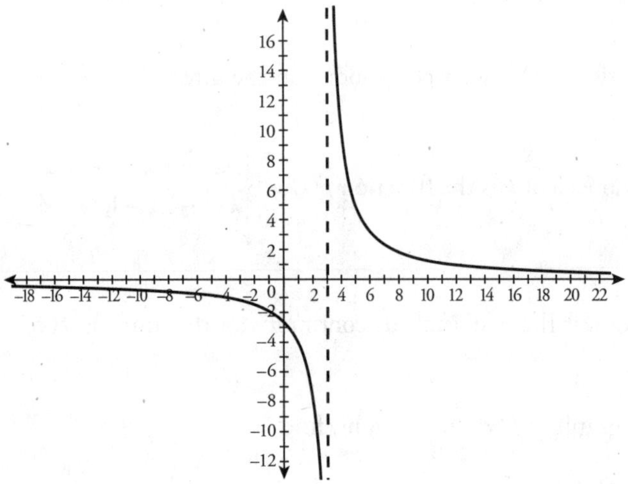

Thus, we can see that there is a vertical asymptote at $x = 3$, where the function has an essential discontinuity.

Let's do another example.

Example 41: Consider the function $f(x)=\frac{20}{(x-4)(x+4)}$. Where are the vertical asymptotes?

If we take the limit as x approaches -4 from the left and from the right, we get $\lim_{x\to -4^-}\frac{20}{(x-4)(x+4)}=+\infty$ and $\lim_{x\to -4^+}\frac{20}{(x-4)(x+4)}=-\infty$. If we take the limit as x approaches 4 from the left and from the right, we get $\lim_{x\to 4^-}\frac{20}{(x-4)(x+4)}=-\infty$ and $\lim_{x\to 4^+}\frac{20}{(x-4)(x+4)}=+\infty$.

Therefore, the function has essential discontinuities at $x = -4$ and $x = 4$, and has vertical asymptotes at $x = -4$ and $x = 4$.

Here is the graph of the function:

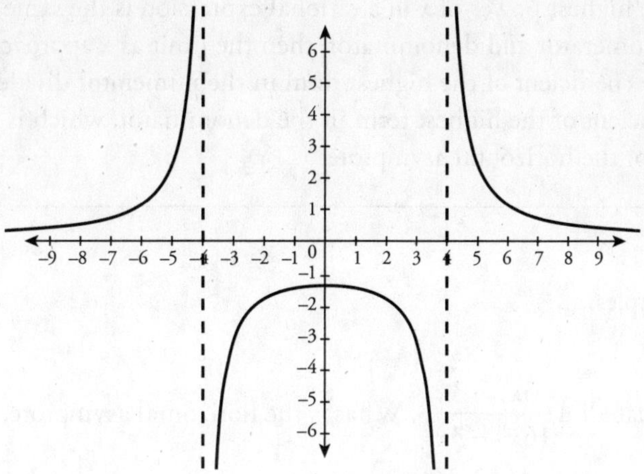

CONNECTING LIMITS AT INFINITY AND HORIZONTAL ASYMPTOTES

Now that we have seen vertical asymptotes, let's look at horizontal asymptotes. The use of the word *asymptote* can be a bit confusing. A vertical asymptote is a line that a function cannot cross because the function is undefined there. A horizontal asymptote describes the end behavior of a function; that is, as the x-values go toward infinity or negative infinity, what do the y-values approach? A function *can* cross a horizontal asymptote. In fact, it can cross a horizontal asymptote infinitely often.

We find the horizontal asymptotes of a function by taking the limit as x approaches infinity (or negative infinity). On the AP Exam, we will almost always see this with rational functions that have polynomials in the numerator and the denominator. Here are the rules for finding the horizontal asymptotes.

(1) If the highest power of x in a rational expression is in the numerator, then the limit as x approaches infinity is infinity and there is no horizontal asymptote.

(2) If the highest power of x in a rational expression is in the denominator, then the limit as x approaches infinity is zero and the horizontal asymptote is the line $y = 0$ (a.k.a. the x-axis).

(3) If the highest power of x in a rational expression is the same in both the numerator and denominator, then the limit as x approaches infinity is the coefficient of the highest term in the numerator divided by the coefficient of the highest term in the denominator, which is the equation of the horizontal asymptote.

Let's do some examples.

Example 42: Evaluate $\lim\limits_{x\to\infty}\dfrac{5x^7-3x}{16x^6-3x^2}$. What is the horizontal asymptote, if any?

Note that the degree of the numerator is greater than the degree of the denominator. Therefore, $\lim\limits_{x\to\infty}\dfrac{5x^7-3x}{16x^6-3x^2}=\infty$ and there is no horizontal asymptote.

Example 43: Evaluate $\lim\limits_{x\to\infty}\dfrac{5x^6-3x}{16x^7-3x^2}$. What is the horizontal asymptote, if any?

Note that the degree of the denominator is greater than the degree of the numerator. Therefore, $\lim\limits_{x\to\infty}\dfrac{5x^6-3x}{16x^7-3x^2}=0$ and the horizontal asymptote is $y = 0$.

Example 44: Evaluate $\lim\limits_{x\to\infty}\dfrac{5x^7-3x}{16x^7-3x^2}$. What is the horizontal asymptote, if any?

Note that the degree of the denominator is equal to the degree of the numerator. Therefore,

$\lim\limits_{x\to\infty}\dfrac{5x^7-3x}{16x^7-3x^2}=\dfrac{5}{16}$ and the horizontal asymptote is $y=\dfrac{5}{16}$.

PRACTICE PROBLEM SET 5

Now try these problems. The answers are at the end of the chapter, starting on page 123.

1. $\lim\limits_{x\to 6}\left(\dfrac{x+2}{x^2-4x-12}\right)=$

2. $\lim\limits_{x\to 7^+}\left(\dfrac{x}{x^2-49}\right)=$

3. $\lim\limits_{x\to 7}\left(\dfrac{x}{x^2-49}\right)=$

4. $\lim\limits_{x\to\infty}\left(\dfrac{x^4-8}{10x^2+25x+1}\right)=$

5. $\lim\limits_{x\to\infty}\left(\dfrac{x^4-8}{10x^4+25x+1}\right)=$

WORKING WITH THE INTERMEDIATE VALUE THEOREM (IVT)

On a few occasions, the AP Exam has had a question involving the Intermediate Value Theorem (IVT). The IVT, among other things, can be used to help figure out if there is a solution to an equation on a particular interval.

The Intermediate Value Theorem guarantees that if a function $f(x)$ is continuous on the interval $[a, b]$ and C is any number between $f(a)$ and $f(b)$, then there is at least one number x in the interval $[a, b]$ such that $f(x) = C$.

Where we usually see the IVT is in the following special case of the IVT.

The Intermediate Value Theorem guarantees that if a function $f(x)$ is continuous on the interval $[a, b]$ and if $f(a)$ and $f(b)$ have opposite signs, then there exists at least one value c in the interval (a, b) such that $f(c) = 0$. What this means is that we are looking for the interval where $f(x)$ is positive at one end of the interval and negative at the other end. This means that there is a value in the interval where the function is zero. We don't necessarily know what the value is, but we know that it exists.

Let's do an example.

Example 45: The Intermediate Value Theorem guarantees a value c, such that $f(c) = 0$ for $f(x) = x^3 + 3x - 7$ on which of the following intervals?

(A) $(-1, 0)$

(B) $(0, 1)$

(C) $(1, 2)$

(D) $(2, 3)$

Note that f is a polynomial and is continuous everywhere. If $f(c) = 0$, then the IVT says that there is a value less than c where f has one sign and a value greater than c where f has the opposite sign.

Let's find the values of $f(x)$ at each of the endpoints:

$$f(-1) = (-1)^3 + 3(-1) - 7 = -11$$
$$f(0) = (0)^3 + 3(0) - 7 = -7$$
$$f(1) = (1)^3 + 3(1) - 7 = -3$$
$$f(2) = (2)^3 + 3(2) - 7 = 7$$
$$f(3) = (3)^3 + 3(3) - 7 = 29$$

Thus, on the interval $(1, 2)$, we have $f(1) < 0$ and $f(2) > 0$. This means that f has a zero somewhere on the interval, which means that the answer is (C).

If we think about what is happening with the function, we can see that it has negative values up to $x = 1$ and positive values after $x = 2$. Therefore, *because f is continuous on the interval, f* must be zero somewhere between 1 and 2 (not necessarily in the middle).

Let's do another example.

Example 46: The Intermediate Value Theorem guarantees a value c, such that $f(c) = 0$ for $f(x) = \sin x$ on which of the following intervals?

(A) $\left(\frac{\pi}{6}, \frac{\pi}{4}\right)$

(B) $\left(\frac{\pi}{4}, \frac{\pi}{3}\right)$

(C) $\left(\frac{\pi}{3}, \frac{\pi}{2}\right)$

(D) $\left(\frac{\pi}{2}, \frac{3\pi}{2}\right)$

Note that sin x is continuous everywhere. If $f(c) = 0$, then the IVT says that there is a value less than c where f has one sign and a value greater than c where f has the opposite sign.

Let's find the values of $f(x)$ at each of the endpoints:

$$f\left(\frac{\pi}{6}\right) = \sin\frac{\pi}{6} = \frac{1}{2}$$

$$f\left(\frac{\pi}{4}\right) = \sin\frac{\pi}{4} = \frac{\sqrt{2}}{2}$$

$$f\left(\frac{\pi}{3}\right) = \sin\frac{\pi}{3} = \frac{\sqrt{3}}{2}$$

$$f\left(\frac{\pi}{2}\right) = \sin\frac{\pi}{2} = 1$$

$$f\left(\frac{3\pi}{2}\right) = \sin\frac{3\pi}{2} = -1$$

Thus, on the interval $\left(\frac{\pi}{2}, \frac{3\pi}{2}\right)$, we have $f\left(\frac{\pi}{2}\right) > 0$ and $f\left(\frac{3\pi}{2}\right) < 0$. This means that f has a zero somewhere on the interval, and this makes the answer (D).

Chapter 3 Drill

The answers are at the end of the chapter, starting on page 125.

1 Mark for Review

Evaluate $\lim_{x \to 2}\left(x^4 - 3x^2 + 5\right)$.

Ⓐ 0

Ⓑ 2

Ⓒ 9

Ⓓ The limit does not exist.

2 Mark for Review

Evaluate $\lim_{x \to 2} \frac{x^2 - x - 2}{x^2 - 8x + 15}$.

Ⓐ 0

Ⓑ $\frac{2}{3}$

Ⓒ 3

Ⓓ The limit does not exist.

3 Mark for Review

Evaluate $\lim_{x \to 7} \frac{x^2 - 3x - 28}{x^2 - 5x - 14}$.

Ⓐ 0

Ⓑ $\frac{11}{9}$

Ⓒ 2

Ⓓ The limit does not exist.

4 Mark for Review

Evaluate $\lim_{x \to \infty} \frac{4x^3 - 8x^2 + 11}{5x^3 + 3x - 9}$.

Ⓐ $-\frac{11}{9}$

Ⓑ 0

Ⓒ $\frac{4}{5}$

Ⓓ The limit does not exist.

5 Mark for Review

Evaluate $\lim_{x \to 0} \frac{4 \sin 3x}{2 \tan 5x}$.

Ⓐ $\frac{6}{5}$

Ⓑ $\frac{3}{5}$

Ⓒ 2

Ⓓ The limit does not exist.

6 Mark for Review

Evaluate $\lim_{x \to 5} \frac{11}{x^2 - 25}$.

Ⓐ 0

Ⓑ $\frac{11}{25}$

Ⓒ ∞

Ⓓ The limit does not exist.

ANSWERS TO PRACTICE PROBLEM SETS

Practice Problem Set 1

1. 13

To find the limit, simply plug in 8 for x: $\lim_{x \to 8}\left(x^2 - 5x - 11\right) = \left(8^2 - (5)(8) - 11\right) = 13$.

2. $\frac{4}{5}$

To find the limit, simply plug in 5 for x: $\lim_{x \to 5}\left(\frac{x+3}{x^2 - 15}\right) = \frac{5+3}{5^2 - 15} = \frac{8}{10} = \frac{4}{5}$.

3. (a) 4; (b) 5; (c) The limit *Does Not Exist.*

(a) Notice that $f(x)$ is a piecewise function, so use the function $f(x) = x^2 - 5$ for all values of x less than or equal to 3. Thus, $\lim_{x \to 3^-} f(x) = 3^2 - 5 = 4$.

(b) Use the function $f(x) = x + 2$ for all values of x greater than 3. Thus, $\lim_{x \to 3^+} f(x) = 3 + 2 = 5$.

(c) In order to evaluate the limit as x approaches 3, find the limit as it approaches 3^+ (from the right) and the limit as it approaches 3^- (from the left). If the two limits approach the same value, or both approach positive infinity or both approach negative infinity, then the limit is that value, or the appropriately signed infinity. If the two limits do not agree, the limit "Does Not Exist." Refer to the solutions in parts (a) and (b) to see that $\lim_{x \to 3^-} f(x) = 4$ and $\lim_{x \to 3^+} f(x) = 5$. Because the two limits are *not* the same, the limit *Does Not Exist.*

4. (a) 4; (b) 4; (c) 4

(a) Notice that $f(x)$ is a piecewise function, so use the function $f(x) = x^2 - 5$ for all values of x less than or equal to 3. Thus, $\lim_{x \to 3^-} f(x) = 3^2 - 5 = 4$.

(b) Use the function $f(x) = x + 1$ for all values of x greater than 3. Thus, $\lim_{x \to 3^+} f(x) = 3 + 1 = 4$.

(c) In order to evaluate the limit as x approaches 3, find the limit as it approaches 3^+ (from the right) and the limit as it approaches 3^- (from the left). If the two limits approach the same value, or both approach positive infinity or both approach negative infinity, then the limit is that value, or the appropriately signed infinity. If the two limits do not agree, the limit "Does Not

Exist." Refer to the solutions in parts (a) and (b) to see that $\lim_{x \to 3^-} f(x) = 4$ and $\lim_{x \to 3^+} f(x) = 4$. Because the two limits are the same, the limit is 4.

5. $\frac{3}{\sqrt{2}}$

Plug in $\frac{\pi}{4}$ for x to get $\lim_{x \to \frac{\pi}{4}} 3\cos x = 3\cos\frac{\pi}{4} = \frac{3}{\sqrt{2}}$.

6. 0

Plug in 0 for x to get $\lim_{x \to 0} 3\frac{x}{\cos x} = 3\frac{0}{\cos 0} = 3\frac{0}{1} = 0$.

Practice Problem Set 2

1. 4

Plug in 3 for x, to get $\frac{0}{0}$, which is indeterminate. When this happens, try to factor the expression in order to get rid of the problem terms. Here, factor the top and get $\lim_{x \to 3}\left(\frac{x^2 - 2x - 3}{x - 3}\right) = \lim_{x \to 3}\frac{(x-3)(x+1)}{x-3}$. Now cancel the term $x - 3$ to get $\lim_{x \to 3}(x+1)$. Notice that you are allowed to cancel the terms because x is not 3 but very close to 3. Now you can plug in 3 for x: $\lim_{x \to 3}(x+1) = 3 + 1 = 4$.

2. 1

Think about what happens when you plug in a value that is very close to 0, but a little bit more. The top and bottom expressions will both be positive and the same value, so you get $\lim_{x \to 0^+}\frac{x}{|x|} = \frac{0^+}{0^+} = 1$.

3. 6

If you plug in 0 for h, you get $\frac{0}{0}$, which is indeterminate. If you expand the expression in the numerator, you get $\lim_{h \to 0}\frac{9 + 6h + h^2 - 9}{h}$. This simplifies to $\lim_{h \to 0}\frac{6h + h^2}{h}$. Next, factor h out of the top expression: $\lim_{h \to 0}\frac{h(6+h)}{h}$. Now, cancel the h and evaluate the limit to get $\lim_{h \to 0}\frac{h(6+h)}{h} = \lim_{h \to 0}(6+h) = 6 + 0 = 6$.

4. $-\frac{1}{x^2}$

Plug in 0 for h and you get $\frac{0}{0}$, which is indeterminate. Combine the two expressions on top with a common denominator and you get $\lim_{h\to 0}\frac{\frac{1}{x+h}-\frac{1}{x}}{h}=\lim_{h\to 0}\frac{\frac{x}{x(x+h)}-\frac{x+h}{x(x+h)}}{h}=\lim_{h\to 0}\frac{\frac{x-(x+h)}{x(x+h)}}{h}$. Simplify the top expression, leaving you with $\lim_{h\to 0}\frac{\frac{-h}{x(x+h)}}{h}$. Next, simplify the expression into $\lim_{h\to 0}\frac{\frac{-h}{x(x+h)}}{h}=\lim_{h\to 0}\frac{-h}{hx(x+h)}$.

Cancel the h to get $\lim_{h\to 0}\frac{-1}{x(x+h)}$. Evaluate the limit to get $\lim_{h\to 0}\frac{-1}{x(x+h)}=\frac{-1}{x(x)}=-\frac{1}{x^2}$.

5. $+\infty$

Think about what happens when you plug in a value that is very close to 6, but a little bit more. The top expression will approach 8. The bottom expression will approach 0, but will be a little bit bigger. Thus, the limit will be $\frac{8}{0^+}$, which is $+\infty$.

6. $-\infty$

Think about what happens when you plug in a value that is very close to 6, but a little bit less. The top expression will approach 8. The bottom expression will approach 0 but will be a little bit less. Thus, the limit will be $\frac{8}{0^-}$, which is $-\infty$.

Practice Problem Set 3

1. 3

Remember Rule No. 1, which says that $\lim_{x\to 0}\frac{\sin x}{x}=1$. To find the limit of its reciprocal, write this as

$\lim_{x\to 0}\frac{1}{\frac{\sin x}{x}}=\frac{1}{\lim_{x\to 0}\frac{\sin x}{x}}=\frac{1}{1}=1$. Plug in 0 for x to get $\lim_{x\to 0}\left(3\frac{x}{\sin x}\right)=(3)(1)=3$.

2. $\frac{7}{5}$

Rewrite the expression as $\lim_{x\to 0}\frac{\tan 7x}{\sin 5x}=\lim_{x\to 0}\frac{\frac{\sin 7x}{\cos 7x}}{\sin 5x}=\lim_{x\to 0}\frac{\sin 7x}{(\cos 7x)(\sin 5x)}$. Remember Rule No. 4, which says that $\lim_{x\to 0}\frac{\sin ax}{\sin bx}=\frac{a}{b}$. Here, $\lim_{x\to 0}\frac{\sin 7x}{(\cos 7x)(\sin 5x)}=\frac{7}{(1)(5)}=\frac{7}{5}$. To evaluate the limit the long way, first divide the numerator and the denominator of the expression by x: $\lim_{x\to 0}\frac{\left(\frac{\sin 7x}{x}\right)}{(\cos 7x)\left(\frac{\sin 5x}{x}\right)}$. Next, multiply the numerator and the denominator of the top expression by 7 and the numerator and the denominator of the bottom expression by 5 to get $\lim_{x\to 0}\frac{\left(\frac{7\sin 7x}{7x}\right)}{(\cos 7x)\left(\frac{5\sin 5x}{5x}\right)}$. Now, evaluate the limit: $\lim_{x\to 0}\frac{(7)(1)}{(1)(5)(1)}=\frac{7}{5}$.

3. The limit *Does Not Exist.*

The value of sin x oscillates between –1 and 1. Thus, as x approaches infinity, sin x does not approach a specific value. Therefore, the limit *Does Not Exist.*

4. 0

Here, as x approaches infinity, $\frac{1}{x}$ approaches 0. Thus, $\lim_{x\to\infty}\sin\frac{1}{x}=\sin 0=0$.

5. $\frac{49}{121}$

Break up the limit into $\lim_{x\to 0}\frac{\sin 7x}{\sin 11x}\frac{\sin 7x}{\sin 11x}$. Remember Rule No. 4, which says that $\lim_{x\to 0}\frac{\sin ax}{\sin bx}=\frac{a}{b}$. Here, $\lim_{x\to 0}\frac{\sin 7x}{\sin 11x}\frac{\sin 7x}{\sin 11x}=\frac{7}{11}\bullet\frac{7}{11}=\frac{49}{121}$. To evaluate the limit the long way, first divide the numerator and the denominator of the expressions by x: $\lim_{x\to 0}\left(\dfrac{\dfrac{\sin 7x}{x}}{\dfrac{\sin 11x}{x}}\dfrac{\dfrac{\sin 7x}{x}}{\dfrac{\sin 11x}{x}}\right)$. Next, multiply the numerator and the denominator of the top expression by 7 and the numerator and the denominator of the bottom expression by 11 to get $\lim_{x\to 0}\left(\dfrac{\dfrac{7\sin 7x}{7x}}{\dfrac{11\sin 11x}{11x}}\dfrac{\dfrac{7\sin 7x}{7x}}{\dfrac{11\sin 11x}{11x}}\right)$. Now, evaluate the limit:

$$\lim_{x\to 0}\left(\dfrac{\dfrac{7\sin 7x}{7x}}{\dfrac{11\sin 11x}{11x}}\dfrac{\dfrac{7\sin 7x}{7x}}{\dfrac{11\sin 11x}{11x}}\right)=\frac{(7)(1)}{(11)(1)}\frac{(7)(1)}{(11)(1)}=\left(\frac{7}{11}\right)\left(\frac{7}{11}\right)=\frac{49}{121}.$$

Practice Problem Set 4

1. Yes. It satisfies all three conditions.

In order for a function $f(x)$ to be continuous at a point $x = c$, it must fulfill *all three* of the following conditions:

Condition 1: $f(c)$ exists.

Condition 2: $\lim_{x\to c} f(x)$ exists.

Condition 3: $\lim_{x\to c} f(x) = f(c)$.

Test each condition.

$f(2)=9$, which satisfies condition 1.

$\lim_{x\to 2^-} f(x)=9$ and $\lim_{x\to 2^+} f(x)=9$, so $\lim_{x\to 2} f(x)=9$, which satisfies condition 2.

$\lim_{x\to 2} f(x)=9=f(2)$, which satisfies condition 3. Therefore, $f(x)$ is continuous at $x = 2$.

2. No. It fails condition 3.

In order for a function $f(x)$ to be continuous at a point $x = c$, it must fulfill *all three* of the following conditions:

Condition 1: $f(c)$ exists.

Condition 2: $\lim_{x\to c} f(x)$ exists.

Condition 3: $\lim_{x\to c} f(x) = f(c)$.

Test each condition.

$f(3) = 29$, which satisfies condition 1.

$\lim_{x\to 3^-} f(x) = 30$ and $\lim_{x\to 3^+} f(x) = 30$, so $\lim_{x\to 3} f(x) = 30$, which satisfies condition 2.

But $\lim_{x\to 3} f(x) \neq f(3)$. Therefore, $f(x)$ is not continuous at $x = 3$ because it fails condition 3.

3. No. It is discontinuous at any odd integral multiple of $\frac{\pi}{2}$.

Recall that $\sec x = \frac{1}{\cos x}$. This means that sec x is undefined at any value where cos $x = 0$, which are the odd multiples of $\frac{\pi}{2}$. Therefore, sec x is not continuous everywhere.

4. No. It is discontinuous at the endpoints of the interval.

Recall that $\sec x = \frac{1}{\cos x}$. This means that sec x is undefined at any value where cos $x = 0$. Also recall that $\cos\frac{\pi}{2} = 0$ and $\cos\left(-\frac{\pi}{2}\right) = 0$. Therefore, sec x is not continuous everywhere on the interval $\left[-\frac{\pi}{2}, \frac{\pi}{2}\right]$.

5. Yes.

Recall that $\sec x = \dfrac{1}{\cos x}$. This means that sec x is undefined at any value where $\cos x = 0$. Also recall that $\cos\dfrac{\pi}{2} = 0$ and $\cos\left(-\dfrac{\pi}{2}\right) = 0$. Therefore, sec x is continuous everywhere on the interval $\left(-\dfrac{\pi}{2}, \dfrac{\pi}{2}\right)$ because the interval does not include the endpoints.

6. The function is continuous for $k = \dfrac{9}{16}$.

In order for a function $f(x)$ to be continuous at a point $x = c$, it must fulfill *all three* of the following conditions:

Condition 1: $f(c)$ exists.

Condition 2: $\lim\limits_{x\to c} f(x)$ exists.

Condition 3: $\lim\limits_{x\to c} f(x) = f(c)$.

Find a value, or values, of k that enables $f(x)$ to satisfy each condition.

Condition 1: $f(4) = 0$.

Condition 2: $\lim\limits_{x\to 4^-} f(x) = 0$ and $\lim\limits_{x\to 4^+} f(x) = 16k - 9$. In order for the limit to exist, the two limits must be the same. If you solve $16k - 9 = 0$, you get $k = \dfrac{9}{16}$.

Condition 3: If you now let $k = \dfrac{9}{16}$, $\lim\limits_{x\to 4} f(x) = 0 = f(4)$. Therefore, the solution is $k = \dfrac{9}{16}$.

7. The removable discontinuity is at $\left(3, \dfrac{11}{5}\right)$.

One of the most common scenarios in which a removable discontinuity occurs is with a rational expression with common factors in the numerator and denominator. In such a case, the graph will have a "hole" whose position can be determined by examining the common factor and canceling it. If you factor $f(x) = \dfrac{x^2 + 5x - 24}{x^2 - x - 6}$, you get $f(x) = \dfrac{(x+8)(x-3)}{(x+2)(x-3)}$. If you cancel the common factor, you get $f(x)=\dfrac{(x+8)}{(x+2)}$. Now, if you plug in $x=3$, you get $f(x)=\dfrac{11}{5}$. Therefore, the removable discontinuity is at $\left(3,\dfrac{11}{5}\right)$.

8. (a) 0; (b) 0; (c) 1; (d) 1; (e) $f(3)$ *Does Not Exist*; (f) a jump discontinuity at $x = -3$, a removable discontinuity at $x = 3$, and an essential discontinuity at $x = 5$.

(a) If you look at the graph, you can see that $\lim_{x\to-\infty} f(x)=0$.

(b) If you look at the graph, you can see that $\lim_{x\to\infty} f(x)=0$.

(c) If you look at the graph, you can see that $\lim_{x\to3^-} f(x)=1$.

(d) If you look at the graph, you can see that $\lim_{x\to3^+} f(x)=1$.

(e) $f(3)$ Does Not Exist.

(f) There are three discontinuities: (1) a jump discontinuity at $x=-3$; (2) a removable discontinuity at $x = 3$; and (3) an essential discontinuity at $x = 5$.

Practice Problem Set 5

1. The limit *Does Not Exist.*

In order to evaluate the limit as x approaches 6, find the limit as it approaches 6^+ (from the right) and the limit as it approaches 6^- (from the left). If the two limits approach the same value, or both approach positive infinity or both approach negative infinity, then the limit is that value, or the appropriately signed infinity. If the two limits do not agree, the limit "Does Not Exist." Here, if you look at the solutions to problems 9 and 10, you find that as x approaches 6^+, the limit is $+\infty$, but as x approaches 6^-, the limit is $-\infty$. Because the two limits are *not* the same, the limit *Does Not Exist.*

2. $+\infty$

Think about what happens when you plug in a value that is very close to 7, but a little bit more. The top expression will approach 7. The bottom expression will approach 0 but will be a little bit positive. Thus, the limit will be $\dfrac{7}{0^+}$, which is $+\infty$.

3. The limit *Does Not Exist.*

In order to evaluate the limit as x approaches 7, find the limit as it approaches 7^+ (from the right) and the limit as it approaches 7^- (from the left). If the two limits approach the same value, or both approach positive infinity or both approach negative infinity, then the limit is that value, or the appropriately signed infinity. If the two limits do not agree, the limit "Does Not Exist." Here, if you look at the solutions to Problem 2, you see that as x approaches 7^+, the limit is $+\infty$. As x approaches 7^-, the top expression will approach 7. The bottom will approach 0, but will be a little bit negative. Thus, the limit will be $\frac{7}{0^-}$, which is $-\infty$. Because the two limits are *not* the same, the limit *Does Not Exist.*

4. $+\infty$

Find the limit as x goes to infinity. Divide the top and bottom by the highest power of x in the expression, which is x^4: $\lim_{x\to\infty}\left(\frac{x^4-8}{10x^2+25x+1}\right)=\lim_{x\to\infty}\left(\frac{\frac{x^4}{x^4}-\frac{8}{x^4}}{\frac{10x^2}{x^4}+\frac{25x}{x^4}+\frac{1}{x^4}}\right)$. Next, simplify the top and bottom: $\lim_{x\to\infty}\left(\frac{1-\frac{8}{x^4}}{\frac{10}{x^2}+\frac{25}{x^3}+\frac{1}{x^4}}\right)$. Now, if you take the limit as x goes to infinity, you get $\lim_{x\to\infty}\left(\frac{1-\frac{8}{x^4}}{\frac{10}{x^2}+\frac{25}{x^3}+\frac{1}{x^4}}\right)=\frac{1-0}{0+0+0}=\infty$.

5. $\frac{1}{10}$

Find the limit as x goes to infinity. Divide the top and bottom by the highest power of x in the expression, which is x^4: $\lim_{x\to\infty}\left(\frac{x^4-8}{10x^4+25x+1}\right)=\lim_{x\to\infty}\left(\frac{\frac{x^4}{x^4}-\frac{8}{x^4}}{\frac{10x^4}{x^4}+\frac{25x}{x^4}+\frac{1}{x^4}}\right)$. Next, simplify the top and bottom: $\lim_{x\to\infty}\left(\frac{1-\frac{8}{x^4}}{10+\frac{25}{x^3}+\frac{1}{x^4}}\right)$. Now, if you take the limit as x goes to infinity, you get $\lim_{x\to\infty}\left(\frac{1-\frac{8}{x^4}}{10+\frac{25}{x^3}+\frac{1}{x^4}}\right)=\frac{1-0}{10+0+0}=\frac{1}{10}$.

ANSWERS TO CHAPTER 3 DRILL

1. **C** Simply plug in $x = 2$: $\lim\limits_{x\to 2}\left(x^4-3x^2+5\right)=(2)^4-3(2)^2+5=16-12+5=9$.

The answer is (C).

2. **A** Plug in $x = 2$: $\lim\limits_{x\to 2}\dfrac{x^2-x-2}{x^2-8x+15}=\dfrac{(2)^2-(2)-2}{(2)^2-8(2)+15}=\dfrac{0}{3}=0$.

The answer is (A).

3. **B** Plug in $x = 7$: $\lim\limits_{x\to 7}\dfrac{(7)^2-3(7)-28}{(7)^2-5(7)-14}=\dfrac{0}{0}$. When you get a limit of $\dfrac{0}{0}$, this does not necessarily mean that the limit does not exist. First, factor the numerator and denominator of the limit: $\lim\limits_{x\to 7}\dfrac{x^2-3x-28}{x^2-5x-14}=\lim\limits_{x\to 7}\dfrac{(x-7)(x+4)}{(x-7)(x+2)}$. Note that when you plug in $x = 7$, the factor $(x - 7)$ is what makes the function zero in the numerator and denominator. Cancel the factors to reduce the limit to $\lim\limits_{x\to 7}\dfrac{(x+4)}{(x+2)}$. Now plug in $x = 7$: $\lim\limits_{x\to 7}\dfrac{(x+4)}{(x+2)}=\dfrac{7+4}{7+2}=\dfrac{11}{9}$.

The answer is (B).

4. **C** Remember that when you are finding the limit at infinity of a rational function, compare the degree of the highest term in the numerator to the degree of the highest term in the denominator. The degrees are both 3, so the limit will be the ratio of the coefficients of the terms, namely $\dfrac{4}{5}$.

The answer is (C).

5. **A** When you have a limit of the form $\lim\limits_{x\to 0}\dfrac{a\sin bx}{c\tan dx}$, the limit is $\dfrac{ab}{cd}$. Here you get $\lim\limits_{x\to 0}\dfrac{4\sin 3x}{2\tan 5x}=\dfrac{(4)(3)}{(2)(5)}=\dfrac{6}{5}$.

The answer is (A).

6. **D** Plug in $x = 5$: $\lim\limits_{x\to 5}\dfrac{11}{x^2-25}=\dfrac{11}{(5)^2-25}=\dfrac{11}{0}$. This is going to give you an infinite limit or the limit doesn't exist. Take the limit from both sides of 5. First, find $\lim\limits_{x\to 5^-}\dfrac{11}{x^2-25}$. The denominator will be just less than zero, so $\lim\limits_{x\to 5^-}\dfrac{11}{x^2-25}=-\infty$. Next, find $\lim\limits_{x\to 5^+}\dfrac{11}{x^2-25}$. Now the denominator will be just greater than zero, so $\lim\limits_{x\to 5^+}\dfrac{11}{x^2-25}=\infty$. The two limits do not match, so the limit does not exist.

The answer is (D).

Chapter 4
Differentiation: Definition and Basic Derivative Rules

DEFINING AVERAGE AND INSTANTANEOUS RATES OF CHANGE AT A POINT

As we discussed in Chapter 3, calculus was invented to help solve the problem of finding change at an instant. There are two types of rates of change that we will be concerned with. The first is called the **Average Rate of Change**. This is the rate that something changes over an interval of time. We can find this with what we call the **Difference Quotient,** and the formula is given in two forms:

$$\frac{f(a+h)-f(a)}{h} \text{ and } \frac{f(x)-f(a)}{x-a}$$

The second type of rate of change is called the **Instantaneous Rate of Change**. This is the rate that something is changing at an instant of time. We can also use the Difference Quotient but this time, we'll use a limit:

$$\lim_{h\to 0}\frac{f(a+h)-f(a)}{h} \text{ and } \lim_{x\to a}\frac{f(x)-f(a)}{x-a}$$

We will explore these formulas later in this chapter. For now, let's do an example.

Example 1: A car leaves its initial location and 3 hours later has traveled 120 kilometers. What is its average rate of change?

Use the first formula for Average Rate of Change. We can call the initial time a, where $a = 0$. Then three hours later would be time $a + 3$ (in other words, $h = 3$). The position at time $a = 0$ is $f(0) = 0$, and the position at time $a + 3$ is $f(3) = 120$. Plugging into the formula, we get $\frac{120-0}{3-0} = 40$ kilometers per hour. If we had used the second Average Rate of Change formula, the initial time would be $a = 0$ and three hours later would be $x = 3$. The position at time $a = 0$ is $f(0) = 0$ and the position at time $x = 3$ is $f(3) = 120$. Plugging into the formula, we get $\frac{120-0}{3-0} = 40$ kilometers per hour. Note that they give the exact same answer. It would be a problem if they didn't!

DEFINING THE DERIVATIVE OF A FUNCTION AND USING DERIVATIVE NOTATION

Deriving the Formula

The best way to understand the definition of the derivative is to start by looking at the simplest continuous function: a line. As you should recall, you can determine the slope of a line by taking two points on that line and plugging them into the slope formula.

$$m = \frac{y_2 - y_1}{x_2 - x_1} \qquad m \text{ stands for slope.}$$

Notice that you can use the coordinates in reverse order and still get the same result. It doesn't matter in which order you do the subtraction as long as you're consistent.

For example, suppose a line goes through the points (3, 7) and (8, 22). First, you subtract the y-coordinates: $(22 - 7) = 15$. Next, subtract the corresponding x-coordinates: $(8 - 3) = 5$. Finally, divide the first number by the second: $\frac{15}{5} = 3$. The result is the slope of the line: $m = 3$.

Let's look at the graph of an arbitrary line. The slope measures the steepness of the line, which looks like the following:

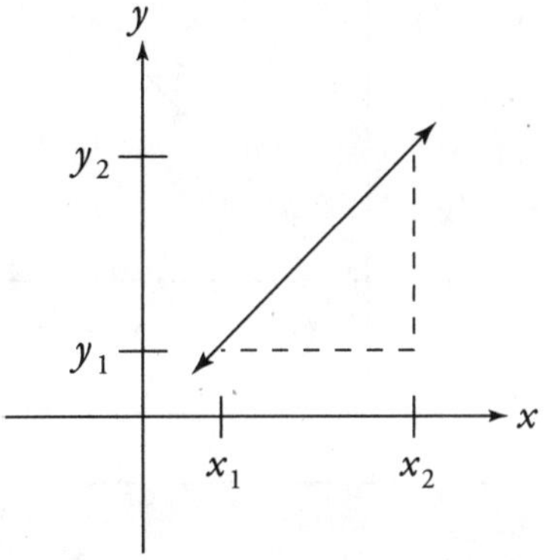

You probably remember your teachers referring to the slope as the "rise" over the "run." The rise is the difference between the y-coordinates, and the run is the difference between the x-coordinates. The slope is the ratio of the two.

Now for a few changes in notation. Instead of calling the x-coordinates x_1 and x_2, we're going to call them x_1 and $x_1 + h$, where h is the difference between the two x-coordinates. Second, instead of using y_1 and y_2, we'll use $f(x_1)$ and $f(x_1 + h)$.

Sometimes, instead of h, some books use Δx.

So now the graph looks like the following:

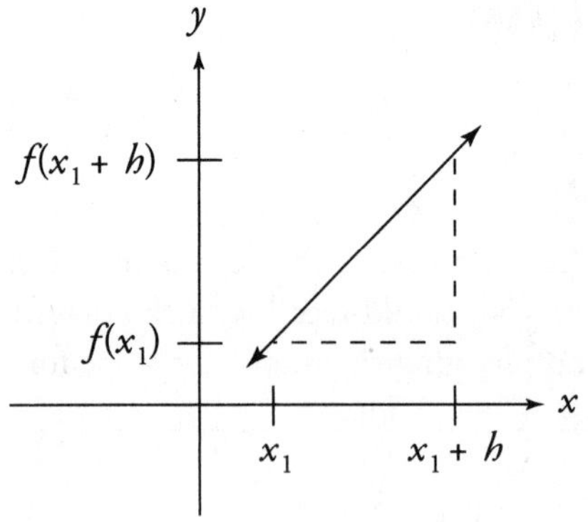

The picture is exactly the same—only the notation has changed.

The Slope of a Curve

Suppose that instead of finding the slope of a line, we wanted to find the slope of a curve. In that case, the slope formula no longer works because the distance from one point to the other is along a curve, not a straight line. But we could find the approximate slope if we took the slope of the line between the two points. This is called the **secant line**.

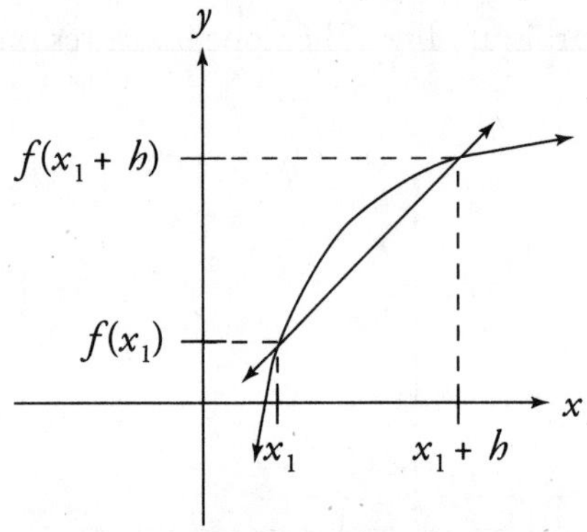

While at first the Difference Quotient may seem like another formula to memorize, keep in mind that it's just another way of writing the slope formula.

The formula for the slope of the secant line is

$$\frac{f(x_1 + h) - f(x_1)}{h}$$

Remember this formula! This is called the **Difference Quotient**.

The Secant and the Tangent

As you can see, the farther apart the two points are, the less the slope of the line corresponds to the slope of the curve.

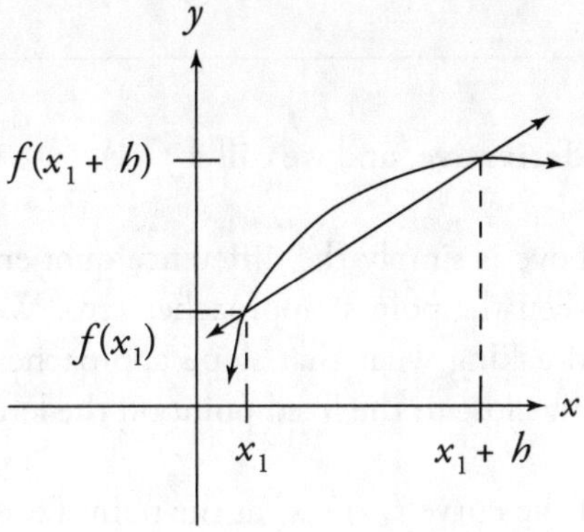

Conversely, the closer the two points are, the more accurate the approximation is.

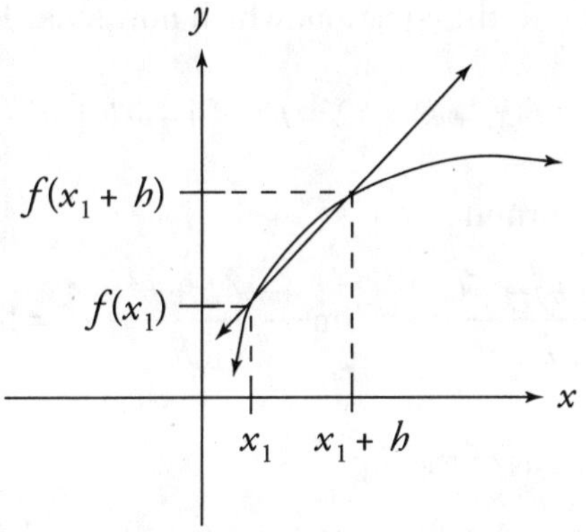

In fact, there is one line, called the **tangent line**, that touches the curve at exactly one point. The slope of the tangent line is equal to the slope of the curve at exactly this point. The objective of using the above formula, therefore, is to shrink h down to an infinitesimally small amount. If we could do that, then the difference between $(x_1 + h)$ and x_1 would be a point.

Keep in mind that there are an infinite number of tangents for any curve because there are any infinite number of points on the curve. We are interested in the slope at a particular point.

Graphically, it looks like the following:

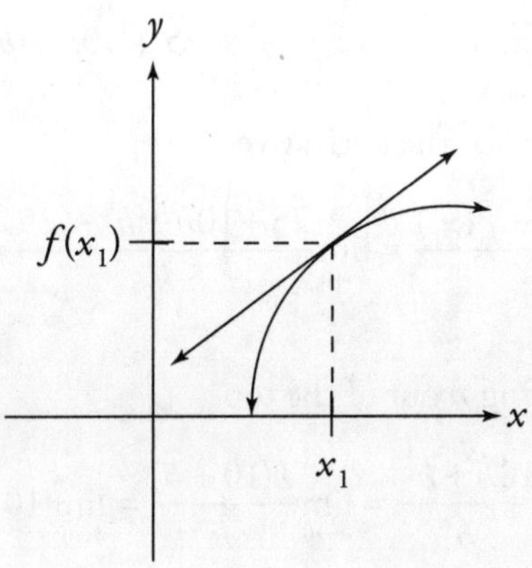

How do we perform this shrinking act? By using the limits we discussed in Chapter 3, we set up a limit during which h approaches zero, like the following:

$$\lim_{h \to 0} \frac{f(x_1 + h) - f(x_1)}{h}$$

This is the **definition of the derivative**, and we call it $f'(x)$.

Notice that the expression above is simply the difference quotient, evaluated at the limit as h (the horizontal distance between the points) approaches zero. We are finding the slope of the line between two points, and asking what that slope approaches as the point on the right is allowed to move infinitesimally close to the fixed point on the left.

Example 2: Find the slope of the curve $f(x) = x^2$ at the point (2, 4).

This means that $x_1 = 2$ and $f(2) = 2^2 = 4$. If we can figure out $f(x_1 + h)$, then we can find the slope. Well, how did we find the value of $f(x)$? We plugged x_1 into the equation $f(x) = x^2$. To find $f(x_1 + h)$, we plug $x_1 + h$ into the equation, which now looks like this:

$$f(x_1 + h) = (2 + h)^2 = 4 + 4h + h^2$$

Now plug this into the slope formula.

$$\lim_{h \to 0} \frac{f(x_1 + h) - f(x_1)}{h} = \lim_{h \to 0} \frac{4 + 4h + h^2 - 4}{h} = \lim_{h \to 0} \frac{4h + h^2}{h}$$

Next, simplify by factoring h out of the top.

$$\lim_{h \to 0} \frac{4h + h^2}{h} = \lim_{h \to 0} \frac{h(4 + h)}{h} = \lim_{h \to 0}(4 + h)$$

Taking the limit as h approaches 0, we get 4. Therefore, the slope of the curve $y = x^2$ at the point (2, 4) is 4. Now we've found the slope of a curve at a certain point, and the notation looks like this: $f'(2) = 4$. Remember this notation!

Example 3: Find the derivative of the equation in Example 2 at the point (5, 25). This means that $x_1 = 5$ and $f(x) = 25$. This time,

$$(x_1 + h)^2 = (5 + h)^2 = 25 + 10h + h^2$$

Now plug this into the formula for the derivative.

$$\lim_{h \to 0} \frac{f(x_1 + h) - f(x_1)}{h} = \lim_{h \to 0} \frac{25 + 10h + h^2 - 25}{h} = \lim_{h \to 0} \frac{10h + h^2}{h}$$

Once again, simplify by factoring h out of the top.

$$\lim_{h \to 0} \frac{10h + h^2}{h} = \lim_{h \to 0} \frac{h(10 + h)}{h} = \lim_{h \to 0}(10 + h)$$

Taking the limit as h goes to 0, you get 10. Therefore, the slope of the curve $y = x^2$ at the point (5, 25) is 10, or $f'(5) = 10$.

Using this pattern, let's forget about the arithmetic for a second and derive a formula.

Example 4: Find the slope of the equation $f(x) = x^2$ at the point $\left(x_1, x_1^2\right)$.

Follow the steps in the last two examples, but instead of using a number, use x_1. This means that $f(x_1) = x_1^2$ and $(x_1 + h)^2 = x_1^2 + 2x_1h + h^2$. Then the derivative is

$$\lim_{h\to 0}\frac{x_1^2 + 2x_1h + h^2 - x_1^2}{h} = \lim_{h\to 0}\frac{2x_1h + h^2}{h}$$

Factor h out of the top.

$$\lim_{h\to 0}\frac{h(2x_1 + h)}{h} = \lim_{h\to 0}(2x_1 + h)$$

Now take the limit as h goes to 0: you get $2x_1$. Therefore, $f'(x_1) = 2x_1$.

This example gives us a general formula for the derivative of this curve. Now we can pick any point, plug it into the formula, and determine the slope at that point. For example, the derivative at the point $x = 7$ is 14. At the point $x = \frac{7}{3}$, the derivative is $\frac{14}{3}$.

Notation

There are several different notations for derivatives in calculus. We'll use two different types interchangeably throughout this book, so get used to them now.

We'll refer to functions three different ways: $f(x)$, u or v, and y. For example, we might write $f(x) = x^3$, $g(x) = x^4$, $h(x) = x^5$. We'll also use notation like $u = \sin x$ and $v = \cos x$. Or we might use $y = \sqrt{x}$. Usually, we pick the notation that causes the least confusion.

The derivatives of the functions will use notation that depends on the function, as shown in the following table:

Function	First Derivative	Second Derivative
$f(x)$	$f'(x)$	$f''(x)$
$g(x)$	$g'(x)$	$g''(x)$
y	y' or $\frac{dy}{dx}$	y'' or $\frac{d^2y}{dx^2}$

Sometimes math books refer to a derivative using either D_x or f_x. We're not going to use either of them.

In addition, if we refer to a derivative of a function in general (for example, $ax^2 + bx + c$), we might enclose the expression in parentheses and use either of the following notations:

$$\left(ax^2 + bx + c\right)' \text{ or } \frac{d}{dx}\left(ax^2 + bx + c\right)$$

PRACTICE PROBLEM SET 6

Now find the derivative of the following expressions. The answers are at the end of the chapter, starting on page 158.

1. $f(x) = 5x$ at $x = 3$

2. $f(x) = 4x$ at $x = -8$

3. $f(x) = 5x^2$ at $x = -1$

4. $f(x) = 8x^2$

5. $f(x) = -10x^2$

6. $f(x) = 20x^2$ at $x = a$

7. $f(x) = 2x^3$ at $x = -3$

8. $f(x) = -3x^3$

9. $f(x) = x^5$

10. $f(x) = 2\sqrt{x}$ at $x = 9$

11. $f(x) = 5\sqrt{2x}$ at $x = 8$

12. $f(x) = \sin x$ at $x = \frac{\pi}{3}$

13. $f(x) = x^2 + x$

14. $f(x) = x^3 + 3x + 2$

15. $f(x) = \frac{1}{x}$

16. $f(x) = ax^2 + bx + c$

ESTIMATING DERIVATIVES OF A FUNCTION AT A POINT

Sometimes we can't find the derivative explicitly, but we can estimate it from some data. We use the formulas for Average Rate of Change and try to make the time interval as small as possible.

The College Board rarely tests this topic, but it sometimes shows up as part of a question in Section II of an exam.

Let's do an example.

Example 5: Suppose that we drop a rock from the roof of a building and we find its height above the ground t seconds later. We find the following data:

Time	Height
0.49 second	44.1584 feet
0.5 second	44 feet
0.51 second	43.8384 feet

Estimate the derivative of the rock's position (its instantaneous velocity) at time $t = 0.5$ second.

We can estimate the derivative by finding $\frac{f(0.5)-f(0.49)}{0.5-0.49}=\frac{44-44.1584}{0.01}=-15.84$ feet per second.

We can also find $\frac{f(0.51)-f(0.5)}{0.51-0.5}=\frac{43.8384-44}{0.01}=-16.16$ feet per second.

A good guess for the instantaneous velocity, or the derivative of the rock's position, is halfway between the two estimates, or –16 feet per second.

Another way that we can estimate a derivative at a point is from the graph of a function. Remember that the derivative is the slope of the tangent line at a point. If we look at a graph, we can guess what the slope of the tangent line is. Let's do an example.

Example 6: Given the graph of the function f below, estimate the derivative at $x = 2$.

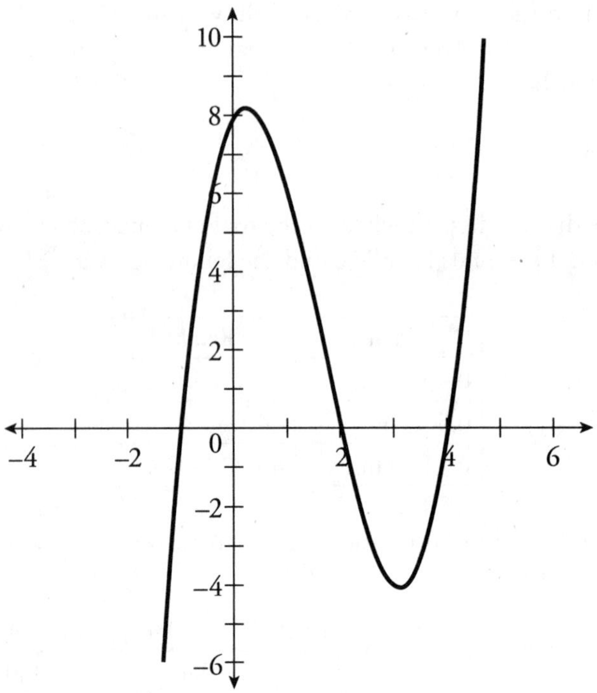

We don't have the equation of this graph, which would make finding the derivative much easier. But we do have some points on the graph that we can use to make an estimate. The function appears to go through the points (0, 8) and (3, −4). If we were to use those points, we would get the slope of the secant line. Let's graph the secant line:

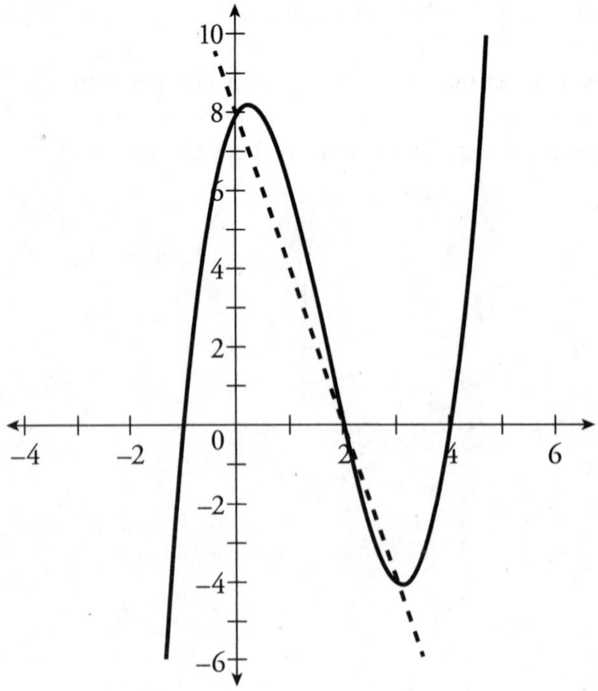

This looks like a pretty good estimate. The slope of the secant line is $\frac{-4-8}{3-0} = -4$, so our estimate for the derivative at $x = 2$ is -4.

Sometimes it's easy to figure out the derivative at a point on a graph.

Example 7: Using the graph from the previous example, find the derivative at $x = 3$.

Notice how the curve has a minimum at $x = 3$. This means that the tangent line would be horizontal there. Let's draw a picture:

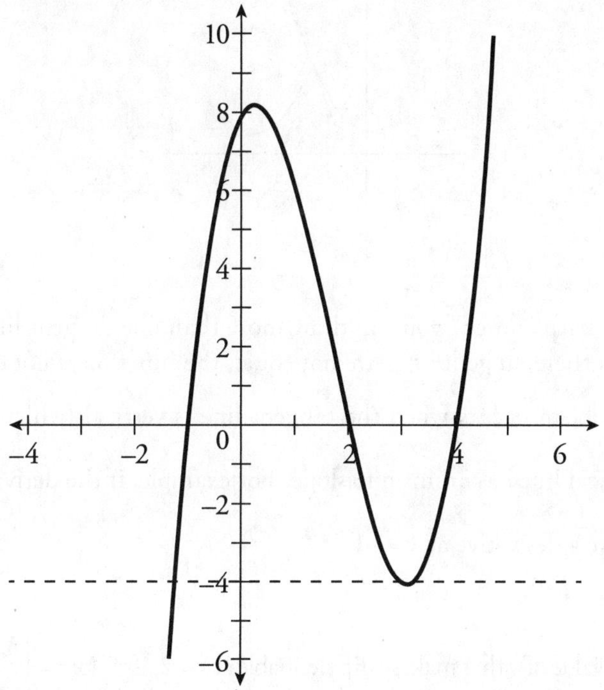

A horizontal line has a slope of zero, so the derivative at $x = 3$ is 0.

CONNECTING DIFFERENTIABILITY AND CONTINUITY: DETERMINING WHEN DERIVATIVES DO AND DO NOT EXIST

One of the important requirements for the differentiability of a function is that the function be continuous. But, even if a function is continuous at a point, the function is not necessarily differentiable there. Check out the graph below.

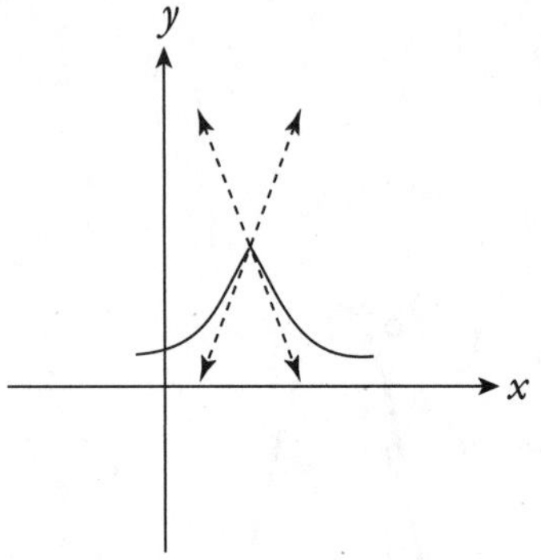

If a function has a "sharp corner," you can draw more than one tangent line at that point, and because the slopes of these tangent lines are not equal, the function is not differentiable there.

Another possible problem occurs when the tangent line is vertical (which can also occur at a cusp) because a vertical line has an infinite slope. For example, if the derivative of a function is $\frac{1}{x+1}$, it doesn't have a derivative at $x = -1$.

Example 8: Find the value of a that makes f differentiable at $x = 2$, if $f'(x) = \begin{cases} 3ax^2 - 4x + 1; x \le 2 \\ -4x^3 + 8ax - 5; x > 2 \end{cases}$.

All we have to do is plug 2 into both pieces of f and set them equal to each other. Then the derivative of the one piece will be the same as the derivative of the other piece. For $x \le 2$, we get $f'(2) = 3a(2)^2 - 4(2) + 1 = 12a - 7$. For $x > 2$, we get $f'(2) = -4(2)^3 + 8a(2) - 5 = 16a - 37$. Set the two pieces equal to each other: $16a - 37 = 12a - 7$. Then $a = \frac{15}{2}$.

APPLYING THE POWER RULE

In the last chapter, you learned how to find a derivative using the definition of the derivative, a process that is very time-consuming and sometimes involves a lot of complex algebra. Fortunately, there's a shortcut to taking derivatives, so you'll never have to use the definition again—except when it's a question on an exam!

The basic technique for taking a derivative is called the **Power Rule**.

> Rule No. 1: If $y = x^n$, then $\frac{dy}{dx} = nx^{n-1}$

That's it. Wasn't that simple? Of course, this and all of the following rules can be derived easily from the definition of the derivative. Look at these next few examples of the Power Rule in action.

Notice that when the power of the function is negative, the power of the derivative is more negative.

Example 9: If $y = x^5$, then $\frac{dy}{dx} = 5x^4$.

Example 10: If $y = x^{20}$, then $\frac{dy}{dx} = 20x^{19}$.

Example 11: If $f(x) = x^{-5}$, then $f'(x) = -5x^{-6}$.

Example 12: If $u = x^{\frac{1}{2}}$, then $\frac{du}{dx} = \frac{1}{2}x^{-\frac{1}{2}}$.

When the power is a fraction, you should be careful to get the subtraction right (you'll often see the powers $\frac{1}{2}$, $\frac{1}{3}$, $\frac{3}{2}$, $-\frac{1}{2}$, and $-\frac{1}{3}$, so be comfortable with subtracting 1 from them).

Example 13: If $y = x^1$, then $\frac{dy}{dx} = 1x^0 = 1$. (Because x^0 is 1!)

Example 14: If $y = x^0$, then $\frac{dy}{dx} = 0 \bullet x^{-1} = 0$.

When the power is 1, the function is linear, and the derivative is a constant. When the power is 0, the function is a constant, and the derivative is 0.

DERIVATIVE RULES: CONSTANT, SUM, DIFFERENCE, AND CONSTANT MULTIPLE

Constant Rule: If $f(x) = k$, where k is a constant, then $f'(x) = 0$. In the other notation, $\frac{d}{dx}(k) = 0$.

Constant Multiple Rule: If you have a constant multiplied by a function, and you would like the derivative of the entire expression, you can "pull the constant out." The derivative will be the constant times the derivative of the function: $[k \cdot f(x)]' = k \cdot f'(x)$. In the other notation, this is $\frac{d}{dx}[k \cdot f(x)] = k \cdot \frac{d}{dx}[f(x)]$.

Note: For future reference, a, b, c, n, and k always stand for constants.

Example 15: If $y = 8x^4$, then $\frac{dy}{dx} = 8 \cdot \frac{d}{dx}x^4 = 8 \cdot 4x^3 = 32x^3$.

Example 16: If $y = 5x^{100}$, then $y' = 500x^{99}$.

Example 17: If $y = -3x^{-5}$, then $\frac{dy}{dx} = 15x^{-6}$.

Example 18: If $f(x) = 7x^{\frac{1}{2}}$, then $f'(x) = \frac{7}{2}x^{-\frac{1}{2}}$.

Example 19: If $y = x\sqrt{15}$, then $\frac{dy}{dx} = \sqrt{15}$.

Example 20: If $y = 12$, then $\frac{dy}{dx} = 0$.

If you have any questions about any of these examples (especially the last two), review the rules. Now for one last rule.

The Sum Rule

> If $y = ax^n + bx^m$, where a and b are constants, then
>
> $$\frac{dy}{dx} = a\left(nx^{n-1}\right) + b\left(mx^{m-1}\right)$$
>
> If $y = ax^n - bx^m$, where a and b are constants, then $\frac{dy}{dx} = a\left(nx^{n-1}\right) - b\left(mx^{m-1}\right)$.

This handy rule works for subtraction too.

Example 21: If $y = 3x^4 + 8x^{10}$, then $\frac{dy}{dx} = 12x^3 + 80x^9$.

Example 22: If $y = 7x^{-4} + 5x^{-\frac{1}{2}}$, then $\frac{dy}{dx} = -28x^{-5} - \frac{5}{2}x^{-\frac{3}{2}}$.

Example 23: If $y = 5x^4(2 - x^3) = 10x^4 - 5x^7$, then $\frac{dy}{dx} = 40x^3 - 35x^6$.

Example 24: If $y = (3x^2 + 5)(x - 1)$, then

$$y = 3x^3 - 3x^2 + 5x - 5 \text{ and } \frac{dy}{dx} = 9x^2 - 6x + 5$$

Example 25: If $y = ax^3 + bx^2 + cx + d$, then $\frac{dy}{dx} = 3ax^2 + 2bx + c$.

After you've worked through Examples 9–25, you should be able to take the derivative of any polynomial with ease.

As you may have noticed from these examples, in calculus you are often asked to convert from fractions and radicals to negative powers and fractional powers. In addition, don't worry if your answer doesn't match any of the answer choices. Because answers to problems are often presented in simplified form, your answer may not be simplified enough.

There are two basic expressions that you'll often be asked to differentiate. You can make your life easier by memorizing the following derivatives:

$$\text{If } y = \frac{k}{x}\text{, then } \frac{dy}{dx} = -\frac{k}{x^2}.$$

$$\text{If } y = k\sqrt{x}\text{, then } \frac{dy}{dx} = \frac{k}{2\sqrt{x}}.$$

Note that the Sum Rule and the Difference Rule are sometimes called the **Addition Rule** and the **Subtraction Rule**, respectively.

PRACTICE PROBLEM SET 7

Find the derivative of each expression and simplify. The answers are at the end of this chapter, starting on page 167.

1. $(4x^2 + 1)^2$

2. $(x^5 + 3x)^2$

3. $11x^7$

4. $18x^3 + 12x + 11$

5. $\frac{1}{2}(x^{12} + 17)$

6. $-\frac{1}{3}(x^9 + 2x^3 - 9)$

7. π^5

8. $-8x^{-8} + 12\sqrt{x}$

9. $6x^{-7} - 4\sqrt{x}$

10. $x^{-5} + \frac{1}{x^8}$

11. $(6x^2 + 3)(12x - 4)$

12. $(3 - x - 2x^3)(6 + x^4)$

13. $e^{10} + \pi^3 - 7$

14. $\left(\frac{1}{x} + \frac{1}{x^2}\right)\left(\frac{4}{x^3} - \frac{6}{x^4}\right)$

15. $\sqrt{x} + \frac{1}{\sqrt{3}}$

16. 0

17. $(x + 1)^3$

18. $\sqrt{x} + \sqrt[3]{x} + \sqrt[3]{x^2}$

19. $x(2x + 7)(x - 2)$

20. $\sqrt{x}\left(\sqrt[3]{x} + \sqrt[5]{x}\right)$

DERIVATIVES OF cos *x*, sin *x*, e^x, AND ln *x*

Derivatives of sin *x* and cos *x*

There are a lot of trigonometry problems in calculus and on the AP Calculus Exam. You'll need to remember your trig formulas, the values of the special angles, and the trig ratios, among other stuff.

In addition, angles are *always* referred to in radians. You can forget all about using degrees.

You should know the derivatives of all six trig functions. The good news is that the derivatives are pretty easy, and all you have to do is memorize them. Because the AP Exam might ask you about this, though, let's use the definition of the derivative to figure out the derivative of sin x.

If $f(x) = \sin x$, then $f(x + h) = \sin(x + h)$.

Substitute this into the definition of the derivative.

$$\lim_{h\to 0}\frac{f(x+h)-f(x)}{h}=\lim_{h\to 0}\frac{\sin(x+h)-\sin x}{h}$$

Remember that $\sin(x + h) = \sin x \cos h + \cos x \sin h$. Now simplify it.

$$\lim_{h\to 0}\frac{\sin x\cos h+\cos x\sin h-\sin x}{h}$$

Next, rewrite this as

$$\lim_{h\to 0}\frac{\sin x(\cos h-1)+\cos x\sin h}{h}=\lim_{h\to 0}\frac{\sin x(\cos h-1)}{h}+\lim_{h\to 0}\frac{\cos x\sin h}{h}$$

Next, use some of the trigonometric limits that you memorized back in Chapter 3. Specifically,

$$\lim_{h\to 0}\frac{(\cos h-1)}{h}=0 \text{ and } \lim_{h\to 0}\frac{\sin h}{h}=1$$

Trigon(line)ometry
If you're unsure of your trig, log on to your free online Student Tools and download the Appendix, which contains a focused review of Prerequisite Math.

This gives you

$$\lim_{h\to 0}\frac{\sin x(\cos h-1)}{h}+\lim_{h\to 0}\frac{\cos x\sin h}{h}=(\sin x)(0)+(\cos x)(1)=\cos x$$

$$\frac{d}{dx}\sin x=\cos x$$

Example 26: Find the derivative of $\sin\left(\frac{\pi}{2}-x\right)$.

$$\frac{d}{dx}\sin\left(\frac{\pi}{2}-x\right)=\cos\left(\frac{\pi}{2}-x\right)(-1)=-\cos\left(\frac{\pi}{2}-x\right)$$

Use some of the rules of trigonometry you remember from last year. Because

$$\sin\left(\frac{\pi}{2}-x\right)=\cos x \text{ and } \cos\left(\frac{\pi}{2}-x\right)=\sin x,$$

you can substitute into the above expression and get

$$\frac{d}{dx}\cos x=-\sin x$$

Log-online-arithms
Remember: if you're not sure about logarithms, you can find a focused, downloadable review of prerequisite mathematics like this online in your free Student Tools.

The Derivative of ln *x*

When you studied logs in the past, you probably concentrated on common logs (that is, those with a base of 10), and avoided natural logarithms (base *e*) as much as possible. Most of what you'll see from now on involves natural logs. In fact, common logs almost never show up in calculus. But that's okay. All you have to do is memorize a bunch of rules, and you'll be fine.

Rule No. 1: If $y = \ln x$, then $\frac{dy}{dx}=\frac{1}{x}$

This rule has a corollary that incorporates the Chain Rule and is actually a more useful rule to memorize.

Rule No. 2: If $y = \ln u$, then $\dfrac{dy}{dx} = \dfrac{1}{u}\dfrac{du}{dx}$

Remember: u is a function of x, and $\dfrac{du}{dx}$ is its derivative.

You'll see how simple this rule is after we try a few examples.

Chain Rule
The Chain says if $f(x) = g(h(x))$, then $f'(x) = g'(h(x))h'(x)$. There will be more on that rule in the next chapter.

Example 27: Find the derivative of $f(x) = \ln(x^3)$.

$$f'(x) = \frac{3x^2}{x^3} = \frac{3}{x}$$

If you recall your rules of logarithms, you could have done this another way.

$$\ln(x^3) = 3\ln x$$

Therefore, $f'(x) = 3\left(\dfrac{1}{x}\right) = \dfrac{3}{x}$.

Example 28: Find the derivative of $f(x) = \ln(5x - 3x^6)$.

$$f'(x) = \frac{(5 - 18x^5)}{(5x - 3x^6)}$$

Example 29: Find the derivative of $f(x) = \ln(\cos x)$.

$$f'(x) = \frac{-\sin x}{\cos x} = -\tan x$$

If you want to find the derivative of ln(function), write a fraction bar and put the function on the bottom and the derivative on the top.

Finding the derivative of a natural logarithm is just a matter of following a simple formula.

The Derivative of e^x

As you'll see in Rule No. 3, the derivative of e^x is probably the easiest thing that you'll ever have to do in calculus.

Rule No. 3: If $y = e^x$, then $\dfrac{dy}{dx} = e^x$

That's not a typo. The derivative is the same as the original function! Incorporating the Chain Rule, we get a good formula for finding the derivative.

Rule No. 4: If $y = e^u$, then $\frac{dy}{dx} = e^u \frac{du}{dx}$

And you were worried that all of this logarithm and exponential stuff was going to be hard!

Example 30: Find the derivative of $f(x) = e^{3x}$.

$$f'(x) = e^{3x}(3) = 3e^{3x}$$

Example 31: Find the derivative of $f(x) = e^{x^3}$.

$$f'(x) = e^{x^3}\left(3x^2\right) = 3x^2e^{x^3}$$

Example 32: Find the derivative of $f(x) = e^{\tan x}$.

$$f'(x) = \left(\sec^2 x\right)e^{\tan x}$$

Example 33: Find the second derivative of $f(x) = e^{x^2}$.

$$f'(x) = 2xe^{x^2}$$
$$f''(x) = 2e^{x^2} + 4x^2e^{x^2}$$

Once again, it's just a matter of following a formula.

The Derivative of $\log_a x$

This derivative is actually a little trickier than the derivative of a natural log. First, if you remember your logarithm rules about change of base, we can rewrite $\log_a x$.

$$\log_a x = \frac{\ln x}{\ln a}$$

Because $\ln a$ is a constant, we can take the derivative.

$$\frac{1}{\ln a}\frac{1}{x}$$

This leads us to our next rule.

Rule No. 5: If $y = \log_a x$, then $\frac{dy}{dx} = \frac{1}{x \ln a}$

Once again, incorporating the Chain Rule gives us a more useful formula.

Rule No. 6: If $y = \log_a u$, then $\frac{dy}{dx} = \frac{1}{u \ln a} \frac{du}{dx}$

Example 34: Find the derivative of $f(x) = \log_{10} x$.

$$f'(x) = \frac{1}{x \ln 10}$$

Note: We refer to $\log_{10} x$ as $\log x$.

Example 35: Find the derivative of $f(x) = \log_8 (x^2 + x)$.

$$f'(x) = \frac{2x+1}{(x^2+x)\ln 8}$$

Example 36: Find the derivative of $f(x) = \log_e x$.

$$f'(x) = \frac{1}{x \ln e} = \frac{1}{x}$$

You can expect this result from Rules 1 and 2 involving natural logs.

The Derivative of a^x

You should recall from your precalculus days that we can rewrite a^x as $e^{x\ln a}$. Keep in mind that $\ln a$ is just a constant, which gives us the next rule.

Rule No. 7: If $y = a^x$, then $\frac{dy}{dx} = \left(e^{x \ln a}\right) \ln a = a^x (\ln a)$

Given the pattern of this chapter, you can guess what's coming: another rule that incorporates the Chain Rule.

Rule No. 8: If $y = a^u$, then $\frac{dy}{dx} = a^u(\ln a)\frac{du}{dx}$

And now, some examples.

Example 37: Find the derivative of $f(x) = 3^x$.

$$f'(x) = 3^x \ln 3$$

Example 38: Find the derivative of $f(x) = 8^{4x^5}$.

$$f'(x) = 8^{4x^5}(20x^4)\ln 8$$

Example 39: Find the derivative of $f(x) = \pi^{\sin x}$.

$$f'(x) = \pi^{\sin x}(\cos x)\ln \pi$$

Finally, here's every nasty teacher's favorite exponential derivative.

Example 40: Find the derivative of $f(x) = x^x$.

First, rewrite this as $f(x) = e^{x \ln x}$. Then, take the derivative.

$$f'(x) = e^{x\ln x}\left(\ln x + \frac{x}{x}\right) = e^{x\ln x}(\ln x + 1) = x^x(\ln x + 1)$$

Would you have thought of that? Remember this trick. It might come in handy! Okay. Ready for some practice? Cover the solutions and get cracking.

Example 41: Find the derivative of $y = 3\ln(5x^2 + 4x)$.

Use Rule No. 2.

$$\frac{dy}{dx} = 3\frac{10x+4}{5x^2+4x} = \frac{30x+12}{5x^2+4x}$$

Example 42: Find the derivative of $f(x) = \ln\left(\sin\left(x^5\right)\right)$.

Use Rule No. 2 in addition to the Chain Rule.

$$f'(x) = \frac{5x^4 \cos\left(x^5\right)}{\sin\left(x^5\right)} = 5x^4 \cot\left(x^5\right)$$

Example 43: Find the derivative of $f(x) = e^{3x^7 - 4x^2}$.

Use Rule No. 4.

$$f'(x) = \left(21x^6 - 8x\right)e^{3x^7 - 4x^2}$$

Example 44: Find the derivative of $f(x) = \log_4(\tan x)$.

Use Rule No. 6.

$$f'(x) = \frac{1}{\ln 4}\frac{\sec^2 x}{\tan x}$$

Example 45: Find the derivative of $y = \log_8 \sqrt{\frac{x^3}{1+x^2}}$.

First, use the rules of logarithms to rewrite the equation.

$$y = \frac{1}{2}\left[3\log_8 x - \log_8\left(1+x^2\right)\right]$$

Now it's much easier to find the derivative.

$$\frac{dy}{dx} = \frac{1}{2}\left[3\frac{1}{x\ln 8}\right] - \frac{1}{2}\left[\frac{2x}{\ln 8\left(1+x^2\right)}\right] = \frac{1}{2\ln 8}\left[\frac{3}{x} - \frac{2x}{\left(1+x^2\right)}\right]$$

Example 46: Find the derivative of $y = 5^{\sqrt{x}}$.

Use Rule No. 8.

$$\frac{dy}{dx} = 5^{\sqrt{x}}\frac{1}{2\sqrt{x}}\ln 5 = \frac{5^{\sqrt{x}}\ln 5}{2\sqrt{x}}$$

Example 47: Find the derivative of $y = \dfrac{e^{x^3}}{5^{\cos x}}$.

Here, you need to use the Quotient Rule and Rules Nos. 4 and 8.

$$\frac{dy}{dx} = \frac{5^{\cos x}\left(3x^2 e^{x^3}\right) - e^{x^3}\left(5^{\cos x}\ln 5(-\sin x)\right)}{\left(5^{\cos x}\right)^2} = 5^{\cos x} e^{x^3} \frac{\left(3x^2\right) + \left(\sin x \ln 5\right)}{5^{2\cos x}} = e^{x^3}\,\frac{3x^2 + \sin x \ln 5}{5^{\cos x}}$$

PRACTICE PROBLEM SET 8

Simplify when possible. The answers are at the end of this chapter, starting on page 171.

1. Find $f'(x)$ if $f(x) = \ln\left(x^4 + 8\right)$.
2. Find $f'(x)$ if $f(x) = \ln\left(3x\sqrt{3+x}\right)$.
3. Find $f'(x)$ if $f(x) = \ln\left(\dfrac{5x^2}{\sqrt{5+x^2}}\right)$.
4. Find $f'(x)$ if $f(x) = e^{x\cos x}$.
5. Find $f'(x)$ if $f(x) = e^{-3x}\sin 5x$.
6. Find $f'(x)$ if $f(x) = e^{\pi x} - \ln e^{\pi x}$.
7. Find $f'(x)$ if $f(x) = \log_{12}\left(x^3\right)$.
8. Find $f'(x)$ if $f(x) = \log\sqrt{10^{3x}}$.
9. Find $f'(x)$ if $f(x) = e^{3x} - 3^{ex}$.
10. Find $f'(x)$ if $f(x) = 10^{\sin x}$.
11. Find $f'(x)$ if $f(x) = \ln\left(10^x\right)$.

THE PRODUCT RULE

Now that you know how to find derivatives of simple polynomials, it's time to get more complicated. What if you had to find the derivative of this?

$$f(x) = (x^3 + 5x^2 - 4x + 1)(x^5 - 7x^4 + x)$$

You could multiply out the expression and take the derivative of each term, like

$$f(x) = x^8 - 2x^7 - 39x^6 + 29x^5 - 6x^4 + 5x^3 - 4x^2 + x$$

And the derivative is

$$f'(x) = 8x^7 - 14x^6 - 234x^5 + 145x^4 - 24x^3 + 15x^2 - 8x + 1$$

Needless to say, this process is messy. Naturally, there's an easier way. When a function involves two terms multiplied by each other, we use the **Product Rule**.

> With the Product Rule, the order of these two operations doesn't matter. It **does** matter with other rules, though, so it helps to use the same order each time.

The Product Rule: If $f(x) = uv$, then $f'(x) = u\frac{dv}{dx} + v\frac{du}{dx}$

To find the derivative of two things multiplied by each other, you multiply the first function by the derivative of the second, and add that to the second function multiplied by the derivative of the first.

Let's use the Product Rule to find the derivative of our example.

$$f'(x) = (x^3 + 5x^2 - 4x + 1)(5x^4 - 28x^3 + 1) + (x^5 - 7x^4 + x)(3x^2 + 10x - 4)$$

If we were to simplify this, we'd get the same answer as before. But here's the best part: we're not going to simplify it. One of the great things about the AP Exam is that when it's difficult to simplify an expression, you almost never have to.

Nonetheless, you'll often need to simplify expressions when you're taking second derivatives, or when you use the derivative in some other equation. Practice simplifying whenever possible.

Example 48: $f(x) = (9x^2 + 4x)(x^3 - 5x^2)$

$$f'(x) = (9x^2 + 4x)(3x^2 - 10x) + (x^3 - 5x^2)(18x + 4)$$

Example 49: $y = \left(\sqrt{x} + 4\sqrt[3]{x}\right)\left(x^5 - 11x^8\right)$

$$y' = \left(\sqrt{x} + 4\sqrt[3]{x}\right)\left(5x^4 - 88x^7\right) + \left(x^5 - 11x^8\right)\left(\frac{1}{2\sqrt{x}} + \frac{4}{3\sqrt[3]{x^2}}\right)$$

Example 50: $y = \left(\frac{1}{x} + \frac{1}{x^2} - \frac{1}{x^3}\right)\left(\frac{1}{x} - \frac{1}{x^3} + \frac{1}{x^5}\right)$

$$y' = \left(\frac{1}{x} + \frac{1}{x^2} - \frac{1}{x^3}\right)\left(-\frac{1}{x^2} + \frac{3}{x^4} - \frac{5}{x^6}\right) + \left(\frac{1}{x} - \frac{1}{x^3} + \frac{1}{x^5}\right)\left(-\frac{1}{x^2} - \frac{2}{x^3} + \frac{3}{x^4}\right)$$

THE QUOTIENT RULE

What happens when you have to take the derivative of a function that is the quotient of two other functions? You guessed it: use the **Quotient Rule**.

> The Quotient Rule: If $f(x) = \frac{u}{v}$, then $f'(x) = \dfrac{v\dfrac{du}{dx} - u\dfrac{dv}{dx}}{v^2}$

In this rule, as opposed to the Product Rule, the order in which you take the derivatives is **very important** because you're subtracting instead of adding. It's always the bottom function times the derivative of the top minus the top function times the derivative of the bottom. Then divide the whole thing by the bottom function squared. A good way to remember this is to say the following:

For the Quotient Rule, remember that order matters!

$$\frac{\text{"}LoDeHi - HiDeLo\text{"}}{(Lo)^2}$$

Here are some more examples.

Example 51: $f(x)=\dfrac{\left(x^5-3x^4\right)}{\left(x^2+7x\right)}$

$$f'(x)=\frac{\left(x^2+7x\right)\left(5x^4-12x^3\right)-\left(x^5-3x^4\right)(2x+7)}{\left(x^2+7x\right)^2}$$

Example 52: $y=\dfrac{\left(x^{-3}-x^{-8}\right)}{\left(x^{-2}+x^{-6}\right)}$

$$\frac{dy}{dx}=\frac{\left(x^{-2}+x^{-6}\right)\left(-3x^{-4}+8x^{-9}\right)-\left(x^{-3}-x^{-8}\right)\left(-2x^{-3}-6x^{-7}\right)}{\left(x^{-2}+x^{-6}\right)^2}$$

We're not going to simplify these, although the Quotient Rule often produces expressions that simplify more readily than those involving the Product Rule. You usually won't have to simplify the derivative using the Quotient Rule unless (1) it's straightforward to do, (2) you need to find the second derivative, or (3) you need to find where the derivative is zero. If this is the case, the AP Exam usually will give you a simple expression to deal with, such as in the example below.

Example 53: $y=\dfrac{3x+5}{5x-3}$

$$\frac{dy}{dx}=\frac{(5x-3)(3)-(3x+5)(5)}{(5x-3)^2}=\frac{(15x-9)-(15x+25)}{(5x-3)^2}=\frac{-34}{(5x-3)^2}$$

FINDING THE DERIVATIVES OF TANGENT, COTANGENT, SECANT, AND/OR COSECANT FUNCTIONS

Now, let's derive the derivatives of the other four trigonometric functions.

Example 54: Find the derivative of $\dfrac{\sin x}{\cos x}$.

Use the Quotient Rule.

$$\frac{d}{dx}\frac{\sin x}{\cos x}=\frac{(\cos x)(\cos x)-(\sin x)(-\sin x)}{(\cos x)^2}=\frac{\cos^2 x+\sin^2 x}{\cos^2 x}=\frac{1}{\cos^2 x}=\sec^2 x$$

Because $\frac{\sin x}{\cos x}$ = tan *x*, you should get

$$\frac{d}{dx}\tan x = \sec^2 x$$

Example 55: Find the derivative of $\frac{\cos x}{\sin x}$.

Use the Quotient Rule.

$$\frac{d}{dx}\frac{\cos x}{\sin x} = \frac{(\sin x)(-\sin x)-(\cos x)(\cos x)}{(\sin x)^2} = \frac{-(\cos^2 x+\sin^2 x)}{\sin^2 x} = -\frac{1}{\sin^2 x} = -\csc^2 x$$

$$\frac{d}{dx}\cot x = -\csc^2 x$$

Example 56: Find the derivative of $\frac{1}{\cos x}$.

Use the Quotient Rule.

$$\frac{d}{dx}\frac{1}{\cos x} = \frac{(\cos x)(0)-1(-\sin x)}{(\cos^2 x)} = \frac{\sin x}{\cos^2 x} = \frac{1}{\cos x}\frac{\sin x}{\cos x} = \sec x\tan x$$

Because $\frac{1}{\cos x}$ = sec *x*, you get

$$\frac{d}{dx}\sec x = \sec x\tan x$$

Example 57: Find the derivative of $\frac{1}{\sin x}$.

You get the idea by now.

$$\frac{d}{dx}\frac{1}{\sin x} = \frac{(\sin x)(0)-1(\cos x)}{(\sin^2 x)} = \frac{-\cos x}{\sin^2 x} = \frac{-1}{\sin x}\frac{\cos x}{\sin x} = -\csc x\cot x$$

Because $\frac{1}{\sin x}$ = csc x, you get

$$\frac{d}{dx}\csc x = -\csc x \cot x$$

There you go. We have now found the derivatives of all six of the trigonometric functions. (You can find a full list of formulas you should know in your free online Student Tools.)

PRACTICE PROBLEM SET 9

Simplify when possible. The answers are at the end of this chapter, starting on page 173.

1. Find $f'(x)$ if $f(x)=\left(\frac{4x^3-3x^2}{5x^7+1}\right)$.

2. Find $f'(x)$ if $f(x)=\left(x^2-4x+3\right)\left(x+1\right)$.

3. Find $f'(x)$ if $f(x)=\left(x+\frac{1}{x}\right)\left(x^2-\frac{1}{x^2}\right)$.

4. Find $f'(x)$ at $x = 2$ if $f(x)=\frac{(x+4)(x-8)}{(x+6)(x-6)}$.

5. Find $f'(x)$ at $x = 1$ if $f(x)=\frac{x^6+4x^3+6}{\left(x^4-2\right)^2}$.

6. Find $f'(x)$ if $f(x)=\frac{x^2-3}{(x-3)}$.

7. Find $f'(x)$ at $x = 1$ if $f(x)=\left(x^4-x^2\right)\left(2x^3+x\right)$.

8. Find $f'(x)$ at $x = 2$ if $f(x) = \dfrac{x^2 + 2x}{x^4 - x^3}$.

9. Find $\dfrac{dy}{dt}$ if $y = (x^6 - 6x^5)(5x^2 + x)$ and $x = \sqrt{t}$.

10. Find $f'(x)$ if $f(x) = \ln(\cot x - \csc x)$.

11. Find $f'(x)$ if $f(x) = \dfrac{\log_4 x}{e^{4x}}$.

12. Find $f'(x)$ if $f(x) = x^5 5^x$.

Chapter 4 Drill

The answers are at the end of the chapter, starting on page 179.

1 Mark for Review

Use the Definition of the Derivative to find $f'(x)$ for $f(x) = 8x - 11$.

(A) -11

(B) 0

(C) 8

(D) $8x$

2 Mark for Review

Use the Definition of the Derivative to find $f'(x)$ for $f(x) = 3x^2 + 7x - 2$.

(A) $6x$

(B) $6x - 2$

(C) $6x + 7$

(D) $3x + 7$

3 Mark for Review

Use the Definition of the Derivative to find $f'(x)$ for $f(x) = x^3 - 4x^2 + 11$.

(A) $3x^2 - 4$

(B) $3x^2 - 4x + 11$

(C) $3x^2 - 8x + 11$

(D) $3x^2 - 8x$

4 Mark for Review

Use the Definition of the Derivative to find $f'(25)$, if $f(x) = 6\sqrt{x}$.

(A) $\frac{3}{5}$

(B) $\frac{3}{25}$

(C) $\frac{6}{25}$

(D) $\frac{6}{5}$

5 Mark for Review

Use the Definition of the Derivative to find $f'(2)$, if $f(x) = \frac{1}{4x^2}$.

(A) $-\frac{1}{16}$

(B) $-\frac{1}{8}$

(C) $\frac{1}{8}$

(D) $\frac{1}{16}$

6 Mark for Review

Find $\frac{dy}{dx}$ if $y = 5x^3 - 7x^2 + 11x + 2$.

(A) $\frac{dy}{dx} = 15x^2 - 14x$

(B) $\frac{dy}{dx} = 35x^2 - 27x + 11$

(C) $\frac{dy}{dx} = 5x^2 - 7x + 11$

(D) $\frac{dy}{dx} = 15x^2 - 14x + 11$

ANSWERS TO PRACTICE PROBLEM SETS

Practice Problem Set 6

1. 5

Find the derivative of a function, $f(x)$, using the definition of the derivative, which is $f'(x)=\lim_{h\to 0}\frac{f(x+h)-f(x)}{h}$. Here, $f(x)=5x$ and $x = 3$. This means that $f(3)=5(3)=15$ and $f(3+h)=5(3+h)=15+5h$. If you now plug these into the definition of the derivative, you get $f'(3)=\lim_{h\to 0}\frac{f(3+h)-f(3)}{h}=\lim_{h\to 0}\frac{15+5h-15}{h}$. This simplifies to $f'(3)=\lim_{h\to 0}\frac{5h}{h}=\lim_{h\to 0}5=5$.

If you noticed that the function is simply the equation of a line, then you would have seen that the derivative is simply the slope of the line, which is 5 everywhere.

2. 4

Find the derivative of a function, $f(x)$, using the definition of the derivative, which is $f'(x)=\lim_{h\to 0}\frac{f(x+h)-f(x)}{h}$. Here, $f(x)=4x$ and $x = -8$. This means that $f(-8)=4(-8)=-32$ and $f(-8+h)=4(-8+h)=-32+4h$. Plug these into the definition of the derivative to get $f'(-8)=\lim_{h\to 0}\frac{f(-8+h)-f(-8)}{h}=\lim_{h\to 0}\frac{-32+4h+32}{h}$. This simplifies to $f'(-8)=\lim_{h\to 0}\frac{4h}{h}=\lim_{h\to 0}4=4$.

If you noticed that the function is simply the equation of a line, then you would have seen that the derivative is simply the slope of the line, which is 4 everywhere.

3. –10

Find the derivative of a function, $f(x)$, using the definition of the derivative, which is $f'(x)=\lim_{h\to 0}\frac{f(x+h)-f(x)}{h}$. Here, $f(x)=5x^2$ and $x=-1$. This means that

$f(-1)=5(-1)^2=5$ and $f(-1+h)=5(-1+h)^2=5(1-2h+h^2)=5-10h+5h^2$. Plug these into the definition of the derivative to get $f'(-1)=\lim\limits_{h\to 0}\frac{f(-1+h)-f(-1)}{h}=\lim\limits_{h\to 0}\frac{5-10h+5h^2-5}{h}$. This simplifies to $f'(-1)=\lim\limits_{h\to 0}\frac{-10h+5h^2}{h}$. Now factor out the h from the numerator and cancel it with the h in the denominator: $f'(-1)=\lim\limits_{h\to 0}\frac{h(-10+5h)}{h}=\lim\limits_{h\to 0}(-10+5h)$. Now take the limit to get $f'(-1)=\lim\limits_{h\to 0}(-10+5h)=-10$.

4. $16x$

Find the derivative of a function, $f(x)$, using the definition of the derivative, which is $f'(x)=\lim\limits_{h\to 0}\frac{f(x+h)-f(x)}{h}$. Here, $f(x)=8x^2$ and $f(x+h)=8(x+h)^2=8(x^2+2xh+h^2)=8x^2+16xh+8h^2$. Plug these into the definition of the derivative to get $f'(x)=\lim\limits_{h\to 0}\frac{f(x+h)-f(x)}{h}=\lim\limits_{h\to 0}\frac{8x^2+16xh+8h^2-8x^2}{h}$. This simplifies to $f'(x)=\lim\limits_{h\to 0}\frac{16xh+8h^2}{h}$. Now factor out the h from the numerator and cancel it with the h in the denominator: $f'(x)=\lim\limits_{h\to 0}\frac{h(16x+8h)}{h}=\lim\limits_{h\to 0}(16x+8h)$. Now take the limit to get $f'(x)=\lim\limits_{h\to 0}(16x+8h)=16x$.

5. $-20x$

Find the derivative of a function, $f(x)$, using the definition of the derivative, which is $f'(x)=\lim\limits_{h\to 0}\frac{f(x+h)-f(x)}{h}$. Here, $f(x)=-10x^2$ and $f(x+h)=-10(x+h)^2=-10(x^2+2xh+h^2)=-10x^2-20xh-10h^2$. Plug these into the definition of the derivative to get $f'(x)=\lim_{h\to 0}\frac{f(x+h)-f(x)}{h}=\lim_{h\to 0}\frac{-10x^2-20xh-10h^2+10x^2}{h}$.

This simplifies to $f'(x)=\lim_{h\to 0}\frac{-20xh-10h^2}{h}$. Now factor out the h from the numerator and cancel it with the h in the denominator: $f'(x)=\lim_{h\to 0}\frac{h(-20x-10h)}{h}=\lim_{h\to 0}(-20x-10h)$. Now take the limit to get $f'(x)=\lim_{h\to 0}(-20x-10h)=-20x$.

6. $40a$

Find the derivative of a function, $f(x)$, using the definition of the derivative, which is $f'(x)=\lim_{h\to 0}\frac{f(x+h)-f(x)}{h}$. Here, $f(x)=20x^2$ and $x=a$. This means that $f(a)=20a^2$ and $f(a+h)=20(a+h)^2=20(a^2+2ah+h^2)=20a^2+40ah+20h^2$. Plug these into the definition of the derivative to get $f'(a)=\lim_{h\to 0}\frac{f(a+h)-f(a)}{h}=\lim_{h\to 0}\frac{20a^2+40ah+20h^2-20a^2}{h}$.

This simplifies to $f'(a)=\lim_{h\to 0}\frac{40ah+20h^2}{h}$. Now factor out the h from the numerator and cancel it with the h in the denominator: $f'(a)=\lim_{h\to 0}\frac{h(40a+20h)}{h}=\lim_{h\to 0}(40a+20h)$. Now take the limit to get $f'(a)=\lim_{h\to 0}(40a+20h)=40a$.

7. 54

Find the derivative of a function, $f(x)$, using the definition of the derivative, which is $f'(x)=\lim_{h\to 0}\frac{f(x+h)-f(x)}{h}$. Here, $f(x)=2x^3$ and $x=-3$. This means that $f(-3)=2(-3)^3=-54$ and

$f(-3+h)=2(-3+h)^3=2\left((-3)^3+3(-3)^2h+3(-3)h^2+h^3\right)=-54+54h-18h^2+2h^3$.

Plug these into the definition of the derivative to get

$f'(-3)=\lim\limits_{h\to 0}\frac{f(-3+h)-f(-3)}{h}=\lim\limits_{h\to 0}\frac{-54+54h-18h^2+2h^3+54}{h}$. This simplifies to $f'(-3)=\lim\limits_{h\to 0}\frac{54h-18h^2+2h^3}{h}$. Now factor out the h from the numerator and cancel it with the h in the denominator: $f'(-3)=\lim\limits_{h\to 0}\frac{h(54-18h+2h^2)}{h}=\lim\limits_{h\to 0}(54-18h+2h^2)$. Now take the limit to get $f'(-3)=\lim\limits_{h\to 0}(54-18h+2h^2)=54$.

8. $-9x^2$

Find the derivative of a function, $f(x)$, using the definition of the derivative, which is $f'(x)=\lim\limits_{h\to 0}\frac{f(x+h)-f(x)}{h}$. Here, $f(x)=-3x^3$. This means that $f(x+h)=-3(x+h)^3=-3(x^3+3x^2h+3xh^2+h^3)=-3x^3-9x^2h-9xh^2-3h^3$.

Plug these into the definition of the derivative to get $f'(x)=\lim\limits_{h\to 0}\frac{f(x+h)-f(x)}{h}=\lim\limits_{h\to 0}\frac{-3x^3-9x^2h-9xh^2-3h^3+3x^3}{h}$. This simplifies to $f'(x)=\lim\limits_{h\to 0}\frac{-9x^2h-9xh^2-3h^3}{h}$. Now factor out the h from the numerator and cancel it with the h in the denominator: $f'(x)=\lim\limits_{h\to 0}\frac{h(-9x^2-9xh-3h^2)}{h}=\lim\limits_{h\to 0}(-9x^2-9xh-3h^2)$. Now take the limit to get $f'(x)=\lim\limits_{h\to 0}(-9x^2-9xh-3h^2)=-9x^2$.

9. $5x^4$

Find the derivative of a function, $f(x)$, using the definition of the derivative, which is $f'(x)=\lim\limits_{h\to 0}\frac{f(x+h)-f(x)}{h}$. Here, $f(x)=x^5$.

This means that $f(x+h)=(x+h)^5=(x^5+5x^4h+10x^3h^2+10x^2h^3+5xh^4+h^5)$.

Plug these into the definition of the derivative to get $f'(x)=\lim\limits_{h\to 0}\frac{f(x+h)-f(x)}{h}=\lim\limits_{h\to 0}\frac{x^5+5x^4h+10x^3h^2+10x^2h^3+5xh^4+h^5-x^5}{h}$.

This simplifies to $f'(x)=\lim\limits_{h\to 0}\dfrac{5x^4h+10x^3h^2+10x^2h^3+5xh^4+h^5}{h}$. Now you can factor out the h from the numerator and cancel it with the h in the denominator: $f'(x)=\lim\limits_{h\to 0}\dfrac{h\left(5x^4+10x^3h+10x^2h^2+5xh^3+h^4\right)}{h}=\lim\limits_{h\to 0}\left(5x^4+10x^3h+10x^2h^2+5xh^3+h^4\right)$.

Now take the limit to get $f'(x)=\lim\limits_{h\to 0}\left(5x^4+10x^3h+10x^2h^2+5xh^3+h^4\right)=5x^4$.

10. $\dfrac{1}{3}$

Find the derivative of a function, $f(x)$, using the definition of the derivative, which is $f'(x)=\lim\limits_{h\to 0}\dfrac{f(x+h)-f(x)}{h}$. Here, $f(x)=2\sqrt{x}$ and $x = 9$. This means that $f(9)=2\sqrt{9}=6$ and $f(9+h)=2\sqrt{9+h}$. Plug these into the definition of the derivative to get $f'(9)=\lim\limits_{h\to 0}\dfrac{f(9+h)-f(9)}{h}=\lim\limits_{h\to 0}\dfrac{2\sqrt{9+h}-6}{h}$. If you now take the limit, you get the indeterminate form $\dfrac{0}{0}$. With polynomials, merely simplify the expression to eliminate this problem. In any derivative of a square root, first multiply the top and the bottom of the expression by the conjugate of the numerator and then simplify. Here the conjugate is $2\sqrt{9+h}+6$. You get $f'(9)=\lim\limits_{h\to 0}\dfrac{2\sqrt{9+h}-6}{h}\times\dfrac{2\sqrt{9+h}+6}{2\sqrt{9+h}+6}$. This simplifies to $f'(9)=\lim\limits_{h\to 0}\dfrac{4(9+h)-36}{h\left(2\sqrt{9+h}+6\right)}=\lim\limits_{h\to 0}\dfrac{36+4h-36}{h\left(2\sqrt{9+h}+6\right)}=\lim\limits_{h\to 0}\dfrac{4h}{h\left(2\sqrt{9+h}+6\right)}$. Now cancel the h in the numerator and the denominator to get $f'(9)=\lim\limits_{h\to 0}\dfrac{4}{\left(2\sqrt{9+h}+6\right)}$. Now take the limit:

$$f'(9)=\lim_{h\to 0}\frac{4}{\left(2\sqrt{9+h}+6\right)}=\frac{4}{\left(2\sqrt{9}+6\right)}=\frac{1}{3}.$$

11. $\frac{5}{4}$

Find the derivative of a function, $f(x)$, using the definition of the derivative, which is $f'(x)=\lim_{h\to 0}\frac{f(x+h)-f(x)}{h}$. Here, $f(x)=5\sqrt{2x}$ and $x = 8$. This means that $f(8)=5\sqrt{16}=20$ and $f(8+h)=5\sqrt{2(8+h)}=5\sqrt{16+2h}$. Plug these into the definition of the derivative to get $f'(8)=\lim_{h\to 0}\frac{f(8+h)-f(8)}{h}=\lim_{h\to 0}\frac{5\sqrt{16+2h}-20}{h}$. If you now take the limit, you get the indeterminate form $\frac{0}{0}$. With polynomials, merely simplify the expression to eliminate this problem. In any derivative of a square root, first multiply the top and the bottom of the expression by the conjugate of the numerator and then simplify. Here the conjugate is $5\sqrt{16+2h}+20$. You get $f'(8)=\lim_{h\to 0}\frac{5\sqrt{16+2h}-20}{h}\times\frac{5\sqrt{16+2h}+20}{5\sqrt{16+2h}+20}$. This simplifies to $f'(8)=\lim_{h\to 0}\frac{25(16+2h)-400}{h\left(5\sqrt{16+2h}+20\right)}=\lim_{h\to 0}\frac{400+50h-400}{h\left(5\sqrt{16+2h}+20\right)}=\lim_{h\to 0}\frac{50h}{h\left(5\sqrt{16+2h}+20\right)}$. Now cancel the h in the numerator and the denominator to get $f'(8)=\lim_{h\to 0}\frac{50}{\left(5\sqrt{16+2h}+20\right)}$. Now, take the limit: $f'(8)=\lim_{h\to 0}\frac{50}{\left(5\sqrt{16+2h}+20\right)}=\frac{50}{(20+20)}=\frac{5}{4}$.

12. $\frac{1}{2}$

Find the derivative of a function, $f(x)$, using the definition of the derivative, which is $f'(x)=\lim_{h\to 0}\frac{f(x+h)-f(x)}{h}$. Here, $f(x)=\sin x$ and $x=\frac{\pi}{3}$. This means that $f\left(\frac{\pi}{3}\right)=\sin\frac{\pi}{3}=\frac{\sqrt{3}}{2}$ and $f\left(\frac{\pi}{3}+h\right)=\sin\left(\frac{\pi}{3}+h\right)$. Plug these into the definition of the derivative to get $f'\left(\frac{\pi}{3}\right)=\lim_{h\to 0}\frac{f\left(\frac{\pi}{3}+h\right)-f\left(\frac{\pi}{3}\right)}{h}=\lim_{h\to 0}\frac{\sin\left(\frac{\pi}{3}+h\right)-\frac{\sqrt{3}}{2}}{h}$. If you now take the limit, you get the indeterminate form $\frac{0}{0}$. You cannot eliminate this problem merely by simplifying the expression the way that you did with a polynomial. Recall the trigonometric formula: $\sin(A+B)=\sin A\cos B+\cos A\sin B$. Here, you can rewrite the top expression as $f'\left(\frac{\pi}{3}\right)=\lim_{h\to 0}\frac{\sin\left(\frac{\pi}{3}+h\right)-\frac{\sqrt{3}}{2}}{h}=\lim_{h\to 0}\frac{\sin\frac{\pi}{3}\cos h+\cos\frac{\pi}{3}\sin h-\frac{\sqrt{3}}{2}}{h}$. Break up the limit into $\lim_{h\to 0}\frac{\sin\frac{\pi}{3}\cos h-\frac{\sqrt{3}}{2}}{h}+\frac{\cos\frac{\pi}{3}\sin h}{h}=\lim_{h\to 0}\frac{\frac{\sqrt{3}}{2}\cos h-\frac{\sqrt{3}}{2}}{h}+\frac{\left(\frac{1}{2}\right)\sin h}{h}$. Next, factor $\frac{\sqrt{3}}{2}$ out of the top of the left-hand expression: $\lim_{h\to 0}\frac{\frac{\sqrt{3}}{2}(\cos h-1)}{h}+\frac{\left(\frac{1}{2}\right)\sin h}{h}$. Now, you can break this into separate limits: $\lim_{h\to 0}\frac{\frac{\sqrt{3}}{2}(\cos h-1)}{h}+\lim_{h\to 0}\frac{\left(\frac{1}{2}\right)\sin h}{h}$. The left-hand limit is $\lim_{h\to 0}\frac{\frac{\sqrt{3}}{2}(\cos h-1)}{h}=\frac{\sqrt{3}}{2}\lim_{h\to 0}\frac{(\cos h-1)}{h}=\frac{\sqrt{3}}{2}\bullet 0=0$.

The right-hand limit is $\frac{1}{2}\lim_{h\to 0}\frac{\sin h}{h}=\frac{1}{2}\bullet 1=\frac{1}{2}$. Therefore, the limit is $\frac{1}{2}$.

13. $2x+1$

Find the derivative of a function, $f(x)$, using the definition of the derivative, which is $f'(x)=\lim_{h\to 0}\frac{f(x+h)-f(x)}{h}$. Here, $f(x)=x^2+x$ and $f(x+h)=(x+h)^2+(x+h)=x^2+2xh+h^2+x+h$. Plug these into the definition of the derivative to get $f'(x)=\lim\limits_{h\to 0}\dfrac{f(x+h)-f(x)}{h}=\lim\limits_{h\to 0}\dfrac{x^2+2xh+h^2+x+h-(x^2+x)}{h}$. This simplifies to $f'(x)=\lim\limits_{h\to 0}\dfrac{2xh+h^2+h}{h}$. Now factor out the h from the numerator and cancel it with the h in the denominator: $f'(x)=\lim\limits_{h\to 0}\dfrac{2xh+h^2+h}{h}=\lim\limits_{h\to 0}\dfrac{h(2x+h+1)}{h}=\lim\limits_{h\to 0}(2x+h+1)$. Now, take the limit to get $f'(x)=\lim\limits_{h\to 0}(2x+h+1)=2x+1$.

14. $3x^2+3$

Find the derivative of a function, $f(x)$, using the definition of the derivative, which is $f'(x)=\lim\limits_{h\to 0}\dfrac{f(x+h)-f(x)}{h}$. Here, $f(x)=x^3+3x+2$ and $f(x+h)=(x+h)^3+3(x+h)+2=x^3+3x^2h+3xh^2+h^3+3x+3h+2$.

Plug these into the definition of the derivative to get $f'(x)=\lim\limits_{h\to 0}\dfrac{f(x+h)-f(x)}{h}=\lim\limits_{h\to 0}\dfrac{x^3+3x^2h+3xh^2+h^3+3x+3h+2-(x^3+3x+2)}{h}$.

This simplifies to $f'(x)=\lim\limits_{h\to 0}\dfrac{3x^2h+3xh^2+h^3+3h}{h}$. Now you can factor out the h from the numerator and cancel it with the h in the denominator: $f'(x)=\lim\limits_{h\to 0}\dfrac{3x^2h+3xh^2+h^3+3h}{h}=\lim\limits_{h\to 0}\dfrac{h(3x^2+3xh+h^2+3)}{h}=\lim\limits_{h\to 0}(3x^2+3xh+h^2+3)$. Now, take the limit to get $f'(x)=\lim\limits_{h\to 0}(3x^2+3xh+h^2+3)=3x^2+3$.

15. $-\dfrac{1}{x^2}$

Find the derivative of a function, $f(x)$, using the definition of the derivative, which is $f'(x)=\lim\limits_{h\to 0}\dfrac{f(x+h)-f(x)}{h}$. Here, $f(x)=\dfrac{1}{x}$ and $f(x+h)=\dfrac{1}{x+h}$. Plug these into the definition of the derivative to get $f'(x)=\lim\limits_{h\to 0}\dfrac{f(x+h)-f(x)}{h}=\lim\limits_{h\to 0}\dfrac{\frac{1}{x+h}-\frac{1}{x}}{h}$.

Notice that if you now take the limit, you get the indeterminate form $\frac{0}{0}$. You cannot eliminate this problem merely by simplifying the expression the way that you did with a polynomial. Here, combine the two terms in the numerator of the expression to get $f'(x)=\lim_{h\to 0}\frac{\frac{1}{x+h}-\frac{1}{x}}{h}=\lim_{h\to 0}\frac{\frac{x}{x(x+h)}-\frac{x+h}{x(x+h)}}{h}=\lim_{h\to 0}\frac{\frac{-h}{x(x+h)}}{h}$. This simplifies to $f'(x)=\lim_{h\to 0}\frac{\frac{-h}{x(x+h)}}{h}=\lim_{h\to 0}\frac{-h}{x(x+h)h}$. Now cancel the factor h in the numerator and the denominator to get $f'(x)=\lim_{h\to 0}\frac{-h}{x(x+h)h}=\lim_{h\to 0}\frac{-1}{x(x+h)}$. Now take the limit: $f'(x)=\lim_{h\to 0}\frac{-1}{x(x+h)}=-\frac{1}{x^2}$.

16. $2ax + b$

Find the derivative of a function, $f(x)$, using the definition of the derivative, which is $f'(x)=\lim_{h\to 0}\frac{f(x+h)-f(x)}{h}$. Here, $f(x) = ax^2 + bx + c$ and $f(x+h)=a(x+h)^2+b(x+h)+c=ax^2+2axh+ah^2+bx+bh+c$.

Plug these into the definition of the derivative to get $f'(x)=\lim_{h\to 0}\frac{f(x+h)-f(x)}{h}=\lim_{h\to 0}\frac{ax^2+2axh+ah^2+bx+bh+c-(ax^2+bx+c)}{h}$. This simplifies to $f'(x)=\lim_{h\to 0}\frac{2axh+ah^2+bh}{h}$. Now you factor out the h from the numerator and cancel it with the h in the denominator: $f'(x)=\lim_{h\to 0}\frac{2axh+ah^2+bh}{h}=\lim_{h\to 0}\frac{h(2ax+ah+b)}{h}=\lim_{h\to 0}(2ax+ah+b)$.

Now take the limit to get $f'(x)=\lim_{h\to 0}(2ax+ah+b)=2ax+b$.

Practice Problem Set 7

1. $64x^3 + 16x$

First, expand $(4x^2 + 1)^2$ to get $16x^4 + 8x^2 + 1$. Now, use the Power Rule to take the derivative of each term. The derivative of $16x^4 = 16(4x^3) = 64x^3$. The derivative of $8x^2 = 8(2x) = 16x$. The derivative of 1 = 0 (because the derivative of a constant is zero). Therefore, the derivative is $64x^3 + 16x$.

2. $10x^9 + 36x^5 + 18x$

First, expand $(x^5 + 3x)^2$ to get $x^{10} + 6x^6 + 9x^2$. Now, use the Power Rule to take the derivative of each term. The derivative of $x^{10} = 10x^9$. The derivative of $6x^6 = 6(6x^5) = 36x^5$. The derivative of $9x^2 = 9(2x) = 18x$. Therefore, the derivative is $10x^9 + 36x^5 + 18x$.

3. $77x^6$

Simply use the Power Rule. The derivative is $11x^7 = 11(7x^6) = 77x^6$.

4. $54x^2 + 12$

Use the Power Rule to take the derivative of each term. The derivative of $18x^3 = 18(3x^2) = 54x^2$. The derivative of $12x$ = 12. The derivative of 11 = 0 (because the derivative of a constant is zero). Therefore, the derivative is $54x^2 + 12$.

5. $6x^{11}$

Use the Power Rule to take the derivative of each term. The derivative of $x^{12} = 12x^{11}$. The derivative of 17 = 0 (because the derivative of a constant is zero). Therefore, the derivative is $\frac{1}{2}(12x^{11}) = 6x^{11}$.

6. $-3x^8 - 2x^2$

Use the Power Rule to take the derivative of each term. The derivative of $x^9 = 9x^8$. The derivative of $2x^3 = 2(3x^2) = 6x^2$. The derivative of $9 = 0$ (because the derivative of a constant is zero). Therefore, the derivative is $-\frac{1}{3}(9x^8 + 6x^2) = -3x^8 - 2x^2$.

7. 0

Don't be fooled by the power. π^5 is a constant, so the derivative is zero.

8. $64x^{-9} + \dfrac{6}{\sqrt{x}}$

Use the Power Rule to take the derivative of each term. The derivative of $-8x^{-8} = -8\left(-8x^{-9}\right) = 64x^{-9}$.

The derivative of $12\sqrt{x} = \dfrac{12}{2\sqrt{x}} = \dfrac{6}{\sqrt{x}}$ (remember the shortcut that we showed you on page 142).

Therefore, the derivative is $64x^{-9} + \dfrac{6}{\sqrt{x}}$.

9. $-42x^{-8} - \dfrac{2}{\sqrt{x}}$

Use the Power Rule to take the derivative of each term. The derivative of $6x^{-7} = 6\left(-7x^{-8}\right) = -42x^{-8}$.

The derivative of $4\sqrt{x} = \dfrac{4}{2\sqrt{x}} = \dfrac{2}{\sqrt{x}}$ (remember the shortcut that we showed you on page 142).

Therefore, the derivative is $-42x^{-8} - \dfrac{2}{\sqrt{x}}$.

10. $\dfrac{-5}{x^6} - \dfrac{8}{x^9}$

Use the Power Rule to take the derivative of each term. The derivative of $x^{-5} = -5x^{-6}$. To find the derivative of $\dfrac{1}{x^8}$, first rewrite it as x^{-8}. The derivative of $x^{-8} = -8x^{-9}$. Therefore, the derivative is $-5x^{-6} - 8x^{-9} = \dfrac{-5}{x^6} - \dfrac{8}{x^9}$.

11. $216x^2 - 48x + 36$

First, expand $(6x^2 + 3)(12x - 4)$ to get $72x^3 - 24x^2 + 36x - 12$. Now, use the Power Rule to take the derivative of each term. The derivative of $72x^3 = 72\left(3x^2\right) = 216x^2$. The derivative of $24x^2 = 24(2x) = 48x$. The derivative of $36x = 36$. The derivative of $12 = 0$ (because the derivative of a constant is zero). Therefore, the derivative is $216x^2 - 48x + 36$.

12. $-6-36x^2+12x^3-5x^4-14x^6$

First, expand $(3-x-2x^3)(6+x^4)$ to get $18-6x-12x^3+3x^4-x^5-2x^7$. Now, use the Power Rule to take the derivative of each term. The derivative of 18 = 0 (because the derivative of a constant is zero). The derivative of 6x = 6. The derivative of $12x^3=12(3x^2)=36x^2$. The derivative of $3x^4=3(4x^3)=12x^3$. The derivative of $x^5=5x^4$. The derivative of $2x^7=2(7x^6)=14x^6$. Therefore, the derivative is $-6-36x^2+12x^3-5x^4-14x^6$.

13. 0

Don't be fooled by the powers. Each term is a constant, so the derivative is zero.

14. $-\frac{16}{x^5}+\frac{10}{x^6}+\frac{36}{x^7}$

First, expand $\left(\frac{1}{x}+\frac{1}{x^2}\right)\left(\frac{4}{x^3}-\frac{6}{x^4}\right)$ to get $\frac{4}{x^4}-\frac{6}{x^5}+\frac{4}{x^5}-\frac{6}{x^6}=\frac{4}{x^4}-\frac{2}{x^5}-\frac{6}{x^6}$. Next, rewrite the terms as $4x^{-4}-2x^{-5}-6x^{-6}$. Now, use the Power Rule to take the derivative of each term. The derivative of $4x^{-4}=4(-4x^{-5})=-16x^{-5}$. The derivative of $2x^{-5}=2(-5x^{-6})=-10x^{-6}$. The derivative of $6x^{-6}=-36x^{-7}$. Therefore, the derivative is $-16x^{-5}+10x^{-6}+36x^{-7}=-\frac{16}{x^5}+\frac{10}{x^6}+\frac{36}{x^7}$.

15. $\frac{1}{2\sqrt{x}}$

Use the Power Rule to take the derivative of each term. The derivative of $\sqrt{x}=\frac{1}{2\sqrt{x}}$ (remember the shortcut that we showed you on page 142). The derivative of $\frac{1}{\sqrt{3}}=0$ (because the derivative of a constant is zero). Therefore, the derivative is $\frac{1}{2\sqrt{x}}$.

16. 0

The derivative of a constant is zero.

17. $3x^2 + 6x + 3$

First, expand $(x+1)^3$ to get $x^3 + 3x^2 + 3x + 1$. Now, use the Power Rule to take the derivative of each term. The derivative of $x^3 = 3x^2$. The derivative of $3x^2 = 3(2x) = 6x$. The derivative of $3x = 3$. The derivative of $1 = 0$ (because the derivative of a constant is zero). Therefore, the derivative is $3x^2 + 6x + 3$.

18. $\frac{1}{2\sqrt{x}} + \frac{1}{3\sqrt[3]{x^2}} + \frac{2}{3\sqrt[3]{x}}$

Use the Power Rule to take the derivative of each term. The derivative of $\sqrt{x} = \frac{1}{2\sqrt{x}}$ (remember the shortcut that we showed you on page 142). Rewrite $\sqrt[3]{x}$ as $x^{\frac{1}{3}}$ and $\sqrt[3]{x^2}$ as $x^{\frac{2}{3}}$. The derivative of $x^{\frac{1}{3}} = \frac{1}{3}x^{-\frac{2}{3}}$. The derivative of $x^{\frac{2}{3}} = \frac{2}{3}x^{-\frac{1}{3}}$. Therefore, the derivative is $\frac{1}{2\sqrt{x}} + \frac{1}{3}x^{-\frac{2}{3}} + \frac{2}{3}x^{-\frac{1}{3}} = \frac{1}{2\sqrt{x}} + \frac{1}{3\sqrt[3]{x^2}} + \frac{2}{3\sqrt[3]{x}}$.

19. $6x^2 + 6x - 14$

First, expand $x(2x+7)(x-2)$ to get $x(2x^2 + 3x - 14) = 2x^3 + 3x^2 - 14x$. Now, use the Power Rule to take the derivative of each term. The derivative of $2x^3 = 2(3x^2) = 6x^2$. The derivative of $3x^2 = 3(2x) = 6x$. The derivative of $14x = 14$. Therefore, the derivative is $6x^2 + 6x - 14$.

20. $\frac{5}{6\sqrt[6]{x}} + \frac{7}{10\sqrt[10]{x^3}}$

First, rewrite the terms as $x^{\frac{1}{2}}\left(x^{\frac{1}{3}} + x^{\frac{1}{5}}\right)$. Next, distribute to get $x^{\frac{1}{2}}\left(x^{\frac{1}{3}} + x^{\frac{1}{5}}\right) = x^{\frac{5}{6}} + x^{\frac{7}{10}}$. Now, use the Power Rule to take the derivative of each term. The derivative of $x^{\frac{5}{6}} = \frac{5}{6}x^{-\frac{1}{6}}$. The derivative of $x^{\frac{7}{10}} = \frac{7}{10}x^{-\frac{3}{10}}$. Therefore, the derivative is $\frac{5}{6}x^{-\frac{1}{6}} + \frac{7}{10}x^{-\frac{3}{10}} = \frac{5}{6\sqrt[6]{x}} + \frac{7}{10\sqrt[10]{x^3}}$.

Practice Problem Set 8

1. $f'(x) = \frac{4x^3}{x^4 + 8}$

   The rule for finding the derivative of $y = \ln u$ is $\frac{dy}{dx} = \frac{1}{u}\frac{du}{dx}$, where u is a function of x. Here, $u = x^4 + 8$. Therefore, the derivative is $f'(x) = \frac{1}{x^4 + 8}(4x^3) = \frac{4x^3}{x^4 + 8}$.

2. $f'(x) = \frac{1}{x} + \frac{1}{6 + 2x}$

   The rule for finding the derivative of $y = \ln u$ is $\frac{dy}{dx} = \frac{1}{u}\frac{du}{dx}$, where u is a function of x. Before finding the derivative, use the laws of logarithms to expand the logarithm. This way, you won't have to use the Product Rule. You get $\ln\left(3x\sqrt{3+x}\right) = \ln 3 + \ln x + \ln\sqrt{3+x} = \ln 3 + \ln x + \frac{1}{2}\ln(3+x)$.

   Now find the derivative: $f'(x) = 0 + \frac{1}{x} + \frac{1}{2}\frac{1}{3+x} = \frac{1}{x} + \frac{1}{6+2x}$.

3. $f'(x) = \frac{2}{x} - \frac{x}{5 + x^2}$

   The rule for finding the derivative of $y = \ln u$ is $\frac{dy}{dx} = \frac{1}{u}\frac{du}{dx}$, where u is a function of x. Before finding the derivative, use the laws of logarithms to expand the logarithm. This way, you won't have to use the Product Rule or the Quotient Rule. You get $\ln\left(\frac{5x^2}{\sqrt{5+x^2}}\right) = \ln 5 + \ln x^2 - \ln\sqrt{5+x^2} = \ln 5 + 2\ln x - \frac{1}{2}\ln\left(5+x^2\right)$. Now find the derivative: $f'(x) = 0 + 2\frac{1}{x} - \frac{1}{2}\frac{1}{5+x^2}(2x) = \frac{2}{x} - \frac{x}{5+x^2}$.

4. $f'(x) = e^{x\cos x}(\cos x - x\sin x)$

   The rule for finding the derivative of $y = e^u$ is $\frac{dy}{dx} = e^u\frac{du}{dx}$, where u is a function of x. Use the Product Rule to find the derivative of the exponent: $f'(x) = e^{x\cos x}(\cos x - x\sin x)$.

5. $f'(x) = -3e^{-3x}\sin 5x + 5e^{-3x}\cos 5x$

The rule for finding the derivative of $y = e^u$ is $\frac{dy}{dx} = e^u \frac{du}{dx}$, where u is a function of x. Use the Product Rule to find the derivative: $f'(x) = e^{-3x}(-3)\sin 5x + e^{-3x}(5\cos 5x)$, which you can rearrange to $f'(x) = -3e^{-3x}\sin 5x + 5e^{-3x}\cos 5x$.

6. $f'(x) = \pi e^{\pi x} - \pi$

The rule for finding the derivative of $y = e^u$ is $\frac{dy}{dx} = e^u \frac{du}{dx}$, where u is a function of x and the rule for finding the derivative of $y = \ln u$ is $\frac{dy}{dx} = \frac{1}{u}\frac{du}{dx}$, where u is a function of x.

You get $f'(x) = e^{\pi x}(\pi) - \frac{1}{e^{\pi x}}\pi e^{\pi x}$. This can be simplified to $f'(x) = \pi e^{\pi x} - \pi$. You might have noticed that $\ln e^{\pi x} = \pi x$, which would have made the derivative a little easier.

7. $f'(x) = \frac{3}{x \ln 12}$

The rule for finding the derivative of $y = \log_a u$ is $\frac{dy}{dx} = \frac{1}{u \ln a}\frac{du}{dx}$, where u is a function of x. Before finding the derivative, use the laws of logarithms to expand the logarithm. You get $f(x) = \log_{12} x^3 = 3\log_{12} x$. Now find the derivative: $f'(x) = \frac{3}{x \ln 12}$.

8. $f'(x) = \frac{3}{2}$

The rule for finding the derivative of $y = \log_a u$ is $\frac{dy}{dx} = \frac{1}{u \ln a}\frac{du}{dx}$, and the rule for finding the derivative of $y = a^u$ is $\frac{dy}{dx} = a^u(\ln a)\frac{du}{dx}$, where u is a function of x. Before finding the derivative, use the laws of logarithms to simplify the logarithm. You get $f(x) = \frac{1}{2}\log 10^{3x}$. Now if you are alert, you will remember that $\log 10^{3x} = 3x$, so this simplifies to $f(x) = \frac{1}{2}3x = \frac{3}{2}x$. The derivative is simply $f'(x) = \frac{3}{2}$.

9. $f'(x) = 3e^{3x} - 3^{ex}(e)(\ln 3)$

The rule for finding the derivative of $y = e^u$ is $\frac{dy}{dx} = e^u \frac{du}{dx}$, and the rule for finding the derivative of $y = a^u$ is $\frac{dy}{dx} = a^u(\ln a)\frac{du}{dx}$, where u is a function of x. You get $f'(x) = 3e^{3x} - 3^{ex}(e)(\ln 3)$.

10. $f'(x) = 10^{\sin x}(\cos x)(\ln 10)$

The rule for finding the derivative of $y = a^u$ is $\frac{dy}{dx} = a^u(\ln a)\frac{du}{dx}$, where u is a function of x. You get $f'(x) = 10^{\sin x}(\cos x)(\ln 10)$.

11. $f'(x) = \ln 10$

Before finding the derivative, use the laws of logarithms to simplify the logarithm. You get $f(x) = \ln(10^x) = x\ln 10$. Now the derivative is simply $f'(x) = \ln 10$.

Practice Problem Set 9

1. $\dfrac{-80x^9 + 75x^8 + 12x^2 - 6x}{(5x^7 + 1)^2}$

Find the derivative using the Quotient Rule, which says that if $f(x) = \frac{u}{v}$, then $f'(x) = \dfrac{v\frac{du}{dx} - u\frac{dv}{dx}}{v^2}$. Here, $f(x) = \dfrac{4x^3 - 3x^2}{5x^7 + 1}$, so $u = 4x^3 - 3x^2$ and $v = 5x^7 + 1$. Using the Quotient Rule, $f'(x) = \dfrac{(5x^7 + 1)(12x^2 - 6x) - (4x^3 - 3x^2)(35x^6)}{(5x^7 + 1)^2}$. This can be simplified to $f'(x) = \dfrac{-80x^9 + 75x^8 + 12x^2 - 6x}{(5x^7 + 1)^2}$.

2. $3x^2 - 6x - 1$

Find the derivative using the Product Rule, which says that if $f(x) = uv$, then $f'(x) = u\frac{dv}{dx} + v\frac{du}{dx}$. Here, $f(x) = (x^2 - 4x + 3)(x + 1)$, so $u = x^2 - 4x + 3$ and $v = x + 1$. Using the Product Rule, $f'(x) = (x^2 - 4x + 3)(1) + (x + 1)(2x - 4)$. This can be simplified to $f'(x) = 3x^2 - 6x - 1$.

3. $3x^2 + 1 + \frac{1}{x^2} + \frac{3}{x^4}$

There are two ways to solve this. Expand the expression first and then take the derivative of each term, or find the derivative using the Product Rule. Do both methods just to see that they both give you the same answer. First, use the Product Rule, which says that if $f(x) = uv$, then $f'(x) = u\frac{dv}{dx} + v\frac{du}{dx}$. Here, $f(x) = \left(x + \frac{1}{x}\right)\left(x^2 - \frac{1}{x^2}\right)$, so $u = \left(x + \frac{1}{x}\right)$ and $v = \left(x^2 - \frac{1}{x^2}\right)$. Using the Product Rule, you get

$f'(x) = \left(x + \frac{1}{x}\right)\left(2x - (-2)x^{-3}\right) + \left(x^2 - \frac{1}{x^2}\right)\left(1 - \frac{1}{x^2}\right) = \left(x + \frac{1}{x}\right)\left(2x + \frac{2}{x^3}\right) + \left(x^2 - \frac{1}{x^2}\right)\left(1 - \frac{1}{x^2}\right)$.

This can be simplified to $f'(x) = \left(2x^2 + \frac{2}{x^2} + 2 + \frac{2}{x^4}\right) + \left(x^2 - 1 - \frac{1}{x^2} + \frac{1}{x^4}\right) = 3x^2 + 1 + \frac{1}{x^2} + \frac{3}{x^4}$.

The other way you could find the derivative is to expand the expression first and then take the derivative. You get $f(x) = \left(x + \frac{1}{x}\right)\left(x^2 - \frac{1}{x^2}\right) = x^3 - \frac{1}{x} + x - \frac{1}{x^3}$. Now take the derivative of each term. You get $f'(x) = 3x^2 + \frac{1}{x^2} + 1 - (-3)x^{-4} = 3x^2 + \frac{1}{x^2} + 1 + \frac{3}{x^4}$. As you can see, the second method is a little quicker, and they both give the same result.

4. $\dfrac{9}{64}$

Find the derivative using the Quotient Rule, which says that if $f(x)=\dfrac{u}{v}$, then $f'(x)=\dfrac{v\dfrac{du}{dx}-u\dfrac{dv}{dx}}{v^2}$. Here, $f(x)=\dfrac{(x+4)(x-8)}{(x+6)(x-6)}$, so $u=(x+4)(x-8)$ and $v=(x+6)(x-6)$.

Before you take the derivative, you can simplify the numerator and denominator of the expression: $f(x)=\dfrac{(x+4)(x-8)}{(x+6)(x-6)}=\dfrac{x^2-4x-32}{x^2-36}$. Now using the Quotient Rule, you get $f(x)=\dfrac{(x^2-36)(2x-4)-(x^2-4x-32)(2x)}{(x^2-36)^2}$. Next, don't simplify. Plug in $x=2$ to get $f(x)=\dfrac{((2)^2-36)(2(2)-4)-((2)^2-4(2)-32)(2(2))}{((2)^2-36)^2}=\dfrac{(-32)(0)-(-36)(4)}{(-32)^2}=\dfrac{9}{64}$.

5. 106

Find the derivative using the Quotient Rule, which says that if $f(x)=\dfrac{u}{v}$, then $f'(x)=\dfrac{v\dfrac{du}{dx}-u\dfrac{dv}{dx}}{v^2}$. You will also need the Chain Rule to take the derivative of the expression in the denominator. The Chain Rule says that if $y=f(g(x))$, then $y'=\left(\dfrac{df(g(x))}{dg}\right)\left(\dfrac{dg}{dx}\right)$. Here, $f(x)=\dfrac{x^6+4x^3+6}{(x^4-2)^2}$, so $u=x^6+4x^3+6$ and $v=(x^4-2)^2$.

You get $f(x)=\dfrac{(x^4-2)^2(6x^5+12x^2)-(x^6+4x^3+6)2(x^4-2)(4x^3)}{(x^4-2)^4}$. Now, don't simplify.

You simply plug in $x = 1$ to get

$$f(x) = \frac{\left((1)^4 - 2\right)^2 \left(6(1)^5 + 12(1)^2\right) - \left((1)^6 + 4(1)^3 + 6\right)2\left((1)^4 - 2\right)\left(4(1)^3\right)}{\left((1)^4 - 2\right)^4} =$$

$$\frac{(1)(18) - (11)2(-1)(4)}{(-1)^4} = 106$$

6. $\dfrac{x^2 - 6x + 3}{(x-3)^2}$

Find the derivative using the Quotient Rule, which says that if $f(x) = \frac{u}{v}$, then $f'(x) = \dfrac{v\frac{du}{dx} - u\frac{dv}{dx}}{v^2}$.

Here, $f(x) = \dfrac{x^2 - 3}{x - 3}$, so $u = x^2 - 3$ and $v = x - 3$. Using the Quotient Rule, $f'(x) = \dfrac{(x-3)(2x) - (x^2 - 3)(1)}{(x-3)^2}$. This can be simplified to $f'(x) = \dfrac{x^2 - 6x + 3}{(x-3)^2}$.

7. 6

Find the derivative using the Product Rule, which says that if $f(x) = uv$, then $f'(x) = u\frac{dv}{dx} + v\frac{du}{dx}$. Here, $f(x) = (x^4 - x^2)(2x^3 + x)$, so $u = x^4 - x^2$ and $v = 2x^3 + x$. Using the Product Rule, $f'(x) = (x^4 - x^2)(6x^2 + 1) + (2x^3 + x)(4x^3 - 2x)$. Now don't simplify. Simply plug in $x = 1$ to get $f'(x) = \left((1)^4 - (1)^2\right)\left(6(1)^2 + 1\right) + \left(2(1)^3 + (1)\right)\left(4(1)^3 - 2(1)\right) = 6$.

8. $-\dfrac{7}{4}$

Find the derivative using the Quotient Rule, which says that if $f(x) = \frac{u}{v}$, then $f'(x) = \dfrac{v\frac{du}{dx} - u\frac{dv}{dx}}{v^2}$. Here, $f(x) = \dfrac{x^2 + 2x}{x^4 - x^3}$, so $u = x^2 + 2x$ and $v = x^4 - x^3$.

Using the Quotient Rule, $f'(x)=\dfrac{(x^4-x^3)(2x+2)-(x^2+2x)(4x^3-3x^2)}{(x^4-x^3)^2}$.

Now, don't simplify. Simply plug in $x=2$ to get

$$f'(x)=\frac{((2)^4-(2)^3)(2(2)+2)-((2)^2+2(2))(4(2)^3-3(2)^2)}{((2)^4-(2)^3)^2}=-\frac{7}{4}.$$

9. $\left(t^3-6t^{\frac{5}{2}}\right)\left(5+\dfrac{1}{2\sqrt{t}}\right)+\left(5t+\sqrt{t}\right)\left(3t^2-15t^{\frac{3}{2}}\right)$

Here, the solution will be much simpler if you first substitute $x=\sqrt{t}$ into the expression for y.

You get $y=\left(t^3-6t^{\frac{5}{2}}\right)\left(5t+\sqrt{t}\right)$. Now, find the derivative using the Product Rule, which says that if $f(x)=uv$, then $f'(x)=u\dfrac{dv}{dx}+v\dfrac{du}{dx}$. Here, $y=\left(t^3-6t^{\frac{5}{2}}\right)\left(5t+\sqrt{t}\right)$, so $u=t^3-6t^{\frac{5}{2}}$ and $v=5t+\sqrt{t}$. Using the Product Rule, $\dfrac{dy}{dt}=\left(t^3-6t^{\frac{5}{2}}\right)\left(5+\dfrac{1}{2\sqrt{t}}\right)+\left(5t+\sqrt{t}\right)\left(3t^2-15t^{\frac{3}{2}}\right)$.

10. $f'(x)=\csc x$

The rule for finding the derivative of $y=\ln u$ is $\dfrac{dy}{dx}=\dfrac{1}{u}\dfrac{du}{dx}$, where u is a function of x.

Here, $u=\cot x-\csc x$. Therefore, the derivative is $f'(x)=\dfrac{1}{\cot x-\csc x}\left(-\csc^2 x+\csc x\cot x\right)$.

This can be simplified to $f'(x)=\dfrac{(-\csc x+\cot x)(\csc x)}{\cot x-\csc x}=\csc x$.

11. $f'(x) = \frac{1}{xe^{4x}\ln 4} - \frac{4\log_4 x}{e^{4x}}$

The rule for finding the derivative of $y = \log_a u$ is $\frac{dy}{dx} = \frac{1}{u \ln a}\frac{du}{dx}$, and the rule for finding the derivative of $y = e^u$ is $\frac{dy}{dx} = e^u \frac{du}{dx}$, where u is a function of x. Use the Quotient Rule to find the derivative: $f'(x) = \frac{\left(e^{4x}\right)\left(\frac{1}{x\ln 4}\right) - \left(\log_4 x\right)\left(e^{4x}\right)(4)}{\left(e^{4x}\right)^2}$. This can be simplified to $f'(x) = \frac{1}{xe^{4x}\ln 4} - \frac{4\log_4 x}{e^{4x}}$.

12. $f'(x) = x^4 5^x (5 + x\ln 5)$

The rule for finding the derivative of $y = a^u$ is $\frac{dy}{dx} = a^u(\ln a)\frac{du}{dx}$, where u is a function of x. Use the Product Rule to find the derivative: $f'(x) = x^5\left(5^x \ln 5\right) + \left(5^x\right)\left(5x^4\right)$, which simplifies to $f'(x) = x^4 5^x (5 + x\ln 5)$.

ANSWERS TO CHAPTER 4 DRILL

1. **C** Find the derivative using the formula $f'(x)=\lim_{h\to 0}\frac{f(x+h)-f(x)}{h}$. You get $f'(x)=\lim_{h\to 0}\frac{8(x+h)-11-(8x-11)}{h}$.

   Simplify the numerator: $\lim_{h\to 0}\frac{8(x+h)-11-(8x-11)}{h}=\lim_{h\to 0}\frac{8x+8h-11-8x+11}{h}=\lim_{h\to 0}\frac{8h}{h}$.

   Cancel h out of the numerator and denominator: $\lim_{h\to 0}\frac{8h}{h}=\lim_{h\to 0}8$.

   And take the limit: $\lim_{h\to 0}8=8$.

   Therefore, the derivative is 8.

   The answer is (C).

2. **C** Find the derivative using the formula $f'(x)=\lim_{h\to 0}\frac{f(x+h)-f(x)}{h}$. You get $f'(x)=\lim_{h\to 0}\frac{3(x+h)^2+7(x+h)-2-(3x^2+7x-2)}{h}$.

   Simplify the numerator: $\lim_{h\to 0}\frac{3(x^2+2xh+h^2)+7x+7h-2-3x^2-7x+2}{h}$

   $\lim_{h\to 0}\frac{3x^2+6xh+3h^2+7x+7h-2-3x^2-7x+2}{h}$

   $\lim_{h\to 0}\frac{6xh+3h^2+7h}{h}$

   Factor h out of the numerator: $\lim_{h\to 0}\frac{h(6x+3h+7)}{h}$.

   Cancel the h in the numerator and the denominator: $\lim_{h\to 0}6x+3h+7$.

   And take the limit: $\lim_{h\to 0}6x+3h+7=6x+7$.

   Therefore, the derivative is $6x+7$.

   The answer is (C).

3. **D** Find the derivative using the formula $f'(x)=\lim_{h\to 0}\frac{f(x+h)-f(x)}{h}$. You get $f'(x)=\lim_{h\to 0}\frac{(x+h)^3-4(x+h)^2+11-(x^3-4x^2+11)}{h}$.

Simplify the numerator: $\lim_{h\to 0}\frac{x^3+3x^2h+3xh^2+h^3-4(x^2+2xh+h^2)+11-x^3+4x^2-11}{h}$

$$\lim_{h\to 0}\frac{x^3+3x^2h+3xh^2+h^3-4x^2-8xh-4h^2+11-x^3+4x^2-11}{h}$$

$$\lim_{h\to 0}\frac{3x^2h+3xh^2+h^3-8xh-4h^2}{h}$$

Factor h out of the numerator: $\lim_{h\to 0}\frac{h(3x^2+3xh+h^2-8x-4h)}{h}$.

Cancel the h in the numerator and the denominator: $\lim_{h\to 0}3x^2+3xh+h^2-8x-4h$.

And take the limit: $\lim_{h\to 0}3x^2+3xh+h^2-8x-4h=3x^2-8x$.

Therefore, the derivative is $3x^2-8x$.

The answer is (D).

4. **A** Find the derivative using the formula $f'(x)=\lim_{h\to 0}\frac{f(x+h)-f(x)}{h}$, and plug in $x = 25$ either after finding the derivative or before. Let's do it afterward.

You get $f'(x)=\lim_{h\to 0}\frac{6\sqrt{(x+h)}-6\sqrt{x}}{h}$.

Multiply the numerator and the denominator by the conjugate of the numerator: $\lim_{h\to 0}\frac{\left(6\sqrt{(x+h)}-6\sqrt{x}\right)}{h}\frac{\left(6\sqrt{(x+h)}+6\sqrt{x}\right)}{\left(6\sqrt{(x+h)}+6\sqrt{x}\right)}$.

Multiply out the numerator: $\lim_{h\to 0}\frac{36(x+h)-36x}{h\left(6\sqrt{(x+h)}+6\sqrt{x}\right)}$.

Simplify the numerator and denominator: $\lim_{h\to 0}\frac{36x+36h-36x}{h\left(6\sqrt{(x+h)}+6\sqrt{x}\right)}=\lim_{h\to 0}\frac{36h}{6h\left(\sqrt{(x+h)}+\sqrt{x}\right)}$.

Cancel the $6h$ in the numerator and the denominator: $\lim_{h\to 0}\frac{6}{\sqrt{(x+h)}+\sqrt{x}}$.

And take the limit: $\dfrac{6}{\sqrt{x+h}+\sqrt{x}}=\dfrac{6}{2\sqrt{x}}=\dfrac{3}{\sqrt{x}}$.

Finally, plug in $x = 25$: $\dfrac{3}{\sqrt{25}}=\dfrac{3}{5}$.

The answer is (A).

5. **A** Find the derivative using the formula $f'(x)=\lim\limits_{h\to 0}\dfrac{f(x+h)-f(x)}{h}$, and plug in $x = 2$ either after finding the derivative or before. Let's do it at the beginning.

You get $f'(x)=\lim\limits_{h\to 0}\dfrac{\dfrac{1}{4(x+h)^2}-\dfrac{1}{4x^2}}{h}$. Plug in $x = 2$: $f'(x)=\lim\limits_{h\to 0}\dfrac{\dfrac{1}{4(2+h)^2}-\dfrac{1}{16}}{h}$.

Combine the two rational expressions in the numerator:

$$\lim_{h\to 0}\frac{\dfrac{1}{4(2+h)^2}-\dfrac{1}{16}}{h}=\lim_{h\to 0}\frac{\dfrac{16}{4(2+h)^2(16)}-\dfrac{4(2+h)^2}{4(2+h)^2 16}}{h}=\lim_{h\to 0}\frac{\dfrac{16-4(2+h)^2}{4(2+h)^2(16)}}{h}$$

Multiply out the numerator: $\lim\limits_{h\to 0}\dfrac{\dfrac{16-4(4+4h+h^2)}{64(2+h)^2}}{h}=\lim\limits_{h\to 0}\dfrac{\dfrac{16-16-16h-4h^2}{64(2+h)^2}}{h}$.

Simplify the numerator and denominator: $\lim\limits_{h\to 0}\dfrac{\dfrac{16-16-16h-4h^2}{64(2+h)^2}}{h}=\lim\limits_{h\to 0}\dfrac{-16h-4h^2}{64h(2+h)^2}$.

Factor $-4h$ out of the numerator: $\lim\limits_{h\to 0}\dfrac{-4h(4+h)}{64h(2+h)^2}$.

Cancel the $4h$ in the numerator and the denominator: $\lim\limits_{h\to 0}\dfrac{-(4+h)}{16(2+h)^2}$.

And take the limit: $\dfrac{-4}{64}=-\dfrac{1}{16}$.

The answer is (A).

6. **D** Use the Power Rule:

$$y = 5x^3 - 7x^2 + 11x + 2$$

$$\frac{dy}{dx} = 15x^2 - 14x + 11$$

The answer is (D).

Chapter 5
Differentiation: Composite, Implicit, and Inverse Functions

THE CHAIN RULE

The most important rule in this chapter (and sometimes the most difficult one) is called the **Chain Rule**. It's used when you're given composite functions—that is, a function inside of another function. You'll always see these on the AP Exam, so it's important to know the Chain Rule cold.

A composite function is usually written as $f(g(x))$.

For example: If $f(x)=\frac{1}{x}$ and $g(x)=\sqrt{3x}$, then $f(g(x))=\frac{1}{\sqrt{3x}}$.

We could also find $g(f(x))=\sqrt{\frac{3}{x}}$.

When finding the derivative of a composite function, we take the derivative of the "outside" function, with the inside function $g(x)$ considered as the variable, leaving the "inside" function alone. Then, we multiply this by the derivative of the "inside" function, with respect to its variable x.

Here is another way to write the Chain Rule.

The Chain Rule: If $y=f(g(x))$, then $y'=f'(g(x))\cdot g'(x)$

For the Chain Rule, always work from the outside in, peeling away layers as with a nesting doll or an onion.

In other words, y' is found by taking the derivative of f, putting $g(x)$ as the input of the derivative function $f'(x)$, and multiplying the result by $g'(x)$.

This rule is tricky, so here are several examples. The last couple incorporate the Product Rule and the Quotient Rule, which we covered in the previous chapter.

Example 1: If $y = (5x^3 + 3x)^5$, then $\frac{dy}{dx} = 5(5x^3 + 3x)^4(15x^2 + 3)$.

We just dealt with the derivative of something to the fifth power, like this:

$$y = (g)^5, \text{ so } \frac{dy}{dg} = 5(g)^4, \text{ where } g = 5x^3 + 3x$$

Then we multiplied by the derivative of g: $(15x^2 + 3)$.

Always do it this way. The process has several successive steps, like peeling away the layers of an onion until you reach the center.

Example 2: If $y = \sqrt{x^3 - 4x}$, then $\frac{dy}{dx} = \frac{1}{2}\left(x^3 - 4x\right)^{-\frac{1}{2}}\left(3x^2 - 4\right)$.

Again, we took the derivative of the outside function and evaluated it using the inside function as input. Then we multiplied by the derivative of the inside function.

Example 3: If $y = \sqrt{\left(x^5 - 8x^3\right)\left(x^2 + 6x\right)}$, then

$$\frac{dy}{dx} = \frac{1}{2}\left[\left(x^5 - 8x^3\right)\left(x^2 + 6x\right)\right]^{-\frac{1}{2}}\left[\left(x^5 - 8x^3\right)(2x + 6) + \left(x^2 + 6x\right)\left(5x^4 - 24x^2\right)\right]$$

Messy, isn't it? That's because we used the Chain Rule and the Product Rule. Now for one with the Chain Rule and the Quotient Rule.

Example 4: If $y = \left(\frac{2x + 8}{x^2 - 10x}\right)^5$, then

$$\frac{dy}{dx} = 5\left[\frac{2x + 8}{x^2 - 10x}\right]^4\left[\frac{\left(x^2 - 10x\right)(2) - (2x + 8)(2x - 10)}{\left(x^2 - 10x\right)^2}\right]$$

Example 5: If $y = \sqrt{5x^3 + x}$, then $\frac{dy}{dx} = \frac{1}{2}\left(5x^3 + x\right)^{-\frac{1}{2}}\left(15x^2 + 1\right)$.

Now we use the Product Rule and the Chain Rule to find the second derivative.

$$\frac{d^2y}{dx^2} = \frac{1}{2}\left(5x^3 + x\right)^{-\frac{1}{2}}(30x) + \left(15x^2 + 1\right)\left[-\frac{1}{4}\left(5x^3 + x\right)^{-\frac{3}{2}}\left(15x^2 + 1\right)\right]$$

You can also simplify this further, if necessary.

There's another representation of the Chain Rule that you need to learn.

> If $y = y(v)$ and $v = v(x)$, then $\frac{dy}{dx} = \frac{dy}{dv}\frac{dv}{dx}$

Example 6: If $y = 8v^2 - 6v$ and $v = 5x^3 - 11x$, then $\frac{dy}{dx} = (16v - 6)(15x^2 - 11)$.

Then substitute for v.

$$\frac{dy}{dx} = \left(16\left(5x^3 - 11x\right) - 6\right)\left(15x^2 - 11\right) = \left(80x^3 - 176x - 6\right)\left(15x^2 - 11\right)$$

> As you can see, these grow quite complex, so we simplify these only as a last resort. If you must simplify, the AP Exam will have only a very simple Chain Rule problem.

PRACTICE PROBLEM SET 10

Simplify when possible. The answers are at the end of this chapter, starting on page 206.

1. Find $f'(x)$ if $f(x)=8\sqrt{\left(x^4-4x^2\right)}$.

2. Find $f'(x)$ if $f(x)=\left(\dfrac{x}{x^2+1}\right)^3$.

3. Find $f'(x)$ if $f(x)=\sqrt[4]{\left(\dfrac{2x-5}{5x+2}\right)}$.

4. Find $f'(x)$ if $f(x)=\left(\dfrac{x}{x+1}\right)^4$.

5. Find $f'(x)$ if $f(x)=\left(x^2+x\right)^{100}$.

6. Find $f'(x)$ if $f(x)=\sqrt{\dfrac{x^2+1}{x^2-1}}$.

7. Find $f'(x)$ if $f(x)=\sqrt{x^4+x^2}$.

8. Find $\dfrac{dy}{dx}$ if $y=u^2-1$ and $u=\dfrac{1}{x-1}$.

9. Find $\dfrac{dy}{dx}$ at $x = 1$ if

   $y=\dfrac{t^2+2}{t^2-2}$ and $t=x^3$.

10. Find $\dfrac{dy}{dx}$ at $x = 1$ if

    $y=\dfrac{1+u}{1+u^2}$ and $u=x^2-1$.

11. Find $\dfrac{du}{dv}$ if $u = y^3$, $y=\dfrac{x}{x+8}$,

    and $x = v^2$.

IMPLICIT DIFFERENTIATION

By now, it should be easy for you to take the derivative of an equation such as $y = 3x^5 - 7x$. If you're given an equation such as $y^2 = 3x^5 - 7x$, you can still figure out the derivative by taking the square root of both sides, which gives you y in terms of x. This is known as finding the derivative **explicitly**. It's messy, but possible.

If you have to find the derivative of $y^2 + y = 3x^5 - 7x$, you don't have an easy way to get y in terms of x, so you can't differentiate this equation using any of the techniques you've learned so far. That's because each of those previous techniques needs to be used on an equation in which y is in terms of x. When you can't isolate y in terms of x (or if isolating y makes taking the derivative a nightmare), it's time to take the derivative **implicitly**.

Implicit differentiation is one of the simpler techniques you need to learn to do in calculus, but for some reason it gives many students trouble. Suppose you have the equation $y^2 = 3x^5 - 7x$. This means that the value of y is a function of the value of x. When we take the derivative, $\frac{dy}{dx}$, we're looking at the rate at which y changes as x changes. Thus, given $y = x^2 + x$, when we write

$$\frac{dy}{dx} = 2x + 1$$

we're saying that "the rate" at which y changes, with respect to how x changes, is $2x + 1$.

Now, suppose you want to find $\frac{dx}{dy}$. As you might imagine,

$$\frac{dx}{dy} = \frac{1}{\frac{dy}{dx}}$$

So here, $\frac{dx}{dy} = \frac{1}{2x+1}$. But notice that this derivative is in terms of x, not y, and you need to find the derivative with respect to y. This derivative is an **implicit** one. When you can't isolate the variables of an equation, you often end up with a derivative that is in terms of both variables.

Another way to think of this is that there is a hidden term in the derivative, $\frac{dx}{dx}$, and when we take the derivative, what we really get is

$$\frac{dy}{dx} = 2x\left(\frac{dx}{dx}\right) + 1\left(\frac{dx}{dx}\right)$$

A fraction that has the same term in its numerator and denominator is equal to 1, so we write

$$\frac{dy}{dx} = 2x(1) + 1(1) = 2x + 1$$

Every time we take a derivative of a term with x in it, we multiply by the term $\frac{dx}{dx}$, but because this is 1, we ignore it. Suppose, however, that we wanted to find out how y changes with respect to t (for time). In that case, we would have

$$\frac{dy}{dt} = 2x\left(\frac{dx}{dt}\right) + 1\left(\frac{dx}{dt}\right)$$

If we wanted to find out how y changes with respect to r, we would have

$$\frac{dy}{dr} = 2x\left(\frac{dx}{dr}\right) + 1\left(\frac{dx}{dr}\right)$$

and if we wanted to find out how y changes with respect to y, we would have

$$\frac{dy}{dy} = 2x\left(\frac{dx}{dy}\right) + 1\left(\frac{dx}{dy}\right) \text{ or } 1 = 2x\frac{dx}{dy} + \frac{dx}{dy}$$

This is how we really do differentiation. Remember the following:

$$\frac{dx}{dy} = \frac{1}{\frac{dy}{dx}}$$

When you can't explicitly write a function in the terms of the variable for which you want to take the derivative, use implicit differentiation.

When you have an equation of x in terms of y, and you want to find the derivative with respect to y, simply differentiate. But if the equation is of y in terms of x, find $\frac{dy}{dx}$ and take its reciprocal to find $\frac{dx}{dy}$. Go back to our original example.

$$y^2 + y = 3x^5 - 7x$$

To take the derivative according to the information in the last paragraph, you get

$$2y\left(\frac{dy}{dx}\right) + 1\left(\frac{dy}{dx}\right) = 15x^4\left(\frac{dx}{dx}\right) - 7\left(\frac{dx}{dx}\right)$$

Notice how each variable is multiplied by its appropriate $\frac{d}{dx}$. Now, remembering that $\frac{dx}{dx} = 1$, rewrite the expression this way: $2y\left(\frac{dy}{dx}\right) + 1\left(\frac{dy}{dx}\right) = 15x^4 - 7$.

Next, factor $\frac{dy}{dx}$ out of the left-hand side: $\frac{dy}{dx}(2y+1)=15x^4-7$.

Isolating $\frac{dy}{dx}$ gives you $\frac{dy}{dx}=\frac{15x^4-7}{(2y+1)}$.

This is the derivative you're looking for. Notice how the derivative is defined in terms of y **and** x. Up until now, $\frac{dy}{dx}$ has been strictly in terms of x. This is why the differentiation is "implicit."

Confused? Let's do a few examples and you will get the hang of it.

Example 7: Find $\frac{dy}{dx}$ if $y^3 - 4y^2 = x^5 + 3x^4$.

Using implicit differentiation, you get

$$3y^2\left(\frac{dy}{dx}\right)-8y\left(\frac{dy}{dx}\right)=5x^4\left(\frac{dx}{dx}\right)+12x^3\left(\frac{dx}{dx}\right)$$

Remember that $\frac{dx}{dx}$ = 1: $\frac{dy}{dx}(3y^2-8y)=5x^4+12x^3$.

After you factor out $\frac{dy}{dx}$, divide both sides by $3y^2 - 8y$.

$$\frac{dy}{dx}=\frac{5x^4+12x^3}{(3y^2-8y)}$$

Note: Now that you understand that the derivative of an x-term with respect to x will always be multiplied by $\frac{dx}{dx}$, and that $\frac{dx}{dx}$ = 1, we won't write $\frac{dx}{dx}$ anymore. You should understand that the term is implied.

Example 8: Find $\frac{dy}{dx}$ if $\sin y^2 - \cos x^2 = \cos y^2 + \sin x^2$.

Use implicit differentiation.

$$\cos y^2\left(2y\frac{dy}{dx}\right)+\sin x^2(2x)=-\sin y^2\left(2y\frac{dy}{dx}\right)+\cos x^2(2x)$$

Then simplify.

$$2y\cos y^2\left(\frac{dy}{dx}\right)+2x\sin x^2=-2y\sin y^2\left(\frac{dy}{dx}\right)+2x\cos x^2$$

Next, put all of the terms containing $\frac{dy}{dx}$ on the left and all of the other terms on the right.

$$2y\cos y^2\left(\frac{dy}{dx}\right)+2y\sin y^2\left(\frac{dy}{dx}\right)=-2x\sin x^2+2x\cos x^2$$

Next, factor out $\frac{dy}{dx}$.

$$\frac{dy}{dx}\left(2y\cos y^2+2y\sin y^2\right)=-2x\sin x^2+2x\cos x^2$$

And isolate $\frac{dy}{dx}$.

$$\frac{dy}{dx}=\frac{-2x\sin x^2+2x\cos x^2}{\left(2y\cos y^2+2y\sin y^2\right)}$$

This can be simplified further to the following:

$$\frac{dy}{dx}=\frac{-x\left(\sin x^2-\cos x^2\right)}{y\left(\cos y^2+\sin y^2\right)}$$

Did you notice the use of the Product Rule to find the derivative of $5xy^2$? The AP Exam loves to make you do this. All of the same differentiation rules that you've learned up until now still apply. We're just adding another technique.

Example 9: Find $\frac{dy}{dx}$ if $3x^2 + 5xy^2 - 4y^3 = 8$.

Implicit differentiation should result in the following:

$$6x+\left[5x\left(2y\frac{dy}{dx}\right)+(5)y^2\right]-12y^2\left(\frac{dy}{dx}\right)=0$$

You can simplify this to

$$6x+10xy\frac{dy}{dx}+5y^2-12y^2\frac{dy}{dx}=0$$

Next, put all of the terms containing $\frac{dy}{dx}$ on the left and all of the other terms on the right.

$$10xy\frac{dy}{dx}-12y^2\frac{dy}{dx}=-6x-5y^2$$

Next, factor out $\frac{dy}{dx}$.

$$\left(10xy-12y^2\right)\frac{dy}{dx}=-6x-5y^2$$

Then, isolate $\frac{dy}{dx}$.

$$\frac{dy}{dx} = \frac{-6x - 5y^2}{\left(10xy - 12y^2\right)}$$

Example 10: Find the derivative of $3x^2 - 4y^2 + y = 9$ at (2, 1).

You need to use implicit differentiation to find $\frac{dy}{dx}$.

$$6x - 8y\left(\frac{dy}{dx}\right) + \left(\frac{dy}{dx}\right) = 0$$

Now, instead of rearranging to isolate $\frac{dy}{dx}$, plug in (2, 1) immediately and solve for the derivative.

$$6(2) - 8(1)\left(\frac{dy}{dx}\right) + \left(\frac{dy}{dx}\right) = 0$$

Simplify: $12 - 7\left(\frac{dy}{dx}\right) = 0$, so $\frac{dy}{dx} = \frac{12}{7}$.

Getting the hang of implicit differentiation yet? We hope so, because these next examples are slightly harder.

Be smart about your problem solving. Just because you can simplify something doesn't mean that you should. In a case like this, plugging into this form of the derivative is more effective.

Example 11: Find the derivative of $\frac{2x - 5y^2}{4y^3 - x^2} = -x$ at (1, 1).

First, cross-multiply.

$$2x - 5y^2 = -x\left(4y^3 - x^2\right)$$

Distribute.

$$2x - 5y^2 = -4xy^3 + x^3$$

Take the derivative.

$$2 - 10y\frac{dy}{dx} = -4x\left(3y^2\frac{dy}{dx}\right) - 4y^3 + 3x^2$$

Do not simplify now. Rather, plug in (1, 1) right away. This will save you from the algebra.

$$2 - 10(1)\frac{dy}{dx} = -4(1)\left(3(1)^2\frac{dy}{dx}\right) - 4(1)^3 + 3(1)^2$$

Now solve for $\frac{dy}{dx}$.

$$2-10\frac{dy}{dx}=-12\frac{dy}{dx}-1$$

$$2\frac{dy}{dx}=-3$$

$$\frac{dy}{dx}=\frac{-3}{2}$$

PRACTICE PROBLEM SET 11

Use implicit differentiation to find the following derivatives. The answers are at the end of this chapter, starting on page 210.

1. Find $\frac{dy}{dx}$ if $x^3 - y^3 = y$.

2. Find $\frac{dy}{dx}$ if $x^2 - 16xy + y^2 = 1$.

3. Find $\frac{dy}{dx}$ at (2, 1) if $\frac{x+y}{x-y} = 3$.

4. Find $\frac{dy}{dx}$ if $16x^2 - 16xy + y^2 = 1$ at (1, 1).

5. Find $\frac{dy}{dx}$ if $x\sin y + y\sin x = \frac{\pi}{2\sqrt{2}}$ at $\left(\frac{\pi}{4}, \frac{\pi}{4}\right)$.

6. Find $\frac{d^2y}{dx^2}$ if $x^2 + 4y^2 = 1$.

7. Find $\frac{d^2y}{dx^2}$ if $\sin x + 1 = \cos y$.

8. Find $\frac{d^2y}{dx^2}$ if $x^2 - 4x = 2y - 2$.

DIFFERENTIATING INVERSE FUNCTIONS

The College Board occasionally asks a question about finding the derivative of an inverse function. To do this, you need to learn only this simple formula.

Suppose we have a function $x = f(y)$ that is defined and differentiable at $y = a$ where $x = c$. Suppose we also know that the $f^{-1}(x)$ exists at $x = c$. Thus, $f(a) = c$ and $f^{-1}(c) = a$. Then, because

$$\frac{dy}{dx} = \frac{1}{\frac{dx}{dy}},$$

$$\left.\frac{d}{dx} f^{-1}(x)\right|_{x=c} = \frac{1}{\left[\frac{d}{dy} f(y)\right]_{y=a}}$$

In short, we can find the derivative of a function's inverse at a particular point by taking the reciprocal of the derivative at that point's corresponding y-value. These examples should help clear up any confusion.

Example 12: If $f(x) = x^2$, find the derivative of $f^{-1}(x)$ at $x = 9$.

First, notice that $f(3) = 9$. One of the most confusing parts of finding the derivative of an inverse function is that when you're asked to find the derivative at a value of x, they're *really* asking you for the derivative of the inverse of the function at the value that *corresponds* to $f(x) = 9$. This is because x-values of the inverse correspond to $f(x)$-values of the original function.

The rule is very simple: when you're asked to find the derivative of $f^{-1}(x)$ at $x = c$, you take the reciprocal of the derivative of $f(x)$ at $x = a$, where $f(a) = c$.

We know that $\frac{d}{dx} f(x) = 2x$. This means that we're going to plug $x = 3$ into the formula (because $f(3) = 9$). This gives us

$$\left.\frac{1}{2x}\right|_{x=3} = \frac{1}{6}$$

We can verify this by finding the inverse of the function first and then taking the derivative. The inverse of the function $f(x) = x^2$ is the function $f^{-1}(x) = \sqrt{x}$. Now we find the derivative and evaluate it at $f(3) = 9$.

$$\frac{d}{dx}\sqrt{x} = \left.\frac{1}{2\sqrt{x}}\right|_{x=9} = \frac{1}{6}$$

Remember the rule: find the value, a, of $f(x)$ that gives you the value of x that the problem asks for. Then plug that value, a, into the reciprocal of the derivative of the *inverse* function.

Example 13: Find a derivative of the inverse of $y = x^3 - 1$ when $y = 7$.

First, we need to find the x-value that corresponds to $y = 7$. A little algebra tells us that this is $x = 2$. Then,

$$\frac{dy}{dx} = 3x^2 \text{ and } \frac{1}{\frac{dy}{dx}} = \frac{1}{3x^2}$$

Therefore, the derivative of the inverse is

$$\left.\frac{1}{3x^2}\right|_{x=2} = \frac{1}{12}$$

Verify it: the inverse of the function $y = x^3 - 1$ is the function $y = \sqrt[3]{x+1}$. The derivative of this latter function is

$$\left.\frac{1}{3\sqrt[3]{(x+1)^2}}\right|_{x=7} = \frac{1}{12}$$

Let's do one more.

Example 14: Find a derivative of the inverse of $y = x^2 + 4$ when $y = 29$.

At $y = 29$, $x = 5$, the derivative of the function is

$$\frac{dy}{dx} = 2x$$

So, a derivative of the inverse is

$$\left.\frac{1}{2x}\right|_{x=5} = \frac{1}{10}$$

Note that $x = -5$ also gives us $y = 29$, so $-\frac{1}{10}$ is also a derivative.

This is all you'll be required to know involving derivatives of inverses. Naturally, there are ways to create harder problems, but the AP Exam stays away from them and sticks to simpler stuff.

Here are some solved problems. Do each problem, cover the answer first, and then check your answer.

Example 15: Find a derivative of the inverse of $f(x) = 2x^3 + 5x + 1$ at $y = 8$.

First, we take the derivative of $f(x)$.

$$f'(x) = 6x^2 + 5$$

A possible value of x is $x = 1$.

Then, we use the formula to find the derivative of the inverse.

$$\frac{1}{f'(1)} = \frac{1}{11}$$

Example 16: Find a derivative of the inverse of $f(x) = 3x^3 - x + 7$ at $y = 9$.

First, take the derivative of $f(x)$.

$$f'(x) = 9x^2 - 1$$

A possible value of x is $x = 1$.

Then, use the formula to find the derivative of the inverse.

$$\frac{1}{f'(1)} = \frac{1}{8}$$

Example 17: Find a derivative of the inverse of $y = \frac{8}{x^3}$ at $y = 1$.

Take the derivative of y.

$$y' = -\frac{24}{x^4}$$

Find the value of x where $y = 1$.

$$1 = \frac{8}{x^3}$$

$$x = 2$$

Use the formula.

$$\frac{1}{\left.\frac{dy}{dx}\right|_{x=2}} = \frac{1}{\left.\left(-\frac{24}{x^4}\right)\right|_{x=2}} = -\frac{2}{3}$$

Here's one more.

Example 18: Find a derivative of the inverse of $y = 2x - x^3$ at $y = 1$.

The derivative of the function is

$$\frac{dy}{dx} = 2 - 3x^2$$

Next, find the value of x where $y = 1$. By inspection, $y = 1$ when $x = 1$.

Then, we use the formula to find the derivative of the inverse.

$$\frac{1}{\left.\frac{dy}{dx}\right|_{x=1}} = \frac{1}{\left.\left(2 - 3x^2\right)\right|_{x=1}} = -1$$

PRACTICE PROBLEM SET 12

Find a derivative of the inverse of each of the following functions. The answers are at the end of this chapter, starting on page 214.

1. $y = x + \frac{1}{x}$ at $y = \frac{17}{4}$; where $x > 1$

2. $y = 3x - 5x^3$ at $y = 2$

3. $y = e^x$ at $y = e$

4. $y = x + x^3$ at $y = -2$

5. $y = 4x - x^3$ at $y = 3$

6. $y = \ln x$ at $y = 0$

DIFFERENTIATING INVERSE TRIGONOMETRIC FUNCTIONS

Now we will learn how to find the derivatives of Inverse Trigonometric Functions.

Suppose you have the equation $\sin y = x$. If you differentiate both sides with respect to x, you get

$$\cos y \frac{dy}{dx} = 1$$

Now divide both sides by $\cos y$.

$$\frac{dy}{dx} = \frac{1}{\cos y}$$

Because $\sin^2 y + \cos^2 y = 1$, we can replace $\cos y$ with $\sqrt{1 - \sin^2 y}$.

$$\frac{dy}{dx} = \frac{1}{\sqrt{1 - \sin^2 y}}$$

Finally, because $x = \sin y$, replace $\sin y$ with x. The derivative equals

$$\frac{1}{\sqrt{1 - x^2}}$$

Now go back to the original equation $\sin y = x$ and solve for y in terms of x: $y = \sin^{-1} x$.

If you differentiate both sides with respect to x, you get

$$\frac{dy}{dx} = \frac{d}{dx}\sin^{-1} x$$

Replace $\frac{dy}{dx}$, and you get the final result.

$$\frac{d}{dx}\sin^{-1} x = \frac{1}{\sqrt{1 - x^2}}$$

The AP Exam rarely tests these derivatives, largely because finding them is just a matter of following a formula. (You're more likely to see them tested as integrals.) Still, make sure you at least know the inverse sine and tangent functions.

This is the derivative of inverse sine. By similar means, you can find the derivatives of all six inverse trig functions. They're not difficult to derive, and they're also not difficult to memorize. The choice is yours. We use u instead of x to account for the Chain Rule.

$$\frac{d}{dx}(\sin^{-1} u) = \frac{1}{\sqrt{1-u^2}}\frac{du}{dx}; \quad -1 < u < 1 \qquad \frac{d}{dx}(\cos^{-1} u) = \frac{-1}{\sqrt{1-u^2}}\frac{du}{dx}; \quad -1 < u < 1$$

$$\frac{d}{dx}(\tan^{-1} u) = \frac{1}{1+u^2}\frac{du}{dx} \qquad \frac{d}{dx}(\cot^{-1} u) = \frac{-1}{1+u^2}\frac{du}{dx}$$

$$\frac{d}{dx}(\sec^{-1} u) = \frac{1}{|u|\sqrt{u^2-1}}\frac{du}{dx}; \quad |u| > 1 \qquad \frac{d}{dx}(\csc^{-1} u) = \frac{-1}{|u|\sqrt{u^2-1}}\frac{du}{dx}; \quad |u| > 1$$

Notice the domain restrictions for inverse sine, cosine, secant, and cosecant.

Example 19: $\frac{d}{dx}\sin^{-1} x^2 = \frac{2x}{\sqrt{1-(x^2)^2}} = \frac{2x}{\sqrt{1-x^4}}$

Example 20: $\frac{d}{dx}\tan^{-1} 5x = \frac{5}{1+(5x)^2} = \frac{5}{1+25x^2}$

Example 21: $\frac{d}{dx}\sec^{-1}(x^2-x) = \frac{2x-1}{|x^2-x|\sqrt{(x^2-x)^2-1}} = \frac{2x-1}{|x^2-x|\sqrt{x^4-2x^3+x^2-1}}$

SELECTING PROCEDURES FOR CALCULATING DERIVATIVES

Now that we have learned how to take a variety of derivatives, let's practice some problems, figuring out which rule to use to find the derivative.

Example 22: Find $\frac{dy}{dx}$ if $y = (6x^3 - 5x^2 + 11x - 23)(2x^4 + 7x^3 - 24x + 100)$.

We have two polynomials being multiplied together. One option is to do the multiplication. Then we would just use the Power Rule on each term. But the multiplying is cumbersome, and there is a good chance we might make a mistake. Instead, we can use the Product Rule.

$$\frac{dy}{dx} = (6x^3 - 5x^2 + 11x - 23)(8x^3 + 21x^2 - 24) + (18x^2 - 10x + 11)(2x^4 + 7x^3 - 24x + 100)$$

Example 23: Find $\frac{dy}{dx}$ if $y = (x^2 - 7x)(3x^3 + 4)$.

Again, we have two polynomials being multiplied together. This time, the multiplication is easy, so let's do it. Then we will just use the Power Rule on each term.

$$y = (x^2 - 7x)(3x^3 + 4) = 3x^5 + 4x^2 - 21x^4 - 28x$$

$$\frac{dy}{dx} = 15x^4 + 8x - 84x^3 - 28$$

No need to use the Product Rule here!

Hint: When we have two terms being multiplied together, use the Product Rule unless it is easy to do the multiplication.

Example 24: Find $f'(x)$ if $f(x) = \frac{x^2 - 4\sin x}{x^3 + 2\cos x}$.

Note that we have two terms being divided, so we will use the Quotient Rule.

$$f(x) = \frac{\left(x^3 + 2\cos x\right)\left(2x - 4\cos x\right) - \left(x^2 - 4\sin x\right)\left(3x^2 - 2\sin x\right)}{\left(x^3 + 2\cos x\right)^2}$$

Example 25: Find $\frac{dy}{dx}$ if $y = \tan^3(1 - x^2)$.

Here, we have functions that are inside of other functions (Composite Functions), so we will use the Chain Rule.

$$\frac{dy}{dx} = 3\tan^2\left(1 - x^2\right)\sec^2\left(1 - x^2\right)\left(-2x\right) = -6x\tan^2\left(1 - x^2\right)\sec^2\left(1 - x^2\right)$$

Hint: If you see a function inside of another function, you will almost certainly need to use the Chain Rule. This is because most of the functions you will see are composite functions.

Example 26: Find $\frac{dy}{dx}$ if $y = \sin^{-1}(6x) + \tan^{-1}(6x)$.

We have inverse trig functions, so we just use the rules to find their derivatives. Note that we will need to use the Chain Rule for each derivative.

$$\frac{dy}{dx} = \frac{1}{\sqrt{1 - (6x)^2}} \bullet 6 + \frac{1}{1 + (6x)^2} \bullet 6 = \frac{6}{\sqrt{1 - 36x^2}} + \frac{6}{1 + 36x^2}$$

Example 27: Find $\frac{dy}{dx}$ if $3x^5 - 4x^2y^3 + 2y^2 = 1$.

Note that we can't rewrite this function as y equals a function of x. We will need to use Implicit Differentiation to find $\frac{dy}{dx}$.

Take the derivative of each term: $15x^4-\left[\left(4x^2\right)\left(3y^2\frac{dy}{dx}\right)+(8x)\left(y^3\right)\right]+4y\frac{dy}{dx}=0.$

Simplify: $15x^4-12x^2y^2\frac{dy}{dx}-8xy^3+4y\frac{dy}{dx}=0.$

Group: $4y\frac{dy}{dx}-12x^2y^2\frac{dy}{dx}=8xy^3-15x^4.$

Factor: $\left(4y-12x^2y^2\right)\frac{dy}{dx}=8xy^3-15x^4.$

Divide: $\frac{dy}{dx}=\frac{8xy^3-15x^4}{4y-12x^2y^2}.$

Example 28: Find $\frac{dy}{dx}$ if $7x^3+2x^4y^2-5y^3=-25$ at the point (1, 2).

Note that we can't rewrite this function as y equals a function of x. We will need to use Implicit Differentiation to find $\frac{dy}{dx}$.

Take the derivative of each term: $21x^2+\left(2x^4\right)\left(2y\frac{dy}{dx}\right)+\left(8x^3\right)\left(y^2\right)-15y^2\frac{dy}{dx}=0.$

Hint: When we have x- and y-terms mixed together, we will need to use Implicit Differentiation. And, if we are finding the derivative at a point, just plug in and avoid the algebra!

In the previous problem, we needed to use algebra to find $\frac{dy}{dx}$. Here, however, we are trying to find $\frac{dy}{dx}$ at the point (1, 2), so we can just plug in. We get

$$21(1)^2+\left(2(1)^4\right)\left(2(2)\frac{dy}{dx}\right)+\left(8(1)^3\right)\left((2)^2\right)-15(2)^2\frac{dy}{dx}=0$$

Simplify: $21+8\frac{dy}{dx}+32-60\frac{dy}{dx}=0$. And solve for $\frac{dy}{dx}$: $\frac{dy}{dx}=\frac{53}{52}$.

Example 29: Find $f'(x)$ if $f(x)=\sqrt{1-\sec^3(\pi x)}$ at $x=1$.

Hint: When we are evaluating derivatives at a point, look to see if terms will become 0 or 1. It will make the calculations much easier!

We have functions inside of each other, so we will need to use the Chain Rule. We get $f'(x)=\frac{1}{2}\left(1-\sec^3(\pi x)\right)^{-\frac{1}{2}}\left(-3\sec^2(\pi x)\right)\sec(\pi x)\tan(\pi x)\pi$.

Now we plug in $x=1$. When we do that, note that one of the terms will be $\tan\pi$. Because $\tan\pi=0$, the derivative is 0.

CALCULATING HIGHER-ORDER DERIVATIVES

This term refers only to taking the derivative of a function more than once. You don't have to stop at the first derivative of a function; you can keep taking derivatives. The derivative of a first derivative is called the second derivative. The derivative of the second derivative is called the third derivative, and so on.

Generally, you'll have to take only first and second derivatives.

Function	First Derivative	Second Derivative
x^6	$6x^5$	$30x^4$
$8\sqrt{x}$	$\frac{4}{\sqrt{x}}$	$-2x^{-\frac{3}{2}}$

Notice how we simplified the derivatives in the latter example? You should be able to do this mentally.

Second Derivatives Using Implicit Differentiation

Sometimes, you'll be asked to find a second derivative implicitly.

Example 30: Find $\frac{d^2y}{dx^2}$ if $y^2 + 2y = 4x^2 + 2x$.

Differentiating implicitly, you get

$$2y\frac{dy}{dx} + 2\frac{dy}{dx} = 8x + 2$$

Remember: When it is required to take a second derivative, the first derivative should be simplified first.

Next, simplify and solve for $\frac{dy}{dx}$.

$$\frac{dy}{dx} = \frac{4x+1}{y+1}$$

Now, it's time to take the derivative again.

$$\frac{d^2y}{dx^2} = \frac{4(y+1) - (4x+1)\left(\frac{dy}{dx}\right)}{(y+1)^2}$$

Finally, substitute for $\frac{dy}{dx}$.

$$\frac{4(y+1) - (4x+1)\left(\frac{4x+1}{y+1}\right)}{(y+1)^2} = \frac{4(y+1)^2 - (4x+1)^2}{(y+1)^3}$$

Try these solved problems without looking at the answers. Then check your work.

Example 31: Find $\frac{dy}{dx}$ if $x^2 + y^2 = 6xy$.

Differentiate with respect to x.

$$2x + 2y\frac{dy}{dx} = 6x\frac{dy}{dx} + 6y$$

Group all of the $\frac{dy}{dx}$ terms on the left and the other terms on the right.

$$2y\frac{dy}{dx} - 6x\frac{dy}{dx} = 6y - 2x$$

Now factor out $\frac{dy}{dx}$.

$$\frac{dy}{dx}(2y - 6x) = 6y - 2x$$

Therefore, the first derivative is the following:

$$\frac{dy}{dx} = \frac{6y - 2x}{2y - 6x} = \frac{3y - x}{y - 3x}$$

Example 32: Find $\frac{dy}{dx}$ if $x - \cos y = xy$.

Differentiate with respect to x.

$$1 + \sin y\frac{dy}{dx} = x\frac{dy}{dx} + y$$

Grouping the terms, you get

$$\sin y\frac{dy}{dx} - x\frac{dy}{dx} = y - 1$$

Now factor out $\frac{dy}{dx}$.

$$\frac{dy}{dx}(\sin y - x) = y - 1$$

The derivative is

$$\frac{dy}{dx} = \frac{y - 1}{\sin y - x}$$

Example 33: Find the derivative of each variable with respect to t of $x^2 + y^2 = z^2$.

$$2x\frac{dx}{dt} + 2y\frac{dy}{dt} = 2z\frac{dz}{dt}$$

Example 34: Find the derivative of each variable with respect to t of $V = \frac{1}{3}\pi r^2 h$.

$$\frac{dV}{dt} = \frac{1}{3}\pi\left(r^2\frac{dh}{dt} + 2r\frac{dr}{dt}h\right)$$

Example 35: Find $\frac{d^2y}{dx^2}$ if $y^2 = x^2 - 2x$.

First, take the derivative with respect to x.

$$2y\frac{dy}{dx} = 2x - 2$$

Then, solve for $\frac{dy}{dx}$.

$$\frac{dy}{dx} = \frac{2x-2}{2y} = \frac{x-1}{y}$$

The second derivative with respect to x becomes

$$\frac{d^2y}{dx^2} = \frac{y(1) - (x-1)\frac{dy}{dx}}{y^2}$$

Now substitute for $\frac{dy}{dx}$ and simplify.

$$\frac{d^2y}{dx^2} = \frac{y - (x-1)\left(\frac{x-1}{y}\right)}{y^2} = \frac{y^2 - (x-1)^2}{y^3}$$

Chapter 5 Drill

The answers are at the end of the chapter, starting on page 218.

1 Mark for Review

Find $\frac{dy}{dx}$ if $y=\tan(4x)$.

(A) $\frac{dy}{dx}=\sec^2(4x)$

(B) $\frac{dy}{dx}=4\sec^2(4x)$

(C) $\frac{dy}{dx}=4\tan(4x)$

(D) $\frac{dy}{dx}=4\cot(4x)$

2 Mark for Review

Find $\frac{dy}{dx}$ if $y=\cos^2\left(x^2\right)$.

(A) $\frac{dy}{dx}=4x\cos\left(x^2\right)$

(B) $\frac{dy}{dx}=\sin^2\left(x^2\right)$

(C) $\frac{dy}{dx}=-4x\cos\left(x^2\right)\sin\left(x^2\right)$

(D) $\frac{dy}{dx}=-4x\sin^2\left(x^2\right)$

3 Mark for Review

Find $\frac{dy}{dx}$ if $y=\frac{4}{\sqrt{x-1}}$.

(A) $\frac{dy}{dx}=-\frac{2}{(x-1)^{\frac{3}{2}}}$

(B) $\frac{dy}{dx}=\frac{2}{(x-1)^{\frac{3}{2}}}$

(C) $\frac{dy}{dx}=-\frac{8}{(x-1)^{\frac{3}{2}}}$

(D) $\frac{dy}{dx}=\frac{8}{(x-1)^{\frac{3}{2}}}$

4 Mark for Review

Find $\frac{dy}{dx}$ if $y=\sqrt{\frac{x^2}{\sec x}}$.

(A) $\frac{dy}{dx}=\frac{1}{2}\left(\frac{2x}{\sec x\tan x}\right)^{-\frac{1}{2}}$

(B) $\frac{dy}{dx}=\left(\frac{2x}{\sec x\tan x}\right)^{\frac{1}{2}}$

(C) $\frac{dy}{dx}=\frac{1}{2}\left(\frac{\sec x(2x)-\left(x^2\right)\sec x\tan x}{\sec^2 x}\right)^{-\frac{1}{2}}$

(D) $\frac{dy}{dx}=\frac{1}{2}\left(\frac{x^2}{\sec x}\right)^{-\frac{1}{2}}\left(\frac{\sec x(2x)-\left(x^2\right)\sec x\tan x}{\sec^2 x}\right)$

5 Mark for Review

Find $\frac{dy}{dx}$ if $x^2+2y^3=x^3-4y^2$.

Ⓐ $\frac{dy}{dx}=\frac{3x^2+2x}{6y^2-8y}$

Ⓑ $\frac{dy}{dx}=\frac{3x^2-2x}{6y^2+8y}$

Ⓒ $\frac{dy}{dx}=\frac{6y^2+8y}{3x^2-2x}$

Ⓓ $\frac{dy}{dx}=\frac{6y^2-8y}{3x^2+2x}$

6 Mark for Review

Find $\frac{dy}{dx}$ if $x^2+3xy^2-y^3=1$.

Ⓐ $\frac{dy}{dx}=\frac{-2x-3y^2}{6xy-3y^2}$

Ⓑ $\frac{dy}{dx}=\frac{2x+3y^2}{6xy-3y^2}$

Ⓒ $\frac{dy}{dx}=2x+6xy-3y^2$

Ⓓ $\frac{dy}{dx}=2x-6xy+3y^2$

7 Mark for Review

Find $\frac{dy}{dx}$ if $4\sin\left(x^2\right)=3y-y^2$ at $\left(\sqrt{\frac{\pi}{6}},\,2\right)$.

Ⓐ $-4\sqrt{\frac{\pi}{6}}$

Ⓑ $4\sqrt{\frac{\pi}{6}}$

Ⓒ $-4\sqrt{\frac{\pi}{2}}$

Ⓓ $4\sqrt{\frac{\pi}{2}}$

8 Mark for Review

Find $\frac{dy}{dx}$ if $xy^2-4x^3y^2=-3$ at (1, 1).

Ⓐ $-\frac{11}{6}$

Ⓑ $-\frac{6}{11}$

Ⓒ $\frac{10}{13}$

Ⓓ $\frac{13}{10}$

9 Mark for Review

Find $\frac{d^2y}{dx^2}$ if $y=y^3-x^2$.

Ⓐ $\frac{\left(3y^2-1\right)^2(2)-\left(24x^2y\right)}{\left(3y^2-1\right)^3}$

Ⓑ $\frac{\left(3y^2-1\right)(2)-\left(24x^2y\right)}{\left(3y^2-1\right)^2}$

Ⓒ $\frac{\left(3y^2-1\right)(2)-\left(12x^2\right)}{\left(3y^2-1\right)^3}$

Ⓓ $\frac{\left(3y^2-1\right)^2(2)-\left(12x^2\right)}{\left(3y^2-1\right)^3}$

10 Mark for Review

Find the derivative of the inverse of $y=5x^3+x-7$ at $y=-7$.

Ⓐ -1

Ⓑ $-\frac{1}{7}$

Ⓒ $\frac{1}{7}$

Ⓓ 1

ANSWERS TO PRACTICE PROBLEM SETS

Practice Problem Set 10

1. $\dfrac{16x^2-32}{\sqrt{(x^2-4)}}$

Find the derivative using the Chain Rule, which says that if $y=f(g(x))$, then $y'=\left(\dfrac{df(g(x))}{dg}\right)\left(\dfrac{dg}{dx}\right)$. Here, $f(x)=8\sqrt{(x^4-4x^2)}$, which can be written as $f(x)=8(x^4-4x^2)^{\frac{1}{2}}$. Using the Chain Rule, $f'(x)=8\left(\dfrac{1}{2}\right)(x^4-4x^2)^{-\frac{1}{2}}(4x^3-8x)$. This can be simplified to $f'(x)=\dfrac{16x^2-32}{\sqrt{(x^2-4)}}$.

2. $\dfrac{3x^2-3x^4}{(x^2+1)^4}$

Find the derivative using the Chain Rule. You will also need the Quotient Rule to take the derivative of the expression inside the parentheses. The Chain Rule says that if $y=f(g(x))$, then $y'=\left(\dfrac{df(g(x))}{dg}\right)\left(\dfrac{dg}{dx}\right)$, and the Quotient Rule says that if $f(x)=\dfrac{u}{v}$, then $f'(x)=\dfrac{v\dfrac{du}{dx}-u\dfrac{dv}{dx}}{v^2}$.

You get $f'(x)=3\left(\dfrac{x}{x^2+1}\right)^2\left(\dfrac{(x^2+1)(1)-x(2x)}{(x^2+1)^2}\right)$. This can be simplified to $f'(x)=\dfrac{3x^2-3x^4}{(x^2+1)^4}$.

3. $\frac{1}{4}\left(\frac{2x-5}{5x+2}\right)^{-\frac{3}{4}}\left(\frac{29}{(5x+2)^2}\right)$

Find the derivative using the Chain Rule. You will also need the Quotient Rule to take the derivative of the expression inside the parentheses. The Chain Rule says that if $y = f(g(x))$, then $y' = \left(\frac{df(g(x))}{dg}\right)\left(\frac{dg}{dx}\right)$, and the Quotient Rule says that if $f(x) = \frac{u}{v}$, then $f'(x) = \frac{v\frac{du}{dx} - u\frac{dv}{dx}}{v^2}$. You get $f'(x) = \frac{1}{4}\left(\frac{2x-5}{5x+2}\right)^{-\frac{3}{4}}\left(\frac{(5x+2)(2)-(2x-5)(5)}{(5x+2)^2}\right)$. This can be simplified to $f'(x) = \frac{1}{4}\left(\frac{2x-5}{5x+2}\right)^{-\frac{3}{4}}\left(\frac{29}{(5x+2)^2}\right)$.

4. $\frac{4x^3}{(x+1)^5}$

Find the derivative using the Chain Rule. You will also need the Quotient Rule to take the derivative of the expression inside the parentheses. The Chain Rule says that if $y = f(g(x))$, then $y' = \left(\frac{df(g(x))}{dg}\right)\left(\frac{dg}{dx}\right)$, and the Quotient Rule says that if $f(x) = \frac{u}{v}$, then $f'(x) = \frac{v\frac{du}{dx} - u\frac{dv}{dx}}{v^2}$. You get $f'(x) = 4\left(\frac{x}{x+1}\right)^3\left(\frac{(x+1)(1)-(x)(1)}{(x+1)^2}\right)$. This can be simplified to $f'(x) = 4\left(\frac{x}{x+1}\right)^3\left(\frac{1}{(x+1)^2}\right) = \frac{4x^3}{(x+1)^5}$.

5. $100\left(x^2+x\right)^{99}(2x+1)$

Find the derivative using the Chain Rule. The Chain Rule says that if $y=f\left(g(x)\right)$, then $y'=\left(\frac{df\left(g(x)\right)}{dg}\right)\left(\frac{dg}{dx}\right)$. You get $f'(x)=100\left(x^2+x\right)^{99}(2x+1)$.

6. $\dfrac{-2x}{\left(x^2+1\right)^{\frac{1}{2}}\left(x^2-1\right)^{\frac{3}{2}}}$

Find the derivative using the Chain Rule. You will also need the Quotient Rule to take the derivative of the expression inside the parentheses. The Chain Rule says that if $y=f\left(g(x)\right)$, then $y'=\left(\frac{df\left(g(x)\right)}{dg}\right)\left(\frac{dg}{dx}\right)$, and the Quotient Rule says that if $f(x)=\frac{u}{v}$, then $f'(x)=\dfrac{v\frac{du}{dx}-u\frac{dv}{dx}}{v^2}$.

You get $f'(x)=\frac{1}{2}\left(\frac{x^2+1}{x^2-1}\right)^{-\frac{1}{2}}\left(\frac{\left(x^2-1\right)(2x)-\left(x^2+1\right)(2x)}{\left(x^2-1\right)^2}\right)$. This can be simplified to $f'(x)=\frac{1}{2}\left(\frac{x^2+1}{x^2-1}\right)^{-\frac{1}{2}}\left(\frac{-4x}{\left(x^2-1\right)^2}\right)=\dfrac{-2x}{\left(x^2+1\right)^{\frac{1}{2}}\left(x^2-1\right)^{\frac{3}{2}}}$.

7. $\dfrac{2x^2+1}{\sqrt{x^2+1}}$

Find the derivative using the Chain Rule, which says that if $y=f\left(g(x)\right)$, then $y'=\left(\frac{df\left(g(x)\right)}{dg}\right)\left(\frac{dg}{dx}\right)$. Here, $f(x)=\sqrt{x^4+x^2}=\left(x^4+x^2\right)^{\frac{1}{2}}$. Using the Chain Rule, $f'(x)=\frac{1}{2}\left(x^4+x^2\right)^{-\frac{1}{2}}\left(4x^3+2x\right)$. This can be simplified to $f'(x)=\dfrac{2x^2+1}{\sqrt{x^2+1}}$.

8. $-\frac{2}{(x-1)^3}$

Find the derivative using the Chain Rule, which says that if $y = y(v)$ and $v = v(x)$, then $\frac{dy}{dx} = \frac{dy}{dv}\frac{dv}{dx}$. Here, $\frac{dy}{du} = 2u$ and $\frac{du}{dx} = -1(x-1)^{-2} = -\frac{1}{(x-1)^2}$. Thus, $\frac{dy}{dx} = (2u)\frac{-1}{(x-1)^2}$ and because $u = \frac{1}{x-1}$, $\frac{dy}{dx} = \left(\frac{2}{x-1}\right)\left(\frac{-1}{(x-1)^2}\right) = -\frac{2}{(x-1)^3}$.

9. –24

Find the derivative using the Chain Rule, which says that if $y = y(v)$ and $v = v(x)$, then $\frac{dy}{dx} = \frac{dy}{dv}\frac{dv}{dx}$. Here, $\frac{dy}{dt} = \frac{(t^2-2)(2t)-(t^2+2)(2t)}{(t^2-2)^2}$ and $\frac{dt}{dx} = 3x^2$. Now, plug $x = 1$ into the derivative. Note that where $x = 1$, $t = (1)^3 = 1$. You get $\frac{dy}{dt} = \frac{((1)^2-2)(2(1))-((1)^2+2)(2(1))}{((1)^2-2)^2} = -8$ and $\frac{dt}{dx} = 3(1)^2 = 3$. Thus, $\frac{dy}{dx} = \frac{dy}{dt}\frac{dt}{dx} = (-8)(3) = -24$.

10. 2

Find the derivative using the Chain Rule, which says that if $y = y(v)$ and $v = v(x)$, then $\frac{dy}{dx} = \frac{dy}{dv}\frac{dv}{dx}$. Here $\frac{dy}{du} = \frac{(1+u^2)(1)-(1+u)(2u)}{(1+u^2)^2}$ and $\frac{du}{dx} = 2x$. Now, plug $x = 1$ into the derivative. Note that, where $x = 1$, $u = 0$. You get $\frac{dy}{du} = \frac{(1+0)(1)-(1+0)(0)}{(1+0)^2} = 1$ and $\frac{du}{dx} = 2$, so $\frac{dy}{dx} = (1)(2) = 2$.

11. $\dfrac{48v^5}{\left(v^2+8\right)^4}$

Find the derivative using the Chain Rule, which says that if $y = y(v)$ and $v = v(x)$, then $\dfrac{dy}{dx} = \dfrac{dy}{dv}\dfrac{dv}{dx}$. Although, in this case, you have $u(y)$, $y(x)$, and $x(v)$, so find $\dfrac{du}{dv}$ by $\dfrac{du}{dv} = \dfrac{du}{dy}\dfrac{dy}{dx}\dfrac{dx}{dv}$. Here, $\dfrac{du}{dy} = 3y^2$, $\dfrac{dy}{dx} = \dfrac{(x+8)(1)-(x)(1)}{(x+8)^2} = \dfrac{8}{(x+8)^2}$, and $\dfrac{dx}{dv} = 2v$. Next, $\dfrac{du}{dv} = \dfrac{du}{dy}\dfrac{dy}{dx}\dfrac{dx}{dv} = \left(3y^2\right)\left(\dfrac{8}{(x+8)^2}\right)(2v)$. Now because $x = v^2$ and $y = \dfrac{x}{x+8} = \dfrac{v^2}{v^2+8}$, you get $\dfrac{du}{dv} = \left(3\dfrac{\left(v^2\right)^2}{\left(v^2+8\right)^2}\right)\left(\dfrac{8}{\left(v^2+8\right)^2}\right)(2v) = \dfrac{48v^5}{\left(v^2+8\right)^4}$.

Practice Problem Set 11

1. $\dfrac{3x^2}{1+3y^2}$

Take the derivative of each term with respect to x: $\left(3x^2\right)\left(\dfrac{dx}{dx}\right) - \left(3y^2\right)\left(\dfrac{dy}{dx}\right) = (1)\left(\dfrac{dy}{dx}\right)$.

Next, because $\dfrac{dx}{dx} = 1$, eliminate that term and get $\left(3x^2\right) - \left(3y^2\right)\left(\dfrac{dy}{dx}\right) = \left(\dfrac{dy}{dx}\right)$.

Next, group the terms containing $\dfrac{dy}{dx}$: $(3x^2) = \left(\dfrac{dy}{dx}\right) + (3y^2)\left(\dfrac{dy}{dx}\right)$.

Factor out the term $\dfrac{dy}{dx}$: $\left(3x^2\right) = \left(\dfrac{dy}{dx}\right)\left(1+3y^2\right)$. Now, isolate $\dfrac{dy}{dx}$: $\dfrac{dy}{dx} = \dfrac{3x^2}{1+3y^2}$.

2. $\dfrac{8y-x}{y-8x}$

Take the derivative of each term with respect to x:

$(2x)\left(\frac{dx}{dx}\right)-16\left[(x)\left(\frac{dy}{dx}\right)+(y)\left(\frac{dx}{dx}\right)\right]+(2y)\left(\frac{dy}{dx}\right)=0$.

Next, because $\frac{dx}{dx}=1$, eliminate that term and distribute the –16 to get $2x-16x\left(\frac{dy}{dx}\right)-16y+2y\left(\frac{dy}{dx}\right)=0$.

Next, group the terms containing $\frac{dy}{dx}$ on one side of the equals sign and the other terms on the other side: $-16x\left(\frac{dy}{dx}\right)+2y\left(\frac{dy}{dx}\right)=16y-2x$.

Factor out the term $\frac{dy}{dx}$: $\left(\frac{dy}{dx}\right)(2y-16x)=16y-2x$. Now isolate $\frac{dy}{dx}$: $\frac{dy}{dx}=\frac{16y-2x}{2y-16x}$, which can be reduced to $\frac{dy}{dx}=\frac{8y-x}{y-8x}$.

3. $\dfrac{1}{2}$

First, cross-multiply so that you don't have to use the Quotient Rule: $x+y=3x-3y$. Next, simplify $4y=2x$, which reduces to $y=\frac{1}{2}x$. Now, take the derivative: $\frac{dy}{dx}=\frac{1}{2}$. Note that just because a problem has the *x*s and *y*s mixed together doesn't mean that you need to use implicit differentiation to solve it!

4. $\dfrac{8}{7}$

Take the derivative of each term with respect to x:

$(32x)\left(\frac{dx}{dx}\right)-16\left[(x)\left(\frac{dy}{dx}\right)+(y)\left(\frac{dx}{dx}\right)\right]+(2y)\left(\frac{dy}{dx}\right)=0$.

Next, because $\frac{dx}{dx}=1$, eliminate that term to get $32x-16x\left(\frac{dy}{dx}\right)-16y+2y\left(\frac{dy}{dx}\right)=0$.

Next, don't simplify. Plug in (1, 1) for x and y: $32(1)-16(1)\left(\frac{dy}{dx}\right)-16(1)+2(1)\left(\frac{dy}{dx}\right)=0$, which simplifies to $16-14\left(\frac{dy}{dx}\right)=0$.

Finally, solve for $\frac{dy}{dx}$: $\frac{dy}{dx}=\frac{8}{7}$.

5. –1

Take the derivative of each term with respect to x: $(x\cos y)\left(\frac{dy}{dx}\right)+(\sin y)\left(\frac{dx}{dx}\right)+(y\cos x)\left(\frac{dx}{dx}\right)+(\sin x)\left(\frac{dy}{dx}\right)=0$.

Next, because $\frac{dx}{dx}=1$, eliminate that term to get $(x\cos y)\left(\frac{dy}{dx}\right)+\sin y+y\cos x+(\sin x)\left(\frac{dy}{dx}\right)=0$. Next, don't simplify. Plug in $\left(\frac{\pi}{4},\frac{\pi}{4}\right)$ for x and y: $\left(\frac{\pi}{4}\cos\frac{\pi}{4}\right)\left(\frac{dy}{dx}\right)+\sin\frac{\pi}{4}+\frac{\pi}{4}\cos\frac{\pi}{4}+\left(\sin\frac{\pi}{4}\right)\left(\frac{dy}{dx}\right)=0$, which simplifies to $\left(\frac{\pi}{4}\frac{1}{\sqrt{2}}\right)\left(\frac{dy}{dx}\right)+\frac{1}{\sqrt{2}}+\frac{\pi}{4}\frac{1}{\sqrt{2}}+\frac{1}{\sqrt{2}}\left(\frac{dy}{dx}\right)=0$. Multiply through by $\sqrt{2}$ to get $\frac{\pi}{4}\frac{dy}{dx}+1+\frac{\pi}{4}+\frac{dy}{dx}=0$.

Finally, solve for $\frac{dy}{dx}$: $\frac{dy}{dx}=-1$.

6. $-\frac{1}{16y^3}$

Take the derivative of each term with respect to x: $(2x)\left(\frac{dx}{dx}\right)+(8y)\left(\frac{dy}{dx}\right)=0$.

Next, because $\frac{dx}{dx}=1$, eliminate that term to get $2x+(8y)\left(\frac{dy}{dx}\right)=0$. Next, isolate $\frac{dy}{dx}$: $\frac{dy}{dx}=-\frac{x}{4y}$. Now, take the derivative again: $\frac{d^2y}{dx^2}=-\frac{(4y)\left(\frac{dx}{dx}\right)-(x)\left(4\frac{dy}{dx}\right)}{16y^2}$.

Next, because $\frac{dx}{dx}=1$ and $\frac{dy}{dx}=-\frac{x}{4y}$, you get $\frac{d^2y}{dx^2}=-\frac{4y-4x\left(-\frac{x}{4y}\right)}{16y^2}$. This can be simplified to $\frac{d^2y}{dx^2}=-\frac{4y+\frac{x^2}{y}}{16y^2}=-\frac{4y^2+x^2}{16y^3}=-\frac{1}{16y^3}$.

7. $\frac{\sin x\sin^2 y-\cos y\cos^2 x}{\sin^3 y}$

Take the derivative of each term with respect to x: $(\cos x)\left(\frac{dx}{dx}\right)=(-\sin y)\left(\frac{dy}{dx}\right)$.

Next, because $\frac{dx}{dx}=1$, eliminate that term to get $\cos x=(-\sin y)\left(\frac{dy}{dx}\right)$. Next, isolate $\frac{dy}{dx}$: $\frac{dy}{dx}=-\frac{\cos x}{\sin y}$. Now, take the derivative again: $\frac{d^2y}{dx^2}=\frac{(\sin y)(\sin x)\left(\frac{dx}{dx}\right)+(\cos x)(\cos y)\left(\frac{dy}{dx}\right)}{\sin^2 y}$.

Next, because $\frac{dx}{dx}=1$ and $\frac{dy}{dx}=-\frac{\cos x}{\sin y}$, you get $\frac{d^2y}{dx^2}=\frac{(\sin y)(\sin x)+(\cos x)(\cos y)\left(-\frac{\cos x}{\sin y}\right)}{\sin^2 y}$.

This can be simplified to $\frac{d^2y}{dx^2}=\frac{\sin x\sin^2 y-\cos y\cos^2 x}{\sin^3 y}$.

8. 1

You can easily isolate y in this equation: $y=\frac{1}{2}x^2-2x+1$. Take the derivative: $\frac{dy}{dx}=x-2$. And take the derivative again: $\frac{d^2y}{dx^2}=1$. Note that just because a problem has the *x*s and *y*s mixed together doesn't mean that you need to use implicit differentiation to solve it!

Practice Problem Set 12

1. $\frac{16}{15}$

First, take the derivative of y: $\frac{dy}{dx} = 1 - \frac{1}{x^2}$. Next, find the value of x where $y = \frac{17}{4}$: $\frac{17}{4} = x + \frac{1}{x}$.

With a little algebra, you should get $x = 4$ or $x = \frac{1}{4}$. Because $x > 1$, you can ignore the second answer. Or, if you are permitted, use your calculator.

Now use the formula for the derivative of the inverse of $f(x)$ (see page 193):

$\left.\frac{d}{dx} f^{-1}(x)\right|_{x=c} = \frac{1}{\left[\frac{d}{dy} f(y)\right]_{y=a}}$, where $f(a) = c$. This formula means that you can find the derivative of the inverse of a function at a value a by taking the reciprocal of the derivative and plugging in the value of x that makes y equal to a.

$$\frac{1}{\left.\frac{dy}{dx}\right|_{x=4}} = \frac{1}{\left.1 - \frac{1}{x^2}\right|_{x=4}} = \frac{16}{15}$$

2. $-\frac{1}{12}$

First, take the derivative of y: $\frac{dy}{dx} = 3 - 15x^2$. Next, find the value of x where $y = 2$: $2 = 3x - 5x^3$.

With a little algebra, you should get $x = -1$. Or, if you are permitted, use your calculator.

Now use the formula for the derivative of the inverse of $f(x)$ (see page 193):

$\left.\frac{d}{dx} f^{-1}(x)\right|_{x=c} = \frac{1}{\left[\frac{d}{dy} f(y)\right]_{y=a}}$, where $f(a) = c$. This formula means that you can find the derivative of the inverse of a function at a value a by taking the reciprocal of the derivative and plugging in the value of x that makes y equal to a.

$$\frac{1}{\left.\frac{dy}{dx}\right|_{x=-1}} = \frac{1}{\left.3-15x^2\right|_{x=-1}} = -\frac{1}{12}$$

3. $\frac{1}{e}$

First, take the derivative of y: $\frac{dy}{dx} = e^x$. Next, find the value of x where $y = e$: $e = e^x$. It should be obvious that $x = 1$.

Now use the formula for the derivative of the inverse of $f(x)$ (see page 193): $\left.\frac{d}{dx} f^{-1}(x)\right|_{x=c} = \frac{1}{\left[\frac{d}{dy} f(y)\right]_{y=a}}$, where $f(a) = c$. This formula means that you can find the derivative of the inverse of a function at a value a by taking the reciprocal of the derivative and plugging in the value of x that makes y equal to a.

$$\frac{1}{\left.\frac{dy}{dx}\right|_{x=1}} = \frac{1}{\left.e^x\right|_{x=1}} = \frac{1}{e}$$

4. $\frac{1}{4}$

First, take the derivative of y: $\frac{dy}{dx} = 1 + 3x^2$. Next, find the value of x where $y = -2$: $-2 = x + x^3$. You should be able to tell by inspection that $x = -1$ is a solution. Or, if you are permitted, use your calculator. Remember that the AP Exam won't give you a problem where it is very difficult to solve for the inverse value of y. If the algebra looks difficult, look for an obvious solution, such as $x = 0$ or $x = 1$ or $x = -1$.

Now use the formula for the derivative of the inverse of $f(x)$: $\left.\frac{d}{dx}f^{-1}(x)\right|_{x=c} = \frac{1}{\left[\frac{d}{dy}f(y)\right]_{y=a}}$,

where $f(a)=c$. This formula means that you can find the derivative of the inverse of a function at

a value a by taking the reciprocal of the derivative and plugging in the value of x that makes y equal

to a: $\frac{1}{\left.\frac{dy}{dx}\right|_{x=-1}} = \frac{1}{\left.1+3x^2\right|_{x=-1}} = \frac{1}{4}$.

5. 1

First, take the derivative of y: $\frac{dy}{dx} = 4 - 3x^2$. Next, find the value of x where $y = 3$: $3 = 4x - x^3$. You should be able to tell by inspection that $x = 1$ is a solution. Or, if you are permitted, use your calculator. Remember that the AP Exam won't give you a problem where it is very difficult to solve for the inverse value of y. If the algebra looks difficult, look for an obvious solution, such as $x = 0$ or $x = 1$ or $x = -1$.

Now use the formula for the derivative of the inverse of $f(x)$: $\left.\frac{d}{dx}f^{-1}(x)\right|_{x=c} = \frac{1}{\left[\frac{d}{dy}f(y)\right]_{y=a}}$,

where $f(a)=c$. This formula means that you can find the derivative of the inverse of a function at

a value a by taking the reciprocal of the derivative and plugging in the value of x that makes y equal

to a: $\frac{1}{\left.\frac{dy}{dx}\right|_{x=1}} = \frac{1}{\left.4-3x^2\right|_{x=1}} = 1$.

6. 1

First, take the derivative of y: $\frac{dy}{dx} = \frac{1}{x}$. Next, find the value of x where $y = 0$: $\ln x = 0$. You should know that $x = 1$ is the solution. Now use the formula for the derivative of the inverse of $f(x)$: $\left.\frac{d}{dx} f^{-1}(x)\right|_{x=c} = \frac{1}{\left[\frac{d}{dy} f(y)\right]_{y=a}}$, where $f(a) = c$. This formula means that you can find the derivative of the inverse of a function at a value a by taking the reciprocal of the derivative and plugging in the value of x that makes y equal to a.

$$\frac{1}{\left.\frac{dy}{dx}\right|_{x=1}} = \frac{1}{\left.\frac{1}{x}\right|_{x=1}} = 1$$

ANSWERS TO CHAPTER 5 DRILL

1. **B** Use the Chain Rule:

$$\frac{dy}{dx} = 4\sec^2(4x)$$

The answer is (B).

2. **C** Use the Chain Rule:

$\frac{dy}{dx} = 2\cos(x^2)(-\sin(x^2))(2x)$, which simplifies to $\frac{dy}{dx} = -4x\cos(x^2)\sin(x^2)$.

The answer is (C).

3. **A** It might help to rewrite the function as $y = 4(x-1)^{-\frac{1}{2}}$. Now use the Chain Rule: $\frac{dy}{dx} = 4\left(-\frac{1}{2}\right)(x-1)^{-\frac{3}{2}}(1)$, which simplifies to $\frac{dy}{dx} = -\frac{2}{(x-1)^{\frac{3}{2}}}$.

The answer is (A).

4. **D** It might help to rewrite the function as $y = \left(\frac{x^2}{\sec x}\right)^{\frac{1}{2}}$. Now use the Chain Rule and the Quotient Rule: $\frac{dy}{dx} = \frac{1}{2}\left(\frac{x^2}{\sec x}\right)^{-\frac{1}{2}}\left(\frac{\sec x(2x) - (x^2)\sec x\tan x}{\sec^2 x}\right)$.

The answer is (D).

5. **B** Take the derivative of each term with respect to *x*. Remember that when you are taking the derivative of a function of *y*, multiply that derivative by $\frac{dy}{dx}$. You get $2x + 6y^2\frac{dy}{dx} = 3x^2 - 8y\frac{dy}{dx}$.

Now isolate $\frac{dy}{dx}$. Group the terms that contain $\frac{dy}{dx}$ on one side of the equals sign and the terms that do not contain $\frac{dy}{dx}$ on the other side: $6y^2\frac{dy}{dx} + 8y\frac{dy}{dx} = 3x^2 - 2x$.

Factor out $\frac{dy}{dx}$: $(6y^2 + 8y)\frac{dy}{dx} = 3x^2 - 2x$.

And divide: $\frac{dy}{dx} = \frac{3x^2 - 2x}{6y^2 + 8y}$.

The answer is (B).

6. **A** Take the derivative of each term with respect to x. Remember that when you are taking the derivative of a function of y, multiply that derivative by $\frac{dy}{dx}$. You get $2x+3x\left(2y\frac{dy}{dx}\right)+3y^2-3y^2\frac{dy}{dx}=0$.

   Now isolate $\frac{dy}{dx}$. Group the terms that contain $\frac{dy}{dx}$ on one side of the equals sign and the terms that do not contain $\frac{dy}{dx}$ on the other side: $6xy\frac{dy}{dx}-3y^2\frac{dy}{dx}=-2x-3y^2$.

   Factor out $\frac{dy}{dx}$: $\left(6xy-3y^2\right)\frac{dy}{dx}=-2x-3y^2$.

   And divide: $\frac{dy}{dx}=\frac{-2x-3y^2}{6xy-3y^2}$.

   The answer is (A).

7. **C** Take the derivative of each term with respect to x. Remember that when you are taking the derivative of a function of y, multiply that derivative by $\frac{dy}{dx}$. You get $4\cos\left(x^2\right)(2x)=3\frac{dy}{dx}-2y\frac{dy}{dx}$.

   Now, simply plug in $\left(\sqrt{\frac{\pi}{6}}, 2\right)$: $4\cos\left(\frac{\pi}{6}\right)\left(2\sqrt{\frac{\pi}{6}}\right)=3\frac{dy}{dx}-4\frac{dy}{dx}$. Now simplify.

$$4\frac{\sqrt{3}}{2}\left(2\sqrt{\frac{\pi}{6}}\right)=-\frac{dy}{dx}$$

$$4\sqrt{3}\left(\sqrt{\frac{\pi}{6}}\right)=-\frac{dy}{dx}$$

$$-4\sqrt{\frac{\pi}{2}}=\frac{dy}{dx}$$

   The answer is (C).

8. **A** Take the derivative of each term with respect to x. Remember that when you are taking the derivative of a function of y, multiply that derivative by $\frac{dy}{dx}$. You get $x\left(2y\frac{dy}{dx}\right)+y^2-4x^3\left(2y\frac{dy}{dx}\right)-12x^2y^2=0$.

   Now, simply plug in (1, 1): $2\frac{dy}{dx}+1-8\frac{dy}{dx}-12=0$.

   This simplifies to

$$-6\frac{dy}{dx}=11$$

$$\frac{dy}{dx}=-\frac{11}{6}$$

   The answer is (A).

9. **A** Take the derivative of each term with respect to x. Remember that when you are taking the derivative of a function of y, multiply that derivative by $\frac{dy}{dx}$. You get $\frac{dy}{dx} = 3y^2\frac{dy}{dx} - 2x$.

Now isolate $\frac{dy}{dx}$. Group the terms that contain $\frac{dy}{dx}$ on one side of the equals sign and the terms that do not contain $\frac{dy}{dx}$ on the other side: $2x = 3y^2\frac{dy}{dx} - \frac{dy}{dx}$.

Factor out $\frac{dy}{dx}$: $2x = (3y^2 - 1)\frac{dy}{dx}$.

And divide: $\frac{dy}{dx} = \frac{2x}{3y^2 - 1}$.

Now find the second derivative, so do implicit differentiation again:

$$\frac{d^2y}{dx^2} = \frac{(3y^2-1)(2) - (2x)\left(6y\frac{dy}{dx}\right)}{(3y^2-1)^2}$$

Next, substitute $\frac{dy}{dx} = \frac{2x}{3y^2-1}$: $\frac{d^2y}{dx^2} = \frac{(3y^2-1)(2) - (12xy)\left(\frac{2x}{3y^2-1}\right)}{(3y^2-1)^2}$. This simplifies to $\frac{d^2y}{dx^2} = \frac{(3y^2-1)^2(2) - (24x^2y)}{(3y^2-1)^3}$.

The answer is (A).

10. **D** First, by inspection, you can see that when $y = -7$, $x = 0$. Now use the formula for the derivative of the inverse. The derivative is $\frac{dy}{dx} = 15x^2 + 1$. Now use the formula to find the derivative of the inverse: $\frac{1}{\left.\frac{dy}{dx}\right|_{x=0}} = \frac{1}{\left.15x^2+1\right|_{x=0}} = 1$.

The answer is (D).

Chapter 6
Contextual Applications of Differentiation

INTERPRETING THE MEANING OF THE DERIVATIVE IN CONTEXT

Now that we have learned how to take derivatives, let's see how the derivative can be used in a variety of different contexts. The derivative can help us find instantaneous velocity. The derivative can help us find a rate of change. The derivative can help us find the slope of a curve at a particular point. Technically, the derivative gives us the slope of the tangent line to a curve at a particular point, but for all intents and purposes, they are the same thing.

Instantaneous velocity is often found in units of distance over time: for example, in miles per hour. Rates of change are often found in units of a quantity over time; for example, gallons per minute. The slope of a curve is usually found in units of the change in y over the change in x. We would expect that for slope.

When we use the notation $\frac{dy}{dx}$, we mean the infinitesimal change in y divided by the infinitesimal change in x. When we use the notation $f'(x)$, we (usually) mean the derivative without thinking of it as a rate of change.

STRAIGHT-LINE MOTION: CONNECTING POSITION, VELOCITY, AND ACCELERATION

This type of question may also show up in an integral calculus problem, and we'll cover that topic in a later chapter.

Almost every AP Exam has a question on position, velocity, or acceleration. It's one of the traditional areas of physics where calculus comes in handy.

If you have a function that gives you the position of an object (usually called a "particle") at a specified time, then the derivative of that function with respect to time is the velocity of the object, and the second derivative is the acceleration. These are usually represented by the following:

Please note that these equations are usually functions of time (t). Typically, t is greater than zero, but it doesn't have to be.

Position: $x(t)$ or sometimes $s(t)$

Velocity: $v(t)$, which is $x'(t)$

Acceleration: $a(t)$, which is $x''(t)$ or $v'(t)$

By the way, speed is the absolute value of velocity.

Example 1: If the position of a particle at a time t is given by the equation $x(t) = t^3 - 11t^2 + 24t$, find the velocity and the acceleration of the particle at time $t = 5$.

First, take the derivative of $x(t)$.

$$x'(t) = 3t^2 - 22t + 24 = v(t)$$

Second, plug in $t = 5$ to find the velocity at that time.

$$v(5) = 3(5^2) - 22(5) + 24 = -11$$

Third, take the derivative of $v(t)$ to find $a(t)$.

$$v'(t) = 6t - 22 = a(t)$$

Finally, plug in $t = 5$ to find the acceleration at that time.

$$a(5) = 6(5) - 22 = 8$$

See the negative velocity? The sign of the velocity is important, because it indicates the direction of the particle. Make sure that you know the following:

When the velocity is negative, the particle is moving to the left.

When the velocity is positive, the particle is moving to the right.

When the velocity and acceleration of the particle have the same signs, the particle's speed is increasing (or speeding up).

When the velocity and acceleration of the particle have opposite signs, the particle's speed is decreasing (or slowing down).

When the velocity is zero and the acceleration is not zero, the particle is momentarily stopped and changing direction.

Example 2: If the position of a particle is given by $x(t) = t^3 - 12t^2 + 36t + 18$, where $t > 0$, find the point at which the particle changes direction.

The derivative is

$$x'(t) = v(t) = 3t^2 - 24t + 36$$

Set it equal to zero and solve for t.

$$x'(t) = 3t^2 - 24t + 36 = 0$$
$$t^2 - 8t + 12 = 0$$
$$(t - 2)(t - 6) = 0$$

So, we know that $t = 2$ or $t = 6$.

You need to check that the acceleration is not 0: $x''(t) = 6t - 24$. This equals 0 at $t = 4$. Therefore, the particle is changing direction at $t = 2$ and $t = 6$.

Example 3: Given the same position function as in Example 2, find the interval of time during which the particle is slowing down.

When $0 < t < 2$ and $t > 6$, the particle's velocity is positive; when $2 < t < 6$, the particle's velocity is negative. You can verify this by graphing the function and seeing when it's above or below the x-axis. Or, try some points in the regions between the roots and outside the roots. Now, we need to determine the same information about the acceleration.

$$a(t) = v'(t) = 6t - 24$$

So the acceleration will be negative when $t < 4$, and positive when $t > 4$.

So we have

Time	Velocity	Acceleration
$0 < t < 2$	Positive	Negative
$2 < t < 4$	Negative	Negative
$4 < t < 6$	Negative	Positive
$t > 6$	Positive	Positive

Whenever the velocity and acceleration have opposite signs, the particle is slowing down. Here, the particle is slowing down during the first two seconds ($0 < t < 2$) and between the fourth and sixth seconds ($4 < t < 6$).

Another typical question you'll be asked is to find the distance a particle has traveled from one time to another. This is the distance that the particle has covered without regard to the sign, not just the displacement. In other words, if the particle had an odometer on it, what would it read? Usually, all you have to do is plug into the position function two times and find the difference.

Example 4: How far does a particle travel between the eighth and tenth seconds if its position function is $x(t) = t^2 - 6t$?

Find $x(10) - x(8) = (100 - 60) - (64 - 48) = 24$.

Be careful about one very important thing: **if the velocity changes sign during the problem's time interval**, you'll get the wrong answer if you simply follow the method in the paragraph above. For example, suppose we had the same position function as above, but we wanted to find the distance that the particle travels from $t = 2$ to $t = 4$.

$$x(4) - x(2) = (-8) - (-8) = 0$$

This is wrong. The particle travels from –8 back to –8, but it hasn't stood still. To fix this problem, divide the time interval into the time when the velocity is negative and the time when the velocity is positive, and add the absolute values of each distance. Here, the velocity is $v(t) = 2t - 6$. The velocity is negative when $t < 3$ and positive when $t > 3$. So we find the absolute value of the distance traveled from $t = 2$ to $t = 3$, and add to that the absolute value of the distance traveled from $t = 3$ to $t = 4$.

Because $x(t) = t^2 - 6t$,

$$|x(3) - x(2)| + |x(4) - x(3)| = |-9 + 8| + |-8 + 9| = 2$$

This is the distance that the particle traveled.

Example 5: Given the position function $x(t) = t^4 - 8t^2$, find the distance that the particle travels from $t = 0$ to $t = 4$.

First, find the first derivative ($v(t) = 4t^3 - 16t$) and set it equal to zero.

$$4t^3 - 16t = 0 \qquad 4t(t^2 - 4) = 0 \qquad t = 0, 2, -2$$

So we need to divide the time interval into $t = 0$ to $t = 2$ and $t = 2$ to $t = 4$.

$$|x(2) - x(0)| + |x(4) - x(2)| = 16 + 144 = 160$$

Here are some solved problems. Do each problem, covering the answer first, then checking your answer.

Example 6: Find the velocity and acceleration of a particle whose position function is $x(t) = 2t^3 - 21t^2 + 60t + 3$, for $t > 0$.

Find the first two derivatives.

$$v(t) = 6t^2 - 42t + 60$$
$$a(t) = 12t - 42$$

Example 7: Given the position function in Example 6, find when the particle's speed is increasing.

First, set $v(t) = 0$.

$$6t^2 - 42t + 60 = 0$$
$$t^2 - 7t + 10 = 0$$
$$(t - 2)(t - 5) = 0$$
$$t = 2,\ t = 5$$

You should be able to determine that the velocity is positive from $0 < t < 2$, negative from $2 < t < 5$, and positive again from $t > 5$.

Now, set $a(t) = 0$.

$$12t - 42 = 0$$
$$t = \frac{7}{2}$$

You should be able to determine that the acceleration is negative from $0 < t < \frac{7}{2}$ and positive from $t > \frac{7}{2}$.

The intervals where the velocity and the acceleration have the same sign are $2 < t < \frac{7}{2}$ and $t > 5$.

Example 8: Given that the position of a particle is found by $x(t) = t^3 - 6t^2 + 1$, $t > 0$, find the distance that the particle travels from $t = 2$ to $t = 5$.

First, find $v(t)$.

$$v(t) = 3t^2 - 12t$$

Second, set $v(t) = 0$ and find the critical values.

$$3t^2 - 12t = 0 \qquad 3t(t - 4) = 0 \qquad t = \{0, 4\}$$

Because the particle changes direction after four seconds, you have to figure out two time intervals separately (from $t = 2$ to $t = 4$ and from $t = 4$ to $t = 5$) and add the absolute values of the distances.

$$|x(4) - x(2)| + |x(5) - x(4)| = |(-31) - (-15)| + |(-24) - (-31)| = 23$$

PRACTICE PROBLEM SET 13

Now try these problems The answers are at the end of this chapter, starting on page 252.

1. Find the velocity and acceleration of a particle whose position function is $x(t) = t^3 - 9t^2 + 24t$, $t > 0$.

2. Find the velocity and acceleration of a particle whose position function is $x(t) = \sin(2t) + \cos(t)$.

3. If the position function of a particle is $x(t) = \sin\left(\frac{t}{2}\right)$, $0 < t < 4\pi$, find when the particle is changing direction.

4. If the position function of a particle is $x(t) = 3t^2 + 2t + 4$, $t > 0$, find the distance that the particle travels from $t = 2$ to $t = 5$.

5. If the position function of a particle is $x(t) = t^2 + 8t$, $t > 0$, find the distance that the particle travels from $t = 0$ to $t = 4$.

6. If the position function of a particle is $x(t) = 2\sin^2 t + 2\cos^2 t$, $t > 0$, find the velocity and acceleration of the particle.

7. If the position function of a particle is $x(t) = t^3 + 8t^2 - 2t + 4$, $t > 0$, find when the particle is changing direction.

8. If the position function of a particle is $x(t) = 2t^3 - 6t^2 + 12t - 18$, $t > 0$, find when the particle is changing direction.

RATES OF CHANGE IN APPLIED CONTEXTS OTHER THAN MOTION

In addition to motion, another place where we use the derivative is when finding the slope of a tangent line to a curve. With the derivative, we can then find the equation for the tangent line.

We can use the derivative to find the rate of change in other contexts. Let's do a couple of examples.

Example 9: The volume of water in a swimming pool at a particular time is given by $V(t) = 8t^2 - 32t + 4$, where V is in gallons and t is in hours. What is the equation for the rate at which the volume is increasing?

Simply take the derivative: $\frac{dV}{dt} = 16t - 32$ gallons per hour. By the way, note that at $t = 2$, the rate that the pool is filling is 0. We will look at that more in subsequent units.

Example 10: The rate that a sample of bacteria is reproducing in a petri dish is given by $B = 64t - 2^t$, where B is the population of bacteria and t is hours. After how many hours will the bacteria stop reproducing?

Take the derivative: $\frac{dB}{dt} = 64 - 2^t \ln 2$. We want to find when the bacteria will stop reproducing, which will be when $\frac{dB}{dt} = 0$. Use your calculator to get $t \approx 6.529$ hours.

INTRODUCTION TO RELATED RATES

In Related Rates problems, we will be looking at situations where the rate of change of something is related to the rate of change of something else. Because the two are related, there is a variable in common to both of them, usually time. Once we have set up these problems, we will need to use Implicit Differentiation and the Chain Rule to solve them.

For example, suppose we are given that the area of an object at time t is given by $A(t) = 6t^2 - 12t$. Then $\frac{dA}{dt}$ gives us the rate of change of the area in terms of time. Or suppose the height of an object at time t is given by $h(t) = -4t^2 + 24t + 50$. Then $\frac{dh}{dt}$ gives us the rate of change of the height of the object.

In order to solve Related Rates problems, we will need to find an equation (or equations) that relates the variables in question and then take the derivative of that equation with respect to time. In the next section, we will learn how to solve Related Rates problems.

SOLVING RELATED RATES PROBLEMS

The idea behind these problems is very simple. In a typical problem, you'll be given an equation relating two or more variables. These variables will change with respect to time, and you'll use derivatives to determine how the rates of change are related.

Example 11: A circular pool of water is expanding at the rate of 16π square inches per second. At what rate is the radius expanding when the radius is 4 inches?

Note: The pool is expanding in square inches per second. We've been given the rate that the area is changing, and we need to find the rate of change of the radius. What equation relates the area of a circle to its radius? $A = \pi r^2$.

Step 1: Set up the equation and take the derivative of this equation with respect to t (time).

$$\frac{dA}{dt} = 2\pi r\frac{dr}{dt}$$

In this equation, $\frac{dA}{dt}$ represents the rate at which the area is changing, and $\frac{dr}{dt}$ is the rate at which the radius is changing. The simplest way to explain this is that whenever you have a variable in an equation (r, for example), the derivative with respect to time $\left(\frac{dr}{dt}\right)$ represents the rate at which that variable is increasing or decreasing.

Step 2: Now we can plug in the values for the rate of change of the area and for the radius. (Never plug in the values until after you have taken the derivative, or you will get nonsense!)

$$16\pi = 2\pi(4)\frac{dr}{dt}$$

Solving for $\frac{dr}{dt}$, we get

$$16\pi = 8\pi\frac{dr}{dt} \text{ and } \frac{dr}{dt} = 2$$

The radius is changing at a rate of 2 inches per second. It's important to note that this is the rate only when the radius is 4 inches. As the circle gets bigger and bigger, the radius will expand at a slower and slower rate.

Example 12: A 25-foot-long ladder is leaning against a wall and sliding toward the floor. If the foot of the ladder is sliding away from the base of the wall at a rate of 15 feet per second, how fast is the top of the ladder sliding down the wall when the top of the ladder is 7 feet from the ground?

Here's another classic Related Rates problem. As always, a picture is worth 1,000 words.

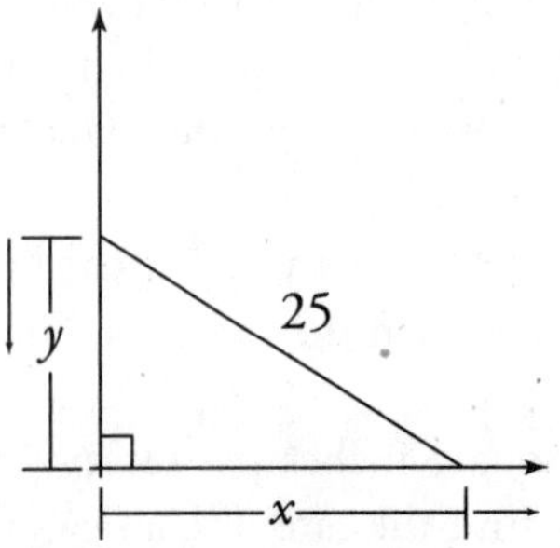

You can see that the ladder forms a right triangle with the wall. Let x stand for the distance from the foot of the ladder to the base of the wall, and let y represent the distance from the top of the ladder to the ground. What's our favorite theorem that deals with right triangles? The Pythagorean Theorem tells us here that $x^2 + y^2 = 25^2$. Now we have an equation that relates the variables to each other.

Now take the derivative of the equation with respect to t.

$$2x\frac{dx}{dt} + 2y\frac{dy}{dt} = 0$$

Just plug in what you know and solve. Because we're looking for the rate at which the vertical distance is changing, we're going to solve for $\frac{dy}{dt}$.

Let's see what we know. We're given the rate at which the ladder is sliding away from the wall: $\frac{dx}{dt} = 15$. The distance from the ladder to the top of the wall is 7 feet ($y = 7$). To find x, use the Pythagorean Theorem. If we plug in $y = 7$ to the equation $x^2 + y^2 = 25^2$, $x = 24$.

Now plug all this information into the derivative equation.

$$2(24)(15) + 2(7)\frac{dy}{dt} = 0$$

$$\frac{dy}{dt} = \frac{-360}{7} \text{ feet per second}$$

Example 13: A spherical balloon is expanding at a rate of 60π cubic inches per second. How fast is the surface area of the balloon expanding when the radius of the balloon is 4 inches?

Step 1: You're given the rate at which the volume is expanding, and you know the equation that relates volume to radius. But you have to relate radius to surface area as well, because you have to find the surface area's rate of change. This means that you'll need the equations for volume and surface area of a sphere.

$$V = \frac{4}{3}\pi r^3$$

$$A = 4\pi r^2$$

You're trying to find $\frac{dA}{dt}$, but A is given in terms of r, so you have to get $\frac{dr}{dt}$ first. Because we know the volume, if we work with the equation that gives us volume in terms of radius, we can find $\frac{dr}{dt}$. From there, work with the other equation to find $\frac{dA}{dt}$. If we take the derivative of the equation with respect to t, we get $\frac{dV}{dt} = 4\pi r^2 \frac{dr}{dt}$. Plugging in for $\frac{dV}{dt}$ and for r, we get $60\pi = 4\pi(4)^2 \frac{dr}{dt}$.

Solving for $\frac{dr}{dt}$, we get

$$\frac{dr}{dt} = \frac{15}{16} \text{ inches per second}$$

Step 2: Now we take the derivative of the other equation with respect to t.

$$\frac{dA}{dt} = 8\pi r \frac{dr}{dt}$$

We can plug in for r and $\frac{dr}{dt}$ from the previous step to get

$$\frac{dA}{dt} = 8\pi(4)\frac{15}{16} = \frac{480\pi}{16} \text{ square inches per second} = 30\pi \text{ square inches per second}$$

Example 14: An underground conical tank, standing on its vertex, is being filled with water at the rate of 18π cubic feet per minute. If the tank has a height of 30 feet and a radius of 15 feet, how fast is the water level rising when the water is 12 feet deep?

This "cone" problem is also typical. The key point to getting these right is knowing that the ratio of the height of a right circular cone to its radius is constant. By telling us that the height of the cone is 30 and the radius is 15, we know that at any level, the height of the water will be twice its radius, or $h = 2r$.

You must find the rate at which the water is rising (the height is changing), or $\frac{dh}{dt}$. Therefore, you want to eliminate the radius from the volume. By substituting $\frac{h}{2} = r$ into the equation for volume, we get

$$V = \frac{1}{3}\pi\left(\frac{h}{2}\right)^2 h = \frac{\pi h^3}{12}$$

Differentiate both sides with respect to t.

$$\frac{dV}{dt} = \frac{\pi}{12}3h^2\frac{dh}{dt}$$

Now we can plug in and solve for $\frac{dh}{dt}$.

$$18\pi = \frac{\pi}{12}3(12)^2\frac{dh}{dt}$$

$$\frac{dh}{dt} = \frac{1}{2} \text{ foot per minute}$$

In order to solve Related Rates problems, you have to be good at determining relationships between variables. Once you figure that out, the rest is a piece of cake. Many of these problems involve geometric relationships, so review the formulas for the volumes and areas of cones, spheres, boxes, and other solids. Once you get the hang of setting up the problems, you'll see that these problems follow the same predictable patterns. Look through these sample problems.

Example 15: A circle is increasing in area at the rate of 16π square inches per second. How fast is the radius increasing when the radius is 2 inches?

Use the expression that relates the area of a circle to its radius: $A = \pi r^2$.

Next, take the derivative of the expression with respect to t.

$$\frac{dA}{dt} = 2\pi r\frac{dr}{dt}$$

Now, plug in $\frac{dA}{dt} = 16\pi$ and $r = 2$.

$$16\pi = 2\pi(2)\frac{dr}{dt}$$

When you solve for $\frac{dr}{dt}$, you'll get $\frac{dr}{dt} = 4$ inches per second.

Example 16: A rocket is rising vertically at a rate of 5400 miles per hour. An observer on the ground is standing 20 miles from the rocket's launch point. How fast (in radians per second) is the angle of elevation between the ground and the observer's line of sight of the rocket increasing when the rocket is at an elevation of 40 miles?

Notice that velocity is given in miles per hour and the answer asks for radians per second. In situations like this one, if you don't correctly convert the units, you may end up missing out on some obtainable points.

First, draw a picture.

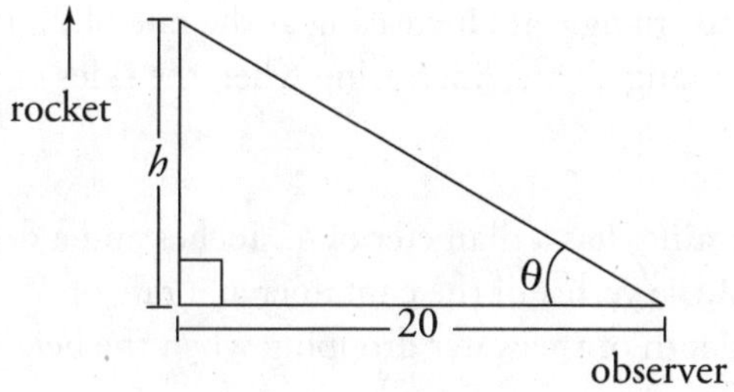

Now, find the equation that relates the angle of elevation to the rocket's altitude.

$$\tan\theta = \frac{h}{20}$$

If we take the derivative of both sides of this expression with respect to *t*, we get

$$\sec^2\theta\frac{d\theta}{dt} = \frac{1}{20}\frac{dh}{dt}$$

We know that $\frac{dh}{dt} = 5{,}400$ miles per hour, but the problem asks for time in seconds, so we need to convert this number. There are 3,600 seconds in an hour, so $\frac{dh}{dt} = \frac{3}{2}$ miles per second. Next, we know that $\tan\theta = \frac{h}{20}$, so when $h = 40$, $\tan\theta = 2$. Because $1 + \tan^2\theta = \sec^2\theta$, we get $\sec^2\theta = 5$.

Plug in the following information:

$$5\frac{d\theta}{dt} = \frac{1}{20}\left(\frac{3}{2}\right) \text{ and } \frac{d\theta}{dt} = \frac{3}{200} \text{ radians per second}$$

PRACTICE PROBLEM SET 14

Now try these problems on your own. The answers are at the end of this chapter, starting on page 254.

1. Oil spilled from a tanker spreads in a circle whose circumference increases at a rate of 40 feet per second. How fast is the area of the spill increasing when the circumference of the circle is 100π feet?

2. A spherical balloon is inflating at a rate of 27π cubic inches per second. How fast is the radius of the balloon increasing when the radius is 3 inches?

3. Cars A and B leave a town at the same time. Car A heads due south at a rate of 80 kilometers per hour and car B heads due west at a rate of 60 kilometers per hour. How fast is the distance between the cars increasing after three hours?

4. The sides of an equilateral triangle are increasing at the rate of 27 inches per second. How fast is the triangle's area increasing when the sides of the triangle are each 18 inches long?

5. An inverted conical container has a diameter of 42 inches and a depth of 15 inches. If water is flowing out of the vertex of the container at a rate of 35π cubic inches per second, how fast is the depth of the water dropping when the height is 5 inches?

6. A boat is being pulled toward a dock by a rope attached to its bow through a pulley on the dock that is 7 feet above the bow. If the rope is hauled in at a rate of 4 feet per second, how fast is the boat approaching the dock when 25 feet of rope is out?

7. A 6-foot-tall woman is walking at the rate of 4 feet per second away from a street lamp that is 24 feet tall. How fast is the length of her shadow changing?

8. The minute hand of a clock is 6 inches long. Starting from noon, how fast is the area of the sector swept out by the minute hand increasing in square inches per minute at any instant?

APPROXIMATING VALUES OF A FUNCTION USING LOCAL LINEARITY AND LINEARIZATION

Equations of Tangent Lines and Normal Lines

Finding the equation of a line tangent to a certain curve at a certain point is a standard calculus problem. This is because, among other things, the derivative is the slope of a tangent line to a curve at a particular point. Thus, we can find the equation of the tangent line to a curve if we have the equation of the curve and the point at which we want to find the tangent line. Then all we have to do is take the derivative of the equation, plug in the *x*-coordinate of the point to find the slope, then use the point and the slope to find the equation of the line. Let's take this one step at a time.

Suppose we have a point (x_1, y_1) and a slope m. Then the equation of the line through that point with that slope is

$$(y - y_1) = m(x - x_1)$$

Next, suppose that we have an equation $y = f(x)$, where (x_1, y_1) satisfies that equation. Then $f'(x_1) = m$, and we can plug all of our values into the equation for a line and get the equation of the tangent line. This is much easier to explain with a simple example.

Example 17: Find the equation of the tangent line to the curve $y = 5x^2$ at the point (3, 45).

First of all, notice that the point satisfies the equation: when $x = 3$, $y = 45$. Now, take the derivative of the equation.

$$\frac{dy}{dx} = 10x$$

Now, if you plug in $x = 3$, you'll get the slope of the curve at that point.

$$\left.\frac{dy}{dx}\right|_{x=3} = 10(3) = 30$$

By the way, the notation for plugging in a point is $|_{x=}$. Learn to recognize it!

Thus, we have the slope and the point, and the equation is

$$(y - 45) = 30(x - 3)$$

It's customary to simplify the equation if it's not too onerous.

$$y = 30x - 45$$

Example 18: Find the equation of the tangent line to $y = x^3 + x^2$ at (3, 36).

The derivative looks like the following:

$$\frac{dy}{dx} = 3x^2 + 2x$$

So, the slope is

$$\left.\frac{dy}{dx}\right|_{x=3} = 3(3)^2 + 2(3) = 33$$

The equation looks like the following:

$$(y - 36) = 33(x - 3), \text{ or } y = 33x - 63$$

Naturally, there are a couple of things that can be done to make problems like this harder. First of all, you can be given only the x-coordinate. Second, the equation can be more difficult to differentiate.

In order to find the y-coordinate, all you have to do is plug the x-value into the equation for the curve and solve for y. Remember this; you'll see it again!

Example 19: Find the equation of the tangent line to $y = \frac{2x+5}{x^2-3}$ at $x = 1$.

First, find the y-coordinate.

$$y(1) = \frac{2(1)+5}{1^2-3} = -\frac{7}{2}$$

Second, take the derivative.

$$\frac{dy}{dx} = \frac{(x^2-3)(2)-(2x+5)(2x)}{(x^2-3)^2}$$

You're probably dreading having to simplify this derivative. Don't waste your time! Plug in $x = 1$ right away.

$$\left.\frac{dy}{dx}\right|_{x=1} = \frac{\left(1^2-3\right)(2)-(2(1)+5)(2(1))}{\left(1^2-3\right)^2} = \frac{-4-14}{4} = -\frac{9}{2}$$

Now, we have a slope and a point, so the equation is

$$y + \frac{7}{2} = -\frac{9}{2}(x-1), \text{ or } 2y = -9x + 2$$

Remember: the slope of a perpendicular line is just the negative reciprocal of the slope of the tangent line.

Sometimes, instead of finding the equation of a tangent line, you will be asked to find the equation of a normal line. A **normal** line is simply the line perpendicular to the tangent line at the same point. You follow the same steps as with the tangent line, but you use the slope that will give you a perpendicular line.

Example 20: Find the equation of the line normal to $y = x^5 - x^4 + 1$ at $x = 2$.

First, find the y-coordinate.

$$y(2) = 2^5 - 2^4 + 1 = 17$$

Second, take the derivative.

$$\frac{dy}{dx} = 5x^4 - 4x^3$$

Third, find the slope at $x = 2$.

$$\left.\frac{dy}{dx}\right|_{x=2} = 5(2)^4 - 4(2)^3 = 48$$

Fourth, take the negative reciprocal of 48, which is $-\frac{1}{48}$.

Finally, the equation becomes

$$y - 17 = -\frac{1}{48}(x - 2)$$

Try these solved problems. Do each problem, covering the answer first, then checking your answer.

Example 21: Find the equation of the tangent line to the graph of $y = 4 - 3x - x^2$ at the point $(2, -6)$.

First, take the derivative of the equation.

$$\frac{dy}{dx} = -3 - 2x$$

Now, plug in $x = 2$ to get the slope of the tangent line.

$$\frac{dy}{dx} = -3 - 2(2) = -7$$

Third, plug the slope and the point into the equation for the line.

$$y - (-6) = -7(x - 2)$$

This simplifies to $y = -7x + 8$.

Example 22: Find the equation of the normal line to the graph of $y = 6 - x - x^2$ at $x = -1$.

Plug $x = -1$ into the original equation to get the y-coordinate.

$$y = 6 + 1 - 1 = 6$$

Once again, take that derivative.

$$\frac{dy}{dx} = -1 - 2x$$

Now plug in $x = -1$ to get the slope of the tangent.

$$\frac{dy}{dx} = -1 - 2(-1) = 1$$

Use the negative reciprocal of the slope in the second step to get the slope of the normal line.

$$m = -1$$

Finally, plug the slope and the point into the equation for the line.

$$y - 6 = -1(x + 1)$$

This simplifies to $y = -x + 5$.

Example 23: Find the equations of the tangent and normal lines to the graph of $y = \frac{10x}{x^2 + 1}$ at the point (2, 4).

This problem will put your algebra to the test. You have to use the Quotient Rule to take the derivative of this mess.

$$\frac{dy}{dx} = \frac{(x^2 + 1)(10) - (10x)(2x)}{(x^2 + 1)^2}$$

Second, plug in $x = 2$ to get the slope of the tangent.

$$\frac{dy}{dx} = \frac{(5)(10) - (20)(4)}{5^2} = -\frac{30}{25} = -\frac{6}{5}$$

Now, plug the slope and the point into the equation for the tangent line.

$$y - 4 = -\frac{6}{5}(x - 2)$$

That simplifies to $6x + 5y = 32$. The equation of the normal line must then be

$$y - 4 = \frac{5}{6}(x - 2)$$

That, in turn, simplifies to $-5x + 6y = 14$.

Example 24: The curve $y = ax^2 + bx + c$ passes through the point (2, 4) and is tangent to the line $y = x + 1$ at (0, 1). Find a, b, and c.

The curve passes through (2, 4), so if you plug in $x = 2$, you'll get $y = 4$. Therefore,

$$4 = 4a + 2b + c$$

Second, the curve also passes through the point (0, 1), so $c = 1$.

Because the curve is tangent to the line $y = x + 1$ at (0, 1), they must both have the same slope at that point. The slope of the line is 1. The slope of the curve is the first derivative.

$$\frac{dy}{dx} = 2ax + b$$

$$\left.\frac{dy}{dx}\right|_{x=0} = 2a(0) + b = b$$

At (0, 1), $\frac{dy}{dx} = b$. Therefore, $b = 1$.

Now that you know b and c, plug them back into the equation from the first step and solve for a.

$$4 = 4a + 2 + 1, \text{ and } a = \frac{1}{4}$$

Example 25: Find the points on the curve $y = 2x^3 - 3x^2 - 12x + 20$ where the tangent is parallel to the x-axis.

The x-axis is a horizontal line, so it has a slope of zero. Therefore, you want to know where the derivative of this curve is zero. Take the derivative.

$$\frac{dy}{dx} = 6x^2 - 6x - 12$$

Set it equal to zero and solve for x. Get accustomed to doing this: it's one of the most common questions in differential calculus.

$$\frac{dy}{dx} = 6x^2 - 6x - 12 = 0$$

$$6(x^2 - x - 2) = 0$$

$$6(x - 2)(x + 1) = 0$$

$$x = 2 \text{ or } x = -1$$

Third, find the y-coordinates of these two points.

$$y = 2(8) - 3(4) - 12(2) + 20 = 0$$

$$y = 2(-1) - 3(1) - 12(-1) + 20 = 27$$

Therefore, the points are (2, 0) and (–1, 27).

PRACTICE PROBLEM SET 15

Now try these problems. The answers are at the end of this chapter, starting on page 258.

1. Find the equation of the tangent to the graph of $y = 3x^2 - x$ at $x = 1$.

2. Find the equation of the tangent to the graph of $y = x^3 - 3x$ at $x = 3$.

3. Find the equation of the tangent to the graph of $y = \dfrac{1}{\sqrt{x^2+7}}$ at $x = 3$.

4. Find the equation of the normal to the graph of $y = \dfrac{x+3}{x-3}$ at $x = 4$.

5. Find the equation of the tangent to the graph of $y = 2x^3 - 3x^2 - 12x + 20$ at $x = 2$.

6. Find the equation of the tangent to the graph of $y = \dfrac{x^2+4}{x-6}$ at $x = 5$.

7. Find the equation of the tangent to the graph of $y = \sqrt{x^3 - 15}$ at $(4, 7)$.

8. Find the values of x where the tangent to the graph of $y = 2x^3 - 8x$ has a slope equal to the slope of $y = x$.

9. Find the equation of the normal to the graph of $y = \dfrac{3x+5}{x-1}$ at $x = 3$.

10. Find the values of x where the normal to the graph of $(x - 9)^2$ is parallel to the y-axis.

11. Find the coordinates where the tangent to the graph of $y = 8 - 3x - x^2$ is parallel to the x-axis.

12. Find the values of a, b, and c where the curves $y = x^2 + ax + b$ and $y = cx + x^2$ have a common tangent line at $(-1, 0)$.

Linearization

A differential is a very small quantity that corresponds to a change in a number. We use the symbol Δx to denote a differential. What are differentials used for? The AP Exam mostly wants you to use them to approximate the value of a function or to find the error of an approximation.

Recall the formula for the definition of the derivative.

$$f'(x) = \lim_{h \to 0} \frac{f(x+h) - f(x)}{h}$$

Replace h with Δx, which also stands for a very small increment of x, and get rid of the limit.

$$f'(x) \approx \frac{f(x+\Delta x) - f(x)}{\Delta x}$$

Notice that this is no longer equal to the derivative, but an approximation of it. If Δx is kept small, the approximation remains fairly accurate. Next, rearrange the equation as follows:

$$f(x+\Delta x) \approx f(x) + f'(x)\Delta x$$

This is our formula for differentials. It says that "the value of a function at x plus a little bit equals the value of the function at x plus the product of the derivative of the function at x and the little bit."

Example 26: Use differentials to approximate $\sqrt{9.01}$.

You can start by letting $x = 9$, $\Delta x = +0.01$, and $f(x) = \sqrt{x}$. Next, we need to find $f'(x)$.

$$f'(x) = \frac{1}{2\sqrt{x}}$$

Now, plug in to the formula.

$$f(x+\Delta x) \approx f(x) + f'(x)\Delta x$$
$$\sqrt{x+\Delta x} \approx \sqrt{x} + \frac{1}{2\sqrt{x}}\Delta x$$

Now, if we plug in $x = 9$ and $\Delta x = +0.01$,

$$\sqrt{9.01} \approx \sqrt{9} + \frac{1}{2\sqrt{9}}(0.01) \approx 3.001666666$$

If you enter $\sqrt{9.01}$ into your calculator, you get 3.001666204. As you can see, our answer is a pretty good approximation. It's not so good, however, when Δx is too big. How big is too big? Good question.

Example 27: Use differentials to approximate $\sqrt{9.5}$.

Let $x = 9, \Delta x = +0.5$, and $f(x) = \sqrt{x}$, and plug in to what you found in Example 26.

$$\sqrt{9.5} \approx \sqrt{9} + \frac{1}{2\sqrt{9}}(0.5) \approx 3.083333333$$

However, $\sqrt{9.5}$ equals 3.082207001 on a calculator. This is good to only two decimal places. As the ratio of $\frac{\Delta x}{x}$ grows larger, the approximation gets less accurate, and we start to get away from the actual value.

There's another approximation formula that you'll need to know for the AP Exam. This formula is used to estimate the error in a measurement or to find the effect on a formula when a small change in measurement is made. The formula is

Note that this equation is simply a rearrangement of $\frac{dy}{dx} = f'(x)$.

$$dy = f'(x)\, dx$$

This notation may look a little confusing. It says that the change in a measurement *dy*, due to a differential *dx*, is found by multiplying the derivative of the equation for *y* by the differential. Let's do an example.

Example 28: The radius of a circle is increased from 3 to 3.04. Estimate the change in area.

Let $A = \pi r^2$. Then our formula says that $dA = A'dr$, where A' is the derivative of the area with respect to r, and $dr = 0.04$ (the change). First, find the derivative of the area: $A' = 2\pi r$. Now, plug in to the formula.

$$dA = 2\pi r\, dr = 2\pi(3)(0.04) = 0.754$$

The actual change in the area is from 9π to 9.2416π, which is approximately 0.759. As you can see, this approximation formula is pretty accurate.

Here are some sample problems involving this differential formula. Try them out, then check your work against the answers directly beneath.

Example 29: Use differentials to approximate $(3.98)^4$.

Let $f(x) = x^4$, $x = 4$, and $\Delta x = -0.02$. Next, find $f'(x)$, which is: $f'(x) = 4x^3$.

Now, plug in to the formula.

$$f(x+\Delta x) \approx f(x) + f'(x)\Delta x$$
$$(x+\Delta x)^4 \approx x^4 + 4x^3\Delta x$$

If you plug in $x = 4$ and $\Delta x = -0.02$, you get

$$(3.98)^4 \approx 4^4 + 4(4)^3(-0.02) \approx 250.88$$

Check $(3.98)^4$ by using your calculator; you should get 250.9182722. Not a bad approximation.

It's natural to ask why we bother learning this technique for approximations when we can just use a calculator. The answer is that this technique comes into play in higher-level calculus courses, where a calculator is of no value.

Example 30: Use differentials to approximate sin 46°.

This is a tricky question. The formula doesn't work if you use degrees. Here's why: Let $f(x) = \sin x$, $x = 45°$, and $\Delta x = 1°$. The derivative is $f'(x) = \cos x$.

If you plug this information into the formula, you get $\sin 46° \approx \sin 45° + \cos 45°(1°) = \sqrt{2}$. You should recognize that this makes no sense for two reasons: (1) the sine of any angle is between -1 and 1; and (2) the answer should be close to $\sin 45° = \frac{1}{\sqrt{2}}$.

What went wrong? You have to use radians! As we mentioned before, angles in calculus problems are measured in radians, not degrees.

Let $f(x) = \sin x$, $x = \frac{\pi}{4}$, and $\Delta x = \frac{\pi}{180}$. Now plug in to the formula.

$$\sin\left(\frac{46\pi}{180}\right) \approx \sin\frac{\pi}{4} + \left(\cos\frac{\pi}{4}\right)\left(\frac{\pi}{180}\right) = 0.7194$$

Example 31: The radius of a sphere is measured to be 4 centimeters with an error of ± 0.01 centimeter. Use differentials to approximate the error in the surface area.

Now it's time for the other differential formula. The formula for the surface area of a sphere is

$$S = 4\pi r^2$$

The formula says that $dS = S'dr$, so first, we find the derivative of the surface area ($S' = 8\pi r$) and plug away.

$$dS = 8\pi r dr = 8\pi(4)(\pm 0.01) = \pm 1.0053$$

This looks like a big error, but given that the surface area of a sphere with radius 4 is approximately 201 square centimeters, the error is quite small.

PRACTICE PROBLEM SET 16

Use the differential formulas in this chapter to solve these problems. The answers are at the end of this chapter, starting on page 263.

1. Approximate $\sqrt{25.02}$.

2. Approximate $\sqrt[3]{63.97}$.

3. Approximate tan 61°.

4. The side of a cube is measured to be 6 inches with an error of ± 0.02 inch. Estimate the error in the volume of the cube.

5. When a spherical ball bearing is heated, its radius increases by 0.01 millimeter. Estimate the change in volume of the ball bearing when the radius is 5 millimeters.

6. A cylindrical tank is constructed to have a diameter of 5 meters and a height of 20 meters. Find the error in the volume if

   (a) the diameter is exact, but the height is 20.1 meters; and

   (b) the height is exact, but the diameter is 5.1 meters.

USING L'HOSPITAL'S RULE FOR DETERMINING LIMITS OF INDETERMINATE FORMS

L'Hospital's Rule is a way to find the limit of certain kinds of expressions that are indeterminate forms. If the limit of an expression results in $\frac{0}{0}$ or $\frac{\infty}{\infty}$, the limit is called "indeterminate" and you can use L'Hospital's Rule to evaluate these expressions.

Remember that you can only use L'Hospital's Rule for indeterminate forms.

If $f(c) = g(c) = 0$, and if $f'(c)$ and $g'(c)$ exist, and if $g'(c) \neq 0$,

then $\lim\limits_{x \to c} \frac{f(x)}{g(x)} = \frac{f'(c)}{g'(c)}$.

Similarly,

If $f(c) = g(c) = \infty$, and if $f'(c)$ and $g'(c)$ exist, and if $g'(c) \neq 0$,

then $\lim\limits_{x \to c} \frac{f(x)}{g(x)} = \frac{f'(c)}{g'(c)}$.

In other words, if the limit of the function gives us an undefined expression, like $\frac{0}{0}$ or $\frac{\infty}{\infty}$, L'Hospital's Rule says we can take the derivative of the top and the derivative of the bottom and see if we get a determinate expression. If not, we can repeat the process.

Example 32: Find $\lim\limits_{x \to 0} \frac{\sin x}{x}$.

First, notice that plugging in 0 results in $\frac{0}{0}$, which is indeterminate. Take the derivative of the top and of the bottom.

$$\lim_{x \to 0} \frac{\cos x}{1}$$

The limit equals 1.

Example 33: Find $\lim\limits_{x \to 0} \dfrac{2x - \sin x}{x}$.

If you plug in 0, you get $\dfrac{0-0}{0}$, which is indeterminate. Now, take the derivative of the top and of the bottom.

$$\lim_{x \to 0} \frac{2 - \cos x}{1}$$

This limit also equals 1.

Example 34: Find $\lim\limits_{x \to 0} \dfrac{\sqrt{4+x} - 2}{2x}$.

Again, plugging in 0 is no help; you get $\dfrac{0}{0}$. Take the derivative of the top and bottom.

$$\lim_{x \to 0} \frac{\frac{1}{2\sqrt{4+x}}}{2} = \frac{1}{8}$$

Example 35: Find $\lim\limits_{x \to 0} \dfrac{\sqrt{4+x} - 2 - \frac{x}{4}}{2x^2}$.

Take the derivative of the top and bottom.

$$\lim_{x \to 0} \frac{\frac{1}{2\sqrt{4+x}} - \frac{1}{4}}{4x}$$

Now if you take the limit, we still get $\dfrac{0}{0}$. So do it again.

$$\lim_{x \to 0} \frac{-\frac{1}{4}(4+x)^{-\frac{3}{2}}}{4} = -\frac{1}{128}$$

Now let's try a couple of $\dfrac{\infty}{\infty}$ forms.

Example 36: Find $\lim\limits_{x \to \infty} \dfrac{5x - 8}{3x + 1}$.

Now, the limit is $\dfrac{\infty}{\infty}$. The derivative of the top and bottom is

$$\lim_{x \to \infty} \frac{5}{3} = \frac{5}{3}$$

Example 37: Find $\lim_{x \to \frac{\pi}{2}} \frac{\sec x}{1 + \sec x}$.

The derivative of the top and bottom is

$$\lim_{x \to \frac{\pi}{2}} \frac{\sec x \tan x}{\sec x \tan x} = 1$$

Example 38: Find $\lim_{x \to 0^+} x \cot x$.

Taking this limit results in $(0)(\infty)$, which is also indeterminate. (But you can't use the rule yet!)

If you rewrite this expression as $\lim_{x \to 0^+} \frac{x}{\tan x}$, it's of the form $\frac{0}{0}$, and we can use L'Hospital's Rule.

$$\lim_{x \to 0^+} \frac{1}{\sec^2 x} = 1$$

That's all that you need to know about L'Hospital's Rule. Just check to see if the limit results in an indeterminate form. If it does, use the rule until you get a determinate form.

Here are some more examples.

Example 39: Find $\lim_{x \to 0} \frac{\sin 8x}{x}$.

First, notice that plugging in 0 gives us an indeterminate result: $\frac{0}{0}$. Now, take the derivative of the top and of the bottom.

$$\lim_{x \to 0} \frac{8 \cos 8x}{1} = 8$$

Example 40: Find $\lim_{x \to \infty} xe^{-2x}$.

First, rewrite this expression as $\frac{x}{e^{2x}}$. Notice that when x nears infinity, the expression becomes $\frac{\infty}{\infty}$, which is indeterminate.

Take the derivative of the top and bottom.

$$\lim_{x \to \infty} \frac{1}{2e^{2x}} = 0$$

Example 41: Find $\lim_{x \to \infty} \frac{5x^3 - 4x^2 + 1}{7x^3 + 2x - 6}$.

We learned this in Chapter 3, remember? Now we'll use L'Hospital's Rule. At first glance, the limit is indeterminate: $\frac{\infty}{\infty}$. Let's take some derivatives.

$$\frac{15x^2 - 8x}{21x^2 + 2}$$

This is still indeterminate, so it's time to take the derivative of the top and bottom again.

$$\frac{30x - 8}{42x}$$

It's still indeterminate! If you try it one more time, you'll get a fraction with no variables: $\frac{30}{42}$, which can be simplified to $\frac{5}{7}$ (as we expected).

Example 42: Find $\lim_{x \to \frac{\pi}{2}} \frac{x - \frac{\pi}{2}}{\cos x}$.

Plugging in $\frac{\pi}{2}$ gives you the indeterminate response of $\frac{0}{0}$. The derivative is

$$-\frac{1}{\sin x}$$

When we take the limit of this expression, we get –1.

PRACTICE PROBLEM SET 17

Now find these limits using L'Hospital's Rule. The answers are at the end of this chapter, starting on page 265.

1. Find $\lim\limits_{x\to 0}\dfrac{\sin 3x}{\sin 4x}$.

2. Find $\lim\limits_{x\to \pi}\dfrac{x-\pi}{\sin x}$.

3. Find $\lim\limits_{x\to 0}\dfrac{x-\sin x}{x^3}$.

4. Find $\lim\limits_{x\to 0}\dfrac{e^{3x}-e^{5x}}{x}$.

5. Find $\lim\limits_{x\to 0}\dfrac{\tan x - x}{\sin x - x}$.

6. Find $\lim\limits_{x\to \infty}\dfrac{x^5}{e^{5x}}$.

7. Find $\lim\limits_{x\to \infty}\dfrac{x^5+4x^3-8}{7x^5-3x^2-1}$.

8. Find $\lim\limits_{x\to 0^+}\dfrac{\ln(\sin x)}{\ln(\tan x)}$.

9. Find $\lim\limits_{x\to 0^+}\dfrac{\cot 2x}{\cot x}$.

10. Find $\lim\limits_{x\to 0^+}\dfrac{x}{\ln(x+1)}$.

Chapter 6 Drill

The answers are at the end of the chapter, starting on page 272.

1 Mark for Review

A 25-ft ladder is sliding down a wall at –6 feet per second. How fast is the bottom of the ladder sliding out when the top of the ladder is 20 feet from the ground?

(A) 2 feet per second

(B) 6 feet per second

(C) 8 feet per second

(D) 15 feet per second

2 Mark for Review

A baseball diamond has right angles at each of the bases, which are 90 feet apart. A baseball player is running from second base to third base at a rate of 30 feet per second. How fast is the distance between the player and home plate changing when the player is halfway from second base to third base?

(A) $-\frac{30}{\sqrt{5}}$ feet per second

(B) –30 feet per second

(C) $-\frac{135}{\sqrt{5}}$ feet per second

(D) –135 feet per second

3 Mark for Review

A spherical balloon is deflating at -24π cubic inches per second. How fast is the surface area of the balloon decreasing when the radius of the balloon is 6 inches?

(A) -6π square inches per second

(B) -8π square inches per second

(C) -16π square inches per second

(D) -24π square inches per second

4 Mark for Review

Find where the velocity of a particle is zero if its position function is $(x)t = t^4 - 2t^2$, where $t > 0$.

(A) $t = \sqrt[4]{t}$

(B) $t = \frac{1}{\sqrt{2}}$

(C) $t = 1$

(D) $t = \sqrt{2}$

5 Mark for Review

If the position function of a particle is $x(t) = 4t^3 - 12t^2 + 18$, $t > 0$, at what time is the particle changing direction?

(A) $t = 0$

(B) $t = \frac{1}{2}$

(C) $t = 1$

(D) $t = 2$

6 Mark for Review

Find the equation of the line tangent to $y = 5x^3 - 20x + 10$ at $x = 2$.

Ⓐ $y - 10 = 10(x - 2)$

Ⓑ $y + 10 = 40(x + 2)$

Ⓒ $y - 10 = 40(x - 2)$

Ⓓ $y - 10 = 90(x + 2)$

7 Mark for Review

Use L'Hospital's Rule to find $\lim_{x \to \infty} \frac{x^{-\frac{4}{3}}}{\sin\left(\frac{1}{x}\right)}$.

Ⓐ $-\infty$

Ⓑ 0

Ⓒ ∞

Ⓓ The limit does not exist.

8 Mark for Review

Find the equation of the line normal to $y = 4\sec(2x)$ at $x = \frac{\pi}{8}$.

Ⓐ $y + 4\sqrt{2} = 8\sqrt{2}\left(x - \frac{\pi}{8}\right)$

Ⓑ $y - 4\sqrt{2} = -8\sqrt{2}\left(x - \frac{\pi}{8}\right)$

Ⓒ $y + 4\sqrt{2} = -\frac{1}{8\sqrt{2}}\left(x + \frac{\pi}{8}\right)$

Ⓓ $y - 4\sqrt{2} = -\frac{1}{8\sqrt{2}}\left(x - \frac{\pi}{8}\right)$

ANSWERS TO PRACTICE PROBLEM SETS

Practice Problem Set 13

1. $v(t)=3t^2-18t+24$; $a(t)=6t-18$

In order to find the velocity function of the particle, simply take the derivative of the position function with respect to t: $\frac{dx}{dt}=v(t)=3t^2-18t+24$. In order to find the acceleration function of the particle, simply take the derivative of the velocity function with respect to t: $\frac{dv}{dt}=6t-18$.

2. $v(t)=2\cos(2t)-\sin(t)$; $a(t)=-4\sin(2t)-\cos(t)$

In order to find the velocity function of the particle, simply take the derivative of the position function with respect to t: $\frac{dx}{dt}=v(t)=2\cos(2t)-\sin(t)$. In order to find the acceleration function of the particle, simply take the derivative of the velocity function with respect to t: $\frac{dv}{dt}=a(t)=-4\sin(2t)-\cos(t)$.

3. $t = \pi$ and $t = 3\pi$

In order to find where the particle is changing direction, find where the velocity of the particle changes sign. The velocity function of the particle is the derivative of the position function: $\frac{dx}{dt}=v(t)=\frac{1}{2}\cos\left(\frac{t}{2}\right)$. Next, set the velocity equal to zero. The solutions are $t = \pi$ and $t = 3\pi$. Actually, there are an infinite number of solutions, but remember that the solution is restricted to $0 < t < 4\pi$. Next, check the sign of the velocity on the intervals $0 < t < \pi$, $\pi < t < 3\pi$, and $3\pi < t < 4\pi$. When $0 < t < \pi$, the velocity is positive, so the particle is moving to the right. When $\pi < t < 3\pi$, the velocity is negative, so the particle is moving to the left. When $3\pi < t < 4\pi$, the velocity is positive, so the particle is moving to the right. Therefore, the particle is changing direction at $t = \pi$ and $t = 3\pi$.

4. The distance is 69.

In order to find the distance that the particle travels, look at the position of the particle at $t = 2$ and at $t = 5$. You also need to see if the particle changes direction anywhere on the interval between the two times. If so, you will need to look at the particle's position at those "turning points" as well. The way to find out if the particle is changing direction is to look at the velocity of the particle, which you can find by taking the derivative of the position function. You get $\frac{dx}{dt}=v(t)=6t+2$. If you set the velocity equal to zero, you get

$t = -\frac{1}{3}$, which is not in the time interval. This means that the velocity doesn't change signs, and thus the particle does not change direction. Now look at the position of the particle on the interval. At $t = 2$, the particle's position is $x = 3(2)^2 + 2(2) + 4 = 20$. At $t = 5$, the particle's position is $x = 3(5)^2 + 2(5) + 4 = 89$. Therefore, the particle travels a distance of 69.

5. The distance is 48.

In order to find the distance that the particle travels, look at the position of the particle at $t = 0$ and at $t = 4$. You also need to see if the particle changes direction anywhere on the interval between the two times. If so, you will need to look at the particle's position at those "turning points" as well. The way to find out if the particle is changing direction is to look at the velocity of the particle, which you can find by taking the derivative of the position function. You get $\frac{dx}{dt} = v(t) = 2t + 8$. If you set the velocity equal to zero, you get $t = -4$, which is not in the time interval. This means that the velocity doesn't change signs, and thus the particle does not change direction. Now look at the position of the particle on the interval. At $t = 0$, the particle's position is $x = (0)^2 + 8(0) = 0$. At $t = 4$, the particle's position is $x = (4)^2 + 8(4) = 48$. Therefore, the particle travels a distance of 48.

6. Velocity is 0; Acceleration is 0

This should not be a surprise because $2\sin^2 t + 2\cos^2 t = 2$, so the position is a constant. This means that the particle is not moving and thus has a velocity and acceleration of 0.

7. $t = \frac{-8 + \sqrt{70}}{3} \approx 0.122$

In order to find where the particle is changing direction, find where the velocity of the particle changes sign. The velocity function of the particle is the derivative of the position function: $\frac{dx}{dt} = v(t) = 3t^2 + 16t - 2$. Next, set the velocity equal to zero. The solutions are $t = \frac{-8 + \sqrt{70}}{3}$ and $t = \frac{-8 - \sqrt{70}}{3}$. You can eliminate the second solution because it is negative and you are restricted to $t > 0$. Next, check the sign of the velocity on the intervals $0 < t < \frac{-8 + \sqrt{70}}{3}$ and $t > \frac{-8 + \sqrt{70}}{3}$. When $0 < t < \frac{-8 + \sqrt{70}}{3}$, the velocity is negative, so the particle is moving to the left. When $t > \frac{-8 + \sqrt{70}}{3}$, the velocity is positive, so the particle is moving to the right. Therefore, the particle is changing direction at $t = \frac{-8 + \sqrt{70}}{3}$.

8. The velocity is never 0, which means that it never changes signs and thus the particle does not change direction.

In order to find where the particle is changing direction, find where the velocity of the particle changes signs. The velocity function of the particle is the derivative of the position function: $\frac{dx}{dt} = v(t) = 6t^2 - 12t + 12$. Next, set the velocity equal to zero. There are no real solutions. If you try a few values, you can see that the velocity is always positive. Therefore, the particle does not change direction.

Practice Problem Set 14

1. 2,000 square feet per second

You are given the rate at which the circumference is increasing, $\frac{dC}{dt} = 40$, and are looking for the rate at which the area is increasing, $\frac{dA}{dt}$. Thus, you need to find a way to relate the area of a circle to its circumference. Recall that the circumference of a circle is $C = 2\pi r$ and that the area is $A = \pi r^2$. You could find C in terms of r and then plug it into the equation for A or work with the equations separately and then relate them. Do both and compare.

Method 1: First, find r in terms of C: $r = \frac{C}{2\pi}$. Now, plug this in for r in the equation for A: $A = \pi\left(\frac{C}{2\pi}\right)^2 = \frac{C^2}{4\pi}$. Next, take the derivative of the equation with respect to t: $\frac{dA}{dt} = \frac{1}{4\pi}(2C)\frac{dC}{dt} = \left(\frac{C}{2\pi}\right)\frac{dC}{dt}$. Next, plug in $C = 100\pi$ and $\frac{dC}{dt} = 40$ and solve: $\frac{dA}{dt} = \left(\frac{100\pi}{2\pi}\right)(40) = 2{,}000$. Therefore, the answer is 2,000 square feet per second.

Method 2: First, take the derivative of C with respect to t: $\frac{dC}{dt} = 2\pi\left(\frac{dr}{dt}\right)$. Next, plug in $\frac{dC}{dt} = 40$ and solve for $\frac{dr}{dt}$: $40 = 2\pi\left(\frac{dr}{dt}\right)$, so $\frac{dr}{dt} = \frac{20}{\pi}$. Next, take the derivative of A with respect to t: $\frac{dA}{dt} = 2\pi r\frac{dr}{dt}$. Now, plug in for $\frac{dr}{dt}$ and r and solve for $\frac{dA}{dt}$. Note that when the circumference is 100π, $r = 50$: $\frac{dA}{dt} = 2\pi(50)\left(\frac{20}{\pi}\right) = 2{,}000$. Therefore, the answer is 2,000 square feet per second.

Which method is better? In this case, they are about the same. Method 1 is going to be more efficient if it is easy to solve for one variable in terms of the other, and it is also easy to take the derivative of the resulting expression. Otherwise, you may prefer to use Method 2 (see Example 13 on page 231).

2. $\frac{3}{4}$ inches per second

You are given the rate at which the volume is increasing, $\frac{dV}{dt} = 27\pi$, and are looking for the rate at which the radius is increasing, $\frac{dr}{dt}$. Thus, you need to find a way to relate the volume of a sphere to its radius. Recall that the volume of a sphere is $V = \frac{4}{3}\pi r^3$. All you have to do is take the derivative of the equation with respect to t: $\frac{dV}{dt} = \frac{4}{3}\pi\left(3r^2\right)\left(\frac{dr}{dt}\right) = 4\pi r^2\frac{dr}{dt}$. Now substitute $\frac{dV}{dt} = 27\pi$ and $r = 3$: $27\pi = 4\pi(3)^2\frac{dr}{dt}$. Solving for $\frac{dr}{dt}$, you get $\frac{dr}{dt} = \frac{3}{4}$ inches per second.

3. 100 kilometers per hour

You are given the rate at which Car A is moving south, $\frac{dA}{dt} = 80$, and the rate at which Car B is moving west, $\frac{dB}{dt} = 60$, and are looking for the rate at which the distance between them is increasing, which you can call $\frac{dC}{dt}$. Note that the directions south and west are at right angles to each other. Thus, the distance that Car A is from the starting point, which you can call A, and the distance that Car B is from the starting point, which you can call B, are the legs of a right triangle, with C as the hypotenuse. You can relate the three distances using the Pythagorean Theorem. Here, because A and B are the legs and C is the hypotenuse, $A^2 + B^2 = C^2$. Now take the derivative of the equation with respect to t: $2A\frac{dA}{dt} + 2B\frac{dB}{dt} = 2C\frac{dC}{dt}$. This simplifies to $A\frac{dA}{dt} + B\frac{dB}{dt} = C\frac{dC}{dt}$. Car A has been driving for 3 hours at 80 kilometers per hour and Car B has been driving for 3 hours at 60 kilometers per hour, so $A = 240$ and $B = 180$. Using the Pythagorean Theorem, $240^2 + 180^2 = C^2$, so $C = 300$. Substitute into the derivative equation: $(240)(80) + (180)(60) = (300)\frac{dC}{dt}$. Solving for C, you get $C = 100$ kilometers per hours.

4. $243\sqrt{3}$ square inches per second

You are given the rate at which the sides of the triangle are increasing, $\frac{ds}{dt} = 27$, and are looking for the rate at which the area is increasing, $\frac{dA}{dt}$. Thus, you need to find a way to relate the area of an equilateral triangle to the length of a side. You know that the area of an equilateral triangle, in terms of its sides, is $A = \frac{s^2\sqrt{3}}{4}$. (If you don't know this formula, memorize it! It will come in very handy in future math problems.) Now take the derivative of this equation with respect to t: $\frac{dA}{dt} = \frac{\sqrt{3}}{4}(2s)\left(\frac{ds}{dt}\right)$. Next, plug $\frac{ds}{dt} = 27$ and s = 18 into the derivative to get $\frac{dA}{dt} = \frac{\sqrt{3}}{4}(36)(27) = 243\sqrt{3}$ square inches per second.

5. $-\frac{5}{7}$ inches per second

You are given the rate at which the water is flowing out of the container, $\frac{dV}{dt} = -35\pi$ (don't forget that this is negative), and are looking for the rate at which the depth of the water is dropping, $\frac{dh}{dt}$. Thus, you need to find a way to relate the volume of a cone to its height. You know that the volume of a cone is $V = \frac{1}{3}\pi r^2 h$. Notice that there is a problem. There is a third variable, r, in the equation. You cannot treat it as a constant the way you did in Problem 4 because as the volume of a cone changes, both its height and radius change. But, you also know that in any cone, the ratio of the radius to the height is a constant. Here, when the radius is 21 (because the diameter is 42), the height is 15. Thus, $\frac{r}{h} = \frac{21}{15} = \frac{7}{5}$. You can now isolate r in this equation: $r = \frac{7h}{5}$. Now plug it into the volume formula to get rid of r: $V = \frac{1}{3}\pi\left(\frac{7h}{5}\right)^2 h$. This simplifies to $V = \frac{49\pi}{75}h^3$. Now take the derivative of this equation with respect to t: $\frac{dV}{dt} = \frac{49\pi}{75}\left(3h^2\right)\frac{dh}{dt}$. Next, plug $\frac{dV}{dt} = -35\pi$ and h = 5 into the derivative to get $-35\pi = \frac{49\pi}{75}\left(3(5)^2\right)\frac{dh}{dt}$. Now solve for $\frac{dh}{dt}$: $\frac{dh}{dt} = -\frac{5}{7}$ inches per second.

6. $-\frac{25}{6}$ feet per second

You are given the rate at which the length of the rope, R, is changing, $\frac{dR}{dt} = -4$, and you are looking for the rate at which the boat, B, is approaching the dock, $\frac{dB}{dt}$. The key to this problem is to realize that the vertical distance from the dock to the bow, the distance from the boat to the dock, and the length of the rope form a right triangle.

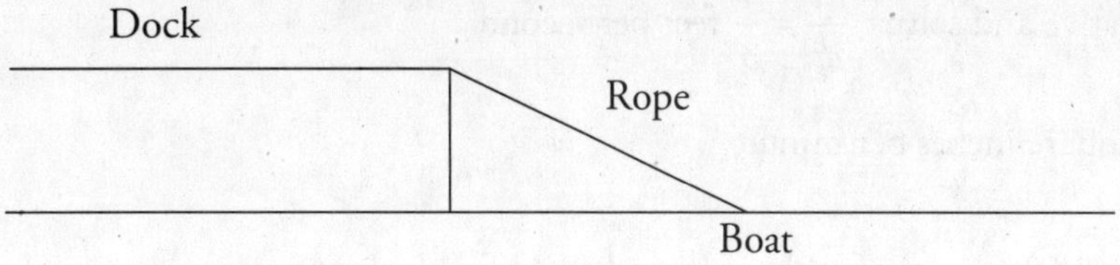

The vertical distance from the dock to the bow is always 7, so, using the Pythagorean Theorem, $7^2 + B^2 = R^2$. Now take the derivative of this equation with respect to t: $2B\frac{dB}{dt} = 2R\frac{dR}{dt}$, which simplifies to $B\frac{dB}{dt} = R\frac{dR}{dt}$. Knowing that $R = 25$, you can use the Pythagorean Theorem to find B: $7^2 + B^2 = 25^2$, so $B = 24$. Now plug $\frac{dR}{dt} = -4$, $R = 25$, and $B = 24$ into the derivative to get $24\frac{dB}{dt} = 25(-4)$. Now solve for $\frac{dB}{dt}$: $\frac{dB}{dt} = -\frac{25}{6}$ feet per second.

7. $\frac{4}{3}$ feet per second

You are given the rate at which the woman is walking away from the street lamp and are looking for the rate at which the length of her shadow is changing. It helps to draw a picture of the situation.

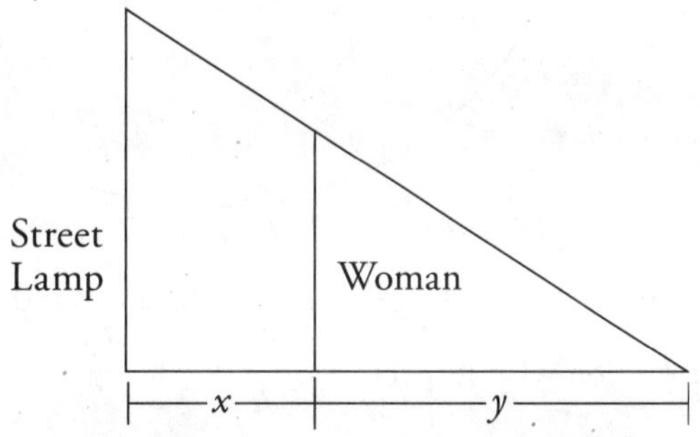

If you label the distance between the woman and the street lamp x and the length of the woman's shadow y, you can use similar triangles to get $\frac{y}{6} = \frac{x+y}{24}$. Cross-multiply and simplify.

$$24y = 6x + 6y$$

$$3y = x$$

Next, take the derivative of the equation with respect to t: $3\frac{dy}{dt} = \frac{dx}{dt}$. Now, plug $\frac{dx}{dt} = 4$ into the derivative and solve: $\frac{dy}{dt} = \frac{4}{3}$ feet per second.

8. $\frac{3\pi}{5}$ square inches per minute

It takes 60 minutes for the minute hand of a clock to make one complete revolution, so the rate at which the angle, θ, formed by the minute hand and noon is increasing, in terms of radians per minute, is $\frac{d\theta}{dt} = \frac{2\pi}{60} = \frac{\pi}{30}$. The area of the sector of a circle is proportional to its central angle, so $\frac{\theta}{2\pi} = \frac{S}{\pi r^2}$, which can be simplified to $S = \frac{1}{2}r^2\theta$.

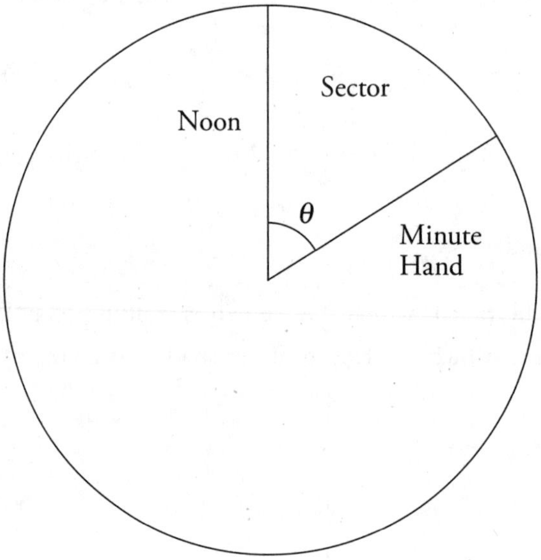

Note that the radius is 6 (the length of the minute hand) and is a constant. Next, take the derivative of the equation with respect to t: $\frac{dS}{dt} = \frac{1}{2}r^2\frac{d\theta}{dt}$.

Finally, substitute $\frac{d\theta}{dt} = \frac{\pi}{30}$ into the derivative: $\frac{dS}{dt} = \frac{1}{2}(6)^2\left(\frac{\pi}{30}\right) = \frac{3\pi}{5}$ square inches per second.

Practice Problem Set 15

1. $y - 2 = 5(x - 1)$

Remember that the equation of a line through a point (x_1, y_1) with slope m is $y - y_1 = m(x - x_1)$. Find the y-coordinate by plugging $x = 1$ into the equation $y = 3x^2 - x$, and the slope by plugging $x = 1$ into the derivative of the equation.

First, find the y-coordinate, y_1: $y = 3(1)^2 - 1 = 2$. This means that the line passes through the point (1, 2).

Next, take the derivative: $\frac{dy}{dx} = 6x - 1$. Now, find the slope, m: $\left.\frac{dy}{dx}\right|_{x=1} = 6(1) - 1 = 5$. Finally, plug in the point $(1, 2)$ and the slope $m = 5$ to get the equation of the tangent line: $y - 2 = 5(x - 1)$.

2. $y - 18 = 24(x - 3)$

Remember that the equation of a line through a point (x_1, y_1) with slope m is $y - y_1 = m(x - x_1)$. Find the y-coordinate by plugging $x = 3$ into the equation $y = x^3 - 3x$, and the slope by plugging $x = 3$ into the derivative of the equation.

First, find the y-coordinate, y_1: $y = (3)^3 - 3(3) = 18$. This means that the line passes through the point $(3, 18)$.

Next, take the derivative: $\frac{dy}{dx} = 3x^2 - 3$. Now, find the slope, m: $\left.\frac{dy}{dx}\right|_{x=3} = 3(3)^2 - 3 = 24$. Finally, plug in the point $(3, 18)$ and the slope $m = 24$ to get the equation of the tangent line: $y - 18 = 24(x - 3)$.

3. $y - \frac{1}{4} = -\frac{3}{64}(x - 3)$

Remember that the equation of a line through a point (x_1, y_1) with slope m is $y - y_1 = m(x - x_1)$. Find the y-coordinate by plugging $x = 3$ into the equation $y = \frac{1}{\sqrt{x^2 + 7}}$, and the slope by plugging $x = 3$ into the derivative of the equation.

First, find the y-coordinate, y_1: $y = \frac{1}{\sqrt{(3)^2 + 7}} = \frac{1}{4}$. This means that the line passes through the point $\left(3, \frac{1}{4}\right)$.

Next, take the derivative: $\frac{dy}{dx} = -\frac{1}{2}(x^2 + 7)^{-\frac{3}{2}}(2x) = -\frac{x}{\left(\sqrt{x^2 + 7}\right)^3}$. Now, find the slope, m: $\left.\frac{dy}{dx}\right|_{x=3} = -\frac{3}{\left(\sqrt{(3)^2 + 7}\right)^3} = -\frac{3}{64}$. Finally, plug in the point $\left(3, \frac{1}{4}\right)$ and the slope $m = -\frac{3}{64}$ to get the equation of the tangent line: $y - \frac{1}{4} = -\frac{3}{64}(x - 3)$.

4. $y-7=\frac{1}{6}(x-4)$

Remember that the equation of a line through a point (x_1, y_1) with slope m is $y-y_1=m(x-x_1)$. Find the y-coordinate by plugging $x = 4$ into the equation $y=\frac{x+3}{x-3}$, and you find the slope by plugging $x = 4$ into the derivative of the equation.

First, find the y-coordinate, y_1: $y=\frac{4+3}{4-3}=7$. This means that the line passes through the point $(4, 7)$. Next, take the derivative: $\frac{dy}{dx}=\frac{(x-3)(1)-(x+3)(1)}{(x-3)^2}=-\frac{6}{(x-3)^2}$. Now, find the slope, m: $\left.\frac{dy}{dx}\right|_{x=4}=-\frac{6}{(4-3)^2}=-6$. However, this is the slope of the *tangent* line. The *normal* line is perpendicular to the tangent line, so its slope will be the negative reciprocal of the tangent line's slope. In this case, the slope of the normal line is $\frac{-1}{-6}=\frac{1}{6}$. Finally, plug in the point (4, 7) and the slope $m=\frac{1}{6}$ to get the equation of the normal line: $y-7=\frac{1}{6}(x-4)$.

5. $y = 0$

Remember that the equation of a line through a point (x_1, y_1) with slope m is $y-y_1=m(x-x_1)$. Find the y-coordinate by plugging $x = 2$ into the equation $y=2x^3-3x^2-12x+20$, and you find the slope by plugging $x = 2$ into the derivative of the equation.

First, find the y-coordinate, y_1: $y=2(2)^3-3(2)^2-12(2)+20=0$. This means that the line passes through the point $(2, 0)$.

Next, take the derivative: $\frac{dy}{dx}=6x^2-6x-12$. Now, find the slope, m: $\left.\frac{dy}{dx}\right|_{x=2}=6(2)^2-6(2)-12=0$. Finally, plug in the point $(2, 0)$ and the slope $m=0$ to get the equation of the tangent line: $y-0=0(x-2)$ or $y=0$.

6. $y + 29 = -39(x - 5)$

Remember that the equation of a line through a point (x_1, y_1) with slope m is $y - y_1 = m(x - x_1)$. Find the y-coordinate by plugging $x = 5$ into the equation $y = \frac{x^2 + 4}{x - 6}$ and the slope by plugging $x = 5$ into the derivative of the equation.

First, find the y-coordinate, y_1: $y = \frac{(5)^2 + 4}{(5) - 6} = -29$. This means that the line passes through the point (5, –29).

Next, take the derivative: $\frac{dy}{dx} = \frac{(x-6)(2x) - (x^2 + 4)(1)}{(x-6)^2} = \frac{x^2 - 12x - 4}{(x-6)^2}$. Now, find the slope, m: $\left.\frac{dy}{dx}\right|_{x=5} = \frac{(5)^2 - 12(5) - 4}{(5-6)^2} = -39$. Finally, plug in the point (5, –29) and the slope $m = -39$ to get the equation of the tangent line: $y + 29 = -39(x - 5)$.

7. $y - 7 = \frac{24}{7}(x - 4)$

Remember that the equation of a line through a point (x_1, y_1) with slope m is $y - y_1 = m(x - x_1)$.

Find the slope by plugging $x = 4$ into the derivative of the equation $y = \sqrt{x^3 - 15}$.

First, take the derivative: $\frac{dy}{dx} = \frac{1}{2}(x^3 - 15)^{-\frac{1}{2}}(3x^2) = \frac{3x^2}{2\sqrt{x^3 - 15}}$. Now, find the slope, m: $\left.\frac{dy}{dx}\right|_{x=4} = \frac{3(4)^2}{2\sqrt{(4)^3 - 15}} = \frac{24}{7}$. Finally, plug in the point $(4, 7)$ and the slope $m = \frac{24}{7}$ to get the equation of the tangent line: $y - 7 = \frac{24}{7}(x - 4)$.

8. $x = \pm\sqrt{\frac{3}{2}}$

The slope of the line $x = y$ is 1, so you want to know where the slope of the tangent line is equal to 1. Find the slope of the tangent line by taking the derivative: $\frac{dy}{dx} = 6x^2 - 8$. Now set the derivative equal to 1: $6x^2 - 8 = 1$. If you solve for x, you get $x = \pm\sqrt{\frac{3}{2}}$.

9. $y - 7 = \frac{1}{2}(x - 3)$

Remember that the equation of a line through a point (x_1, y_1) with slope m is $y - y_1 = m(x - x_1)$. Find the y-coordinate by plugging $x = 3$ into the equation $y = \frac{3x+5}{x-1}$ and the slope by plugging $x = 3$ into the derivative of the equation.

First, find the y-coordinate, y_1: $y = \frac{3(3)+5}{(3)-1} = 7$. This means that the line passes through the point $(3, 7)$.

Next, take the derivative: $\frac{dy}{dx} = \frac{(x-1)(3)-(3x+5)(1)}{(x-1)^2} = \frac{-8}{(x-1)^2}$. Now, find the slope, m: $\left.\frac{dy}{dx}\right|_{x=3} = \frac{-8}{(3-1)^2} = -2$. However, this is the slope of the *tangent* line. The *normal* line is perpendicular to the tangent line, so its slope will be the negative reciprocal of the tangent line's slope. In this case, the slope of the normal line is $\frac{-1}{-2} = \frac{1}{2}$. Finally, plug in the point $(3, 7)$ and the slope $m = \frac{1}{2}$ to get the equation of the normal line: $y - 7 = \frac{1}{2}(x - 3)$.

10. $x = 9$

A line that is parallel to the y-axis has an infinite (or undefined) slope. In order to find where the normal line has an infinite slope, first take the derivative to find the slope of the tangent line: $\frac{dy}{dx} = 2(x-9)(1) = 2x - 18$. Next, because the normal line is perpendicular to the tangent line, the slope of the normal line is the negative reciprocal of the slope of the tangent line: $m = \frac{-1}{2x-18}$.

Now, find where the slope is infinite. This is simply where the denominator of the slope is zero: $x = 9$.

11. $\left(-\frac{3}{2}, \frac{41}{4}\right)$

A line that is parallel to the x-axis has a zero slope. In order to find where the tangent line has a zero slope, first take the derivative: $\frac{dy}{dx} = -3 - 2x$. Now find where the slope is zero. The derivative $-3 - 2x = 0$ at $x = -\frac{3}{2}$. Now find the y-coordinate, which you get by plugging $x = -\frac{3}{2}$ into the equation for y: $8 - 3\left(-\frac{3}{2}\right) - \left(-\frac{3}{2}\right)^2 = \frac{41}{4}$. Therefore, the answer is $\left(-\frac{3}{2}, \frac{41}{4}\right)$.

12. $a = 1$, $b = 0$, and $c = 1$

The two equations will have a common tangent line where they have the same slope, which you find by taking the derivative of each equation. The derivative of the first equation is $\frac{dy}{dx} = 2x + a$. The derivative of the second equation is $\frac{dy}{dx} = c + 2x$. Setting the two derivatives equal to each other, you get $a = c$. Each equation will pass through the point $(-1, 0)$. If you plug $(-1, 0)$ into the first equation, you get $0 = (-1)^2 + a(-1) + b$, which simplifies to $a - b = 1$. If you plug $(-1, 0)$ into the second equation, you get $0 = c(-1) + (-1)^2$, which simplifies to $c = 1$. Now find the values for a, b, and c. You get $a = 1$, $b = 0$, and $c = 1$.

Practice Problem Set 16

1. 5.002

Recall the differential formula used for approximating the value of a function: $f(x + \Delta x) \approx f(x) + f'(x)\Delta x$. Approximate the value of $\sqrt{25.02}$ by using $f(x) = \sqrt{x}$ with $x = 25$ and $\Delta x =$ 0.02. First, find $f'(x)$: $f'(x) = \frac{1}{2\sqrt{x}}$. Now, you plug into the formula: $f(x + \Delta x) \approx \sqrt{x} + \frac{1}{2\sqrt{x}}\Delta x$. If you plug in $x = 25$ and $\Delta x = 0.02$, you get $\sqrt{25 + 0.02} \approx \sqrt{25} + \frac{1}{2\sqrt{25}}(0.02)$. If you evaluate this, you get $\sqrt{25.02} \approx 5 + \frac{1}{10}(0.02) = 5.002$.

2. 3.999375

Recall the differential formula used for approximating the value of a function: $f(x + \Delta x) \approx f(x) + f'(x)\Delta x$. Approximate the value of $\sqrt{63.97}$ by using $f(x)=\sqrt[3]{x}$ with $x = 64$ and $\Delta x = -0.03$. First, find $f'(x)$: $f'(x)=\frac{1}{3}x^{-\frac{2}{3}}=\frac{1}{3\sqrt[3]{x^2}}$. Now, plug into the formula: $f(x+\Delta x)\approx\sqrt[3]{x}+\frac{1}{3\sqrt[3]{x^2}}\Delta x$. If you plug in $x = 64$ and $\Delta x = -0.03$, you get $\sqrt[3]{64-0.03}\approx\sqrt[3]{64}+\frac{1}{3\sqrt[3]{64^2}}(-0.03)$. If you evaluate this, you get $\sqrt[3]{63.97}\approx 4+\frac{1}{48}(-0.03)=3.999375$.

3. 1.802

Recall the differential formula used for approximating the value of a function: $f(x + \Delta x) \approx f(x) + f'(x)\Delta x$. Approximate the value of tan 61°. Be careful! Whenever working with trigonometric functions, it is *very* important to work with radians, *not* degrees! Remember that $60° = \frac{\pi}{3}$ radians and $1° = \frac{\pi}{180}$ radians, so use $f(x) = \tan x$ with $x=\frac{\pi}{3}$ and $\Delta x=\frac{\pi}{180}$. First, find $f'(x)$: $f'(x) = \sec^2 x$. Now, plug into the formula: $f(x+\Delta x)\approx\tan x+\sec^2 x\,\Delta x$. If you plug in $x=\frac{\pi}{3}$ and $\Delta x=\frac{\pi}{180}$, you get $\tan\left(\frac{\pi}{3}+\frac{\pi}{180}\right)\approx\tan\left(\frac{\pi}{3}\right)+\sec^2\left(\frac{\pi}{3}\right)\left(\frac{\pi}{180}\right)$. If you evaluate this, you get $\tan\left(\frac{61\pi}{180}\right)\approx\sqrt{3}+(2)^2\left(\frac{\pi}{180}\right)\approx 1.802$.

4. ±2.16 cubic inches

Recall the formula used to approximate the error in a measurement: $dy = f'(x)\,dx$. Approximate the error in the volume of a cube when you know that it has a side of length 6 inches with an error of ±0.02 inches, where $V(x) = x^3$ (the volume of a cube of side x) with $dx = \pm 0.02$. Find the derivative of the volume: $V'(x) = 3x^2$. Now plug into the formula: $dV = 3x^2 dx$. If you plug in $x = 6$ and $dx = \pm 0.02$, you get $dV = 3(6)^2(\pm 0.02) = \pm 2.16$.

5. $\pi \approx 3.142$ cubic millimeters

Recall the formula used when we want to approximate the error in a measurement: $dy = f'(x)\,dx$. Approximate the increase in the volume of a sphere when you know that it has a radius of length

5 millimeters with an increase of 0.01 millimeters, where $V(r)=\frac{4}{3}\pi r^3$ (the volume of a sphere of radius r) with $dr = 0.01$. Find the derivative of the volume: $V'(r)=\frac{4}{3}\pi(3r^2)=4\pi r^2$. Now plug into the formula: $dV = 4\pi r^2\, dr$. If you plug in $r = 5$ and $dr = 0.01$, you get $dV = 4\pi(5)^2(0.01) = \pi \approx 3.142$.

6. (a) 1.963 cubic meters; (b) 15.708 cubic meters

(a) Recall the formula used to approximate the error in a measurement: $dy = f'(x)\, dx$. Approximate the error in the volume of a cylinder when you know that it has a diameter of length 5 meters (which means that its radius is 2.5 meters) and its height is 20 meters, with an error in the height of 0.1 meters, where $V = \pi r^2 h$ with $dh = 0.1$. Note that the radius is exact, so when you take the derivative, treat only the height as a variable. Find the derivative of the volume: $V' = \pi r^2$. Now plug into the formula: $dV = \pi r^2\, dh$. If you plug in $r = 2.5$, $h = 20$, and $dh = 0.1$, you get $dV = \pi(2.5)^2(0.1) = 0.625\pi \approx 1.963$.

(b) Approximate the error in the volume of a cylinder when you know that it has a diameter of length 5 meters (which means that its radius is 2.5 meters) and its height is 20 meters, with an error in the diameter of 0.1 meters (which means that the error in the radius is 0.05 meters), where $V = \pi r^2 h$ with $dr = 0.05$. Note that the height is exact, so when you take the derivative, treat only the radius as a variable. Find the derivative of the volume: $V' = 2\pi rh$. Now plug into the formula: $dV = 2\pi rh\, dr$. If you plug in $r = 2.5$, $h = 20$, and $dr = 0.05$, you get $dV = 2\pi(2.5)(20)(0.05) = 5\pi \approx 15.708$.

Practice Problem Set 17

1. $\frac{3}{4}$

Recall L'Hospital's Rule: If $f(c)=g(c)=0$, or if $f(c)=g(c)=\infty$, and if $f'(c)$ and $g'(c)$ exist, and if $g'(c)\neq 0$, then $\lim_{x\to c}\frac{f(x)}{g(x)}=\lim_{x\to c}\frac{f'(x)}{g'(x)}$. Here, $f(x)=\sin 3x$ and $g(x)=\sin 4x$, and $\sin 0 = 0$. This means that you can use L'Hospital's Rule to find the limit. Take the derivative of the numerator and the denominator: $\lim_{x\to 0}\frac{\sin 3x}{\sin 4x}=\lim_{x\to 0}\frac{3\cos 3x}{4\cos 4x}$. If you take the new limit, you get $\lim_{x\to 0}\frac{3\cos 3x}{4\cos 4x}=\frac{3\cos 0}{4\cos 0}=\frac{(3)(1)}{(4)(1)}=\frac{3}{4}$.

2. -1

Recall L'Hospital's Rule: If $f(c) = g(c) = 0$, or if $f(c) = g(c) = \infty$, and if $f'(c)$ and $g'(c)$ exist, and if $g'(c) \neq 0$, then $\lim_{x \to c} \frac{f(x)}{g(x)} = \lim_{x \to c} \frac{f'(x)}{g'(x)}$. Here, $f(x) = x - \pi$ and $g(x) = \sin x$. Notice that $x - \pi = 0$ when $x = \pi$, and that $\sin \pi = 0$. This means that you can use L'Hospital's Rule to find the limit. Take the derivative of the numerator and the denominator: $\lim_{x \to \pi} \frac{x - \pi}{\sin x} = \lim_{x \to \pi} \frac{1}{\cos x}$. Taking the new limit, you get $\lim_{x \to \pi} \frac{1}{\cos x} = \frac{1}{\cos \pi} = \frac{1}{-1} = -1$.

3. $\frac{1}{6}$

Recall L'Hospital's Rule: If $f(c) = g(c) = 0$, or if $f(c) = g(c) = \infty$, and if $f'(c)$ and $g'(c)$ exist, and if $g'(c) \neq 0$, then $\lim_{x \to c} \frac{f(x)}{g(x)} = \lim_{x \to c} \frac{f'(x)}{g'(x)}$. Here, $f(x) = x - \sin x$ and $g(x) = x^3$. Notice that $x - \sin x = 0$ when $x = 0$, and that $0^3 = 0$. This means that you can use L'Hospital's Rule to find the limit. Take the derivative of the numerator and the denominator: $\lim_{x \to 0} \frac{x - \sin x}{x^3} = \lim_{x \to 0} \frac{1 - \cos x}{3x^2}$. If you take the new limit, you get $\lim_{x \to 0} \frac{1 - \cos x}{3x^2} = \frac{1 - \cos 0}{3(0)^2} = \frac{1 - 1}{0} = \frac{0}{0}$. But this is still indeterminate, so what now? Use L'Hospital's Rule again! Take the derivative of the numerator and the denominator: $\lim_{x \to 0} \frac{1 - \cos x}{3x^2} = \lim_{x \to 0} \frac{\sin x}{6x}$. If you take the limit, you get $\lim_{x \to 0} \frac{\sin x}{6x} = \frac{\sin 0}{6(0)} = \frac{0}{0}$. Use L'Hospital's Rule one more time to get $\lim_{x \to 0} \frac{\sin x}{6x} = \lim_{x \to 0} \frac{\cos x}{6}$. Now, if you take the limit, you get $\lim_{x \to 0} \frac{\cos x}{6} = \frac{\cos 0}{6} = \frac{1}{6}$. Notice that if L'Hospital's Rule results in an indeterminate form, you can use the rule again and again (but not infinitely often).

4. -2

Recall L'Hospital's Rule: If $f(c) = g(c) = 0$, or if $f(c) = g(c) = \infty$, and if $f'(c)$ and $g'(c)$ exist, and if $g'(c) \neq 0$, then $\lim_{x \to c} \frac{f(x)}{g(x)} = \lim_{x \to c} \frac{f'(x)}{g'(x)}$. Here, $f(x) = e^{3x} - e^{5x}$ and $g(x) = x$.

Notice that $e^{3x} - e^{5x} = 0$ when $x = 0$, and that the denominator is obviously zero at $x = 0$. This means that you can use L'Hospital's Rule to find the limit. Take the derivative of the numerator and the denominator: $\lim_{x\to 0}\frac{e^{3x}-e^{5x}}{x} = \lim_{x\to 0}\frac{3e^{3x}-5e^{5x}}{1}$. If you take the new limit, you get $\lim_{x\to 0}\frac{3e^{3x}-5e^{5x}}{1} = \frac{3e^{0}-5e^{0}}{1} = \frac{3(1)-5(1)}{1} = -2$.

5. -2

Recall L'Hospital's Rule: If $f(c) = g(c) = 0$, or if $f(c) = g(c) = \infty$, and if $f'(c)$ and $g'(c)$ exist, and if $g'(c) \neq 0$, then $\lim_{x\to c}\frac{f(x)}{g(x)} = \lim_{x\to c}\frac{f'(x)}{g'(x)}$. Here, $f(x) = \tan x - x$ and $g(x) = \sin x - x$.

Notice that $\tan x - x = 0$ and $\sin x - x = 0$ when $x = 0$. This means that you can use L'Hospital's Rule to find the limit. Take the derivative of the numerator and the denominator: $\lim_{x\to 0}\frac{\tan x - x}{\sin x - x} = \lim_{x\to 0}\frac{\sec^2 x - 1}{\cos x - 1}$. If you take the new limit, you get $\lim_{x\to 0}\frac{\sec^2 x - 1}{\cos x - 1} = \frac{\sec^2(0)-1}{\cos(0)-1} = \frac{1-1}{1-1} = \frac{0}{0}$. But this is still indeterminate, so what now?

Use L'Hospital's Rule again! Take the derivative of the numerator and the denominator: $\lim_{x\to 0}\frac{\sec^2 x - 1}{\cos x - 1} = \lim_{x\to 0}\frac{2\sec x(\sec x\tan x)}{-\sin x} = \lim_{x\to 0}\frac{2\sec^2 x\tan x}{-\sin x}$. If you take the limit, you get $\lim_{x\to 0}\frac{2\sec^2(0)\tan(0)}{-\sin(0)} = \frac{0}{0}$. At this point, you might be getting nervous as the derivatives start to get messier. Try using trigonometric identities to simplify the limit. If you rewrite the numerator in terms of sin x and cos x, you get $\frac{2\sec^2 x\tan x}{-\sin x} = \frac{2\left(\frac{1}{\cos^2 x}\right)\left(\frac{\sin x}{\cos x}\right)}{-\sin x}$. Simplify this to

$\dfrac{2\left(\dfrac{1}{\cos^2 x}\right)\left(\dfrac{\sin x}{\cos x}\right)}{-\sin x} = -2\dfrac{\dfrac{\sin x}{\cos^3 x}}{\sin x}$. This simplifies to $-2\dfrac{\sin x}{\cos^3 x}\dfrac{1}{\sin x} = \dfrac{-2}{\cos^3 x}$. Now, if you take the limit, you get $\lim\limits_{x \to 0}\dfrac{-2}{\cos^3 x} = \dfrac{-2}{\cos^3(0)} = \dfrac{-2}{1^3} = -2$.

Note that you could have used trigonometric identities on either of the first two limits as well. Remember that when you have a limit that is an indeterminate form, you can sometimes use algebra or trigonometric identities (or both) to simplify the limit. Sometimes this will get rid of the problem.

6. 0

Recall L'Hospital's Rule: If $f(c) = g(c) = 0$, or if $f(c) = g(c) = \infty$, and if $f'(c)$ and $g'(c)$ exist, and if $g'(c) \neq 0$, then $\lim\limits_{x \to c}\dfrac{f(x)}{g(x)} = \lim\limits_{x \to c}\dfrac{f'(x)}{g'(x)}$. Here, $f(x) = x^5$ and $g(x) = e^{5x}$.

Notice that $x^5 = \infty$ when $x = \infty$, and that $e^{\infty} = \infty$. This means that you can use L'Hospital's Rule to find the limit. Take the derivative of the numerator and the denominator: $\lim\limits_{x \to \infty}\dfrac{x^5}{e^{5x}} = \lim\limits_{x \to \infty}\dfrac{5x^4}{5e^{5x}} = \lim\limits_{x \to \infty}\dfrac{x^4}{e^{5x}}$. If you take the new limit, you get $\lim\limits_{x \to 0}\dfrac{x^4}{e^{5x}} = \dfrac{\infty}{\infty}$. But this is still indeterminate, so what now? Use L'Hospital's Rule again! Take the derivative of the numerator and the denominator: $\lim\limits_{x \to \infty}\dfrac{x^4}{e^{5x}} = \lim\limits_{x \to \infty}\dfrac{4x^3}{5e^{5x}}$. You need to use L'Hospital's Rule until the x-term in the numerator is gone. Take the derivative: $\lim\limits_{x \to \infty}\dfrac{4x^3}{5e^{5x}} = \lim\limits_{x \to \infty}\dfrac{12x^2}{25e^{5x}}$. And again: $\lim\limits_{x \to \infty}\dfrac{12x^2}{25e^{5x}} = \lim\limits_{x \to \infty}\dfrac{24x}{125e^{5x}}$. And one more time: $\lim\limits_{x \to \infty}\dfrac{24x}{125e^{5x}} = \lim\limits_{x \to \infty}\dfrac{24}{625e^{5x}}$. Now, if you take the limit, you get $\lim\limits_{x \to \infty}\dfrac{24}{625e^{5x}} = 0$.

Notice that if L'Hospital's Rule results in an indeterminate form, you can use the rule again and again (but not infinitely often).

7. $\frac{1}{7}$

Recall L'Hospital's Rule: If $f(c) = g(c) = 0$, or if $f(c) = g(c) = \infty$, and if $f'(c)$ and $g'(c)$ exist, and if $g'(c) \neq 0$, then $\lim_{x \to c} \frac{f(x)}{g(x)} = \lim_{x \to c} \frac{f'(x)}{g'(x)}$. Here, $f(x) = x^5 + 4x^3 - 8$ and $g(x) = 7x^5 - 3x^2 - 1$. Notice that $x^5 + 4x^3 - 8 = \infty$ and $7x^5 - 3x^2 - 1 = \infty$ when $x = \infty$. This means that you can use L'Hospital's Rule to find the limit. Take the derivative of the numerator and the denominator: $\lim_{x \to \infty} \frac{x^5 + 4x^3 - 8}{7x^5 - 3x^2 - 1} = \lim_{x \to \infty} \frac{5x^4 + 12x^2}{35x^4 - 6x}$. If you take the new limit, you get $\lim_{x \to \infty} \frac{5x^4 + 12x^2}{35x^4 - 6x} = \frac{\infty}{\infty}$. But this is still indeterminate, so what now? Use L'Hospital's Rule again! Take the derivative of the numerator and the denominator: $\lim_{x \to \infty} \frac{5x^4 + 12x^2}{35x^4 - 6x} = \lim_{x \to \infty} \frac{20x^3 + 24x}{140x^3 - 6}$. You need to use L'Hospital's Rule again. In fact, each time you use the rule, you reduce the power of the *x*-terms in the numerator and denominator, and you need to keep doing so until the *x*-terms are gone. Take the derivative: $\lim_{x \to \infty} \frac{20x^3 + 24x}{140x^3 - 6} = \lim_{x \to \infty} \frac{60x^2 + 24}{420x^2}$. And again, $\lim_{x \to \infty} \frac{60x^2 + 24}{420x^2} = \lim_{x \to \infty} \frac{120x}{840x}$. Now take the limit: $\lim_{x \to \infty} \frac{120x}{840x} = \lim_{x \to \infty} \frac{1}{7} = \frac{1}{7}$. Notice that if L'Hospital's Rule results in an indeterminate form, you can use the rule again and again (but not infinitely often).

8. 1

Recall L'Hospital's Rule: If $f(c) = g(c) = 0$, or if $f(c) = g(c) = \infty$, and if $f'(c)$ and $g'(c)$ exist, and if $g'(c) \neq 0$, then $\lim_{x \to c} \frac{f(x)}{g(x)} = \lim_{x \to c} \frac{f'(x)}{g'(x)}$. Here, $f(x) = \ln(\sin x)$ and $g(x) = \ln(\tan x)$, and both of these approach infinity as x approaches 0 from the right. This means that you can use L'Hospital's Rule to find the limit. Take the derivative of the numerator and the denominator: $\lim_{x\to 0^+}\frac{\ln(\sin x)}{\ln(\tan x)} = \lim_{x\to 0^+}\frac{\frac{\cos x}{\sin x}}{\frac{\sec^2 x}{\tan x}}$. Use trigonometric identities to simplify the limit: $\frac{\frac{\cos x}{\sin x}}{\frac{\sec^2 x}{\tan x}} = \frac{\cos x}{\sin x}\frac{\tan x}{\sec^2 x} = \frac{\cos x}{\sin x}\frac{\frac{\sin x}{\cos x}}{\sec^2 x} = \frac{1}{\sec^2 x}$. Now, if you take the new limit, you get $\lim_{x\to 0^+}\frac{1}{\sec^2 x} = 1$.

9. $\frac{1}{2}$

Recall L'Hospital's Rule: If $f(c) = g(c) = 0$, or if $f(c) = g(c) = \infty$, and if $f'(c)$ and $g'(c)$ exist, and if $g'(c) \neq 0$, then $\lim_{x\to c}\frac{f(x)}{g(x)} = \lim_{x\to c}\frac{f'(x)}{g'(x)}$. Here, $f(x) = \cot 2x$ and $g(x) = \cot x$, and both of these approach infinity as x approaches 0 from the right. This means that you can use L'Hospital's Rule to find the limit. Take the derivative of the numerator and the denominator: $\lim_{x\to 0^+}\frac{\cot 2x}{\cot x} = \lim_{x\to 0^+}\frac{-2\cot 2x\csc 2x}{-\csc^2 x}$. This seems to be worse than what you started with. Instead, try using trigonometric identities to simplify the limit: $\frac{\cot 2x}{\cot x} = \frac{\frac{\cos 2x}{\sin 2x}}{\frac{\cos x}{\sin x}} = \frac{\cos 2x}{\sin 2x}\frac{\sin x}{\cos x} = \frac{\cos^2 x - \sin^2 x}{2\sin x\cos x}\frac{\sin x}{\cos x}$. Notice that the problem is with sin x as x approaches zero (because it becomes zero), not with cos x. As long as you are multiplying the numerator and denominator by sin x, you are going to get an indeterminate form. So, thanks to trigonometric identities, you can eliminate the problem term: $\frac{\cos^2 x - \sin^2 x}{2\sin x\cos x}\frac{\sin x}{\cos x} = \frac{\cos^2 x - \sin^2 x}{2\cos^2 x}$. If you take the limit of this expression, it is not indeterminate. You get $\lim\limits_{x\to 0^+}\dfrac{\cos^2 x-\sin^2 x}{2\cos^2 x}=\dfrac{\cos^2 0-\sin^2 0}{2\cos^2 0}=\dfrac{1^2-0}{2\left(1^2\right)}=\dfrac{1}{2}$. Notice that you didn't need to use L'Hospital's Rule here. You should bear in mind that just because a limit is indeterminate, it does not mean that the best way to evaluate it is with L'Hospital's Rule.

10. 1

Recall L'Hospital's Rule: If $f(c)=g(c)=0$, or if $f(c)=g(c)=\infty$, and if $f'(c)$ and $g'(c)$ exist, and if $g'(c)\neq 0$, then $\lim\limits_{x\to c}\dfrac{f(x)}{g(x)}=\lim\limits_{x\to c}\dfrac{f'(x)}{g'(x)}$. Here, $f(x)=x$ and $g(x)=\ln(x+1)$, and both of these approach zero as x approaches 0 from the right. This means that you can use L'Hospital's Rule to find the limit. Take the derivative of the numerator and the denominator: $\lim\limits_{x\to 0^+}\dfrac{1}{\frac{1}{x+1}}=\lim\limits_{x\to 0^+}(x+1)$. Now, if you take the new limit, you get $\lim\limits_{x\to 0^+}(x+1)=1$.

ANSWERS TO CHAPTER 6 DRILL

1. **C** First, draw a picture:

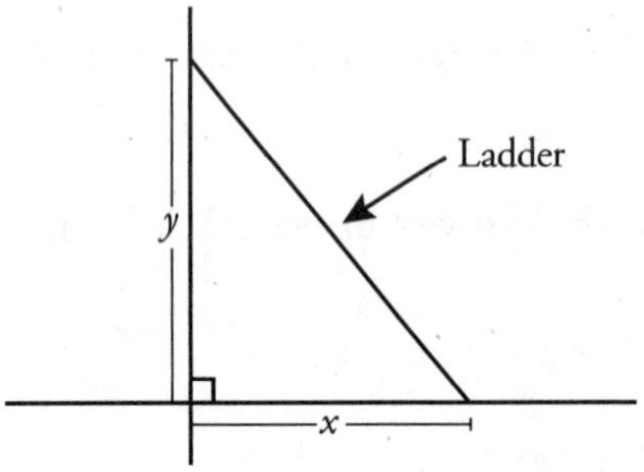

Call the distance from the top of the ladder to the ground y and the distance from the wall to the bottom of the ladder x. The relationship between x and y is $x^2 + y^2 = 25^2$ (Pythagorean Theorem). You are given $\frac{dy}{dt} = -6$ and are looking for $\frac{dx}{dt}$ when $y = 20$.

Take the derivative with respect to t: $2x\frac{dx}{dt} + 2y\frac{dy}{dt} = 0$, which simplifies to $x\frac{dx}{dt} + y\frac{dy}{dt} = 0$. You have $y = 20$ and $\frac{dy}{dt}$, so in order to find $\frac{dx}{dt}$, you will need to find x. Plugging into the original equation, you get $x^2 + (20)^2 = 25^2$, so $x = 15$. Now plug this into the derivative and solve for $\frac{dx}{dt}$: $15\frac{dx}{dt} + (20)(-6) = 0$.

Therefore, $\frac{dx}{dt} = \frac{120}{15} = 8$ feet per second.

The answer is (C).

2. **A** First, draw a picture:

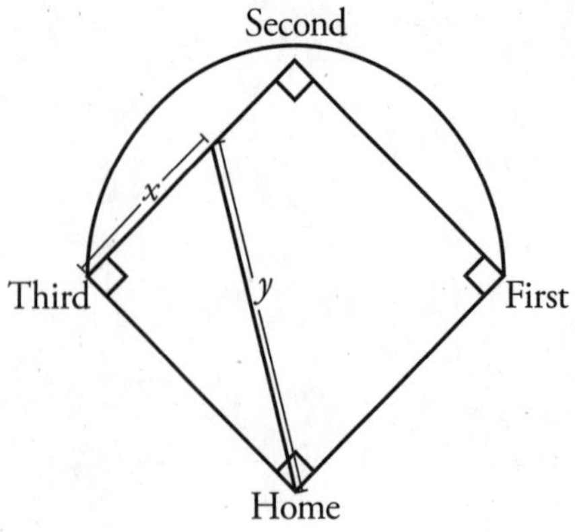

Call the distance from the runner to third base x and the distance from the runner to home plate y. Then $\frac{dx}{dt} = -30$ (it's negative because the distance is shrinking), and you are looking for $\frac{dy}{dt}$ when $x = 45$ (halfway from second base to third base). The relationship between x and y is $x^2 + 90^2 = y^2$ (Pythagorean Theorem). Take the derivative with respect to t: $2x\frac{dx}{dt} = 2y\frac{dy}{dt}$, which simplifies to $x\frac{dx}{dt} = y\frac{dy}{dt}$. Find y next. You have $45^2 + 90^2 = y^2$, so $y = 45\sqrt{5}$. Plugging everything in, you get $(45)(-30) = \left(45\sqrt{5}\right)\frac{dy}{dt}$. Thus $\frac{dy}{dt} = -\frac{30}{\sqrt{5}}$ feet per second.

The answer is (A).

3. **B** You are given $\frac{dV}{dt} = -24\pi$ and are looking for $\frac{dS}{dt}$ when $r = 6$. The volume of a sphere is $V = \frac{4}{3}\pi r^3$ and the surface area is $S = 4\pi r^2$. You could solve the volume for r and plug it into the surface area equation to get S in terms of V, but this is messy. It's easier to work with the equations separately. First, take the derivative of V with respect to t: $\frac{dV}{dt} = \frac{4}{3}\pi\left(3r^2\frac{dr}{dt}\right) = 4\pi r^2\frac{dr}{dt}$. Next, plug in to solve for $\frac{dr}{dt}$:

$$-24\pi = 4\pi(6)^2\frac{dr}{dt}$$

$$\frac{dr}{dt} = -\frac{1}{6}$$

Next, take the derivative of S with respect to t: $\frac{dS}{dt} = 8\pi r\frac{dr}{dt}$. Now plug in $r = 6$ and $\frac{dr}{dt} = -\frac{1}{6}$: $\frac{dS}{dt} = 8\pi(6)\frac{-1}{6} = -8\pi$ square inches per second.

The answer is (B).

4. **C** The velocity function is simply the first derivative: $v(t) = 4t^3 - 4t$.
Set this equal to zero and solve for t:

$$4t^3 - 4t = 0$$

$$4t(t^2 - 1) = 0$$

$$t = 0 \text{ and } \pm 1$$

Because $t > 0$, the only valid answer is $t = 1$.
The answer is (C).

5. **D** The particle is changing direction at the time that its velocity is zero and the velocity changes sign there. The first derivative of the position function gives the velocity function: $x'(t) = v(t) = 12t^2 - 24t$.

Set this equal to zero and solve for t: $12t^2 - 24t = 0$, and $t(12t - 24) = 0$, so $t = 0$ or $t = 2$.

Throw out $t = 0$, so the particle is changing direction at $t = 2$. Verify this by making sure that the velocity changes sign at $t = 2$. At $t = 1$, $v(1)=12(1)^2-24(1)<0$, and at $t = 3$, $v(3)=12(3)^2-24(3)>0$, so the position is indeed changing direction at $t = 2$.

The answer is (D).

6. **C** First, find the y-coordinate: $y(2)=5(2)^3-20(2)+10=10$.

Second, take the derivative: $\frac{dy}{dx}=15x^2-20$. Plug in $x = 2$ to get the slope of the tangent line: $\frac{dy}{dx}=15(2)^2-20=40$.

Using the slope and a point, the equation is $y-10=40(x-2)$.

The answer is (C).

7. **B** Because $\lim\limits_{x\to\infty} x^{-\frac{4}{3}}=0$ and $\lim\limits_{x\to\infty}\sin\left(\frac{1}{x}\right)=0$, the limit is an indeterminate form $\frac{0}{0}$. Try to find the limit using L'Hospital's Rule. Take the derivative of the numerator and the denominator:

$$\lim_{x\to\infty}\frac{x^{-\frac{4}{3}}}{\sin\left(\frac{1}{x}\right)}=\lim_{x\to\infty}\frac{-\frac{4}{3}x^{-\frac{7}{3}}}{\left(-\frac{1}{x^2}\right)\cos\left(\frac{1}{x}\right)}.$$

This simplifies to $\lim\limits_{x\to\infty}\frac{-\frac{4}{3}x^{-\frac{7}{3}}}{\left(-\frac{1}{x^2}\right)\cos\left(\frac{1}{x}\right)}=\lim\limits_{x\to\infty}\frac{\frac{4}{3}x^{-\frac{1}{3}}}{\cos\left(\frac{1}{x}\right)}$. Now when taking the limit, you get $\lim\limits_{x\to\infty}\frac{\frac{4}{3}x^{-\frac{1}{3}}}{\cos\left(\frac{1}{x}\right)}=\frac{0}{1}=0$.

The answer is (B).

8. **D** First, find the y-coordinate: $y\left(\frac{\pi}{8}\right)=4\sec\left(\frac{\pi}{4}\right)=4\sqrt{2}$.

Second, take the derivative: $\frac{dy}{dx}=8\sec(2x)\tan(2x)$. Plug in $x=\frac{\pi}{8}$ to get the slope of the tangent line: $\frac{dy}{dx}=8\sec\left(\frac{\pi}{4}\right)\tan\left(\frac{\pi}{4}\right)=8\sqrt{2}$. Thus, the slope of the normal line is the negative reciprocal of the slope of the tangent line: $-\frac{1}{8\sqrt{2}}$.

Using the slope and a point, the equation is $y-4\sqrt{2}=-\frac{1}{8\sqrt{2}}\left(x-\frac{\pi}{8}\right)$.

The answer is (D).

Chapter 7
Analytical Applications of Differentiation

USING THE MEAN VALUE THEOREM

In the beginning of Chapter 4, we discussed the difference between average rate of change and instantaneous rate of change. While this difference is important, they are actually closely linked though the Mean Value Theorem.

> In order for the Mean Value Theorem for Derivatives to work, the curve must be continuous on the interval <u>and</u> at the endpoints.

If $y = f(x)$ is continuous on the interval $[a, b]$, and is differentiable everywhere on the interval (a, b), then there is at least one number c between a and b such that

$$f'(c) = \frac{f(b) - f(a)}{b - a}$$

In other words, there's some point in the interval where the slope of the tangent line equals the slope of the secant line that connects the endpoints of the interval. (The function has to be continuous at the endpoints of the interval, but it doesn't have to be differentiable at the endpoints.) You can see this graphically in the following figure:

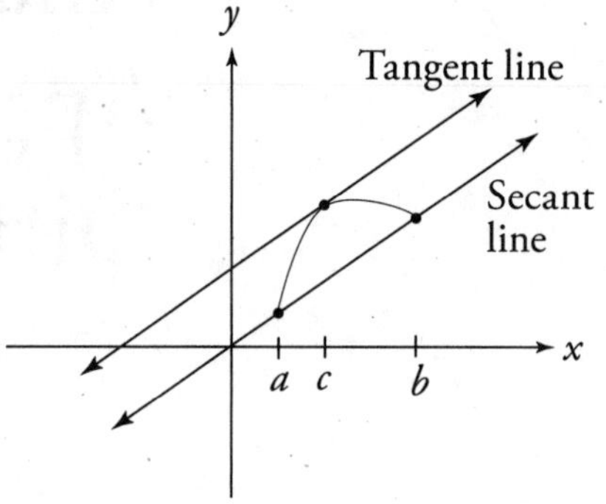

Example 1: Suppose you have the function $f(x) = x^2$, and you're looking at the interval [1, 3]. The Mean Value Theorem for derivatives (this is often abbreviated MVTD) states that there is some number c such that

$$f'(c) = \frac{3^2 - 1^2}{3 - 1} = 4$$

Because $f'(x) = 2x$, plug in c for x and solve: $2c = 4$ so $c = 2$. Notice that 2 is in the interval. This is what the MVTD predicted! If you don't get a value for c within the interval, something went wrong; either the function is not continuous and differentiable in the required interval, or you made a mistake.

Example 2: Consider the function $f(x) = x^3 - 12x$ on the interval [–2, 2]. The MVTD states that there is a c such that

$$f'(c) = \frac{\left(2^3 - 24\right) - \left((-2)^3 + 24\right)}{2 - (-2)} = -8$$

Then $f'(c) = 3c^2 - 12 = -8$ and $c = \pm\frac{2}{\sqrt{3}}$ (which is approximately ±1.155).

Notice that there are two values of c that satisfy the MVTD. That's allowed. In fact, there can be infinitely many values, depending on the function.

Example 3: Consider the function $f(x) = \frac{1}{x}$ on the interval [–2, 2].

Follow the MVTD.

$$f'(c) = \frac{\frac{1}{2} - \left(-\frac{1}{2}\right)}{2 - (-2)} = \frac{1}{4}$$

Then

$$f'(c) = \frac{-1}{c^2} = \frac{1}{4}$$

There is no value of c that will satisfy this equation! We expected this. Why? Because $f(x)$ is not continuous at $x = 0$, which is in the interval. Suppose the interval had been [1, 3], eliminating the discontinuity. The result would have been

$$f'(c) = \frac{\frac{1}{3} - (1)}{3 - 1} = -\frac{1}{3} \text{ and } f'(c) = \frac{-1}{c^2} = -\frac{1}{3};\ c = \pm\sqrt{3}$$

$c = -\sqrt{3}$ is not in the interval, but $c = \sqrt{3}$ is. The answer is $c = \sqrt{3}$.

Example 4: Consider the function $f(x) = x^2 - x - 12$ on the interval [–3, 4].

Follow the MVTD.

$$f'(c) = \frac{0 - 0}{7} = 0 \text{ and } f'(c) = 2c - 1 = 0\text{, so } c = \frac{1}{2}$$

In this last example, you discovered where the derivative of the equation equaled zero. This is going to be the single most common problem you'll work with in differential calculus. So now, we've got an important tip for you.

When you don't know what to do, take the derivative of the equation and set it equal to zero!

Remember this advice for the rest of AP Calculus.

Rolle's Theorem

Now let's learn Rolle's Theorem, which is a special case of the MVTD.

If $y = f(x)$ is continuous on the interval $[a, b]$, and is differentiable everywhere on the interval (a, b), and if $f(a) = f(b)$, then there is at least one number c between a and b such that $f'(c) = 0$.

Graphically, this means that a continuous, differentiable curve has a horizontal tangent between any two points where it crosses the x-axis.

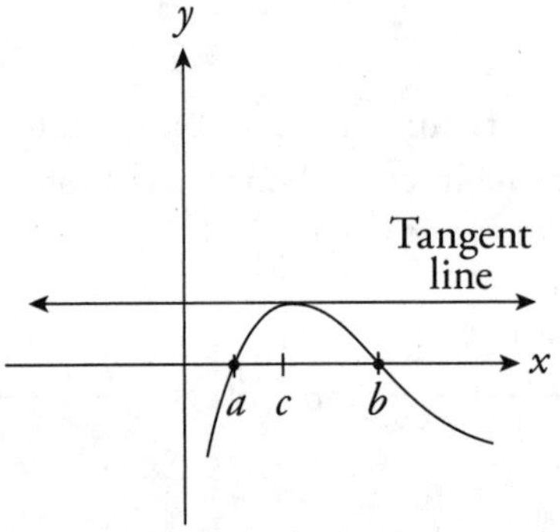

Example 4 was an example of Rolle's Theorem, but let's do another.

Example 5: Consider the function $f(x) = \frac{x^2}{2} - 6x$ on the interval [0, 12].

First, show that

$$f(0) = \frac{0}{2} - 6(0) = 0 \text{ and } f(12) = \frac{144}{2} - 6(12) = 0$$

Then find

$$f'(x) = x - 6, \text{ so } f'(c) = c - 6$$

If you set this equal to zero (remember what we told you!), you get $c = 6$. This value of c falls in the interval, so the theorem holds for this example.

As with the MVTD, you'll run into problems with the theorem when the function is not continuous and differentiable over the interval. This is where you need to look out for a trap set by the College Board. Otherwise, just follow what we did here and you won't have any trouble with either Rolle's Theorem or the MVTD. Try these example problems, and cover the responses until you check your work.

Example 6: Find the values of c that satisfy the MVTD for $f(x) = x^2 + 2x - 1$ on the interval [0, 1].

First, find $f(0)$ and $f(1)$.

$$f(0) = 0^2 + 2(0) - 1 = -1 \text{ and } f(1) = 1^2 + 2(1) - 1 = 2$$

Then,

$$\frac{2-(-1)}{1-0} = \frac{3}{1} = 3 = f'(c)$$

Next, find $f'(x)$.

$$f'(x) = 2x + 2$$

Thus, $f'(c) = 2c + 2 = 3$, and $c = \frac{1}{2}$.

Example 7: Find the values of c that satisfy the MVTD for $f(x) = x^3 + 1$ on the interval [1, 2].

Find $f(1) = 1^3 + 1 = 2$ and $f(2) = 2^3 + 1 = 9$. Then,

$$\frac{9-2}{2-1} = 7 = f'(c)$$

Next, $f'(x) = 3x^2$, so $f'(c) = 3c^2 = 7$ and $c = \pm\sqrt{\frac{7}{3}}$.

Notice that there are two answers for c, but only one of them is in the interval. The answer is $c = \sqrt{\frac{7}{3}}$.

Example 8: Find the values of c that satisfy the MVTD for $f(x) = x + \frac{1}{x}$ on the interval [–4, 4].

First, because the function is not continuous on the interval, there may not be a solution for c. Let's show that this is true. Find $f(-4) = -4 - \frac{1}{4} = -\frac{17}{4}$ and $f(4) = 4 + \frac{1}{4} = \frac{17}{4}$. Then,

$$\frac{\frac{17}{4} - \left(-\frac{17}{4}\right)}{4-(-4)} = \frac{17}{16} = f'(c)$$

Next, $f'(x) = 1 - \frac{1}{x^2}$. Therefore, $f'(c) = 1 - \frac{1}{c^2} = \frac{17}{16}$.

There's no solution to this equation.

Example 9: Find the values of c that satisfy Rolle's Theorem for $f(x) = x^4 - x$ on the interval [0, 1].

Show that $f(0) = 0^4 - 0 = 0$ and that $f(1) = 1^4 - 1 = 0$.

Next, find $f'(x) = 4x^3 - 1$. By setting $f'(c) = 4c^3 - 1 = 0$ and solving, you'll see that $c = \sqrt[3]{\frac{1}{4}}$, which is in the interval.

PRACTICE PROBLEM SET 18

Now try these problems. The answers are at the end of this chapter, starting on page 328.

1. Find the values of c that satisfy the MVTD for $f(x) = 3x^2 + 5x - 2$ on the interval [–1, 1].
2. Find the values of c that satisfy the MVTD for $f(x) = x^3 + 24x - 16$ on the interval [0, 4].
3. Find the values of c that satisfy the MVTD for $f(x) = \frac{6}{x} - 3$ on the interval [1, 2].
4. Find the values of c that satisfy the MVTD for $f(x) = \frac{6}{x} - 3$ on the interval [–1, 2].
5. Find the values of c that satisfy Rolle's Theorem for $f(x) = x^2 - 8x + 12$ on the interval [2, 6].
6. Find the values of c that satisfy Rolle's Theorem for $f(x) = x(1 - x)$ on the interval [0, 1].
7. Find the values of c that satisfy Rolle's Theorem for $f(x) = 1 - \frac{1}{x^2}$ on the interval [–1, 1].
8. Find the values of c that satisfy Rolle's Theorem for $f(x) = x^{\frac{2}{3}} - x^{\frac{1}{3}}$ on the interval [0, 1].

EXTREME VALUE THEOREM, GLOBAL VERSUS LOCAL EXTREMA, AND CRITICAL POINTS

There are theorems in calculus that will let us draw conclusions about the behavior of a function on an interval without knowing the precise location of that behavior. One of these is the **Extreme Value Theorem**. This theorem says that if a function f is continuous on the interval $[a, b]$, then the Extreme Value Theorem guarantees that f has at least one maximum value and at least one minimum value on the interval $[a, b]$.

These maximum and minimum values are called **extrema** and they come in two varieties—**absolute** (sometimes called **global**) and **local** (sometimes called **relative**). An absolute maximum (or minimum) on an interval means that there is no other value of f that is higher (or lower) than the absolute maximum (or minimum). Look at the graph below. Point B is a local maximum because it is the highest value of the function in its immediate neighborhood. In other words, if you look right around $x = 0$, the highest value of the function is at B. On the other hand, point D is an absolute maximum because it is the highest value of the function for all values of x in the domain of the function. Similarly, point C is a local minimum and point A is an absolute minimum.

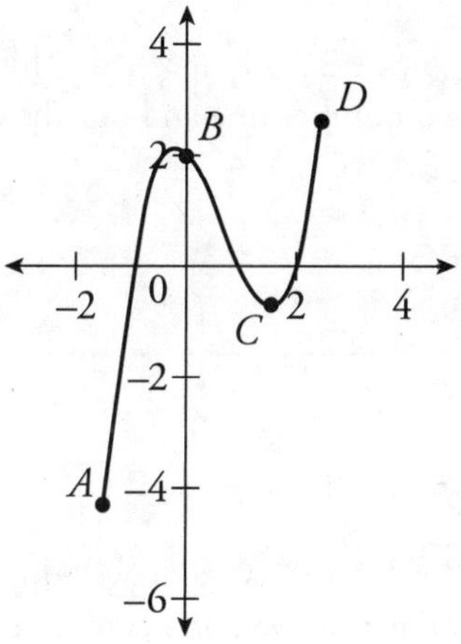

Later in this chapter, we will learn how to find the exact values of extrema.

The points where the extrema could exist are called the **critical points** of the function and we will also learn how to find those later in this chapter.

DETERMINING INTERVALS ON WHICH A FUNCTION IS INCREASING OR DECREASING

Notice how in the graph on the previous page, the function is increasing (going up) as we go from point *A* to point *B*, and then it reaches a maximum value and decreases as we go from point *B* to point *C*. We can use the first derivative to figure out on which intervals (if any) a function is increasing or decreasing. It's quite simple. If the first derivative is positive at a point, then the function is increasing there. If the first derivative is negative at a point, then the function is decreasing there. If the first derivative is zero at a point (or the function is undefined there), then the function is neither increasing nor decreasing at that point. To put it in math speak, if $f'(a) > 0$ at $x = a$, then f is increasing at a. And if $f'(a) < 0$ at $x = a$, then f is decreasing at a.

Example 10: Given $f(x) = x^2 - 8x + 6$, on what interval(s) is f increasing? Decreasing?

All we have to do is take the derivative and find out where it is positive or negative. We get $f'(x) = 2x - 8$. How do we find out where it is positive or negative? Simple. First we set the derivative of the function equal to zero. The values that we get are called the **critical values**. We will then do a **sign test** on the critical values to help us find the intervals.

Set the derivative equal to zero: $f'(x) = 2x - 8$ at $x = 4$. This means that $x = 4$ is the critical value. Now for the sign test. Make a number line and put the critical value on the number line.

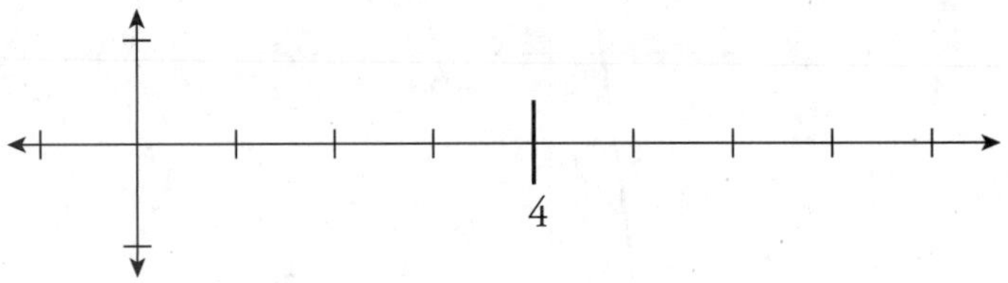

Hint. Any time we can use 0 for a sign test, we should do so. It's easy to plug in!

Now pick a value to the left of 4. Any value will do, so let's use 0. Find the *sign* of the derivative at that value. Note that we don't care what the actual value of the derivative is, only whether it's positive or negative. We get $f'(0) = 2(0) - 8 = -8$. It's negative, so we can conclude that the derivative will be negative for all values of x less than 4: that is, on the interval $(-\infty, 4)$. Let's indicate this on the number line:

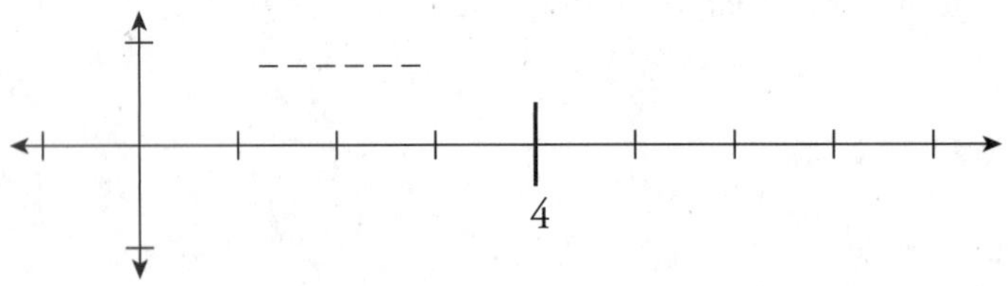

We use the negative signs to show us that the derivative is negative on the interval $(-\infty, 4)$.

Now pick a value to the right of 4. Any value will do, so let's use 5. Find the *sign* of the derivative at that value. We get $f'(5) = 2(5) - 8 = 2$. It's positive, so we can conclude that the derivative will be positive for all values of x greater than 4: that is, on the interval $(4, \infty)$. Let's indicate this on the number line:

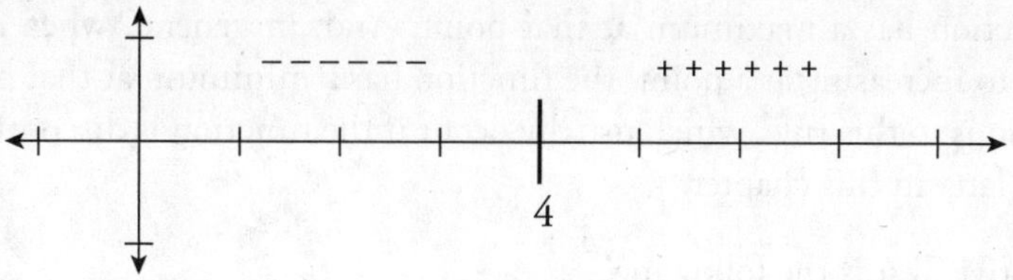

See how this sign test helps us figure out where the function is increasing or decreasing? The function $f(x) = x^2 - 8x + 6$ is thus increasing on the interval $(4, \infty)$ and decreasing on the interval $(-\infty, 4)$. Note that at $x = 4$, the function is neither increasing nor decreasing. We will discuss this more in the next section on determining relative extrema.

Example 11: Given $f(x) = x^3 - 6x^2 + 9x - 5$, on what interval(s) is f increasing? Decreasing?

First, let's find the derivative: $f'(x) = 3x^2 - 12x + 9$. Set it equal to zero: $f'(x) = 3x^2 - 12x + 9 = 0$. Divide through by 3: $x^2 - 4x + 3 = 0$. Factor: $(x - 3)(x - 1) = 0$. Therefore, the derivative is 0 at $x = 1$ and at $x = 3$. These are the critical values. Next, let's do a sign test. First, make a number line and put the critical values on the line:

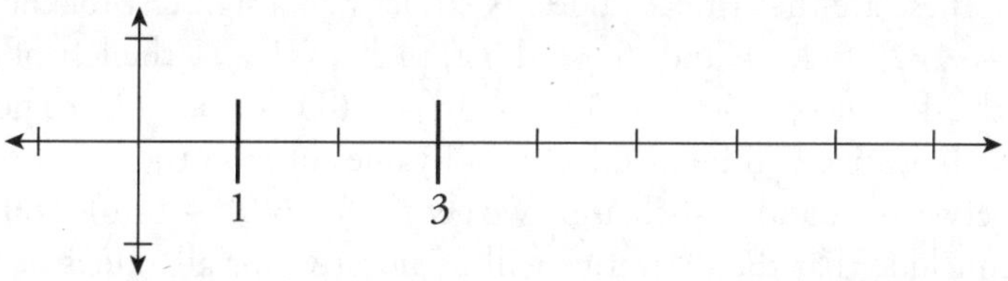

Note that we have three intervals this time, namely $(-\infty, 1)$, $(1, 3)$, and $(3, \infty)$. Let's do a sign test for each of the three intervals. First, pick a value to the left of 1. Let's use 0. Find the sign of the derivative at 0: $f'(0) - 3(0)^2 - 12(0) + 9 = 9$. It's positive, so we can conclude that the derivative will be positive for all values of x on the interval $(-\infty, 1)$. Next, let's test a value between 1 and 3, such as 2. We get $f'(2) = 3(2)^2 - 12(2) + 9 = -3$. It's negative, so we can conclude that the derivative will be negative for all values of x on the interval $(1, 3)$. Finally, let's test a value to the right of 3, such as 4: $f'(4) = 3(4)^2 - 12(4) + 9 = 9$. It's positive, so we can conclude that the derivative will be positive for all values of x on the interval $(3, \infty)$. Here's our sign chart:

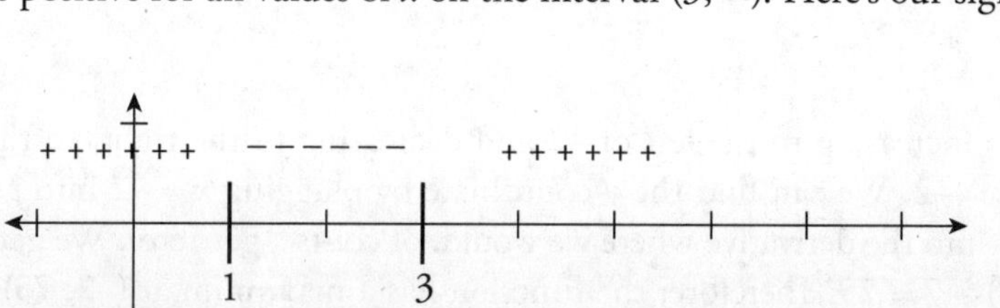

The function $f(x) = x^3 - 6x^2 + 9x - 5$ is thus increasing on the intervals $(-\infty, 1)$ and $(3, \infty)$, and it is decreasing on the interval $(1, 3)$. Note that at $x = 1$ and $x = 3$, the function is neither increasing nor decreasing.

USING THE FIRST DERIVATIVE TEST TO DETERMINE RELATIVE (LOCAL) EXTREMA

As we saw in the previous section, we can use a sign test to figure out where a function is increasing or decreasing. In general, when a function shifts from increasing to decreasing at a point, the function has a maximum at that point. And, in general, when a function shifts from decreasing to increasing at a point, the function has a minimum at that point. There are a couple of exceptions to this rule, which usually occur if the function is discontinuous there. We will get to those later in this chapter.

The First Derivative Test is the following:

> If $f(x)$ is a continuous function and $f'(x) = 0$ and $x = a$, then if $f'(x) > 0$ to the left of $x = a$, and if $f'(x) < 0$ to the right of $x = a$, then f has a relative maximum at $x = a$. If $f'(x) < 0$ to the left of $x = a$ and if $f'(x) > 0$ to the right of $x = a$, then f has a relative minimum at $x = a$.

Example 12: Use the First Derivative Test to find any relative maxima or minima of the function $f(x) = 2x^3 - 9x^2 - 60x + 7$.

First, find the derivative: $f'(x) = 6x^2 - 18x - 60$. Factor: $f'(x) = 6(x - 5)(x + 2)$. Thus, $f'(x) = 0$ at $x = 5$ and $x = -2$. These are the critical values. Next, let's do a sign test for the derivative at the three intervals: $(-\infty, -2)$, $(-2, 5)$, and $(5, \infty)$. First, pick a value to the left of -2. Let's use -3. Find the sign of the derivative at -3: $f'(-3) = 6(-3)^2 - 18(-3) - 60 = 48$. It's positive, so we can conclude that the derivative will be positive for all values of x on the interval $(-\infty, -2)$. Next, let's test a value between -2 and 5, such as 0. We get $f'(0) = 6(0)^2 - 18(0) - 60 = -60$. It's negative, so we can conclude that the derivative will be negative for all values of x on the interval $(-2, 5)$. Finally, let's test a value to the right of 5, such as 6: $f'(6) = 6(6)^2 - 18(6) - 60 = 48$. It's positive, so we can conclude that the derivative will be positive for all values of x on the interval $(5, \infty)$. Here's our sign chart:

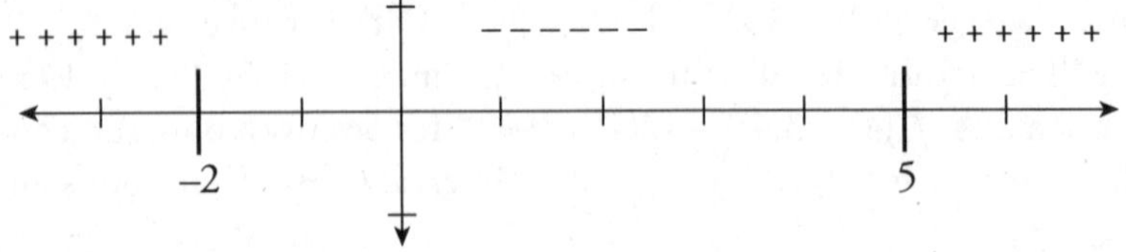

The function is increasing to the left of -2 and decreasing to the right, so the function has a maximum at $x = -2$. We can find the y-coordinate by plugging $x = -2$ into $f(x) = 2x^3 - 9x^2 - 60x + 7$ (NOT into the derivative where we would, of course, get zero). We get $f(-2) = 2(-2)^3 - 9(-2)^2 - 60(-2) + 7 = 75$. Therefore, the function has a maximum at $(-2, 75)$. The function is decreasing to the left of 5 and increasing to the right, so the function has a minimum at $x = 5$. Plug $x = 5$ into $f(x) = 2x^3 - 9x^2 - 60x + 7$: $f(5) = 2(5)^3 - 9(5)^2 - 60(5) + 7 = -268$. Therefore, the function has a minimum at $(5, -268)$.

Example 13: Use the First Derivative Test to find any relative maxima or minima of the function $f(x) = 2\sin\left(\frac{\pi}{4}x\right)$ on the interval [1, 11].

Note: Don't confuse coordinate parentheses with interval notation. In interval notation, square brackets include endpoints and parentheses do not. For example, the interval $2 \le x \le 4$ is written [2, 4] and the interval $2 < x < 4$ is written (2, 4).

First, find the derivative: $f'(x) = \frac{\pi}{2}\cos\left(\frac{\pi}{4}x\right)$. Next, find where the derivative is zero. Note that we are only interested in where the derivative is zero on the interval [1, 11]. Cosine is zero at $x = \frac{\pi}{2}, \frac{3\pi}{2}, \frac{5\pi}{2}, \ldots$. Thus, $f'(x) = 0$ at $x = 2$, $x = 6$, and $x = 10$. These are the critical values. Next, let's sign test the derivative. We have four intervals: (1, 2), (2, 6), (6, 10), and (10, 11). Let's do a sign test for each of the intervals. First, pick a value between 1 and 2. Let's use $\frac{3}{2}$. Find the sign of the derivative at $\frac{3}{2}$: $f'\left(\frac{3}{2}\right) = \frac{\pi}{2}\cos\left(\frac{3\pi}{8}\right) > 0$. It's positive, so we can conclude that the derivative will be positive for all values of x on the interval (1, 2). Next, let's test a value between 2 and 6, such as 4: $f'(4) = \frac{\pi}{2}\cos(\pi) < 0$. It's negative, so we can conclude that the derivative will be negative for all values of x on the interval (2, 6). Next, let's test a value between 6 and 10, such as 8: $f'(8) = \frac{\pi}{2}\cos(2\pi) > 0$. It's positive, so we can conclude that the derivative will be positive for all values of x on the interval (6, 10). Finally, let's test a value between 10 and 11, such as $\frac{21}{2}$: $f'\left(\frac{21}{2}\right) = \frac{\pi}{2}\cos\left(\frac{21\pi}{8}\right) < 0$, so we can conclude that the derivative will be negative for all values of x on the interval (10, 11). Here's our sign chart:

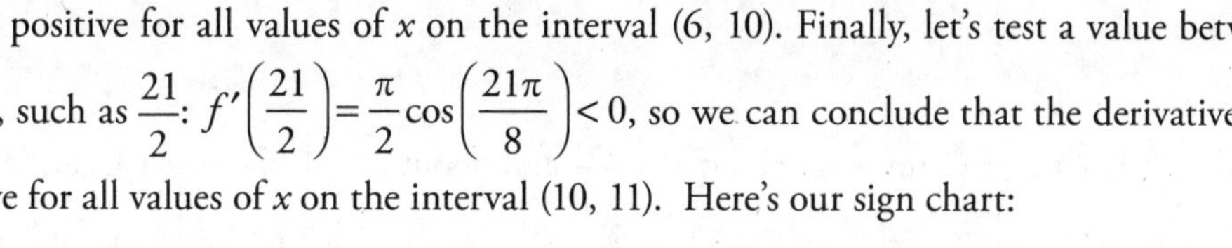

We can see that f has a maximum at $x = 2$ and $x = 10$, and a minimum at $x = 6$. We can find the y-coordinates by plugging the x-values into f. We get

$$f(2) = 2\sin\left(\frac{\pi}{2}\right) = 2$$

$$f(6) = 2\sin\left(\frac{3\pi}{2}\right) = -2$$

$$f(10) = 2\sin\left(\frac{5\pi}{2}\right) = 2$$

Thus, the maxima are at (2, 2) and (10, 2), and the minimum is at (6, –2).

USING THE CANDIDATES TEST TO DETERMINE ABSOLUTE (GLOBAL) EXTREMA

So far, we have been using the first derivative to find relative (local) maxima and minima. Now we will find absolute (global) maxima and minima. What's the difference? Look at the graph below:

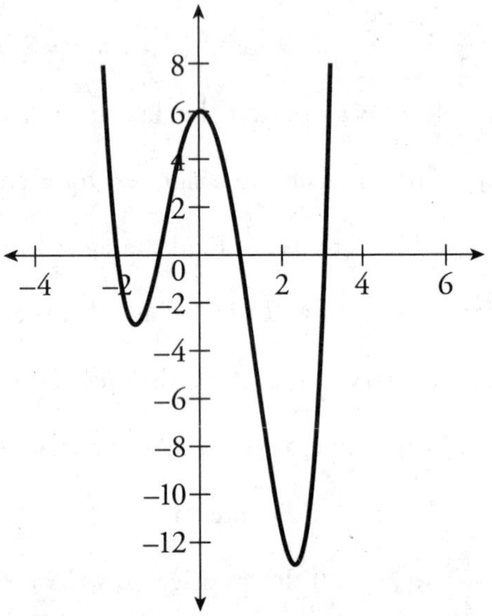

Notice that it has two minima—at about $x = -\frac{3}{2}$ and about $x = 2$ (the exact locations aren't important). Both of those are relative minima because they are the lowest points locally. The minimum at $x = 2$ is lower than the minimum at $x = -\frac{3}{2}$, so it is also the absolute minimum.

There is a local maximum at $x = 0$, but that isn't the absolute maximum of the function. In fact, unless the domain is restricted, there is no absolute maximum because the ends of the function just keep going up to infinity.

So how do we figure out the absolute maximum or minimum of a function? Simple.

> Take the critical values and the endpoints of the function (if there are any) and plug them into the function. The highest value is the absolute maximum, and the lowest value is the absolute minimum.

Example 14: Find the relative and absolute extrema of the function $f(x) = 2x^3 + 3x^2 - 36x + 10$ on the interval $[-5, 5]$.

First, take the derivative: $f'(x) = 6x^2 + 6x - 36$. Set the derivative equal to zero: $6x^2 + 6x - 36 = 0$ and divide through by 6: $x^2 + x - 6 = 0$. Factor: $(x + 3)(x - 2) = 0$. The critical values are $x = -3$ and $x = 2$. Let's do a sign test:

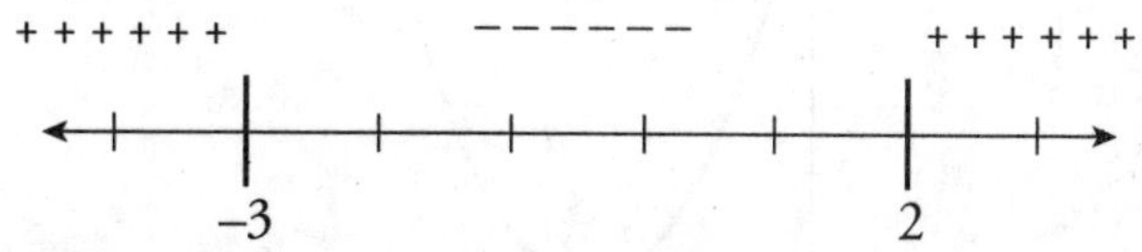

We can see that there is a maximum at $x = -3$ and a minimum at $x = 2$. But are those relative or absolute extrema? As we said above, we can find out by plugging these values and the endpoints into $f(x) = 2x^3 + 3x^2 - 36x + 10$. The highest value of f is the absolute maximum and the lowest value is the absolute minimum. Let's make a table (you can use your calculator):

x	$f(x)$
–5	15
–3	91
2	–34
5	155

We can see that (–3, 91) is a relative maximum and (5, 155) is the absolute maximum, and that (–5, 15) is a relative minimum and (2, –34) is the absolute minimum.

That's all there is to finding absolute extrema. Next, let's learn about concavity.

DETERMINING CONCAVITY OF FUNCTIONS OVER THEIR DOMAINS

Now that we have seen how to use the first derivative of a function, let's work with the second derivative. The first derivative tells us whether a function is increasing ($f'(x) > 0$), decreasing ($f'(x) < 0$), or neither ($f'(x) = 0$ or $f(x)$ does not exist at the point). The second derivative tells us the **concavity** of a function: that is the curvature. If the second derivative of a function is positive at a point, it means that the derivative is increasing there, and the function is **concave up** at that point. In the figure below, the function is concave up at the point $x = 2$. In fact, it is concave up everywhere.

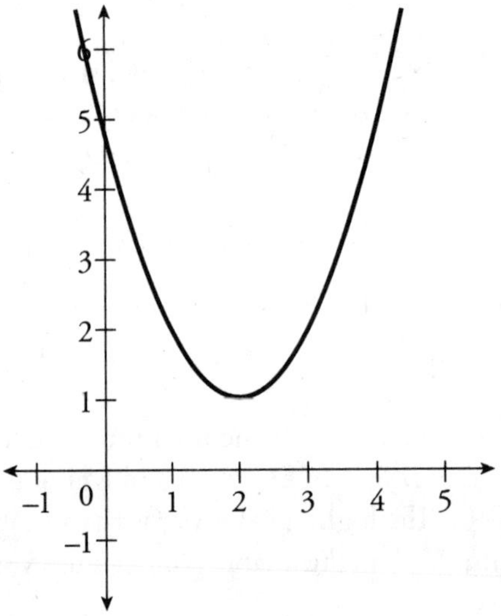

If the second derivative of a function is negative at a point, it means that the derivative is decreasing there and that the function is **concave down** at that point. In the figure below, the function is concave down at the point $x = 2$. In fact, it is concave down everywhere.

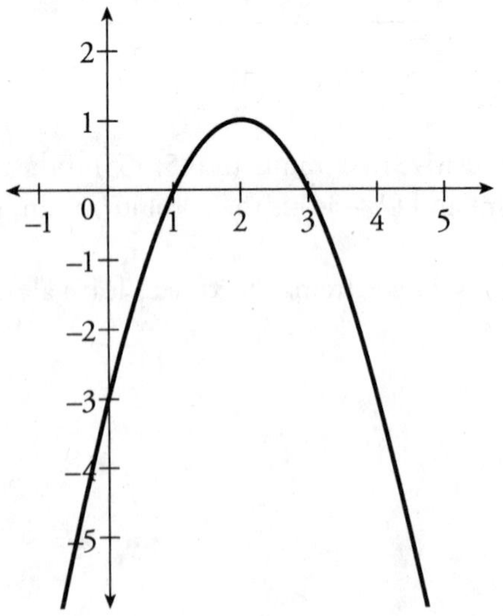

If the second derivative is zero at a point, or the function does not exist there, then there is no concavity at that point.

Example 15: Given $f(x) = -2x^4 + 4x^3 + 24x^2 + 20x + 4$, find the intervals where the function is concave up or concave down, and any points of inflection.

First, we need to find the second derivative. The first derivative is $f'(x) = -8x^3 + 12x^2 + 48x + 20$ and the second derivative is $f''(x) = -24x^2 + 24x + 48$. Set the second derivative equal to zero and solve: $-24x^2 + 24x + 48 = 0$.

$$x^2 - x - 2 = 0$$

$$(x - 2)(x + 1) = 0$$

$$x = 2 \text{ or } x = -1$$

Now, just as with the first derivative, let's do a sign test:

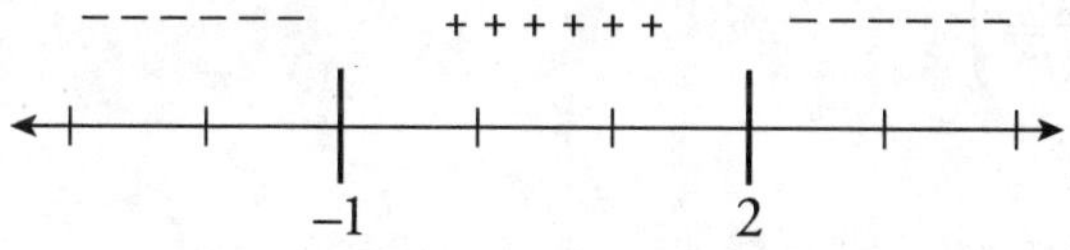

Thus, the function is concave down on the intervals $(-\infty, -1)$ and $(2, \infty)$ and concave up on the interval $(-1, 2)$.

Notice how the function changes concavity at $x = -1$ and again at $x = 2$, where the second derivative is zero. These are the x-coordinates of what are called **points of inflection**. We can find the y-coordinates by plugging into the original equation. We get $f(-1) = 2$ and $f(2) = 140$. Therefore, the points of inflection are $(-1, 2)$ and $(2, 140)$.

Let's do another example.

Example 16: Given $f(x) = e^{-x^2}$, find the intervals where the function is concave up or concave down, and any points of inflection.

First, we need to find the second derivative. The first derivative is $f'(x) = -2xe^{-x^2}$ and the second derivative is $f''(x) = -2x\left(-2xe^{-x^2}\right) - 2e^{-x^2} = \left(4x^2 - 2\right)e^{-x^2}$. Set the second derivative equal to zero and solve: $\left(4x^2 - 2\right)e^{-x^2} = 0$. Note that e^{-x^2} is always positive, so we just need to find where $(4x^2 - 2)$ is zero.

$$4x^2 - 2 = 0$$

$$x = \pm\frac{1}{\sqrt{2}}$$

Now let's do a sign test:

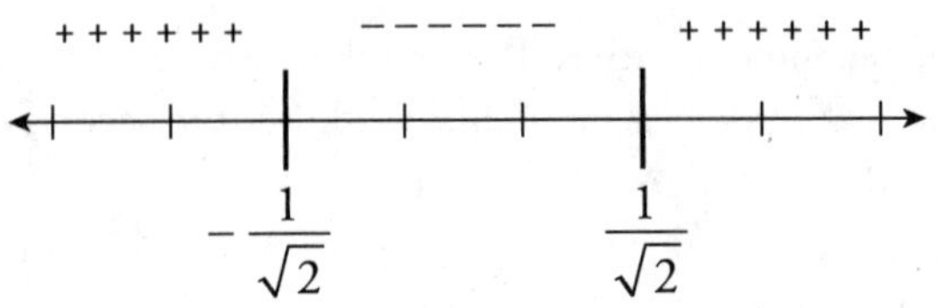

Thus, the function is concave up on the intervals $\left(-\infty, -\frac{1}{\sqrt{2}}\right)$ and $\left(\frac{1}{\sqrt{2}}, \infty\right)$ and concave down on the interval $\left(-\frac{1}{\sqrt{2}}, \frac{1}{\sqrt{2}}\right)$.

Notice how the function changes concavity at $x = -\frac{1}{\sqrt{2}}$ and again at $x = \frac{1}{\sqrt{2}}$, where the second derivative is zero. We can find the y-coordinates by plugging into the original equation. We get $f\left(-\frac{1}{\sqrt{2}}\right) = e^{-\frac{1}{2}}$ and $f\left(\frac{1}{\sqrt{2}}\right) = e^{-\frac{1}{2}}$. Therefore, the points of inflection are $\left(\frac{-1}{\sqrt{2}}, e^{-\frac{1}{2}}\right)$ and $\left(\frac{1}{\sqrt{2}}, e^{-\frac{1}{2}}\right)$.

USING THE SECOND DERIVATIVE TEST TO DETERMINE EXTREMA

Because the second derivative gives us the concavity of a function, we can use it to determine whether a critical point is the location of a relative maximum (or minimum). Let's look again at the curves from the previous section:

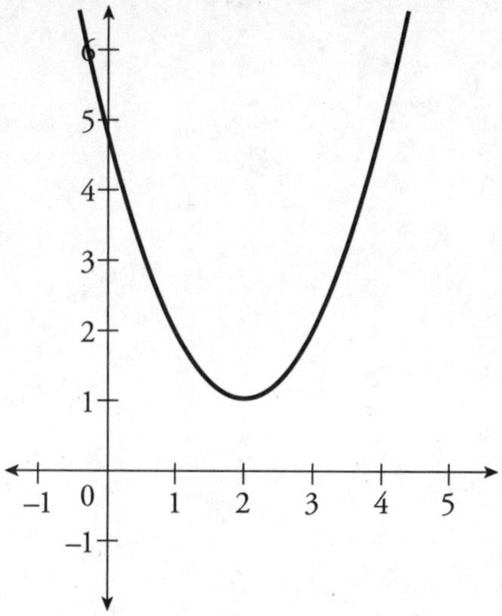

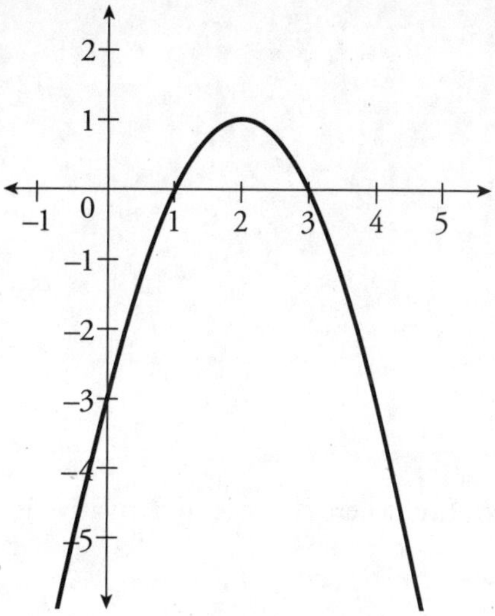

In the curve on the left, we have a minimum at $x = 2$ and in the curve on the right, we have a maximum at $x = 2$. At a relative minimum, we have positive concavity and at a relative maximum, we have negative concavity. We can use this information to create the **Second Derivative Test**:

> Given a continuous function f, with at least a first and second derivative: If $f'(a) = 0$, then if $f''(a) > 0$, the function has a minimum at the point $(a, f(a))$; if $f''(a) < 0$, the function has a maximum at the point $(a, f(a))$.

We now have two ways to determine whether a critical point is a relative maximum or minimum (or neither). We can use the sign test with the first derivative or we can use the Second Derivative Test. When it's easy to find a second derivative, we recommend using the Second Derivative Test. When it's difficult to find a second derivative, then sign testing is better. Also, sometimes we can trivially know the sign of the second derivative (for example, $f(x) = x^2$), in which case we know right away if a critical point is a relative maximum or minimum.

Example 17: Given $f(x) = 4x^3 - 5x^2 - 8x - 24$, find any relative extrema, point(s) of inflection, the intervals where f is increasing or decreasing, and the intervals where f is concave up or concave down.

Let's take the first and second derivatives: $f'(x) = 12x^2 - 10x - 8$ and $f''(x) = 24x - 10$.

Next, let's find the critical values. Set the first derivative equal to zero and solve. We get

$$12x^2 - 10x - 8 = 0$$

$$6x^2 - 5x - 4 = 0$$

$$(3x - 4)(2x + 1) = 0$$

$$x = \frac{4}{3} \text{ or } x = -\frac{1}{2}$$

Next, find where the second derivative is zero:

$$24x - 10 = 0$$

$$x = \frac{5}{12}$$

Use the Second Derivative Test to determine the relative maximum and relative minimum. We get

$f''\left(\frac{4}{3}\right) = 24\left(\frac{4}{3}\right) - 10 = 22$, so there is a relative minimum at $x = \frac{4}{3}$.

$f''\left(-\frac{1}{2}\right) = 24\left(-\frac{1}{2}\right) - 10 = -22$, so there is a relative maximum at $x = -\frac{1}{2}$.

Next, we can use the sign test to find the intervals where the function is increasing or decreasing. Test the first derivative:

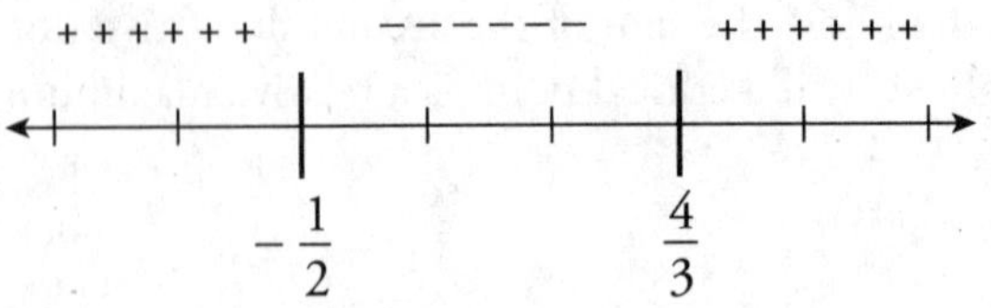

Test the second derivative:

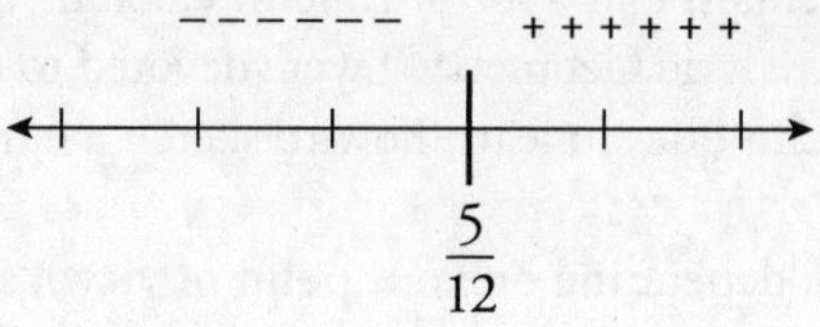

We can see that we have a point of inflection at $x = \frac{5}{12}$.

Finally, let's find the y-coordinates of the relevant points. You may want to use a calculator.

x	$f(x)$
$-\frac{1}{2}$	–21.75
$\frac{4}{3}$	–34.074
$\frac{5}{12}$	–27.912

Therefore, we have a relative maximum at $\left(-\frac{1}{2}, -21.75\right)$, a relative minimum at $\left(\frac{4}{3}, -34.074\right)$, and a point of inflection at $\left(\frac{5}{12}, -27.912\right)$. The function is increasing on the intervals $\left(-\infty, -\frac{1}{2}\right)$ and $\left(\frac{4}{3}, \infty\right)$, is decreasing on the interval $\left(-\frac{1}{2}, \frac{4}{3}\right)$, is concave down on the interval $\left(-\infty, \frac{5}{12}\right)$, and is concave up on the interval $\left(\frac{5}{12}, \infty\right)$.

In the next section, we will put all of this together to learn how to graph functions.

SKETCHING GRAPHS OF FUNCTIONS AND THEIR DERIVATIVES

Another topic on which students spend a lot of time in calculus is curve sketching. In the old days, whole courses (called "Analytic Geometry") were devoted to the subject, and students had to master a wide variety of techniques to learn how to sketch a curve accurately.

Fortunately (or unfortunately, depending on your point of view), students no longer need to be as good at analytic geometry. There are two reasons for this: (1) The AP Exam tests only a few types of curves; and (2) you can use a graphing calculator. Because of the calculator, you can get an idea of the shape of the curve, and all you need to do is find important points to label the graph. We use calculus to find some of these points.

When it's time to sketch a curve, we'll show you a four-part analysis that'll give you all the information you need.

STEP 1

Step 1: Test the Function

Find where $f(x) = 0$. This tells you the function's x-intercepts (or roots). By setting $x = 0$, we can determine the y-intercepts. Then, find any horizontal and/or vertical asymptotes.

STEP 2

Step 2: Test the First Derivative

Find where $f'(x) = 0$. This tells you the critical points. We can determine whether the curve is rising or falling, as well as where the maxima and minima are. It's also possible to determine if the curve has any points where it's nondifferentiable.

STEP 3

Step 3: Test the Second Derivative

Find where $f''(x) = 0$. This shows you where any points of inflection are. (These are points where the graph of a function changes concavity.) Then we can determine where the graph curves upward and where it curves downward.

STEP 4

Step 4: Test End Behavior

Look at what the general shape of the graph will be, based on the values of y for very large values of $\pm x$. Using this analysis, we can always come up with a sketch of a curve.

(1) When $f'(x) > 0$, the curve is rising; when $f'(x) < 0$, the curve is falling; when $f'(x) = 0$, the curve is at a critical point.

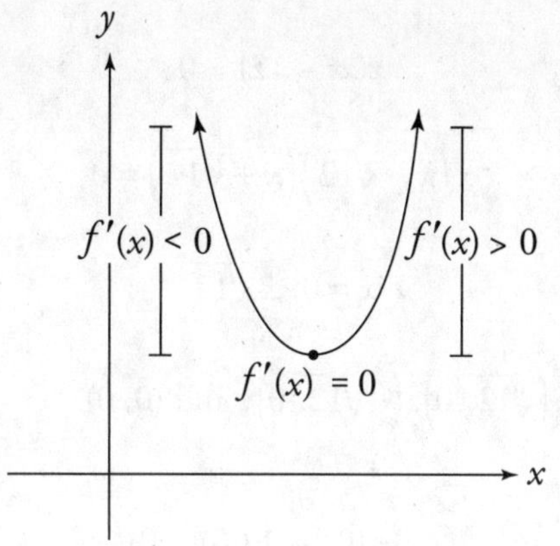

(2) When $f''(x) > 0$, the curve is "concave up"; when $f''(x) < 0$, the curve is "concave down"; and when $f''(x) = 0$, the curve is at a point of inflection.

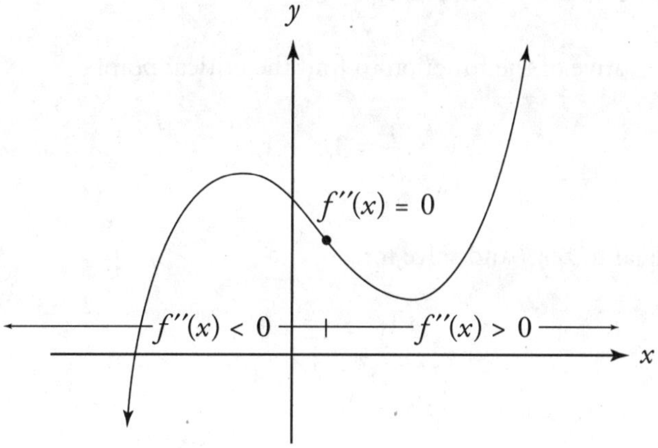

(3) The y-coordinates of each critical point are found by plugging the x-value into the *original* equation.

Example 18: Sketch the equation $y = x^3 - 12x$.

Step 1: Find the x-intercepts.

$$x^3 - 12x = 0$$

$$x(x^2 - 12) = 0$$

$$x\left(x-\sqrt{12}\right)\left(x+\sqrt{12}\right)=0$$

$$x = 0, \pm\sqrt{12}$$

The curve has x-intercepts at $\left(\sqrt{12},\ 0\right)$, $\left(-\sqrt{12},\ 0\right)$, and (0, 0).

Next, find the y-intercepts.

$$y = (0)^3 - 12(0) = 0$$

The curve has a y-intercept at (0, 0).

There are no asymptotes because there's no place where the curve is undefined (you won't have asymptotes for curves that are polynomials).

Step 2: Take the derivative of the function to find the critical points.

$$\frac{dy}{dx} = 3x^2 - 12$$

Set the derivative equal to zero, and solve for x.

$$3x^2 - 12 = 0$$
$$3(x^2 - 4) = 0$$
$$3(x - 2)(x + 2) = 0$$

So $x = 2, -2$.

Next, plug $x = 2, -2$ into the original equation to find the y-coordinates of the critical points.

$$y = (2)^3 - 12(2) = -16$$
$$y = (-2)^3 - 12(-2) = 16$$

Thus, we have critical points at (2, –16) and (–2, 16).

Step 3: Now, take the second derivative to find any points of inflection.

$$\frac{d^2y}{dx^2} = 6x$$

This equals zero at $x = 0$. We already know that when $x = 0$, $y = 0$, so the curve has a point of inflection at (0, 0).

Now, plug the critical values into the second derivative to determine whether each is a maximum or a minimum: $f''(2) = 6(2) = 12$. This is positive, so the curve has a minimum at (2, –16), and the curve is concave up at that point: $f''(-2) = 6(-2) = -12$. This value is negative, so the curve has a maximum at (–2, 16) and the curve is concave down there.

Armed with this information, we can now plot the graph.

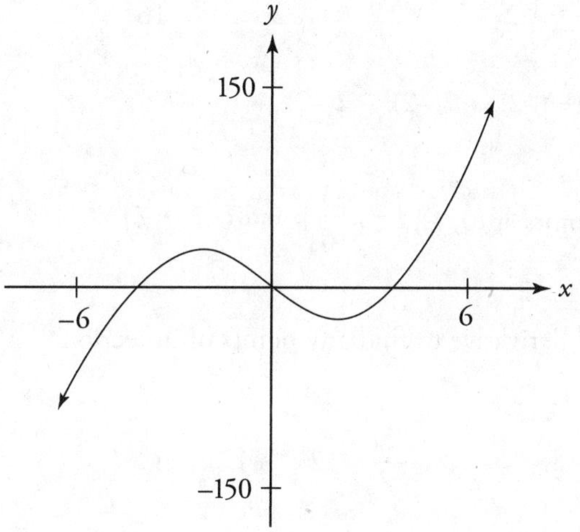

Example 19: Sketch the graph of $y = x^4 + 2x^3 - 2x^2 + 1$.

Step 1: First, let's find the x-intercepts.

$$x^4 + 2x^3 - 2x^2 + 1 = 0$$

If the equation doesn't factor easily, it's best not to bother to find the function's roots.

The good news is that if the roots aren't easy to find, the College Board won't ask you to find them, or you can find them with your calculator.

Next, let's find the y-intercepts.

$$y = (0)^4 + 2(0)^3 - 2(0)^2 + 1$$

The curve has a y-intercept at (0, 1).

There are no vertical asymptotes because there is no place where the curve is undefined.

Step 2: Now we take the derivative to find the critical points.

$$\frac{dy}{dx} = 4x^3 + 6x^2 - 4x$$

Set the derivative equal to zero.

$$4x^3 + 6x^2 - 4x = 0$$
$$2x(2x^2 + 3x - 2) = 0$$
$$2x(2x - 1)(x + 2) = 0$$
$$x = 0, \frac{1}{2}, -2$$

Next, plug these three values into the original equation to find the y-coordinates of the critical points. We already know that when $x = 0$, $y = 1$.

When $x = \frac{1}{2}$, $y = \left(\frac{1}{2}\right)^4 + 2\left(\frac{1}{2}\right)^3 - 2\left(\frac{1}{2}\right)^2 + 1 = \frac{13}{16}$

When $x = -2$, $y = (-2)^4 + 2(-2)^3 - 2(-2)^2 + 1 = -7$

Thus, we have critical points at $(0, 1)$, $\left(\frac{1}{2}, \frac{13}{16}\right)$, and $(-2, -7)$.

Step 3: Take the second derivative to find any points of inflection.

$$\frac{d^2y}{dx^2} = 12x^2 + 12x - 4$$

Set this equal to zero.

$$12x^2 + 12x - 4 = 0$$
$$3x^2 + 3x - 1 = 0$$
$$x = \frac{-3 \pm \sqrt{21}}{6} \approx 0.26, -1.26$$

Therefore, the curve has points of inflection at $x = \frac{-3 \pm \sqrt{21}}{6}$.

Now solve for the y-coordinates.

$$(0.26, 0.90) \text{ and } (-1.26, -3.66)$$

We can now plug the critical values into the second derivative to determine whether each is a maximum or a minimum.

$$12(0)^2 + 12(0) - 4 = -4$$

This is negative, so the curve has a maximum at (0, 1); the curve is concave down there.

$$12\left(\frac{1}{2}\right)^2+12\left(\frac{1}{2}\right)-4=5$$

This is positive, so the curve has a minimum at $\left(\frac{1}{2},\frac{13}{16}\right)$; the curve is concave up there.

$$12(-2)^2+12(-2)-4=20$$

This is positive, so the curve has a minimum at (–2, –7) and the curve is also concave up there.

Let's sketch the graph.

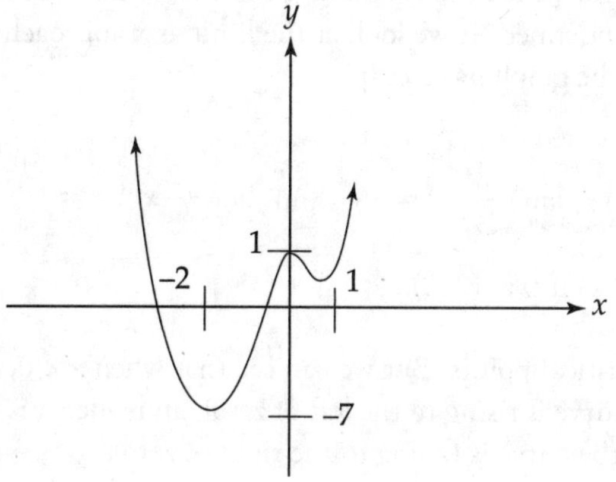

Finding a Cusp

If the derivative of a function approaches ∞ from one side of a point and –∞ from the other, and if the function is continuous at that point, then the curve has a "cusp" at that point. In order to find a cusp, you need to look at points where the first derivative is undefined, as well as where it's zero.

Example 20: Sketch the graph of $y=2-x^{\frac{2}{3}}$.

Step 1: Find the *x*-intercepts.

$$2-x^{\frac{2}{3}}=0$$

$$x^{\frac{2}{3}}=2 \qquad x=\pm 2^{\frac{3}{2}}=\pm 2\sqrt{2}$$

The *x*-intercepts are at $\left(\pm 2\sqrt{2},\ 0\right)$.

Next, find the *y*-intercepts.

$$y=2-(0)^{\frac{2}{3}}=2$$

The curve has a y-intercept at (0, 2).

There are no asymptotes because there is no place where the curve is undefined.

Step 2: Now, take the derivative to find the critical points.

$$\frac{dy}{dx} = -\frac{2}{3}x^{-\frac{1}{3}}$$

What's next? You guessed it! Set the derivative equal to zero.

$$-\frac{2}{3}x^{-\frac{1}{3}} = 0$$

There are no values of x for which the equation is zero. But here's the new stuff to deal with: at $x = 0$, the derivative is undefined. If we look at the limit as x approaches 0 from both sides, we can determine whether the graph has a cusp.

$$\lim_{x \to 0^+} -\frac{2}{3}x^{-\frac{1}{3}} = -\infty \quad \text{and} \quad \lim_{x \to 0^-} -\frac{2}{3}x^{-\frac{1}{3}} = \infty$$

Therefore, the curve has a cusp at (0, 2).

There aren't any other critical points. But we can see that when $x < 0$, the derivative is positive (which means that the curve is rising to the left of zero), and when $x > 0$, the derivative is negative (which means that the curve is falling to the right of zero).

Step 3: Now, we take the second derivative to find any points of inflection.

$$\frac{d^2y}{dx^2} = \frac{2}{9}x^{-\frac{4}{3}}$$

Again, there's no x-value where this is zero. In fact, the second derivative is positive at all values of x except 0. Therefore, the graph is concave up everywhere.

Let's sketch the graph.

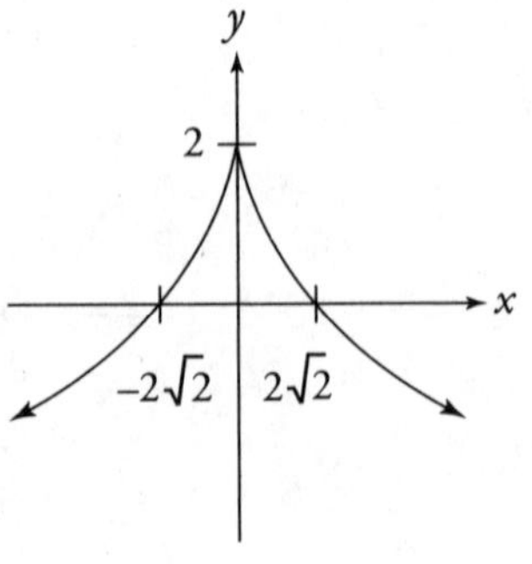

There's one other type of graph you should know about: a rational function. In order to graph a rational function, you need to know how to find that function's asymptotes.

How to Find Asymptotes

A line $y = c$ is a horizontal asymptote of the graph of $y = f(x)$ if

$$\lim_{x\to\infty} f(x) = c \text{ or if } \lim_{x\to-\infty} f(x) = c$$

A line $x = k$ is a vertical asymptote of the graph of $y = f(x)$ if

$$\lim_{x\to k^+} f(x) = \pm\infty \text{ or if } \lim_{x\to k^-} f(x) = \pm\infty$$

Example 21: Sketch the graph of $y = \dfrac{3x}{x+2}$.

Step 1: Find the x-intercepts. A fraction can be equal to zero only when its numerator is equal to zero (provided that the denominator is not also zero there). All we have to do is set $3x = 0$, and you get $x = 0$. Thus, the graph has an x-intercept at (0, 0). Note: this is also the y-intercept.

Next, look for asymptotes. The denominator is undefined at $x = -2$, and if we take the left- and right-hand limits of the function, we see the following:

$$\lim_{x\to-2^+} \frac{3x}{x+2} = -\infty \text{ and } \lim_{x\to-2^-} \frac{3x}{x+2} = \infty$$

The curve has a vertical asymptote at $x = -2$.

If we take $\lim_{x\to\infty} \dfrac{3x}{x+2} = 3$ and $\lim_{x\to-\infty} \dfrac{3x}{x+2} = 3$, the curve has a horizontal asymptote at $y = 3$.

Step 2: Now, take the derivative to figure out the critical points.

$$\frac{dy}{dx} = \frac{(x+2)(3)-(3x)(1)}{(x+2)^2} = \frac{6}{(x+2)^2}$$

There are no values of x that make the derivative equal to zero. Because the numerator is 6 and the denominator is squared, the derivative will always be positive (the curve is always rising). You should note that the derivative is undefined at $x = -2$, but you already know that there's an asymptote at $x = -2$, so you don't need to examine this point further.

Step 3: Now, it's time for the second derivative.

$$\frac{d^2y}{dx^2} = \frac{-12}{(x+2)^3}$$

This is never equal to zero. The expression is positive when $x < -2$, so the graph is concave up when $x < -2$. The second derivative is negative when $x > -2$, so it's concave down when $x > -2$.

Let's sketch the graph.

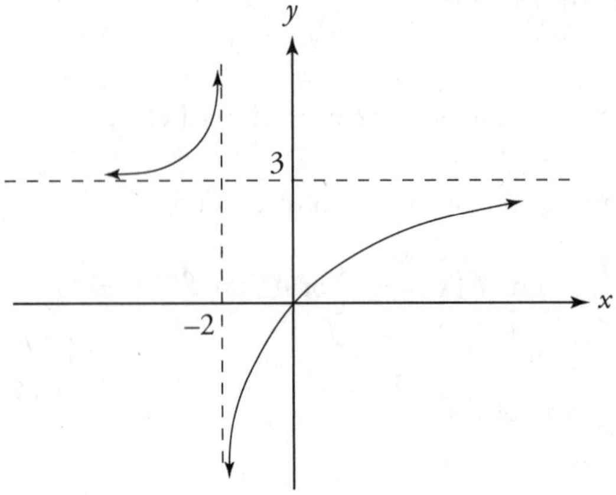

Now it's time to practice some problems. Do each problem, covering the answer first, then check your answer.

Example 22: Sketch the graph of $y = x^3 - 9x^2 + 24x - 10$. Plot all extrema, points of inflection, and asymptotes.

Follow the three steps.

First, see if the x-intercepts are easy to find. This is a cubic equation that isn't easily factored. So skip this step.

Next, find the y-intercepts by setting $x = 0$.

$$y = (0)^3 - 9(0)^2 + 24(0) - 10 = -10$$

The curve has a y-intercept at $(0, -10)$.

There are no asymptotes, because the curve is a simple polynomial.

Next, find the critical points using the first derivative.

$$\frac{dy}{dx} = 3x^2 - 18x + 24$$

Set the derivative equal to zero and solve for x.

$$3x^2 - 18x + 24 = 0$$
$$3(x^2 - 6x + 8) = 0$$
$$3(x - 4)(x - 2) = 0$$
$$x = 2, 4$$

Plug $x = 2$ and $x = 4$ into the original equation to find the y-coordinates of the critical points.

When $x = 2$, $y = 10$
When $x = 4$, $y = 6$

Thus, we have critical points at (2, 10) and (4, 6).

In our third step, the second derivative indicates any points of inflection.

$$\frac{d^2y}{dx^2} = 6x - 18$$

This equals zero at $x = 3$.

Next, plug $x = 3$ into the original equation to find the y-coordinate of the point of inflection, which is at (3, 8). Plug the critical values into the second derivative to determine whether each is a maximum or a minimum.

$$6(2) - 18 = -6$$

This is negative, so the curve has a maximum at (2, 10), and the curve is concave down there.

$$6(4) - 18 = 6$$

This is positive, so the curve has a minimum at (4, 6), and the curve is concave up there.

Let's sketch the graph.

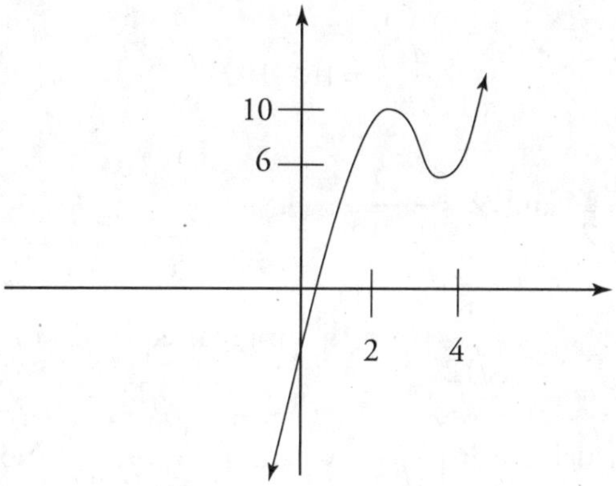

Example 23: Sketch the graph of $y = 8x^2 - 16x^4$. Plot all extrema, points of inflection, and asymptotes.

Factor the polynomial.

$$8x^2\left(1 - 2x^2\right) = 0$$

Solving for x, we get $x = 0$ (a double root), $x = \frac{1}{\sqrt{2}}$, and $x = -\frac{1}{\sqrt{2}}$.

Find the y-intercepts: when $x = 0$, $y = 0$.

There are no asymptotes because the curve is a simple polynomial.

Find the critical points using the first derivative.

$$\frac{dy}{dx} = 16x - 64x^3$$

Set the derivative equal to zero and solve for x. You get $x = 0$, $x = \frac{1}{2}$, and $x = -\frac{1}{2}$.

Next, plug $x = 0$, $x = \frac{1}{2}$, and $x = -\frac{1}{2}$ into the original equation to find the y-coordinates of the critical points.

When $x = 0$, $y = 0$.

When $x = \frac{1}{2}$, $y = 1$.

When $x = -\frac{1}{2}$, $y = 1$.

Thus, there are critical points at (0, 0), $\left(\frac{1}{2}, 1\right)$, and $\left(-\frac{1}{2}, 1\right)$. Take the second derivative to find any points of inflection.

$$\frac{d^2y}{dx^2} = 16 - 192x^2$$

This equals zero at $x = \frac{1}{\sqrt{12}}$ and $x = -\frac{1}{\sqrt{12}}$.

Next, plug $x = \frac{1}{\sqrt{12}}$ and $x = -\frac{1}{\sqrt{12}}$ into the original equation to find the y-coordinates of the points of inflection, which are at $\left(\frac{1}{\sqrt{12}}, \frac{5}{9}\right)$ and $\left(-\frac{1}{\sqrt{12}}, \frac{5}{9}\right)$. Now determine whether the points are maxima or minima.

At $x = 0$, we have a minimum; the curve is concave up there.

At $x = \frac{1}{2}$, it's a maximum, and the curve is concave down.

At $x = -\frac{1}{2}$, it's also a maximum (still concave down).

Let's sketch the graph.

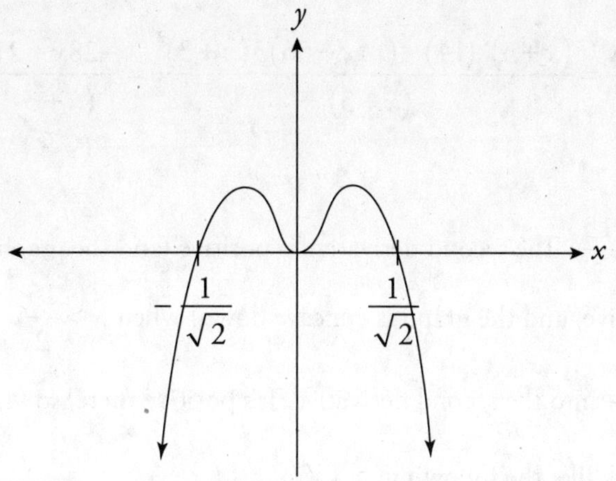

Example 24: Sketch the graph of $y=\left(\frac{x-4}{x+3}\right)^2$. Plot all extrema, points of inflection, and asymptotes.

This should seem rather routine by now.

Find the x-intercepts by setting the numerator equal to zero; $x = 4$. The graph has an x-intercept at (4, 0). (It's a double root.)

Next, find the y-intercept by plugging in $x = 0$.

$$y=\frac{16}{9}$$

The denominator is undefined at $x = -3$, so there's a vertical asymptote at that point.

Look at the limits.

$$\lim_{x\to\infty}\left(\frac{x-4}{x+3}\right)^2=1 \text{ and } \lim_{x\to-\infty}\left(\frac{x-4}{x+3}\right)^2=1$$

The curve has a horizontal asymptote at $y = 1$.

It's time for the first derivative.

$$\frac{dy}{dx}=2\left(\frac{x-4}{x+3}\right)\frac{(x+3)(1)-(x-4)(1)}{(x+3)^2}=\frac{14x-56}{(x+3)^3}$$

The derivative is zero when $x = 4$, and the derivative is undefined at $x = -3$. (There's an asymptote there, so we can ignore the point. If the curve were *defined* at $x = -3$, then it would be a critical point, as you'll see in the next example.)

Now for the second derivative.

$$\frac{d^2y}{dx^2} = \frac{(x+3)^3(14)-(14x-56)3(x+3)^2}{(x+3)^6} = \frac{-28x+210}{(x+3)^4}$$

This is zero when $x = \frac{15}{2}$. The second derivative is positive (and the graph is concave up) when $x < \frac{15}{2}$, and it's negative (and the graph is concave down) when $x > \frac{15}{2}$.

We can now plug $x = 4$ into the second derivative. It's positive there, so (4, 0) is a minimum.

Your graph should look like the following:

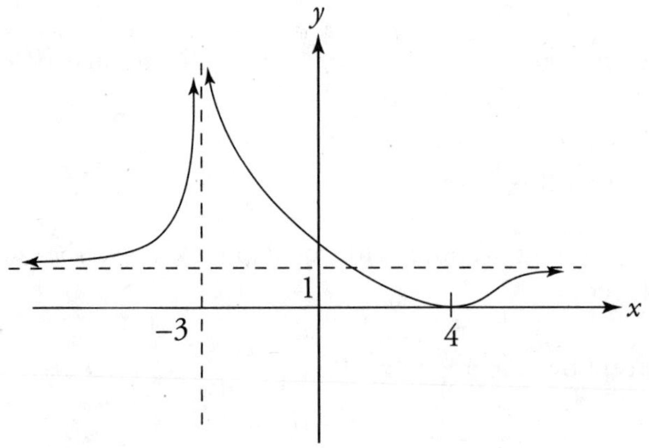

Example 25: Sketch the graph of $y = (x-4)^{\frac{2}{3}}$. Plot all extrema, points of inflection, and asymptotes.

By inspection, the x-intercept is at $x = 4$.

Next, find the y-intercepts. When $x = 0$, $y = \sqrt[3]{16} \approx 2.52$.

No asymptotes exist because there's no place where the curve is undefined.

The first derivative is

$$\frac{dy}{dx} = \frac{2}{3}(x-4)^{-\frac{1}{3}}$$

Set it equal to zero.

$$\frac{2}{3}(x-4)^{-\frac{1}{3}} = 0$$

This can never equal zero. But at $x = 4$, the derivative is undefined, so this is a critical point. If you look at the limit as x approaches 4 from both sides, you can see if there's a cusp.

$$\lim_{x \to 4^+} \frac{2}{3}(x-4)^{-\frac{1}{3}} = \infty \text{ and } \lim_{x \to 4^-} \frac{2}{3}(x-4)^{-\frac{1}{3}} = -\infty$$

The curve has a cusp at (4, 0).

There were no other critical points. But we can see that when $x > 4$, the derivative is positive and the curve is rising; when $x < 4$, the derivative is negative and the curve is falling.

The second derivative is

$$\frac{d^2y}{dx^2} = -\frac{2}{9}(x-4)^{-\frac{4}{3}}$$

No value of x can set this equal to zero. In fact, the second derivative is negative at all values of x except 4. Therefore, the graph is concave down everywhere.

Your graph should look like the following:

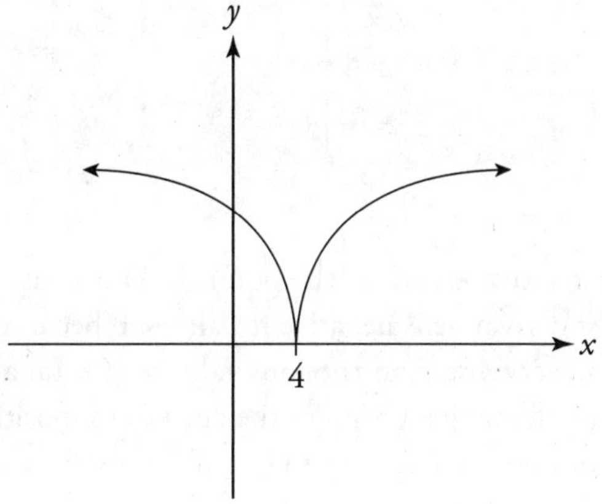

CONNECTING A FUNCTION, ITS FIRST DERIVATIVE, AND ITS SECOND DERIVATIVE

Sometimes, we are given the graph of a function and we are asked to graph the derivative. We do this by analyzing the sign of the first and second derivatives at various places on the graph and then sketching a graph of the derivative from that information. Let's start with something simple. Suppose we have the graph of $y = f(x)$ below and we are asked to sketch the graph of its derivative.

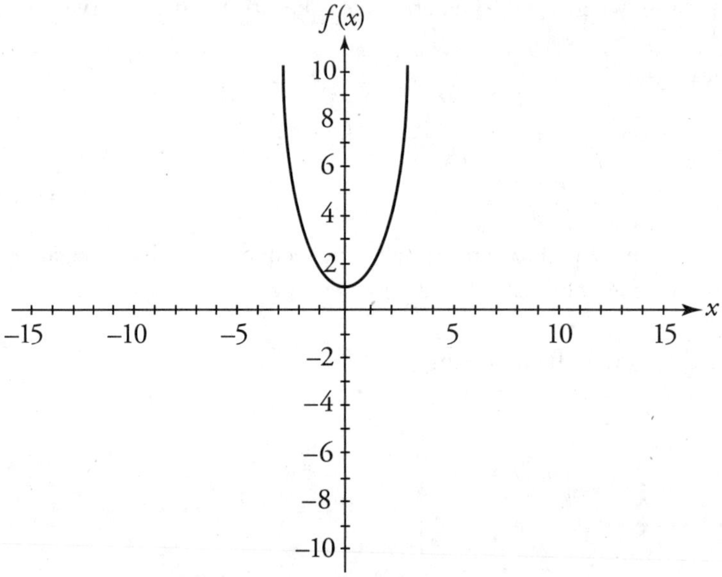

First, note that the first derivative is zero at the point (0, 1) because the tangent line is horizontal there. Next, the first derivative is negative for all $x < 0$ because the tangent lines to the curve have negative slopes everywhere on the interval $(-\infty, 0)$. Finally, the first derivative is positive for all $x > 0$ because the tangent lines to the curve have positive slopes everywhere on the interval $(0, \infty)$. Furthermore, notice that the graph is always concave up, so the second derivative is always positive. Since the second derivative is the derivative of the first derivative, the graph of first derivative is always increasing. Now we can make a graph of the derivative. It will go through the origin (because the derivative is 0 at $x = 0$), it will be negative on the interval $(-\infty, 0)$, it will be positive on the interval $(0, \infty)$, and it will always be increasing.

The graph looks something like the following:

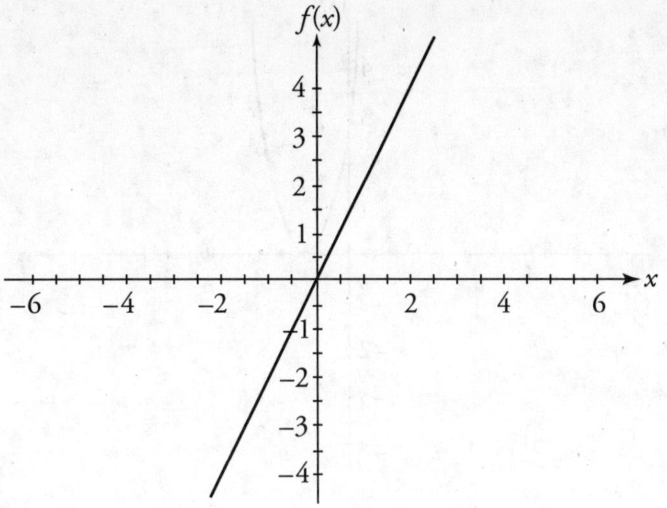

Note that it's not important for your graph to be exact. All we are doing here is sketching the derivative. The important parts of the graph are where the derivative is positive, negative, and zero.

Now let's try something a little harder. Suppose we have the following graph and we are asked to sketch the graph of the derivative:

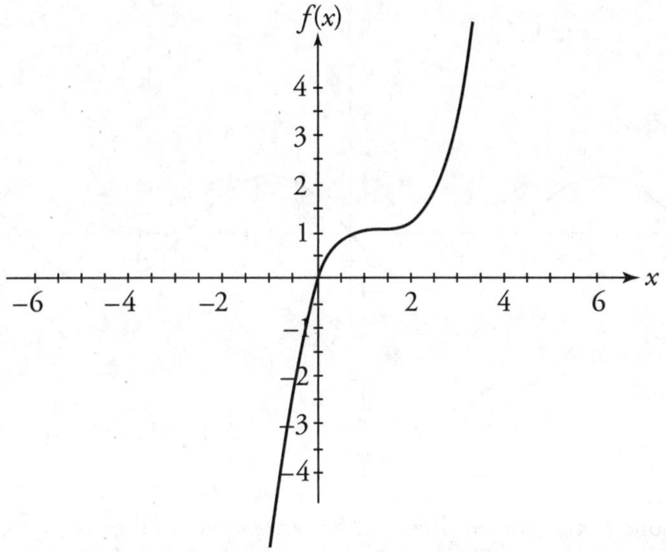

Maxima and minima on the graph of f become zeros on the graph of f'.

Notice that the tangent line looks as if it's horizontal at $x = 1$. This means that the graph of the derivative is zero there. Next, the curve is increasing for all other values of x, so the graph of the derivative will be positive. However, since the graph does not reach a local extrema at $x = 1$, this is actually an inflection point. Notice that the graph is concave down to the left of $x = 1$. That means the second derivative is negative and thus the first derivative is decreasing to the left of $x = 1$. Conversely, the graph is concave up to the right of $x = 1$, so the derivative is increasing to the right of $x = 1$.

If we sketch the derivative, we get something like the following:

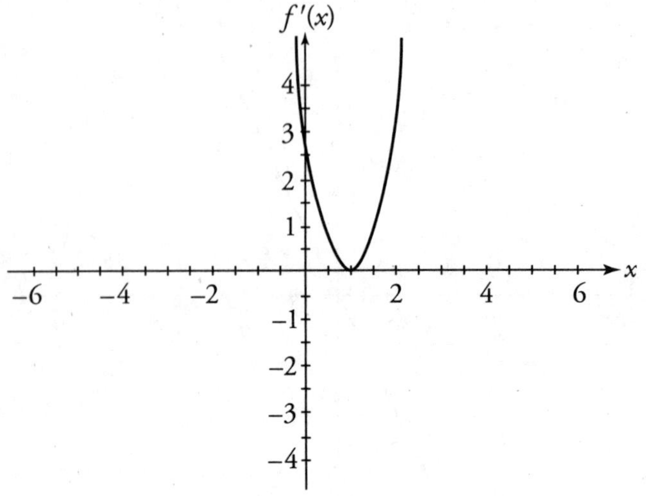

Here is an important one to understand. Suppose we have the graph of $y = \sin x$.

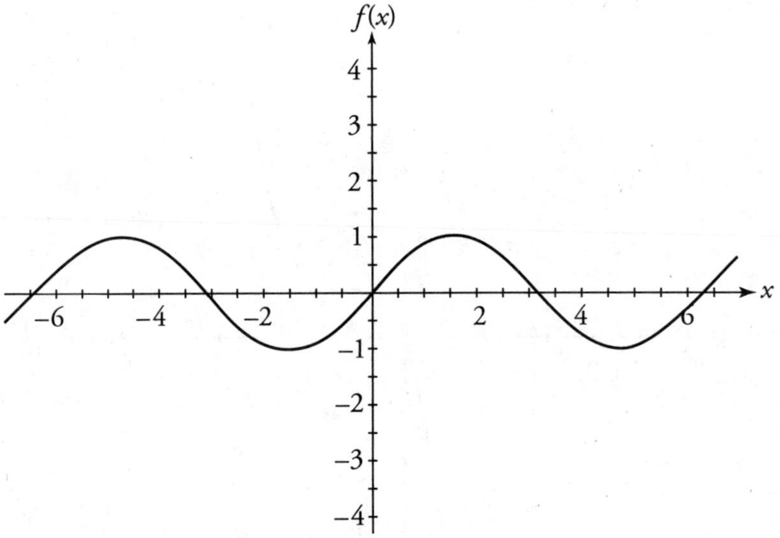

Notice that the slope of the tangent line will be horizontal at all of the maxima and minima of the graph. Because the slope of a horizontal line is zero, this means that the derivative will be zero at those values $\left(\pm\frac{\pi}{2}, \frac{3\pi}{2}, \ldots\right)$. Next, notice that the slope of the curve is about 1 as the curve goes through the origin. This should make sense if you recall that $\lim_{x \to 0} \frac{\sin x}{x} = 1$. The slope of the curve is about –1 as the curve goes through $x = \pi$, and so on. Also, notice that the graph has inflection points at $x = -2\pi, -\pi, 0, \pi$, and 2π, so the second derivative is 0 and the derivative has critical points at each of the x-values. More specifically, the concavity (and thus the second derivative) goes from positive to negative at $x = -\pi$ and π and from negative to positive at $x = -2\pi$, 0, and 2π. Therefore, the derivative must have a local maximum and $x = -\pi$ and π and a local minimum at $x = -2\pi$, 0, and 2π.

If we now sketch the derivative, it looks something like the following:

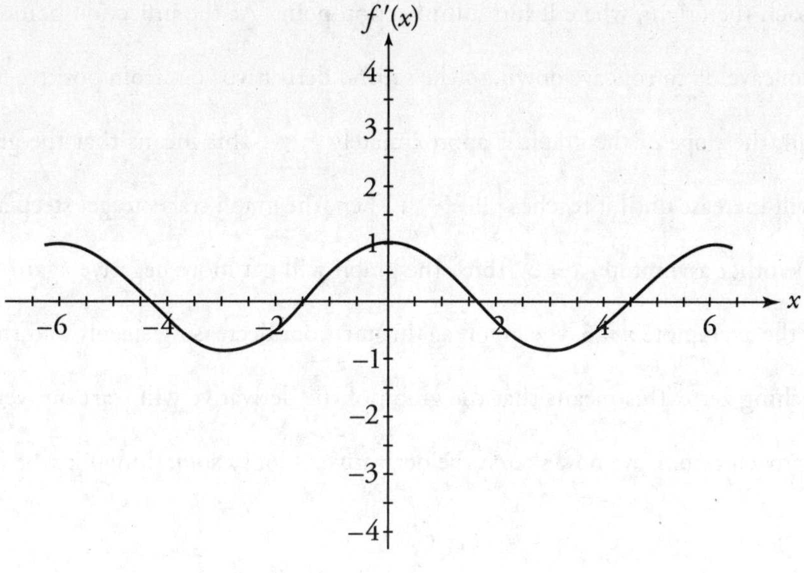

Notice that this is the graph of $y = \cos x$. This is because the derivative of $\sin x$ is $\cos x$.

Now let's do a hard one. Suppose we have the following graph:

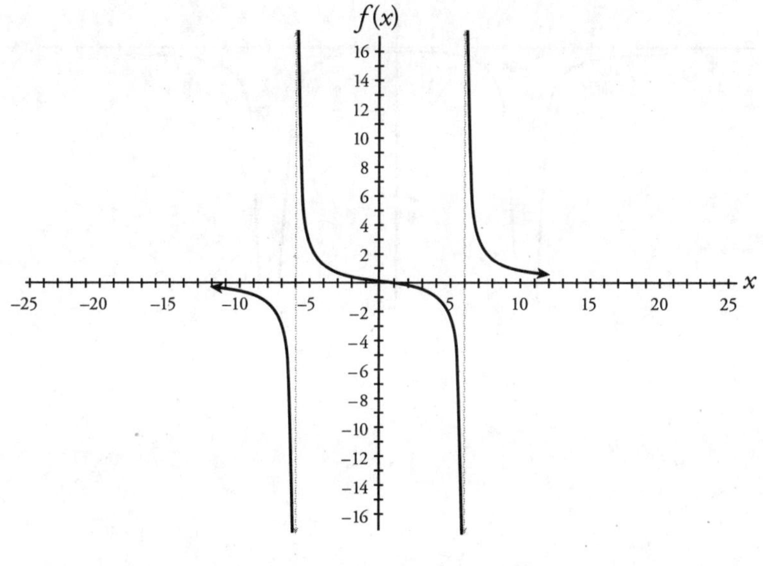

First, notice that we have two vertical asymptotes at $x = 6$ and $x = -6$. This means that the graph of the derivative will also have vertical asymptotes at $x = 6$ and $x = -6$. Next, notice that the curve is always decreasing. This means that the graph of the derivative will always be negative. Moving from left to right, the graph starts out close to flat, so the derivative will be close to zero. Then, the graph is concave down, so the derivative will be decreasing. Then, we have the asymptote $x = -6$. Next, the graph begins decreasing very steeply and starts to flatten out as we approach the origin, where it hits an inflection point. At the inflection point, the graph goes from concave up to concave down, so the second derivative goes from positive to negative. At the origin, the slope of the graph is approximately $-\frac{1}{2}$. This means that the graph of the derivative will increase until it reaches $\left(0, -\frac{1}{2}\right)$. Then, the graph starts to get steep again as we approach the other asymptote, $x = 6$. Thus, the graph will get more negative again. Finally, to the right of the asymptote $x = 6$, the graph again starts out decreasing steeply and then flattens out, approaching zero. This means that the graph of the derivative will start out very negative and will approach zero. If we now sketch the derivative, it looks something like the following:

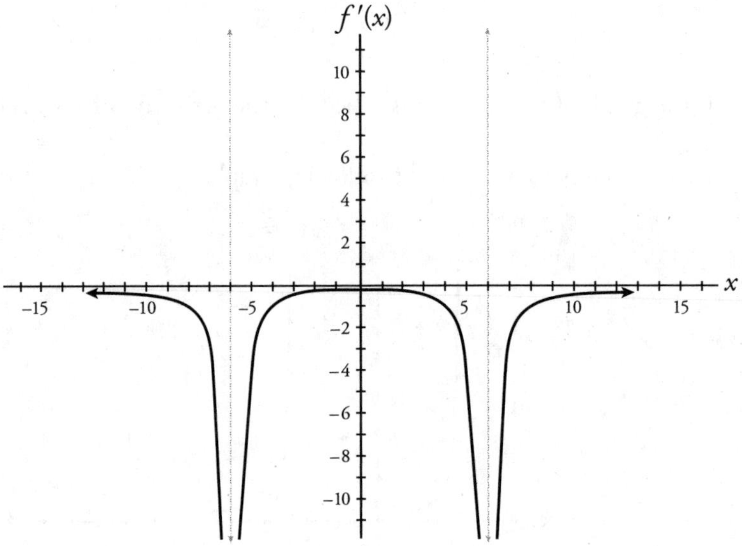

PRACTICE PROBLEM SET 19

Sketch the following graphs. The answers are at the end of the chapter, starting on page 332.

1. $y = x^3 - 9x - 6$

2. $y = -x^3 - 6x^2 - 9x - 4$

3. $y = (x^2 - 4)(9 - x^2)$

4. $y = \dfrac{x-3}{x+8}$

5. $y = \dfrac{x^2-4}{x-3}$

6. $y = 3 + x^{\frac{2}{3}}$

7. $y = x^{\frac{2}{3}}\left(3 - 2x^{\frac{1}{3}}\right)$

INTRODUCTION TO OPTIMIZATION PROBLEMS

Now that we have seen how to use the derivative to graph functions or to interpret the graphs of functions, let's use the derivative to solve some word problems.

One of the most common applications of the derivative is to find a maximum or minimum value of a function. These values can be called extreme values, optimal values, or critical points. Each of these problems involves the same, very simple principle.

A maximum or a minimum of a function occurs at a point where the derivative of a function is zero, or where the derivative fails to exist.

At a point where the first derivative equals zero, the curve has a horizontal tangent line, at which point it could be reaching either a "peak" (maximum) or a "valley" (minimum).

If the derivative of a function is zero at a certain point, it is usually a maximum or minimum—but not always.

There are two different kinds of maxima and minima: relative and absolute. A **relative** or **local** maximum or minimum means that the curve has a horizontal tangent line at that point, but it is not necessarily the highest or lowest value that the function attains. In the figure to the right, the two indicated points are relative maxima/minima.

There are a few exceptions to every rule. This rule is no different.

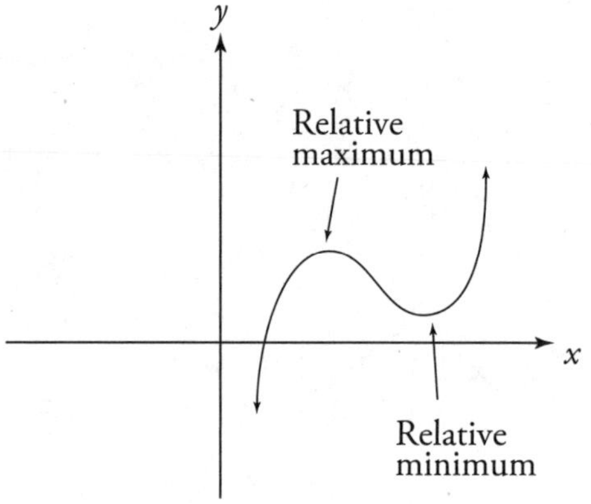

An **absolute** maximum or minimum occurs either at a critical point or an end point. In the following figure, the two indicated points are absolute maxima/minima. A relative maximum can also be an absolute maximum.

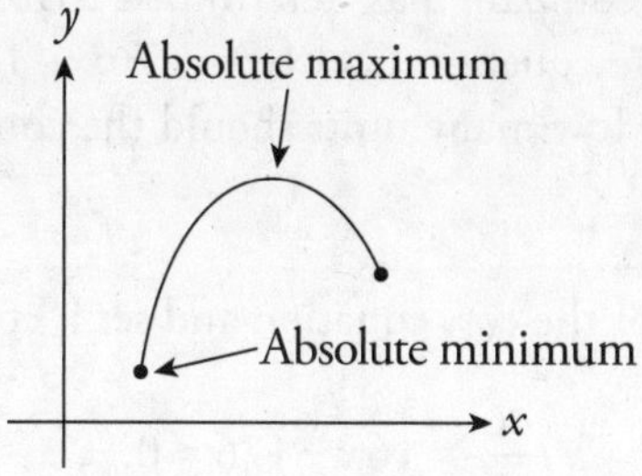

A typical word problem will ask you to find a maximum or a minimum value of a function, as it pertains to a certain situation. Sometimes you're given the equation; other times, you have to figure it out for yourself. Once you have the equation, you find its derivative and set it equal to zero. The values you get are called critical values. That is, if $f'(c) = 0$ or $f'(c)$ does not exist, then c is a critical value. Then, test these values to determine whether each value is a maximum or a minimum. The simplest way to do this is with the Second Derivative Test.

If a function has a critical value at $x = c$, then that value is a relative maximum if $f''(c) < 0$ and it is a relative minimum if $f''(c) > 0$.

If the second derivative is also zero at $x = c$, then the point is neither a maximum nor a minimum, but it could be a point of inflection.

Extreme Learning

The Extreme Value Theorem says that if a function is continuous on the interval $[a, b]$, then somewhere in the interval (a, b), there exists a maximum or minimum value of $f(c)$ between $f(a)$ and $f(b)$ inclusive. In other words, if the function is continuous in some interval, then you always have a maximum or minimum somewhere in the interval, or the curve is flat.

SOLVING OPTIMIZATION PROBLEMS

Let's use the tools that we have developed in this chapter to solve some word problems. Many students struggle with these problems so you may want to work through them more than once.

Example 26: Find the minimum value on the curve $y = ax^2$, if $a > 0$.

Take the derivative and set it equal to zero.

$$\frac{dy}{dx} = 2ax = 0$$

The first derivative is equal to zero at $x = 0$. By plugging 0 back into the original equation, we can solve for the y-coordinate of the minimum (the y-coordinate is also 0, so the point is at the origin).

In order to determine if this is a maximum or a minimum, take the second derivative.

$$\frac{d^2y}{dx^2} = 2a$$

Because a is positive, the second derivative is positive and the critical point we obtained from the first derivative is a minimum point. Had a been negative, the second derivative would have been negative, and a maximum would have occurred at the critical point.

Example 27: A manufacturing company has determined that the total cost of producing an item can be determined from the equation $C = 8x^2 - 176x + 1800$, where x is the number of units that the company makes. How many units should the company manufacture in order to minimize the cost?

Once again, take the derivative of the cost equation and set it equal to zero.

$$\frac{dC}{dx} = 16x - 176 = 0$$

$$x = 11$$

This tells us that 11 is a critical point of the equation. Now we need to figure out if this is a maximum or a minimum using the second derivative.

$$\frac{d^2C}{dx^2} = 16$$

Because 16 is always positive, any critical value is going to be a minimum. Therefore, the company should manufacture 11 units in order to minimize its cost.

Example 28: A rocket is fired into the air, and its height in meters at any given time t can be calculated using the formula $h(t) = 1600 + 196t - 4.9t^2$. Find the maximum height of the rocket and the time at which it occurs.

Take the derivative and set it equal to zero.

$$\frac{dh}{dt} = 196 - 9.8t$$

$$t = 20$$

Now that we know 20 is a critical point of the equation, use the Second Derivative Test.

$$\frac{d^2h}{dt^2} = -9.8$$

This is always negative, so any critical value is a maximum. To determine the maximum height of the rocket, plug $t = 20$ into the equation.

$$h(20) = 1{,}600 + 196(20) - 4.9(20^2) = 3{,}560 \text{ meters}$$

> The technique is always the same: (a) take the derivative of the equation; (b) set it equal to zero; and (c) use the Second Derivative Test to show that it is a maximum or minimum.

The hardest part of these word problems is when you have to set up the equation yourself. The following is a classic AP problem:

Example 29: Max wants to make a box with no lid from a rectangular sheet of cardboard that is 18 inches by 24 inches. The box is to be made by cutting a square of side x from each corner of the sheet and folding up the sides (see figure below). Find the value of x that maximizes the volume of the box.

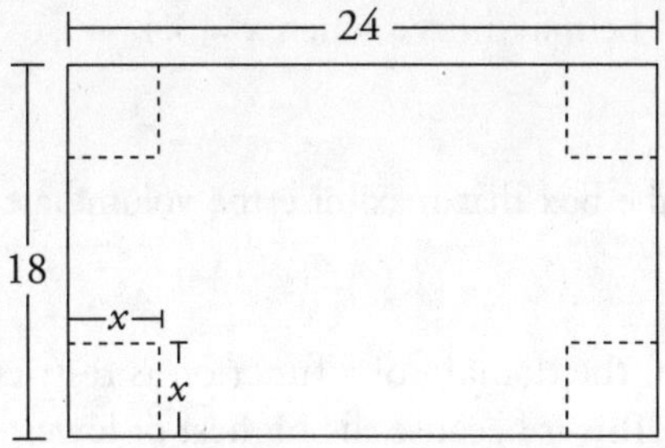

After we cut out the squares of side x and fold up the sides, the dimensions of the box will be

width: $18 - 2x$
length: $24 - 2x$
depth: x

Using the formula for the volume of a rectangular prism, which is the length times the width times the height, we can get an equation for the volume in terms of x.

$$V = x(18 - 2x)(24 - 2x)$$

Multiply the terms together (and be careful with your algebra).

$$V = x(18 - 2x)(24 - 2x) = 4x^3 - 84x^2 + 432x$$

Now take the derivative.

$$\frac{dV}{dx} = 12x^2 - 168x + 432$$

Set the derivative equal to zero, and solve for x.

$$12x^2 - 168x + 432 = 0$$

$$x^2 - 14x + 36 = 0$$

$$x = \frac{14 \pm \sqrt{196 - 144}}{2} = 7 \pm \sqrt{13} \approx 3.4, 10.6$$

Common sense tells us that you can't cut out two square pieces that measure 10.6 inches to a side (the sheet's only 18 inches wide!), so the maximizing value has to be 3.4 inches. Here's the Second Derivative Test, just to be sure.

$$\frac{d^2V}{dx^2} = 24x - 168$$

At $x = 3.4$,

$$\frac{d^2V}{dx^2} = -86.4$$

So the volume of the box will be maximized when $x = 3.4$.

Therefore, the dimensions of the box that maximize the volume are approximately: 11.2 inches × 17.2 inches × 3.4 inches.

Sometimes, particularly when the domain of a function is restricted, you have to test the endpoints of the interval as well. This is because the highest or lowest value of a function may be at an endpoint of that interval; the critical value you obtained from the derivative might be just a local maximum or minimum. For the purposes of the AP Exam, however, endpoints are considered separate from critical values.

Example 30: Find the absolute maximum and minimum values of $y = x^3 - x$ on the interval $[-3, 3]$.

Take the derivative and set it equal to zero.

$$\frac{dy}{dx} = 3x^2 - 1 = 0$$

Solve for x.

$$x = \pm\frac{1}{\sqrt{3}}$$

Test the critical points.

$$\frac{d^2y}{dx^2} = 6x$$

At $x = \frac{1}{\sqrt{3}}$, we have a minimum. At $x = -\frac{1}{\sqrt{3}}$, we have a maximum.

$$\text{At } x = -\frac{1}{\sqrt{3}},\ y = -\frac{1}{3\sqrt{3}} + \frac{1}{\sqrt{3}} = \frac{2}{3\sqrt{3}} \approx 0.385.$$

$$\text{At } x = \frac{1}{\sqrt{3}},\ y = \frac{1}{3\sqrt{3}} - \frac{1}{\sqrt{3}} = -\frac{2}{3\sqrt{3}} \approx -0.385.$$

Now it's time to check the endpoints of the interval.

At $x = -3$, $y = -24$.
At $x = 3$, $y = 24$.

We can see that the function actually has a *lower* value at $x = -3$ than at its "minimum" when $x = \frac{1}{\sqrt{3}}$. Similarly, the function has a *higher* value at $x = 3$ than at its "maximum" of $x = -\frac{1}{\sqrt{3}}$. This means that the function has a "local minimum" at $x = \frac{1}{\sqrt{3}}$ and an "absolute minimum" when $x = -3$. And, the function has a "local maximum" at $x = -\frac{1}{\sqrt{3}}$ and an "absolute maximum" at $x = 3$.

Example 31: A rectangle is to be inscribed in a semicircle with radius 4, with one side on the semicircle's diameter. What is the largest area this rectangle can have?

Let's look at this on the coordinate axes. The equation for a circle of radius 4, centered at the origin, is $x^2 + y^2 = 16$; a semicircle has the equation $y = \sqrt{16 - x^2}$. Our rectangle can then be expressed as a function of *x*, where the height is $\sqrt{16 - x^2}$ and the base is $2x$. See the following figure:

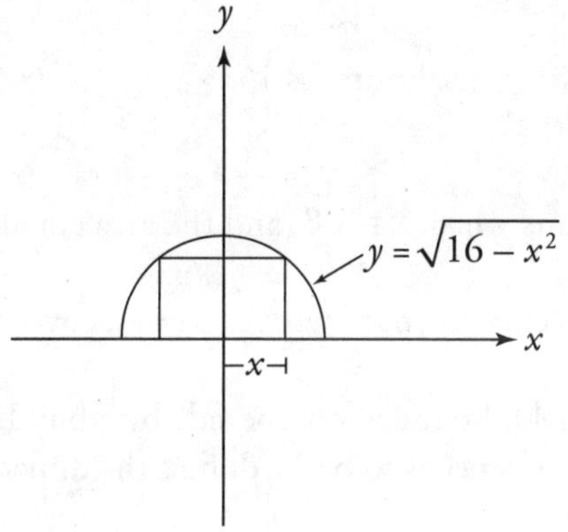

The area of the rectangle, which is the base times the height, is $A = 2x\sqrt{16 - x^2}$. Let's take the derivative of the area.

$$\frac{dA}{dx} = 2\sqrt{16 - x^2} - \frac{2x^2}{\sqrt{16 - x^2}}$$

The derivative is not defined at $x = \pm 4$. Setting the derivative equal to zero, we get

$$2\sqrt{16-x^2} - \frac{2x^2}{\sqrt{16-x^2}} = 0$$

$$2\sqrt{16-x^2} = \frac{2x^2}{\sqrt{16-x^2}}$$

$$2\left(16-x^2\right) = 2x^2$$

$$32 - 2x^2 = 2x^2$$

$$32 = 4x^2$$

$$x = \pm\sqrt{8}$$

If you're wondering why we don't use the negative root, it's because there is no such thing as a negative area.

Note that the domain of this function is $-4 \leq x \leq 4$, so these numbers serve as endpoints of the interval. Let's compare the critical values and the endpoints.

When $x = -4$, $y = 0$ and the area is 0.

When $x = 4$, $y = 0$ and the area is 0.

When $x = \sqrt{8}$, $y = \sqrt{8}$ and the area is 16.

Thus, the maximum area occurs when $x = \sqrt{8}$ and the area equals 16.

Try some of these solved problems on your own. As always, cover the answers as you work.

Example 32: A rectangular field, bounded on one side by a building, is to be fenced in on the other three sides. If 3000 feet of fence is to be used, find the dimensions of the largest field that can be fenced in.

First, let's make a rough sketch of the situation.

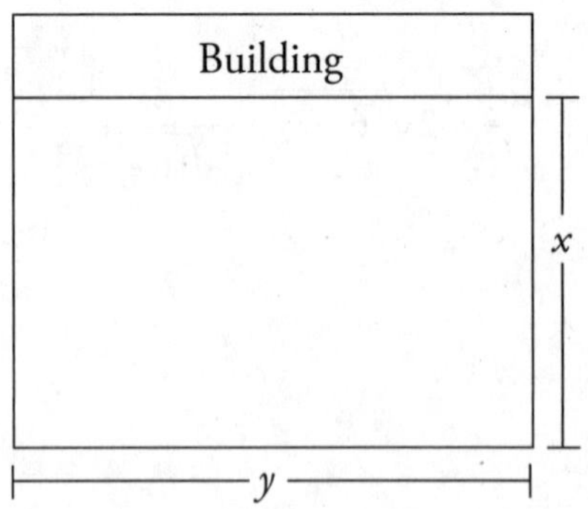

If we call the length of the field y and the width of the field x, the formula for the area of the field becomes

$$A = xy$$

The perimeter of the fencing is equal to the sum of two widths and the length.

$$2x + y = 3{,}000$$

Now solve this second equation for y:

$$y = 3{,}000 - 2x$$

When you plug this expression into the formula for the area, you get a formula for A in terms of x.

$$A = x(3{,}000 - 2x) = 3{,}000x - 2x^2$$

Next, take the derivative, set it equal to zero, and solve for x.

$$\frac{dA}{dx} = 3{,}000 - 4x = 0$$

$$x = 750$$

Let's check to make sure it's a maximum. Find the second derivative.

$$\frac{d^2A}{dx^2} = -4$$

Because we have a negative result, $x = 750$ is a maximum. Finally, if we plug in $x = 750$ and solve for y, we find that $y = 1{,}500$. The largest field will measure 750 feet by 1,500 feet.

Example 33: A poster is to contain 100 square inches of picture surrounded by a 4-inch margin at the top and bottom and a 2-inch margin on each side. Find the overall dimensions that will minimize the total area of the poster.

First, make a sketch.

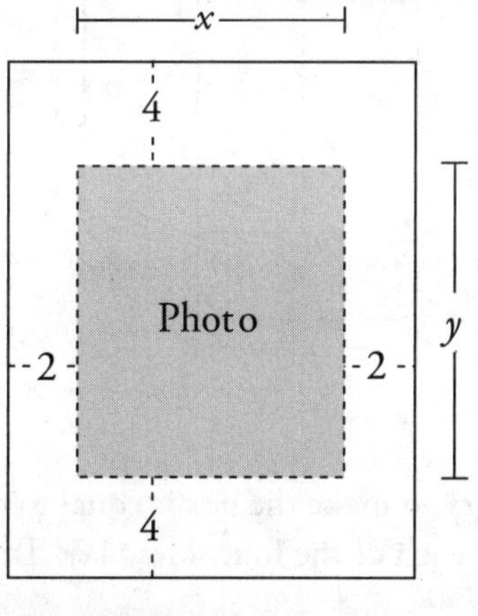

Let the area of the picture be $xy = 100$. The total area of the poster is $A = (x + 4)(y + 8)$. Then, expand the equation.

$$A = xy + 4y + 8x + 32$$

Substitute $xy = 100$ and $y = \frac{100}{x}$ into the area equation, and we get

$$A = 132 + \frac{400}{x} + 8x$$

Now take the derivative and set it equal to zero.

$$\frac{dA}{dx} = 8 - \frac{400}{x^2} = 0$$

Solving for x, we find that $x = \sqrt{50}$. Now solve for y by plugging $x = \sqrt{50}$ into the area equation: $y = 2\sqrt{50}$. Then check that these dimensions give us a minimum.

$$\frac{d^2A}{dx^2} = \frac{800}{x^3}$$

This is positive when x is positive, so the minimum area occurs when $x = \sqrt{50}$. Thus, the overall dimensions of the poster are $4 + \sqrt{50}$ inches by $8 + 2\sqrt{50}$ inches.

Example 34: An open-top box with a square bottom and rectangular sides is to have a volume of 256 cubic inches. Find the dimensions that require the minimum amount of material.

First, make a sketch of the situation.

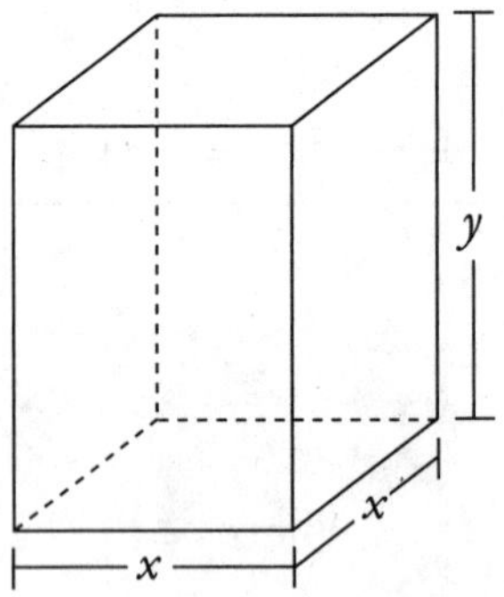

The amount of material necessary to make the box is equal to the surface area, which is the area of the base, x^2, plus the areas of each of the four sides, $4xy$. This gives us

$$S = x^2 + 4xy$$

The formula for the volume of the box is $x^2y = 256$.

If we solve the latter equation for y, $y = \frac{256}{x^2}$, and plug it into the former equation, we get

$$S = x^2 + 4x\frac{256}{x^2} = x^2 + \frac{1{,}024}{x}$$

Now take the derivative and set it equal to zero.

$$\frac{dS}{dx} = 2x - \frac{1{,}024}{x^2} = 0$$

If we solve this for x, we get $x^3 = 512$ and $x = 8$. Solving for y, we get $y = 4$.

Check that these dimensions give us a minimum.

$$\frac{d^2A}{dx^2} = 2 + \frac{2{,}048}{x^3}$$

This is positive when x is positive, so the minimum surface area occurs when $x = 8$. The dimensions of the box should be 8 inches by 8 inches by 4 inches.

Example 35: Find the point on the curve $y = \sqrt{x}$ that is a minimum distance from the point (4, 0).

First, make that sketch.

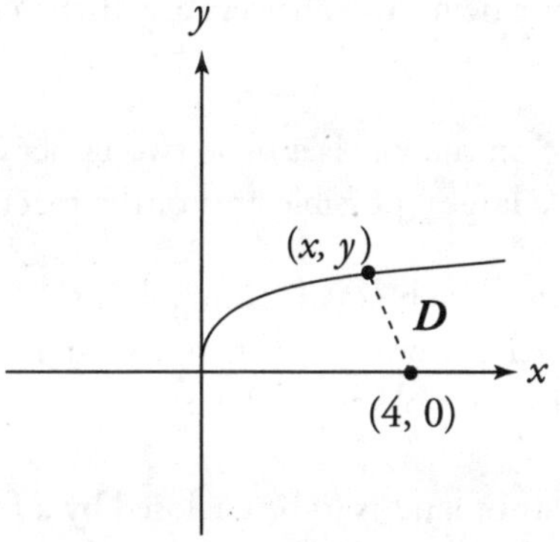

The distance formula is $D = \sqrt{(x_1 - x_2)^2 + (y_1 - y_2)^2}$, and if you square both sides, you get $D^2 = (x_1 - x_2)^2 + (y_1 - y_2)^2$.

Using the distance formula, we get

$$D^2 = (x - 4)^2 + (y - 0)^2 = x^2 - 8x + 16 + y^2$$

Because $y = \sqrt{x}$,

$$D^2 = x^2 - 8x + 16 + x = x^2 - 7x + 16$$

Next, let $L = D^2$. We can do this because the minimum value of D^2 will occur at the same value of x as the minimum value of D. Therefore, it's simpler to minimize D^2 rather than D (because we won't have to take a square root).

$$L = x^2 - 7x + 16$$

Now, take the derivative and set it equal to zero.

$$\frac{dL}{dx} = 2x - 7 = 0$$

$$x = \frac{7}{2}$$

Solving for y, we get $y = \sqrt{\frac{7}{2}}$.

Finally, because $\frac{d^2L}{dx^2} = 2$, the point $\left(\frac{7}{2}, \sqrt{\frac{7}{2}}\right)$ is the minimum distance from the point (4, 0).

PRACTICE PROBLEM SET 20

Now try these problems on your own. The answers are at the end of this chapter, starting on page 340.

1. A rectangle has its base on the x-axis and its two upper corners on the parabola $y = 12 - x^2$. What is the largest possible area of the rectangle?

2. An open rectangular box is to be made from a 9- by 12-inch piece of tin by cutting squares of side x inches from the corners and folding up the sides. What should x be to maximize the volume of the box?

3. A 384-square-meter plot of land is to be enclosed by a fence and divided into two equal parts by another fence parallel to one pair of sides. What dimensions of the outer rectangle will minimize the amount of fence used?

4. What is the radius of a cylindrical soda can with a volume of 512 cubic inches that will use the minimum material?

5. A swimmer is at a point 500 meters from the closest point on a straight shoreline. She needs to reach a cottage located 1800 meters downshore from the closest point. If she swims at 4 meters per second and she walks at 6 meters per second, how far from the cottage should she come ashore so as to arrive at the cottage in the shortest time?

6. Find the closest point on the curve $x^2 + y^2 = 1$ to the point (2, 1).

7. A window consists of an open rectangle topped by a semicircle and is to have a perimeter of 288 inches. Find the radius of the semicircle that will maximize the area of the window.

8. The range of a projectile is $R = \dfrac{v_0^{\,2} \sin 2\theta}{g}$, where v_0 is its initial velocity, g is the acceleration due to gravity and is a constant, and θ is its firing angle. Find the angle that maximizes the projectile's range.

EXPLORING BEHAVIORS OF IMPLICIT RELATIONS

Just as we can find critical points for explicitly defined functions, so too can we find them for implicit functions. In other words, a point on an implicitly defined function where the derivative does not exist or is undefined is a critical point of the function.

Note: This topic is rarely tested on the AP Exam.

Example 36: Given $3x^2 + 5y^2 - 2x + 2y = x$, find any critical points.

First, let's take the derivative implicitly: $6x + 10y\dfrac{dy}{dx} - 2 + 2\dfrac{dy}{dx} = 1$. Now let's isolate $\dfrac{dy}{dx}$. Group: $10y\dfrac{dy}{dx} + 2\dfrac{dy}{dx} = 3 - 6x$. Factor: $(10y + 2)\dfrac{dy}{dx} = 3 - 6x$. Divide: $\dfrac{dy}{dx} = \dfrac{3 - 6x}{10y + 2}$. Now let's find where the derivative is zero or undefined. First, the derivative will be zero where the numerator is zero, which is at $x = \dfrac{1}{2}$. Second, the derivative will be undefined where the denominator is zero, which is at $y = -\dfrac{1}{5}$.

At the point where $x = \dfrac{1}{2}$, $3\left(\dfrac{1}{2}\right)^2 + 5y^2 - 2\left(\dfrac{1}{2}\right) + 2y = \left(\dfrac{1}{2}\right)$, so $5y^2 + 2y - \dfrac{3}{4} = 0$ and thus $y = 0.236$ or $y = -0.636$.

At the point where $y = -\dfrac{1}{5}$, $3x^2 + 5\left(-\dfrac{1}{5}\right)^2 - 2x + 2\left(-\dfrac{1}{5}\right) = x$, so $3x^2 - 3x - \dfrac{1}{5} = 0$ and thus $x = 1.063$ or $x = -0.063$.

Therefore, the critical points are $\left(\dfrac{1}{2}, 0.236\right)$, $\left(\dfrac{1}{2}, -0.636\right)$, $\left(1.063, -\dfrac{1}{5}\right)$, and $\left(-0.063, -\dfrac{1}{5}\right)$.

Chapter 7 Drill

The answers are at the end of the chapter, starting on page 347.

1 Mark for Review

A rectangle has its base on the x-axis and its two upper corners on the semicircle $y = \sqrt{200 - x^2}$. Find the value of x that maximizes the area of the rectangle.

(A) 0

(B) 5

(C) 10

(D) $\sqrt{200}$

2 Mark for Review

An open rectangular box with a square base is to have a volume of 108 cubic inches. Find the dimensions of the box that minimize its surface area (base by height).

(A) 3 inches by 6 inches

(B) 6 inches by 6 inches

(C) 3 inches by 3 inches

(D) 6 inches by 3 inches

3 Mark for Review

Find the coordinates of the maximum and minimum (if they exist) of $y = x^3 - 3x^2 - 24x + 6$.

	Minimum	Maximum
(A)	(–4, –10)	(2, 46)
(B)	(4, –74)	(–2, 34)
(C)	(1, –20)	None
(D)	(1, –20)	(2, 46)

4 Mark for Review

At what value(s) of x does the curve $y = x^3 - 12x^2 + 120$ have a point (or points) of inflection?

(A) (4, –8)

(B) (0, 120)

(C) (8, –136)

(D) (–4, –136)

5 Mark for Review

On what interval is the curve $y = \sin x + \sqrt{3}\cos x$ decreasing on $[0, \pi]$?

(A) $\left(0, \frac{\pi}{6}\right)$

(B) $\left(\frac{\pi}{6}, \pi\right)$

(C) $\left(\frac{\pi}{2}, 1\right)$

(D) The function is always increasing.

6 Mark for Review

Find the value of c that satisfies the MVTD for $f(x) = x^2 - 6x + 5$ on the interval [3, 8].

(A) 5

(B) $\frac{11}{2}$

(C) 11

(D) There is no value of c.

7 Mark for Review

Find the value of c that satisfies the MVTD for $f(x)=\frac{2}{x-4}$ on the interval [2, 6].

(A) –2

(B) 6

(C) 8

(D) There is no value of c.

8 Mark for Review

Find all of the values of c that satisfy Rolle's Theorem for $f(x)=x^4-5x^2+4$ on the interval [–3, 3].

(A) $c=0$

(B) $c=\pm\sqrt{\frac{5}{2}}$

(C) $c=0, \pm 2$

(D) $c=0, \pm\sqrt{\frac{5}{2}}$

ANSWERS TO PRACTICE PROBLEM SETS

Practice Problem Set 18

1. $c = 0$

The Mean Value Theorem says that if $f(x)$ is continuous on the interval $[a, b]$ and is differentiable everywhere on the interval (a, b), then there exists at least one number c on the interval (a, b) such that $f'(c) = \frac{f(b) - f(a)}{b - a}$. Here, the function is $f(x) = 3x^2 + 5x - 2$ and the interval is $[-1, 1]$. Thus, the Mean Value Theorem says that $f'(c) = \frac{\left(3(1)^2 + 5(1) - 2\right) - \left(3(-1)^2 + 5(-1) - 2\right)}{(1+1)}$. This simplifies to $f'(c) = 5$. Next, find $f'(c)$. The derivative of $f(x)$ is $f'(x) = 6x + 5$, so $f'(c) = 6c + 5$. Now, solve for c: $6c + 5 = 5$ and $c = 0$. Note that 0 is in the interval $(-1, 1)$, just as expected.

2. $c = \frac{4}{\sqrt{3}}$

The Mean Value Theorem says that if $f(x)$ is continuous on the interval $[a, b]$ and is differentiable everywhere on the interval (a, b), then there exists at least one number c on the interval (a, b) such that $f'(c) = \frac{f(b) - f(a)}{b - a}$. Here, the function is $f(x) = x^3 + 24x - 16$ and the interval is $[0, 4]$. Thus, the Mean Value Theorem says that $f'(c) = \frac{\left((4)^3 + 24(4) - 16\right) - \left((0)^3 + 24(0) - 16\right)}{(4-0)}$. This simplifies to $f'(c) = 40$. Next, find $f'(c)$ from the equation. The derivative of $f(x)$ is $f'(x) = 3x^2 + 24$, so $f'(c) = 3c^2 + 24$. Now, solve for c: $3c^2 + 24 = 40$ and $c = \pm\frac{4}{\sqrt{3}}$. Note that $\frac{4}{\sqrt{3}}$ is in the interval $(0, 4)$, but $-\frac{4}{\sqrt{3}}$ is *not* in the interval. Thus, the answer is only $c = \frac{4}{\sqrt{3}}$. It's *very* important to check that the answers you get for c fall in the given interval when doing Mean Value Theorem problems.

3. $c = \sqrt{2}$

The Mean Value Theorem says that if $f(x)$ is continuous on the interval $[a, b]$ and is differentiable everywhere on the interval (a, b), then there exists at least one number c on the interval (a, b) such that $f'(c) = \frac{f(b) - f(a)}{b - a}$. Here, the function is $f(x) = \frac{6}{x} - 3$ and the interval is [1, 2]. Thus, the Mean Value Theorem says that $f'(c) = \frac{\left(\frac{6}{2} - 3\right) - \left(\frac{6}{1} - 3\right)}{(2 - 1)}$. This simplifies to $f'(c) = -3$. Next, find $f'(c)$ from the equation. The derivative of $f(x)$ is $f'(x) = -\frac{6}{x^2}$, so $f'(c) = -\frac{6}{c^2}$. Now, solve for c: $-\frac{6}{c^2} = -3$ and $c = \pm\sqrt{2}$. Note that $c = \sqrt{2}$ is in the interval (1, 2), but $-\sqrt{2}$ is *not* in the interval. Thus, the answer is only $c = \sqrt{2}$. It's *very* important to check that the answers you get for c fall in the given interval when doing Mean Value Theorem problems.

4. *No Solution.*

The Mean Value Theorem says that if $f(x)$ is continuous on the interval $[a, b]$ and is differentiable everywhere on the interval (a, b), then there exists at least one number c on the interval (a, b) such that $f'(c) = \frac{f(b) - f(a)}{b - a}$. Here, the function is $f(x) = \frac{6}{x} - 3$ and the interval is [–1, 2]. Note that the function is *not* continuous on the interval. It has an essential discontinuity (vertical asymptote) at $x = 0$. Thus, the Mean Value Theorem does not apply on the interval, and there is no solution.

Suppose you applied the theorem anyway. You would get $f'(c)=\dfrac{\left(\dfrac{6}{2}-3\right)-\left(\dfrac{6}{-1}-3\right)}{(2+1)}$.

This simplifies to $f'(c) = 3$. Next, find $f'(c)$ from the equation. The derivative of $f(x)$ is $f'(x)=-\dfrac{6}{x^2}$, so $f'(c)=-\dfrac{6}{c^2}$. Now, solve for c: $-\dfrac{6}{c^2}=3$. This has no real solution. Therefore, remember that it's *very* important to check that the function is continuous and differentiable everywhere on the given interval (it does not have to be differentiable at the endpoints) when doing Mean Value Theorem problems. If it is not, then the theorem does not apply.

5. $c = 4$

Rolle's Theorem says that if $f(x)$ is continuous on the interval $[a, b]$ and is differentiable everywhere on the interval (a, b), and if $f(a) = f(b) = 0$, then there exists at least one number c on the interval (a, b) such that $f'(c) = 0$. Here, the function is $f(x) = x^2 - 8x + 12$ and the interval is $[2, 6]$. First, check if the function is equal to zero at both of the endpoints: $f(6) = (6)^2 - 8(6) + 12 = 0$ and $f(2) = (2)^2 - 8(2) + 12 = 0$. Next, take the derivative to find $f'(c)$: $f'(x) = 2x - 8$, so $f'(c) = 2c - 8$. Now, solve for c: $2c - 8 = 0$ and $c = 4$. Note that 4 is in the interval $(2, 6)$, just as expected.

6. $c=\dfrac{1}{2}$

Rolle's Theorem says that if $f(x)$ is continuous on the interval $[a, b]$ and is differentiable everywhere on the interval (a, b), and if $f(a) = f(b) = 0$, then there exists at least one number c on the interval (a, b) such that $f'(c) = 0$. Here, the function is $f(x) = x(1 - x)$ and the interval is $[0, 1]$. First, check if the function is equal to zero at both of the endpoints: $f(0) = (0)(1 - 0) = 0$ and $f(1) = (1)(1 - 1) = 0$. Next, take the derivative to find $f'(c)$: $f'(x) = 1 - 2x$, so $f'(c) = 1 - 2c$. Now, solve for c: $1 - 2c = 0$ and $c=\dfrac{1}{2}$. Note that $\dfrac{1}{2}$ is in the interval $(0, 1)$, just as expected.

7. *No Solution.*

Rolle's Theorem says that if $f(x)$ is continuous on the interval $[a, b]$ and is differentiable everywhere on the interval (a, b), and if $f(a) = f(b) = 0$, then there exists at least one number c on the interval (a, b) such that $f'(c) = 0$. Here, the function is $f(x) = 1 - \frac{1}{x^2}$ and the interval is $[-1, 1]$. Note that the function is *not* continuous on the interval. It has an essential discontinuity (vertical asymptote) at $x = 0$. Thus, Rolle's Theorem does not apply on the interval, and there is no solution.

Suppose you applied the theorem anyway. First, check if the function is equal to zero at both of the endpoints: $f(1) = 1 - \frac{1}{(1)^2} = 0$ and $f(-1) = 1 - \frac{1}{(-1)^2} = 0$. Next, take the derivative to find $f'(c)$: $f'(x) = \frac{2}{x^3}$, so $f'(c) = \frac{2}{c^3}$. This has no solution.

Therefore, remember that it's *very* important to check that the function is continuous and differentiable everywhere on the given interval (it does not have to be differentiable at the endpoints) when doing Rolle's Theorem problems. If it is not, then the theorem does not apply.

8. $c = \frac{1}{8}$

Rolle's Theorem says that if $f(x)$ is continuous on the interval $[a, b]$ and is differentiable everywhere on the interval (a, b), and if $f(a) = f(b) = 0$, then there exists at least one number c on the interval (a, b) such that $f'(c) = 0$. Here, the function is $f(x) = x^{\frac{2}{3}} - x^{\frac{1}{3}}$ and the interval is $[0, 1]$. First, check if the function is equal to zero at both of the endpoints: $f(0) = (0)^{\frac{2}{3}} - (0)^{\frac{1}{3}} = 0$ and $f(1) = (1)^{\frac{2}{3}} - (1)^{\frac{1}{3}} = 0$. Next, take the derivative to find $f'(c)$: $f(x) = \frac{2}{3}x^{-\frac{1}{3}} - \frac{1}{3}x^{-\frac{2}{3}} = \frac{2}{3\sqrt[3]{x}} - \frac{1}{3\sqrt[3]{x^2}}$, so $f(c) = \frac{2}{3\sqrt[3]{c}} - \frac{1}{3\sqrt[3]{c^2}}$. Now, solve for c: $\frac{2}{3\sqrt[3]{c}} - \frac{1}{3\sqrt[3]{c^2}} = 0$ and $c = \frac{1}{8}$. Note that $\frac{1}{8}$ is in the interval $(0, 1)$, just as expected.

Practice Problem Set 19

1. Minimum at $\left(\sqrt{3}, -6\sqrt{3}-6\right)$; Maximum at $\left(-\sqrt{3}, 6\sqrt{3}-6\right)$; Point of inflection at (0, –6)

First, find the y-intercept. Set $x = 0$ to get $y = (0)^3 - 9(0) - 6 = -6$. Therefore, the y-intercept is (0, –6). Next, find any critical points using the first derivative. The derivative is $\frac{dy}{dx} = 3x^2 - 9$. If you set this equal to zero and solve for x, you get $x = \pm\sqrt{3}$. Plug $x = \sqrt{3}$ and $x = -\sqrt{3}$ into the original equation to find the y-coordinates of the critical points. When $x = \sqrt{3}$, $y = \left(\sqrt{3}\right)^3 - 9\left(\sqrt{3}\right) - 6 = -6\sqrt{3} - 6$. When $x = -\sqrt{3}$, $y = \left(-\sqrt{3}\right)^3 - 9\left(-\sqrt{3}\right) - 6 = 6\sqrt{3} - 6$. Thus, the critical points are $\left(\sqrt{3}, -6\sqrt{3}-6\right)$ and $\left(-\sqrt{3}, 6\sqrt{3}-6\right)$. Next, take the second derivative to find any points of inflection. The second derivative is $\frac{d^2y}{dx^2} = 6x$, which is equal to zero at $x = 0$. Note that this is the y-intercept (0, –6), which you already found, so there is a point of inflection at (0, –6). Next, determine if each critical point is maximum, minimum, or something else. If you plug $x = \sqrt{3}$ into the second derivative, the value is obviously positive, so $\left(\sqrt{3}, -6\sqrt{3}-6\right)$ is a minimum. If you plug $x = -\sqrt{3}$ into the second derivative, the value is obviously negative, so $\left(-\sqrt{3}, 6\sqrt{3}-6\right)$ is a maximum. Now, you can draw the curve. It looks like the following:

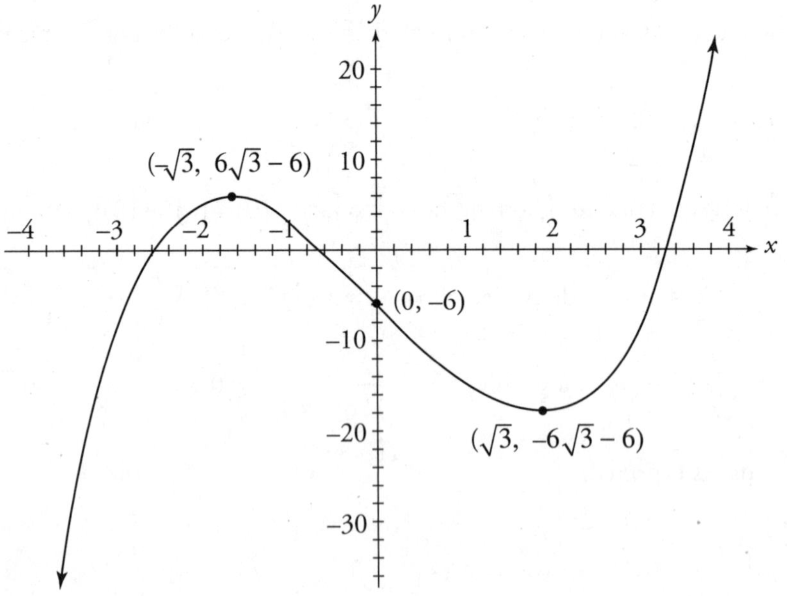

2. Minimum at (–3, –4); Maximum at (–1, 0); Point of inflection at (–2, –2)

First, find the y-intercept. Set $x = 0$ to get $y = -(0)^3 - 6(0)^2 - 9(0) - 4 = -4$. Therefore, the y-intercept is (0, –4). Next, find any critical points using the first derivative. The derivative is $\frac{dy}{dx} = -3x^2 - 12x - 9$. If you set this equal to zero and solve for x, you get $x = -1$ and $x = -3$. Plug $x = -1$ and $x = -3$ into the original equation to find the y-coordinates of the critical points. When $x = -1$, $y = -(-1)^3 - 6(-1)^2 - 9(-1) - 4 = 0$. When $x = -3$, $y = -(-3)^3 - 6(-3)^2 - 9(-3) - 4 = -4$. Thus, the critical points are (–1, 0) and (–3, –4). Next, take the second derivative to find any points of inflection. The second derivative is $\frac{d^2y}{dx^2} = -6x - 12$, which is equal to zero at $x = -2$. Plug $x = -2$ into the original equation to find the y-coordinate: $y = -(-2)^3 - 6(-2)^2 - 9(-2) - 4 = -2$, so there is a point of inflection at (–2, –2). Next, determine if each critical point is maximum, minimum, or something else. If you plug $x = -1$ into the second derivative, the value is negative, so (–1, 0) is a maximum. If you plug $x = -3$ into the second derivative, the value is positive, so (–3, –4) is a minimum. Now, you can draw the curve. It looks like the following:

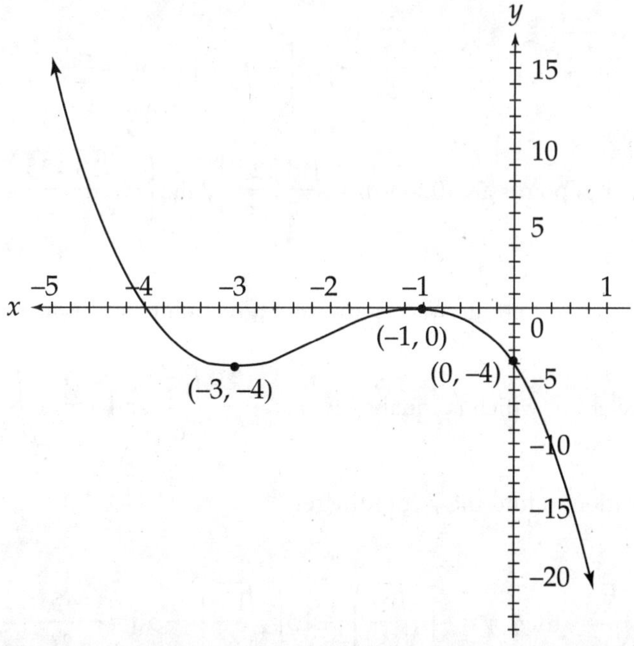

3. Minimum at (0, –36); Maxima at $\left(\sqrt{\frac{13}{2}}, \frac{25}{4}\right)$ and $\left(-\sqrt{\frac{13}{2}}, \frac{25}{4}\right)$; points of inflection at $\left(\sqrt{\frac{13}{6}}, -\frac{451}{36}\right)$ and $\left(-\sqrt{\frac{13}{6}}, -\frac{451}{36}\right)$

First, you can easily see that the graph has x-intercepts at $x = \pm 2$ and $x = \pm 3$. Next, before you take the derivative, multiply the two terms so you don't have to use the Product Rule. You get $y = -x^4 + 13x^2 - 36$. Now we can take the derivative: $\frac{dy}{dx} = -4x^3 + 26x$. Next, set the derivative equal to zero to find the critical points. There are three solutions: $x = 0$, $x = \sqrt{\frac{13}{2}}$, and $x = -\sqrt{\frac{13}{2}}$.

Plug these values into the original equation to find the y-coordinates of the critical points. When $x = 0$, $y = -(0)^4 + 13(0)^2 - 36 = -36$. When $x = \sqrt{\frac{13}{2}}$, $y = -\left(\sqrt{\frac{13}{2}}\right)^4 + 13\left(\sqrt{\frac{13}{2}}\right)^2 - 36 = \frac{25}{4}$.

When $x = -\sqrt{\frac{13}{2}}$, $y = -\left(-\sqrt{\frac{13}{2}}\right)^4 + 13\left(-\sqrt{\frac{13}{2}}\right)^2 - 36 = \frac{25}{4}$.

Thus, the critical points are (0, –36), $\left(\sqrt{\frac{13}{2}}, \frac{25}{4}\right)$, and $\left(-\sqrt{\frac{13}{2}}, \frac{25}{4}\right)$.

Next, take the second derivative to find any points of inflection. The second derivative is $\frac{d^2y}{dx^2} = -12x^2 + 26$, which is equal to zero at $x = \sqrt{\frac{13}{6}}$ and $x = -\sqrt{\frac{13}{6}}$. Plug these values into the original equation to find the y-coordinates.

When $x = \sqrt{\frac{13}{6}}$, then $y = -\left(\sqrt{\frac{13}{6}}\right)^4 + 13\left(\sqrt{\frac{13}{6}}\right)^2 - 36 = -\frac{451}{36}$.

When $x = -\sqrt{\frac{13}{6}}$, then $y = -\left(-\sqrt{\frac{13}{6}}\right)^4 + 13\left(-\sqrt{\frac{13}{6}}\right)^2 - 36 = -\frac{451}{36}$. So there are points of inflection at $\left(\sqrt{\frac{13}{6}}, -\frac{451}{36}\right)$ and $\left(-\sqrt{\frac{13}{6}}, -\frac{451}{36}\right)$. Next, determine if each critical point is maximum, minimum, or something else. If you plug $x = 0$ into the second derivative, the value is positive, so (0, –36) is a minimum. If you plug $x = \sqrt{\frac{13}{2}}$ into the second derivative, the value is negative, so $\left(\sqrt{\frac{13}{2}}, \frac{25}{4}\right)$ is a maximum. If you plug $x = -\sqrt{\frac{13}{2}}$ into the second derivative, the value is negative, so $\left(-\sqrt{\frac{13}{2}}, \frac{25}{4}\right)$ is a maximum. Now, you can draw the curve. It looks like the following:

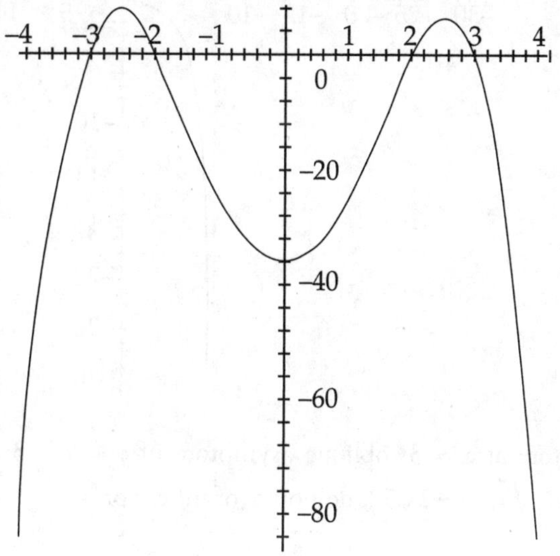

4. Vertical asymptote at $x = -8$; Horizontal asymptote at $y = 1$; No maxima, minima, or points of inflection

First, notice that there is an x-intercept at $x = 3$ and that the y-intercept is $\left(0, -\frac{3}{8}\right)$. Next, take the derivative: $\frac{dy}{dx} = \frac{(x+8)(1)-(x-3)(1)}{(x+8)^2} = \frac{11}{(x+8)^2}$. Next, set the derivative equal to zero to find the critical points. There is no solution. Next, take the second derivative: $\frac{d^2y}{dx^2} = -\frac{22}{(x+8)^3}$. If you set this equal to zero, there is also no solution. Therefore, there are no maxima, minima, or points of inflection. Note that the first derivative is always positive. This means that the curve is always increasing. Also notice that the second derivative changes sign from positive to negative at $x = -8$. This means that the curve is concave up for values of x less than $x = -8$ and concave down for values of x greater than $x = -8$.

Now, you can draw the curve. It looks like the following:

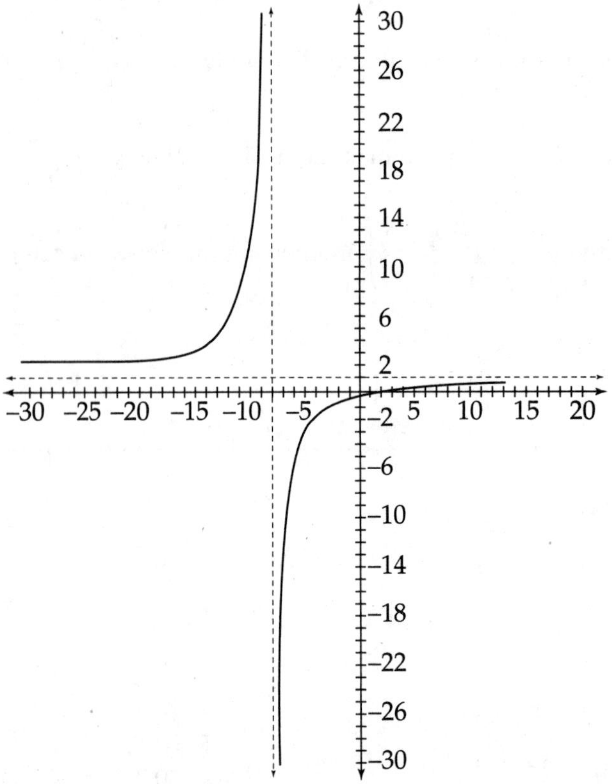

5. Vertical asymptote at $x = 3$; oblique asymptote of $y = x + 3$; maximum at $\left(3+\sqrt{5},\ 6+2\sqrt{5}\right)$; minimum at $\left(3-\sqrt{5},\ 6-2\sqrt{5}\right)$; no points of inflection

First, notice that there are x-intercepts at $x = \pm 2$ and that the y-intercept is $\left(0, \frac{4}{3}\right)$. There is a vertical asymptote at $x = 3$. There is no horizontal asymptote, but notice that the degree of the numerator of the function is 1 greater than the denominator. This means that there is an oblique (slant) asymptote. You can find this by dividing the denominator into the numerator and looking at the quotient. You get

$$\begin{array}{r} x+3+\dfrac{5}{x-3} \\ x-3\overline{\big)\ x^2+0x-4} \end{array}$$

This means that as $x \to \pm\infty$, the function will behave like the function $y = x + 3$. This means that there is an oblique asymptote of $y = x + 3$. Next, take the derivative: $\frac{dy}{dx} = \frac{(x-3)(2x)-(x^2-4)(1)}{(x-3)^2} = \frac{x^2-6x+4}{(x-3)^2} = 1 - \frac{5}{(x-3)^2}$. Next, set the derivative equal to zero to find the critical points. There are two solutions: $x = 3+\sqrt{5}$ and $x = 3-\sqrt{5}$. Plug these values into the original equation to find the y-coordinates of the critical points. When $x = 3+\sqrt{5}$, $y = \dfrac{\left(3+\sqrt{5}\right)^2 - 4}{\left(3+\sqrt{5}\right)-3} = 6+2\sqrt{5}$. When $x = 3-\sqrt{5}$, $y = \dfrac{\left(3-\sqrt{5}\right)^2 - 4}{\left(3-\sqrt{5}\right)-3} = 6-2\sqrt{5}$. Thus, the critical points are $\left(3+\sqrt{5}, 6+2\sqrt{5}\right)$ and $\left(3-\sqrt{5}, 6-2\sqrt{5}\right)$. Next, take the second derivative: $\dfrac{d^2y}{dx^2} = \dfrac{10}{(x-3)^3}$. If you set this equal to zero, there is no solution. Therefore, there is no point of inflection. But, notice that the second derivative changes sign from negative to positive at $x = 3$. This means that the curve is concave down for values of x less than $x = 3$ and concave up for values of x greater than $x = 3$. Next, determine if each critical point is maximum, minimum, or something else. If you plug $x = 3+\sqrt{5}$ into the second derivative, the value is positive, so $\left(3+\sqrt{5}, 6+2\sqrt{5}\right)$ is a minimum. If you plug $x = 3-\sqrt{5}$ into the second derivative, the value is positive, so $\left(3-\sqrt{5}, 6-2\sqrt{5}\right)$ is a maximum.

Now, you can draw the curve. It looks like the following:

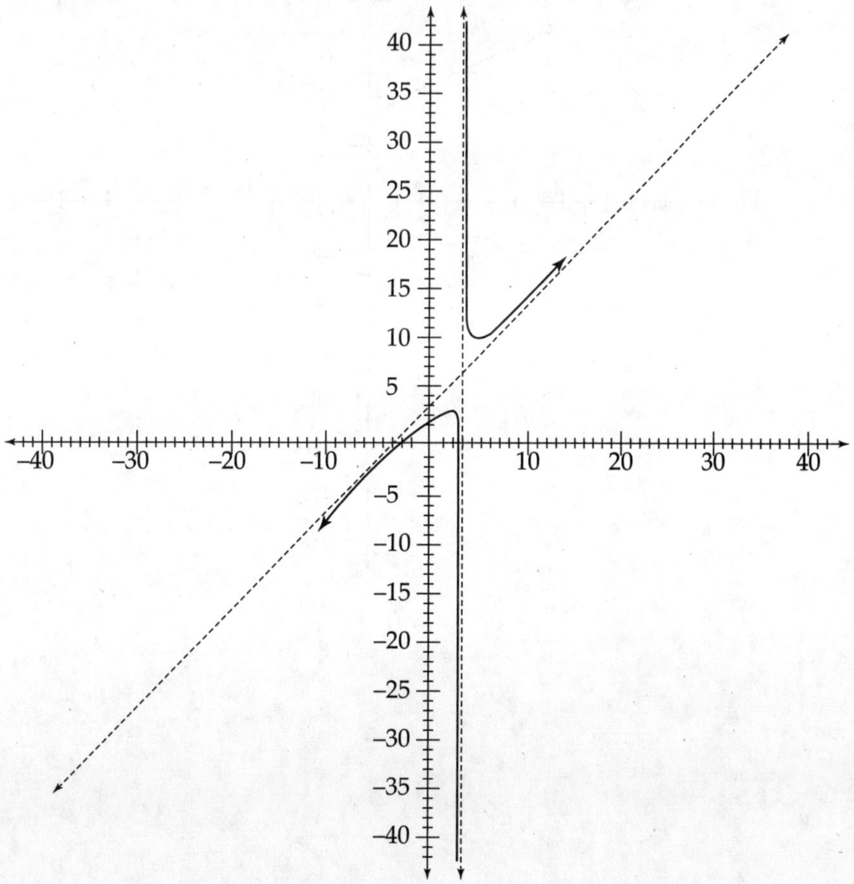

6. Minimum at (0, 3); no points of inflection; cusp at (0, 3)

First, notice that the function is always positive. There is a y-intercept at (0, 3). There is no x-intercept. Next, take the derivative: $\frac{dy}{dx} = \frac{2}{3}x^{-\frac{1}{3}}$. If you set the derivative equal to zero, there is no solution. But, notice that the derivative is not defined at $x = 0$. This means that the function has either a vertical tangent or a cusp at $x = 0$. You'll be able to determine which after you take the second derivative. Notice also that the derivative is negative for $x < 0$ and positive for $x > 0$. Therefore, the curve is decreasing for $x < 0$ and increasing for $x > 0$. Next, take the second derivative: $\frac{dy}{dx} = -\frac{2}{9}x^{-\frac{4}{3}}$. If you set this equal to zero, there is no solution. The second derivative is always negative, which means that the curve is always concave down and that the curve has a cusp at $x = 0$. Note that if it had switched concavity there, then $x = 0$ would be a vertical tangent.

Now, you can draw the curve. It looks like the following:

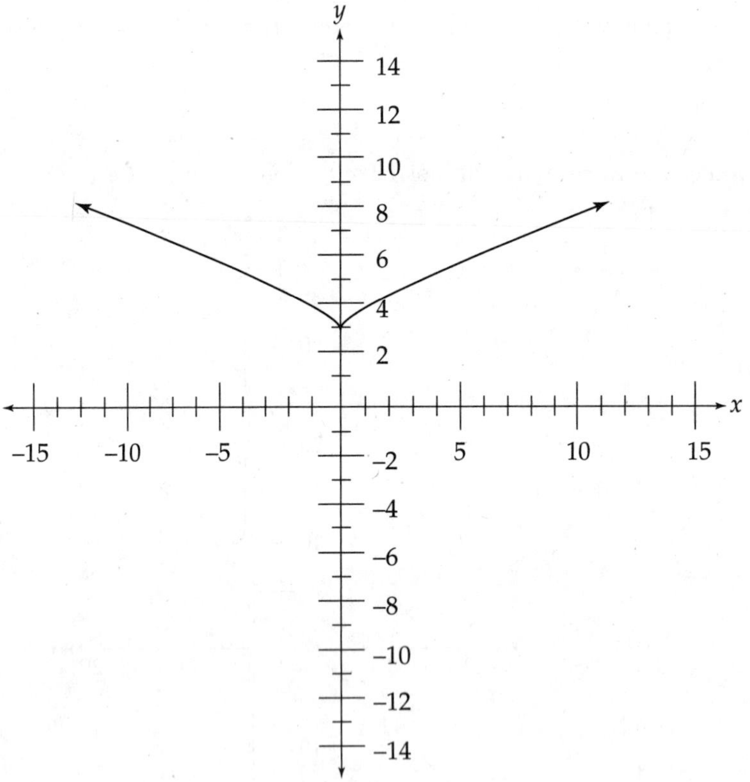

7. Maximum at (1, 1); no point of inflection; cusp at (0, 0)

First, notice that the curve has x-intercepts at $x = 0$ and $x = \frac{27}{8}$ and that there is a y-intercept at (0, 0). Before you take the derivative, expand the expression. That way, you won't have to use the Product Rule. You get $y = 3x^{\frac{2}{3}} - 2x$. Next, take the derivative: $\frac{dy}{dx} = 2x^{-\frac{1}{3}} - 2$. Next, set the derivative equal to zero to find the critical points. There is one solution: $x = 1$. Plug this value into the original equation to find the y-coordinate of the critical point: when $x = 1$, $y = 3(1)^{\frac{2}{3}} - 2(1) = 1$. Thus, the critical point is (1, 1). But, notice that the derivative is not defined at $x = 0$. This means that the function has either a vertical tangent or a cusp at $x = 0$. You'll be able to determine which after you take the second derivative. Notice also that the derivative is negative for $x < 0$ and for $x > 1$ and positive for $0 < x < 1$. Therefore, the curve is decreasing for $x < 0$ and for $x > 1$, and increasing for $0 < x < 1$. Next, take the second derivative: $\frac{d^2y}{dx^2} = -\frac{2}{3}x^{-\frac{4}{3}}$. If you set this equal to zero, there is no solution. Therefore, there is no point of inflection. The second derivative is always negative, which means that the curve is always concave down and that the curve has a cusp at $x = 0$. Note that if it had switched concavity there, then $x = 0$ would be a vertical tangent. Next, determine if each critical point is maximum, minimum, or something else. If you plug $x = 1$ into the second derivative, the value is negative, so (1, 1) is a maximum.

Now, you can draw the curve. It looks like the following:

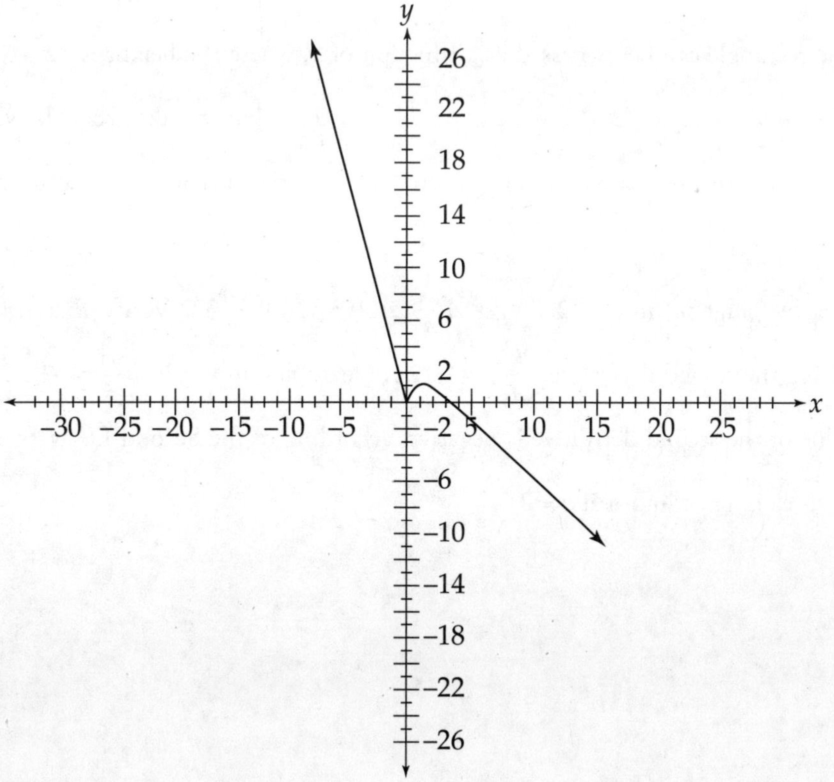

Practice Problem Set 20

1. The area is 32.

First, draw a picture.

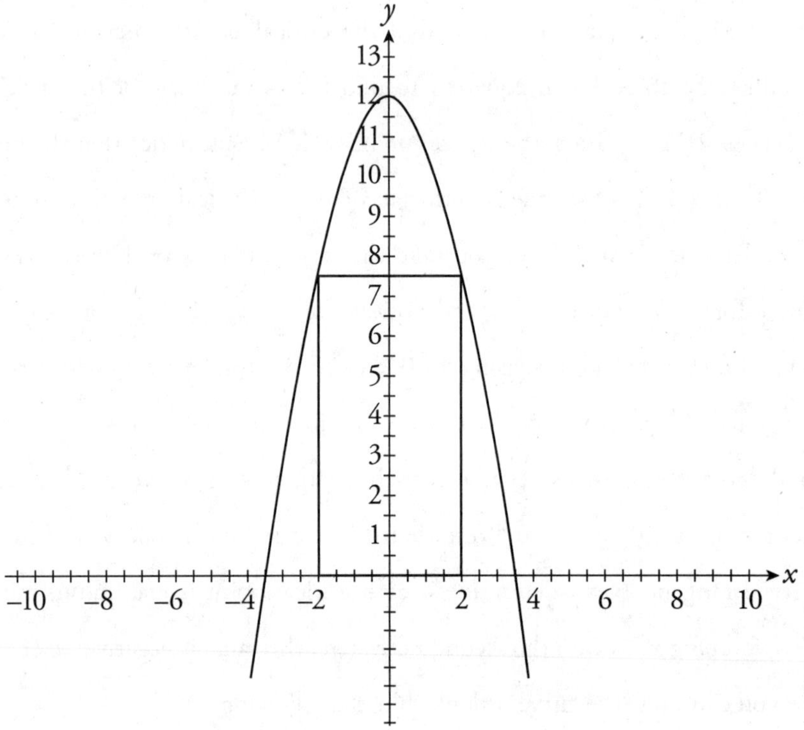

The rectangle can be expressed as a function of x, where the height is $12 - x^2$ and the base is $2x$. The area is $A = 2x(12 - x^2) = 24x - 2x^3$. Now, take the derivative: $\frac{dA}{dx} = 24 - 6x^2$. Next, set the derivative equal to zero: $24 - 6x^2 = 0$. Solve for x to get $x = \pm 2$. A negative answer doesn't make any sense in this case, so use the solution $x = 2$. Then find the area by plugging in $x = 2$ to get $A = 24(2) - 2(2)^3 = 32$. Verify that this is a maximum by taking the second derivative: $\frac{d^2A}{dx^2} = -12x$. Next, plug in $x = 2$: $\frac{d^2A}{dx^2} = -12(2) = -24$. Because the value of the second derivative is negative, according to the Second Derivative Test (see page 291), the area is a maximum at $x = 2$.

2. $x = \dfrac{7-\sqrt{13}}{2} \approx 1.697$ inches

First, draw a picture.

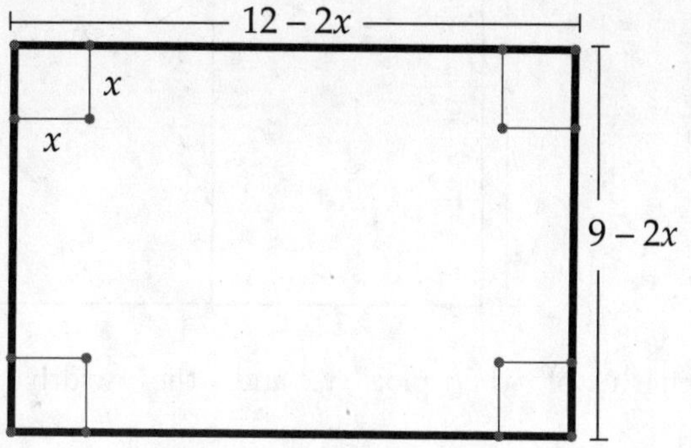

After cutting out the squares of side x and folding the sides, the dimensions of the box will be width = $9 - 2x$, length = $12 - 2x$, and depth = x.

Using the formula for the volume of a rectangular prism, you can get an equation for the volume of the box in terms of x: $V = x(9 - 2x)(12 - 2x) = 108x - 42x^2 + 4x^3$.

Now, take the derivative: $\dfrac{dV}{dx} = 108 - 84x + 12x^2$. Next, set the derivative equal to zero: $108 - 84x + 12x^2 = 0$. Solving for x, you get $x = \dfrac{7 \pm \sqrt{13}}{2} \approx 5.303$ or 1.697. You can't cut two squares of length 5.303 inches from a side of length 9 inches, so you can get rid of that answer. Therefore, the answer must be $x = \dfrac{7-\sqrt{13}}{2} \approx 1.697$ inches. Verify that this is a maximum by taking the second derivative: $\dfrac{d^2V}{dx^2} = -84 + 24x$. Next, plug in $x = 1.697$ to get approximately $\dfrac{d^2V}{dx^2} = -84 + 24(1.697) = -43.272$. Because the value of the second derivative is negative, according to the Second Derivative Test (see page 291), the volume is a maximum at $x = \dfrac{7-\sqrt{13}}{2} \approx 1.697$ inches.

3. 16 meters by 24 meters

First, draw a picture.

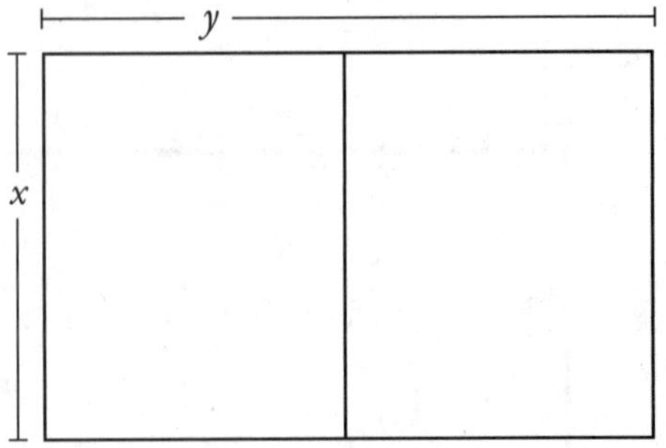

Calling the length of the plot y and the width x, the area of the plot is $A = xy = 384$. The perimeter is $P = 3x + 2y$. So, if you want to minimize the length of the fence, you need to minimize the perimeter of the plot. If you solve the area equation for y, $y = \frac{384}{x}$. Now substitute this for y in the perimeter equation: $P = 3x + 2\left(\frac{384}{x}\right) = 3x + \frac{768}{x}$. Now take the derivative of P: $\frac{dP}{dx} = 3 - \frac{768}{x^2}$. Solving for x, $x = \pm 16$. A negative answer doesn't make any sense in this case, so use the solution $x = 16$ meters. Now solve for y: $y = \frac{384}{16} = 24$ meters.

Verify that this is a minimum by taking the second derivative: $\frac{d^2P}{dx^2} = \frac{1536}{x^3}$. Next, plug in $x = 16$ to get $\frac{d^2P}{dx^2} = \frac{1536}{16^3} = \frac{3}{8}$. Because the value of the second derivative is positive, according to the Second Derivative Test (see page 291), the perimeter is a minimum at $x = 16$.

4. Radius is $\sqrt[3]{\frac{256}{\pi}}$ inches.

First, draw a picture.

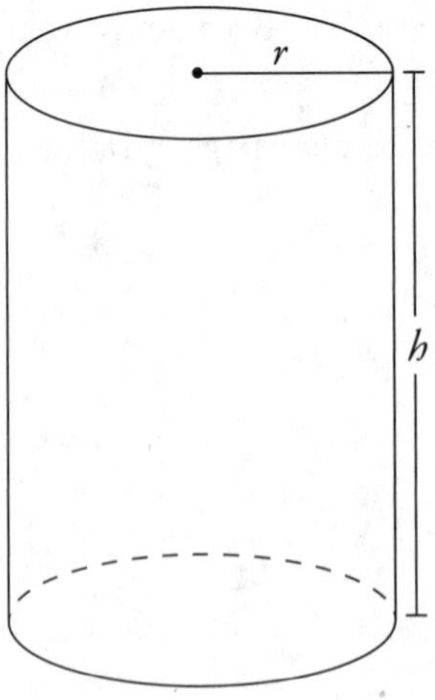

The volume of a cylinder is $V = \pi r^2 h = 512$. The material used for the can is the surface area of the cylinder (don't forget the ends!): $S = 2\pi rh + 2\pi r^2$. If you solve the volume equation for h, you get $h = \frac{512}{\pi r^2}$. Substitute this for h in the surface area equation: $S = 2\pi r\left(\frac{512}{\pi r^2}\right) + 2\pi r^2 = \frac{1024}{r} + 2\pi r^2$. Now take the derivative of S: $\frac{dS}{dr} = -\frac{1024}{r^2} + 4\pi r$. Solving this for r, $r = \sqrt[3]{\frac{256}{\pi}}$ inches. Verify that this is a minimum by taking the second derivative: $\frac{d^2S}{dx^2} = \frac{2048}{r^3} + 4\pi$. Next, plug in $\sqrt[3]{\frac{256}{\pi}}$ and you can see that the value of the second derivative is positive. Therefore, according to the Second Derivative Test (see page 291), the perimeter is a minimum at $r = \sqrt[3]{\frac{256}{\pi}}$.

5. 1,352.786 meters

Think about the situation. If the swimmer swims the whole distance to the cottage, she will be traveling the entire time at her slowest speed. If she swims straight to shore first, minimizing her swimming distance, she will be maximizing her running distance. Therefore, there should be a point, somewhere between the cottage and the point on the shore directly opposite her, where the swimmer should come on land to switch from swimming to running to get to the cottage in the shortest time.

Draw a picture.

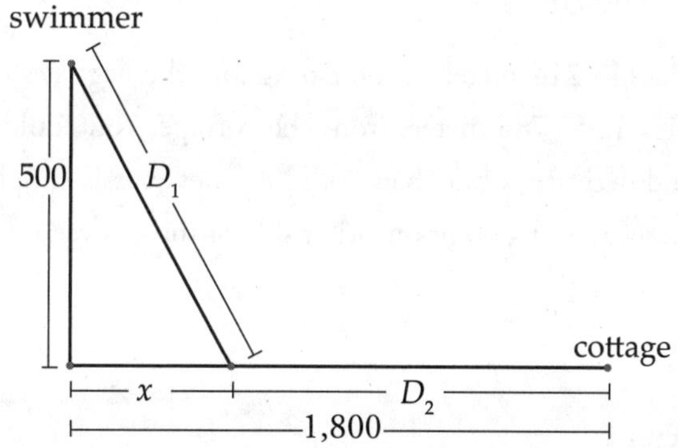

Let x be the distance from the point on the shore directly opposite the swimmer to the point where she comes on land. There are two distances to consider. The first is the diagonal distance that the swimmer swims. This distance, D_1, is $D_1 = \sqrt{500^2 + x^2} = \sqrt{250{,}000 + x^2}$. The second distance, D_2, is simply $D_2 = 1{,}800 - x$. Remember that *rate* × *time* = *distance.*

Use this formula to find the total time that the swimmer needs. The time for the swimmer to travel D_1 is $T_1 = \frac{D_1}{4} = \frac{\sqrt{250{,}000 + x^2}}{4}$ (because she swims at 4 meters per second) and the time for the swimmer to travel D_2 is $T_2 = \frac{D_2}{6} = \frac{1{,}800 - x}{6} = 300 - \frac{x}{6}$. Therefore, the total time is $T = \frac{\sqrt{250{,}000 + x^2}}{4} + 300 - \frac{x}{6}$. Now simply take the derivative: $\frac{dT}{dx} = \frac{1}{4}\left(\frac{1}{2}\right)\left(250{,}000 + x^2\right)^{-\frac{1}{2}}(2x) - \frac{1}{6} = \frac{x}{4\sqrt{250{,}000 + x^2}} - \frac{1}{6}$. Next, set this equal to zero and solve: $\frac{x}{4\sqrt{250{,}000 + x^2}} - \frac{1}{6} = 0$. The best way to solve this is to move the $\frac{1}{6}$ to the other side of the equals sign and cross-multiply:

$$\frac{x}{4\sqrt{250{,}000 + x^2}} = \frac{1}{6}$$

$$6x = 4\sqrt{250{,}000 + x^2}$$

Next, square both sides: $36x^2 = 16(250{,}000 + x^2)$.

Simplify: $20x^2 = 4{,}000{,}000$.

Solve for x: $x = \pm 447.214$ meters. (You can ignore the negative answer.) Therefore, she should land $1{,}800 - 447.214 = 1{,}352.786$ meters from the cottage. You could verify that this is a minimum by taking the second derivative, but that would be messy. It is simpler to use your calculator to check the sign of the derivative at a point on either side of the answer, or to graph the equation for the time.

6. $\left(\frac{2}{\sqrt{5}}, \frac{1}{\sqrt{5}}\right)$

First, draw a picture.

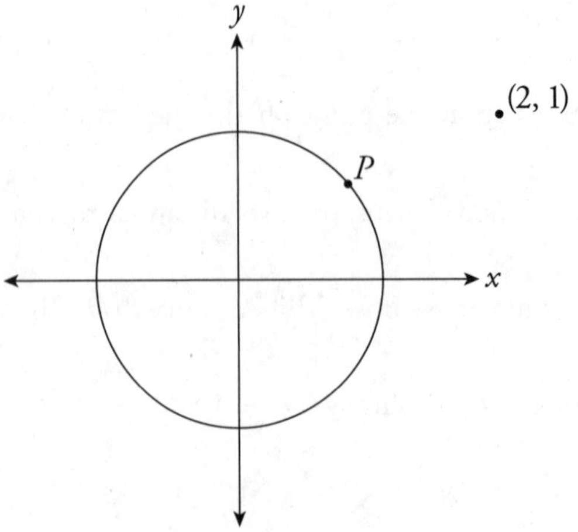

Find an expression for the distance from the point P to the point (2, 1) and then minimize the distance. Calling the coordinates of P (x, y), you can find the distance to (2, 1) using the distance formula: $D^2 = (x - 2)^2 + (y - 1)^2$. Next, just as in Example 35 (page 323), let $L = D^2$ and minimize L: $L = (x - 2)^2 + (y - 1)^2 = x^2 - 4x + 4 + y^2 - 2y + 1$.

Because $x^2 + y^2 = 1$, substitute for y: $L = x^2 - 4x + 4 + (1 - x^2) - 2\sqrt{1 - x^2} + 1$, which simplifies to $L = -4x + 6 - 2\sqrt{1 - x^2}$. Next, take the derivative: $\frac{dL}{dx} = -4 - 2\left(\frac{1}{2}\right)(1 - x^2)^{-\frac{1}{2}}(-2x) = -4 + \frac{2x}{\sqrt{1 - x^2}}$. Next, set the derivative equal to zero: $-4 + \frac{2x}{\sqrt{1 - x^2}} = 0$. The best way to solve this is to move the 4 to the other side of the equals sign and cross-multiply.

$$\frac{2x}{\sqrt{1 - x^2}} = 4$$

$$2x = 4\sqrt{1 - x^2}$$

Next, simplify and square both sides: $x^2 = 4(1 - x^2)$. Now, solve to get $x = \pm\frac{2}{\sqrt{5}}$. Next, find the y-coordinate: $y = \pm\frac{1}{\sqrt{5}}$. There are thus four possible answers, but if you look at the picture, the answer is obviously the point $\left(\frac{2}{\sqrt{5}}, \frac{1}{\sqrt{5}}\right)$. You could verify that this is a minimum by taking the second derivative, but that will be messy. It is simpler to use your calculator to check the sign of the derivative at a point on either side of the answer, or to graph the equation for the distance.

7. $r = \frac{288}{4 + \pi}$ inches

First, draw a picture.

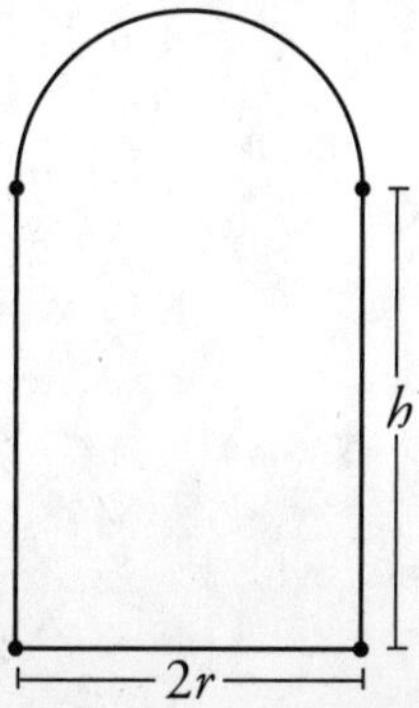

Call the width of the window $2r$. Notice that this is the diameter of the semicircle. Call the height of the window h. The area of the rectangular portion of the window is $2rh$, and the perimeter is $2r + 2h$. The area of the semicircular portion of the window is $\frac{\pi r^2}{2}$, and the perimeter is $\frac{2\pi r}{2} = \pi r$. Therefore, the area of the window is $A = 2rh + \frac{\pi r^2}{2}$, and the perimeter is $2r + 2h + \pi r = 288$. You can use the equation for the perimeter to eliminate a variable from the equation for the area. Isolate h: $h = 144 - r - \frac{\pi r}{2}$. Now substitute for h in the equation for the area: $A = 2r\left(144 - r - \frac{\pi r}{2}\right) + \frac{\pi r^2}{2} = 288r - 2r^2 - \frac{\pi r^2}{2}$. Next, take the derivative: $\frac{dA}{dr} = 288 - 4r - \pi r$. Setting this equal to zero and solving for r, you get $r = \frac{288}{4+\pi}$ inches. You can verify that this is a maximum by taking the second derivative: $\frac{d^2A}{dr^2} = -4 - \pi$. You can see that the value of the second derivative is negative. Therefore, according to the Second Derivative Test (see page 291), the area is a maximum at $r = \frac{288}{4+\pi}$ inches.

8. $\theta = \frac{\pi}{4}$ radians (or 45 degrees)

Simply take the derivative: $\frac{dR}{d\theta} = \frac{v_0^{\,2}}{g}(2\cos 2\theta)$. Note that v_0 and g are constants, so the only variable you need to take the derivative with respect to is θ. Now set the derivative equal to zero: $\frac{v_0^{\,2}}{g}(2\cos 2\theta) = 0$. Although this has an infinite number of solutions, you are interested only in values of θ between 0 and $\frac{\pi}{2}$ radians. The value of θ that makes the derivative zero is $\theta = \frac{\pi}{4}$ radians (or 45 degrees). You can verify that this is a maximum by taking the second derivative: $\frac{d^2R}{d\theta^2} = \frac{v_0^{\,2}}{g}(-4\sin 2\theta)$. You can see that the value of the second derivative is negative at $\theta = \frac{\pi}{4}$. Therefore, according to the Second Derivative Test (see page 291), the range is a maximum at $\theta = \frac{\pi}{4}$ radians.

ANSWERS TO CHAPTER 7 DRILL

1. **C** First, draw a picture:

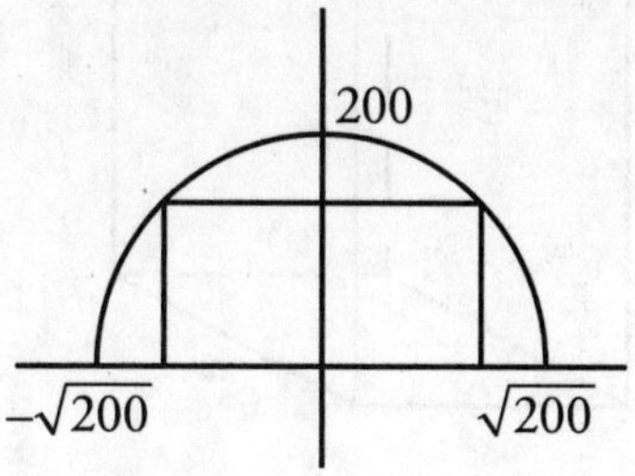

Note that the base of the rectangle is $2x$ and the height of the rectangle is y. Thus, the area of the rectangle is $A = 2xy$. Because $y = \sqrt{200 - x^2}$, you can substitute, and the area is $A = 2x\sqrt{200 - x^2}$.

Take the derivative: $\frac{dA}{dx} = 2\sqrt{200 - x^2} + 2x\left(\frac{1}{2}(200 - x^2)^{-\frac{1}{2}}(-2x)\right) = 2\sqrt{200 - x^2} + \frac{-2x^2}{\sqrt{200 - x^2}}$.

Set the derivative equal to zero and solve for x:

$$2\sqrt{200 - x^2} + \frac{-2x^2}{\sqrt{200 - x^2}} = 0$$

$$2\sqrt{200 - x^2} = \frac{2x^2}{\sqrt{200 - x^2}}$$

$$2(200 - x^2) = 2x^2$$

$$200 - x^2 = x^2$$

$$200 = 2x^2$$

$$x = \pm 10$$

Throw out the negative answer; the maximum is at $x = 10$.

The answer is (C).

2. **D** First, draw a picture:

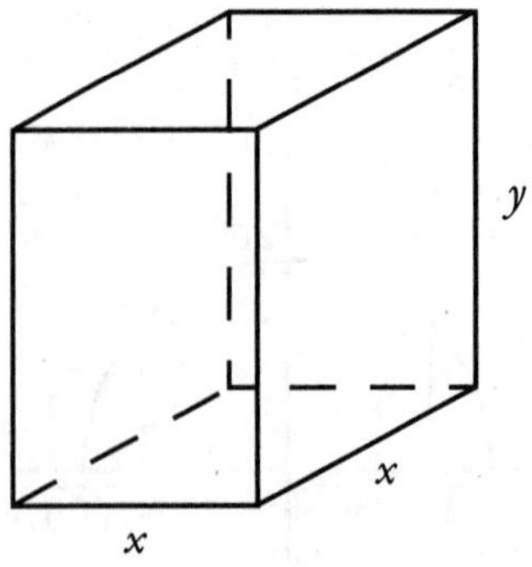

Note that the area of the base is x^2 and the height is *y*, so the volume of the box is $V = x^2 y = 108$ and the surface area is $A = x^2 + 4xy$. If you solve the volume for *y*, you get $y = \frac{108}{x^2}$. Then plug this into the area formula to get the area in terms of *x* only: $A = x^2 + 4x\left(\frac{108}{x^2}\right) = x^2 + \frac{432}{x}$.

Take the derivative: $\frac{dA}{dx} = 2x - \frac{432}{x^2}$.

Set it equal to zero and solve for *x*:

$$2x - \frac{432}{x^2} = 0$$

$$2x^3 = 432$$

$$x^3 = 216$$

$$x = 6 \text{ inches and thus } y = \frac{108}{6^2} = 3 \text{ inches}$$

The answer is (D).

3. **B** First, take the derivative: $\frac{dy}{dx} = 3x^2 - 6x - 24$. Set the derivative equal to zero and solve for *x*:

$$3x^2 - 6x - 24 = 0$$

$$x^2 - 2x - 8 = 0$$

$$(x-4)(x+2) = 0$$

$$x = 4 \text{ and } -2$$

Now that you have the critical values, you can find the *y*-coordinates by plugging the critical values back into the original equation. You get $y = 4^3 - 3(4)^2 - 24(4) + 6 = -74$.

$$y = (-2)^3 - 3(-2)^2 - 24(-2) + 6 = 34$$

Finally, determine which gives the maximum and which gives the minimum by using the Second Derivative Test. The second derivative is $\frac{d^2y}{dx^2} = 6x - 6$.

At $x = 4$, the second derivative is $6(4)-6>0$, so (4, –74) is at a minimum.

At $x = -2$, the second derivative is $6(-2)-6<0$, so (–2, 34) is at a maximum.

The answer is (B).

4. **A** First, find the second derivative.

$$\frac{dy}{dx} = 3x^2 - 24x$$

$$\frac{d^2y}{dx^2} = 6x - 24$$

Now, set the second derivative equal to zero and solve for x: $6x - 24 = 0$, and $x = 4$.

Using the x-coordinate of the point of inflection, find the y-coordinate by plugging x into the original equation. You get $y = (4)^3 - 12(4)^2 + 120 = -8$.

Therefore, the point of inflection is at (4, –8).

The answer is (A).

5. **B** First, take the derivative: $\frac{dy}{dx} = \cos x - \sqrt{3}\sin x$. Next, set the derivative equal to zero and solve for x:

$$\cos x - \sqrt{3}\sin x = 0$$

$$\cos x = \sqrt{3}\sin x$$

$$\frac{1}{\sqrt{3}} = \frac{\sin x}{\cos x} = \tan x$$

$$x = \frac{\pi}{6}$$

Now figure out the interval where the function is increasing or decreasing. Put the critical value on a number line:

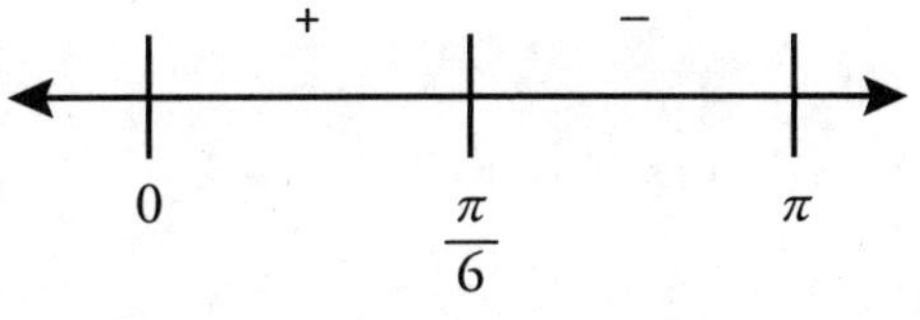

Test values in each interval to find the sign of the derivative. Try a value between $x = 0$ and $x = \frac{\pi}{6}$:
$\frac{dy}{dx} = \cos\frac{\pi}{12} - \sqrt{3}\sin\frac{\pi}{12} > 0$.

Now try a value between $x = \frac{\pi}{6}$ and $x = \pi$: $\frac{dy}{dx} = \cos\frac{\pi}{2} - \sqrt{3}\sin\frac{\pi}{2} < 0$.

Thus, the function is increasing on the interval $\left(0, \frac{\pi}{6}\right)$ and decreasing on $\left(\frac{\pi}{6}, \pi\right)$.

The answer is (B).

6. **B** First, find $f(3)$ and $f(8)$: $f(3)=3^2-6(3)+5=-4$ and $f(8)=8^2-6(8)+5=21$. Then $\frac{21-(-4)}{8-3}=5=f'(c)$. Next, find $f'(x)$: $f'(x)=2x-6$. Then, according to the MVTD: $2c-6=5$, so $c=\frac{11}{2}$.

The answer is (B).

7. **D** First, because the function is not continuous on the interval, there may not be a solution. Prove that this is true. First, find $f(2)$ and $f(6)$: $f(2)=\frac{2}{2-4}=-1$ and $f(6)=\frac{2}{6-4}=1$. Then $\frac{1-(-1)}{6-2}=\frac{1}{2}=f'(c)$. Next, find $f'(x)$: $f'(x)=\frac{-2}{(x-4)^2}$. Then, according to the MVTD: $\frac{-2}{(c-4)^2}=\frac{1}{2}$. This gives $(c-4)^2=-4$, which has no solution.

The answer is (D).

8. **D** First, find $f(-3)$ and $f(3)$: $f(-3)=(-3)^4-5(-3)^2+4=40$ and $f(3)=(3)^4-5(3)^2+4=40$. Next, find $f'(x)$: $f'(x)=4x^3-10x$. By setting $f'(c)=4c^3-10c=0$, you get $c(4c^2-10)=0$, and c = 0 or $c=\pm\sqrt{\frac{5}{2}}$, all of which are in the interval.

The answer is (D).

Chapter 8
Integration and Accumulation of Change

EXPLORING ACCUMULATIONS OF CHANGE

Now we will be moving from finding and using derivatives of functions (**Differential Calculus**) to finding and using integrals (**Integral Calculus**). We'll also be using a new symbol $\int$, which stands for integration and, among other things, represents the antiderivative.

Many students consider this half of calculus to be the more challenging part, but don't worry! We'll get you through it, starting with how to find the reverse of a derivative, the integral, along with a whole bunch of new fun things.

First, we will learn about the **definite integral**. It can be used to give the area of the region between a function and the x-axis, and it gives us the accumulation of change in that function.

Example 1: Look at the region below.

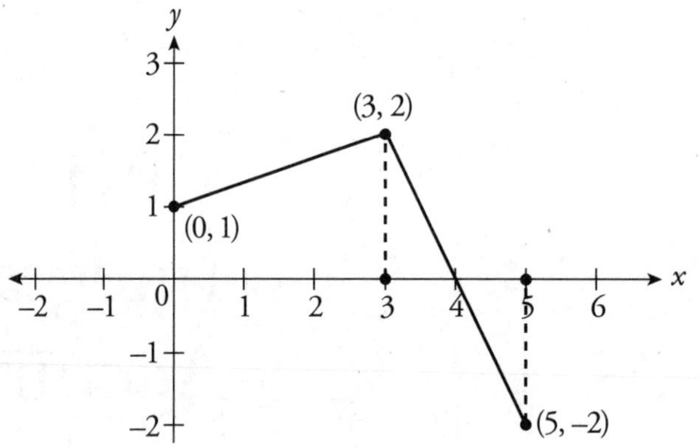

The definite integral gives us the area of the region between the function and the x-axis. We will learn the notation for this a little further into this chapter. We can find the area using geometry. It helps to divide the region into three pieces. The first piece, from $x = 0$ to $x = 3$, is a trapezoid with an area of $\frac{1}{2}(1+2)(3)=\frac{9}{2}$; the second region is a triangle with an area of $\frac{1}{2}(1)(2)=1$; and the third region is a triangle with an area of $\frac{1}{2}(1)(2)=1$. Notice that the third region is below the x-axis, so we consider this region negative. Therefore, the total area of the region is $\frac{9}{2}$ square units.

The units for the region are found by multiplying the units along the x-axis by the units along the y-axis. So, for example, if the graph were of velocity (y-axis) versus time (x-axis), where velocity was measured in meters per second and time was measured in seconds, then the units would be $\frac{\text{m}}{\text{s}} \times \text{s} = \text{m}$. This should make sense because the accumulation of change for an object moving at a given velocity for a given time should be the total distance, in this case measured in meters.

Notice that some of the accumulation can be positive and some can be negative. For example, one could have liquid flowing into a container, increasing the volume in the container, or flowing out of a container, decreasing the volume in the container.

APPROXIMATING AREAS WITH RIEMANN SUMS

Area Under a Curve

It's time to learn one of the most important uses of the integral: finding the area under a curve. First, here's a little background about how to find the area without using integration.

Suppose you have to find the area under the curve $y = x^2 + 2$ from $x = 1$ to $x = 3$. The graph of the curve looks like the following:

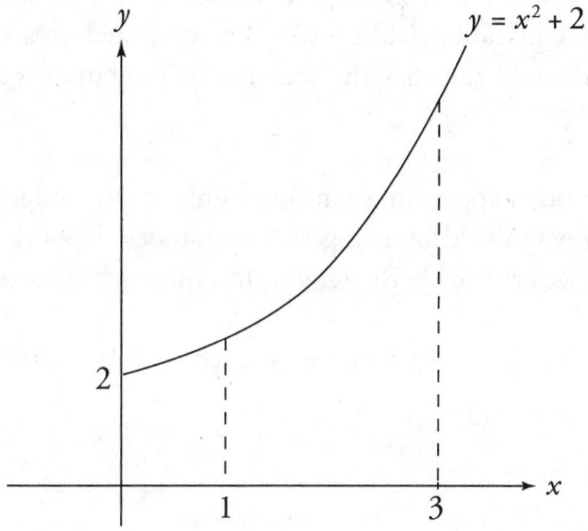

Nothing you've learned in geometry thus far has taught you how to find the area of something like this, but you have learned how to find the area of a rectangle, though, and we're going to use rectangles to approximate the area between the curve and the x-axis.

Let's divide the region into two rectangles, one from $x = 1$ to $x = 2$ and one from $x = 2$ to $x = 3$, where the top of each rectangle comes just under the curve.

It looks like the following:

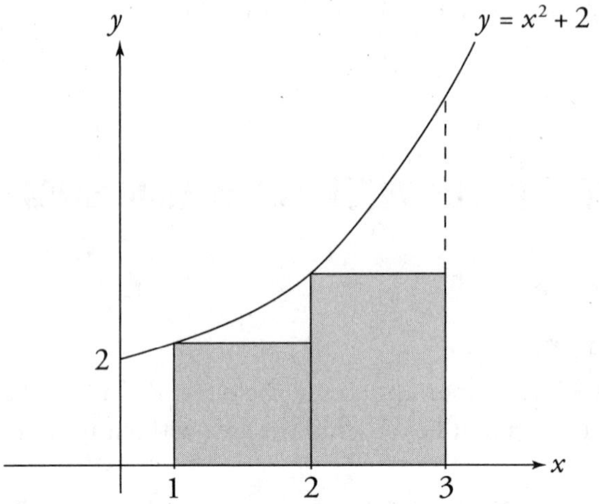

Notice that the width of each rectangle is 1. The height of the left rectangle is found by plugging 1 into the equation $y = x^2 + 2$ (yielding 3); the height of the right rectangle is found by plugging 2 into the same equation (yielding 6). The combined area of the two rectangles is $(1)(3) + (1)(6) = 9$. So we could say that the area under the curve is approximately 9 square units.

Naturally, this is a pretty rough approximation that significantly underestimates the area. Look at how much of the area we missed by using two rectangles! How do you suppose we could make the approximation better? Divide the region into more, thinner rectangles.

This time, cut up the region into four rectangles, each with a width of $\frac{1}{2}$. It looks like the following:

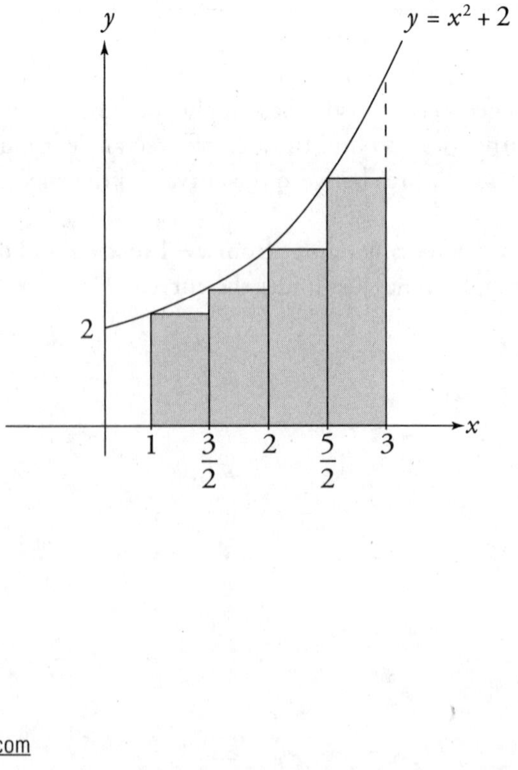

Now find the height of each rectangle the same way as before. Notice that the values we use are the left endpoints of each rectangle. The heights of the rectangles are, respectively,

$$(1)^2+2=3;\ \left(\frac{3}{2}\right)^2+2=\frac{17}{4};\ (2)^2+2=6;\text{ and}\left(\frac{5}{2}\right)^2+2=\frac{33}{4}$$

Now multiply each height by the width of $\frac{1}{2}$ and add up the areas.

$$\left(\frac{1}{2}\right)(3)+\left(\frac{1}{2}\right)\left(\frac{17}{4}\right)+\left(\frac{1}{2}\right)(6)+\left(\frac{1}{2}\right)\left(\frac{33}{4}\right)=\frac{43}{4}$$

This is a much better approximation of the area, but there's still a lot of space that isn't accounted for. We're still underestimating the area. The rectangles need to be thinner. But before we do that, let's do something else.

Notice how each of the rectangles is inscribed in the region. Suppose we used circumscribed rectangles instead—that is, we could determine the height of each rectangle by the higher of the two y-values, not the lower. The following graph shows what the region would look like:

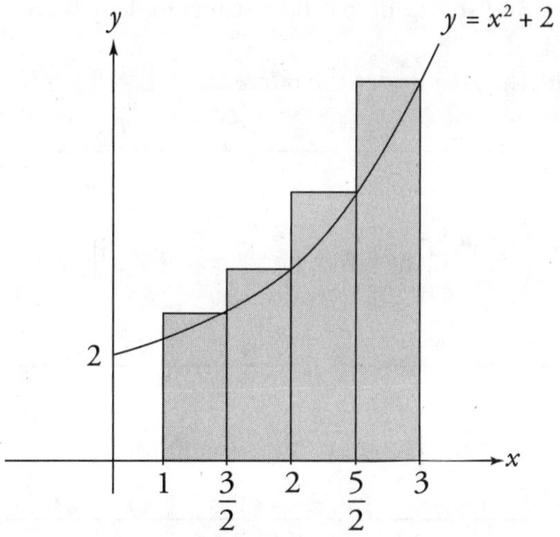

To find the area of the rectangles, we would still use the width of $\frac{1}{2}$, but the heights would change. The heights of each rectangle are now found by plugging in the right endpoint of each rectangle.

$$\left(\frac{3}{2}\right)^2+2=\frac{17}{4};(2)^2+2=6;\left(\frac{5}{2}\right)^2+2=\frac{33}{4};\text{ and }(3)^2+2=11$$

Once again, multiply each height by the width of $\frac{1}{2}$ and add up the areas.

$$\left(\frac{1}{2}\right)\left(\frac{17}{4}\right)+\left(\frac{1}{2}\right)(6)+\left(\frac{1}{2}\right)\left(\frac{33}{4}\right)+\left(\frac{1}{2}\right)(11)=\frac{59}{4}$$

The area under the curve using four left endpoint rectangles is $\frac{43}{4}$, and the area using four right endpoint rectangles is $\frac{59}{4}$, so why not average the two? This gives us $\frac{51}{4}$, which is a better approximation of the area.

Now that we've found the area using the rectangles a few times, let's turn the method into a formula. Call the left endpoint of the interval a and the right endpoint of the interval b, and set the number of rectangles we use equal to n. So the width of each rectangle is $\frac{b-a}{n}$. The height of the first inscribed rectangle is y_0, the height of the second rectangle is y_1, the height of the third rectangle is y_2, and so on, up to the last rectangle, which is y_{n-1}. If we use the left endpoint of each rectangle, the area under the curve is

$$\left(\frac{b-a}{n}\right)\left[y_0+y_1+y_2+y_3+\ldots+y_{n-1}\right]$$

If we use the right endpoint of each rectangle, then the formula is

$$\left(\frac{b-a}{n}\right)\left[y_1+y_2+y_3+\ldots+y_n\right]$$

Remember how we said that we could make the approximation better by making more, thinner rectangles? By letting n approach infinity, we create an infinite number of rectangles that are infinitesimally thin. We can do this by using limits. The formula for "left-endpoint" rectangles becomes

$$\lim_{n\to\infty}\left(\frac{b-a}{n}\right)\left[y_0+y_1+y_2+y_3+\ldots+y_{n-1}\right]$$

Note: Sometimes these are called "inscribed" and "circumscribed" rectangles, but that restricts the use of the formula. It's more exact to evaluate these rectangles using right and left endpoints.

For "right-endpoint" rectangles, the formula becomes

$$\lim_{n\to\infty}\left(\frac{b-a}{n}\right)\left[y_1+y_2+y_3+\ldots+y_n\right]$$

We could also find the area using the midpoint of each interval. Let's once again use the above example, dividing the region into four rectangles. The region would look like the following:

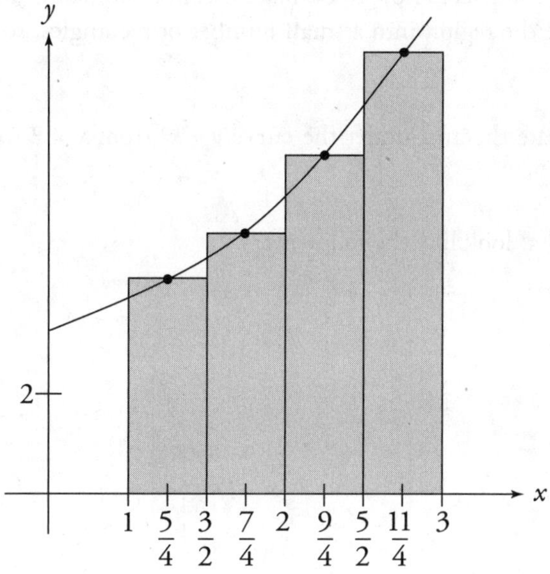

To find the area of the rectangles, we would still use the width of $\frac{1}{2}$, but the heights would now be found by plugging the midpoint of each interval into the equation.

$$\left(\frac{5}{4}\right)^2+2=\frac{57}{16};\left(\frac{7}{4}\right)^2+2=\frac{81}{16};\left(\frac{9}{4}\right)^2+2=\frac{113}{16};\left(\frac{11}{4}\right)^2+2=\frac{153}{16}$$

Once again, multiply each height by the width of $\frac{1}{2}$ and add up the areas.

$$\left(\frac{1}{2}\right)\left(\frac{57}{16}\right)+\left(\frac{1}{2}\right)\left(\frac{81}{16}\right)+\left(\frac{1}{2}\right)\left(\frac{113}{16}\right)+\left(\frac{1}{2}\right)\left(\frac{153}{16}\right)=\frac{404}{32}=\frac{101}{8}=12.625$$

The general formula for approximating the area under a curve using midpoints is

$$\left(\frac{b-a}{n}\right)\left[y_{\frac{1}{2}} + y_{\frac{3}{2}} + y_{\frac{5}{2}} + \ldots + y_{\frac{2n-1}{2}}\right]$$

(Note: The fractional subscript means to evaluate the function at the number halfway between each integral pair of values of n.)

When you take the limit of this infinite sum, you get the integral. Actually, we write the integral as

$$\int_1^3 \left(x^2 + 2\right)dx$$

This is called a definite integral, and it means that we're finding the area under the curve $x^2 + 2$ from $x = 1$ to $x = 3$. (We'll discuss how to evaluate this in a moment.) On the AP Exam, you'll only be asked to divide the region into a small number of rectangles, so it won't be very hard. Let's do an example.

Example 2: Approximate the area under the curve $y = x^3$ from $x = 2$ to $x = 3$ using four left-endpoint rectangles.

Draw four rectangles that look like the following:

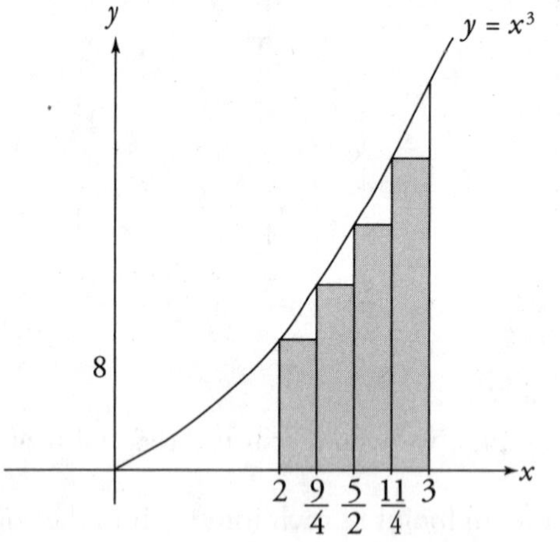

The width of each rectangle is $\frac{1}{4}$. The heights of the rectangles are

$$2^3, \left(\frac{9}{4}\right)^3, \left(\frac{5}{2}\right)^3, \text{ and } \left(\frac{11}{4}\right)^3$$

Therefore, the area is

$$\left(\frac{1}{4}\right)\left(2^3\right) + \left(\frac{1}{4}\right)\left(\frac{9}{4}\right)^3 + \left(\frac{1}{4}\right)\left(\frac{5}{2}\right)^3 + \left(\frac{1}{4}\right)\left(\frac{11}{4}\right)^3 = \frac{893}{64} \approx 13.953$$

Example 3: Repeat Example 2 using four right-endpoint rectangles.

Now draw four rectangles that look like the following:

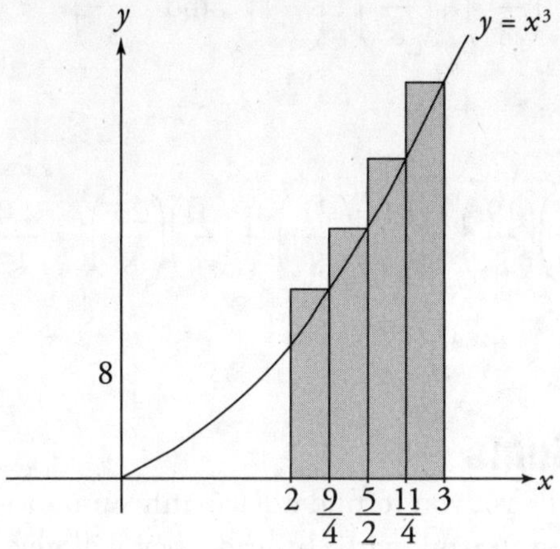

The width of each rectangle is still $\frac{1}{4}$, but the heights of the rectangles are now

$$\left(\frac{9}{4}\right)^3, \left(\frac{5}{2}\right)^3, \left(\frac{11}{4}\right)^3, \text{ and } \left(3^3\right)$$

The area is now

$$\left(\frac{1}{4}\right)\left(\frac{9}{4}\right)^3 + \left(\frac{1}{4}\right)\left(\frac{5}{2}\right)^3 + \left(\frac{1}{4}\right)\left(\frac{11}{4}\right)^3 + \left(\frac{1}{4}\right)\left(3^3\right) = \frac{1{,}197}{64} \approx 18.703$$

Example 4: Repeat Example 2 using four midpoint rectangles.

Now draw four rectangles that look like the following:

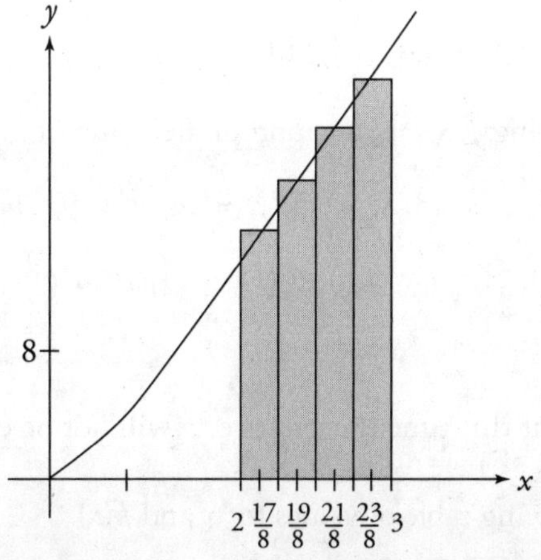

The width of each rectangle is still $\frac{1}{4}$, but the heights of the rectangles are now

$$\left(\frac{17}{8}\right)^3, \left(\frac{19}{8}\right)^3, \left(\frac{21}{8}\right)^3, \text{ and } \left(\frac{23}{8}\right)^3$$

The area is now

$$\left(\frac{1}{4}\right)\left(\frac{17}{8}\right)^3 + \left(\frac{1}{4}\right)\left(\frac{19}{8}\right)^3 + \left(\frac{1}{4}\right)\left(\frac{21}{8}\right)^3 + \left(\frac{1}{4}\right)\left(\frac{23}{8}\right)^3 = \frac{2{,}075}{128} \approx 16.211$$

Tabular Riemann Sums

Sometimes the AP Exam will ask you to find a Riemann sum, or to approximate an integral, but won't give you a function to work with. Instead, they will give you a table of values for x and $f(x)$. These are quite simple to evaluate. All you do is use the right-hand or left-hand sum formula, plugging in the appropriate values for $f(x)$. One thing you should watch out for is that sometimes the x-values are not evenly spaced, so make sure that you use the correct values for the widths of the rectangles. Let's do an example.

Example 5: Suppose we are given the following table of values for x and $f(x)$.

x	2	4	6	8	10	12
$f(x)$	10	13	15	14	9	3

Use a *right-hand* Riemann sum with 5 subintervals indicated by the data in the table to approximate $\int_2^{12} f(x)\,dx$.

Recall that the formula for finding the area under the curve using the right endpoints is $\left(\frac{b-a}{n}\right)[y_1 + y_2 + y_3 + \ldots + y_n]$. Here, the width of each rectangle is 2. We find the height of each rectangle by evaluating $f(x)$ at the appropriate value of x, the right endpoint of each interval on the x-axis. Here, $y_1 = 13$, $y_2 = 15$, $y_3 = 14$, $y_4 = 9$, and $y_5 = 3$. Therefore, we can approximate the integral with $\int_2^{12} f(x)\,dx = (2)(13) + (2)(15) + (2)(14) + (2)(9) + (2)(3) = 108$.

Let's do another example, but this time the values of x will not be evenly spaced on the x-axis.

Example 6: Given the following table of values for x and $f(x)$:

x	0	2	5	11	19	22	23
$f(x)$	4	6	16	18	22	29	50

Use a *left-hand* Riemann sum with 6 subintervals indicated by the data in the table to approximate $\int_0^{23} f(x)\,dx$.

Recall that the formula for finding the area under the curve using the left endpoints is $\left(\frac{b-a}{n}\right)[y_0 + y_1 + y_2 + \ldots + y_{n-1}]$. This formula assumes that the *x*-values are evenly spaced but they aren't here, so we will replace the values of $\left(\frac{b-a}{n}\right)$ with the appropriate widths of each rectangle. The width of the first rectangle is 2 – 0 = 2; the second width is 5 – 2 = 3; the third is 11 – 5 = 6; the fourth is 19 – 11 = 8; the fifth is 22 – 19 = 3; and the sixth is 23 – 22 = 1. We find the height of each rectangle by evaluating $f(x)$ at the appropriate value of *x*, the left endpoint of each interval on the *x*-axis. Here, $y_0 = 4$, $y_1 = 6$, $y_2 = 16$, $y_3 = 18$, $y_4 = 22$, and $y_5 = 29$. Therefore, we can approximate the integral with $\int_0^{23} f(x)\,dx = (2)(4) + (3)(6) + (6)(16) + (8)(18) + (3)(22) + (1)(29) = 361$.

That's all there is to approximating the area under a curve using rectangles. Now let's learn how to find the area using the Trapezoid Rule.

The Trapezoid Rule

There's another approximation method that's even better than the rectangle method. Essentially, all you do is divide the region into trapezoids instead of rectangles. Let's use the problem that we did at the beginning of the chapter.

We get a picture that looks like the following:

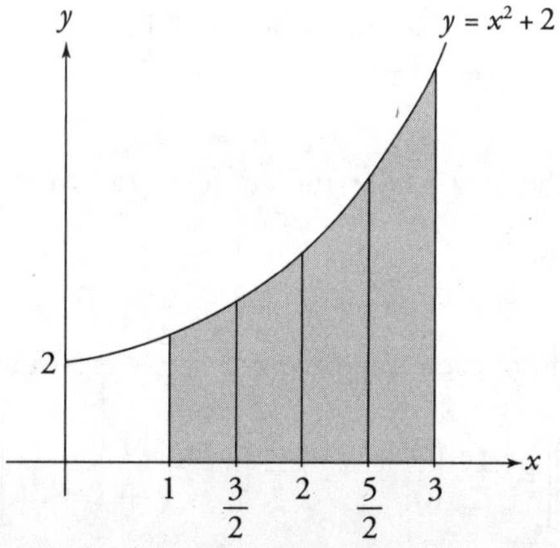

As you should recall from geometry, the formula for the area of a trapezoid is

$$\frac{1}{2}(b_1 + b_2)h$$

(Note: b_1 and b_2 are the two bases of the trapezoid.) Notice that each of the shapes is a trapezoid on its side, so the height of each trapezoid is the length of the interval $\frac{1}{2}$, and the bases are the y-values that correspond to each x-value. We found these earlier in the rectangle example; they are, in order: 3, $\frac{17}{4}$, 6, $\frac{33}{4}$, and 11. We can find the area of each trapezoid and add them up.

$$\frac{1}{2}\left(3+\frac{17}{4}\right)\left(\frac{1}{2}\right)+\frac{1}{2}\left(\frac{17}{4}+6\right)\left(\frac{1}{2}\right)+\frac{1}{2}\left(6+\frac{33}{4}\right)\left(\frac{1}{2}\right)+\frac{1}{2}\left(\frac{33}{4}+11\right)\left(\frac{1}{2}\right)=$$

$$\frac{51}{4} \text{ or } 12.75$$

Recall that the actual value of the area is $\frac{38}{3}$ or 12.67; the Trapezoid Rule gives a pretty good approximation.

Notice how each trapezoid shares a base with the trapezoid next to it, except for the end ones. This enables us to simplify the formula for using the Trapezoid Rule. Each trapezoid has a height equal to the length of the interval divided by the number of trapezoids we use. If the interval is from $x = a$ to $x = b$, and the number of trapezoids is n, then the height of each trapezoid is $\frac{b-a}{n}$. Then our formula is

$$\left(\frac{1}{2}\right)\left(\frac{b-a}{n}\right)\left[y_0+2y_1+2y_2+2y_3+\ldots+2y_{n-2}+2y_{n-1}+y_n\right]$$

This is all you need to know about the Trapezoid Rule. Just follow the formula, and you won't have any problems. Let's do one more example.

Example 7: Approximate the area under the curve $y = x^3$ from $x = 2$ to $x = 3$ using four inscribed trapezoids.

Following the rule, the height of each trapezoid is $\frac{3-2}{4}=\frac{1}{4}$. Thus, the approximate area is

$$\left(\frac{1}{2}\right)\left(\frac{1}{4}\right)\left[2^3+2\left(\frac{9}{4}\right)^3+2\left(\frac{5}{2}\right)^3+2\left(\frac{11}{4}\right)^3+3^3\right]=\frac{1{,}045}{64}$$

Compare this answer to the actual value we found earlier—it's pretty close!

Example 8: Approximate the area under the curve $y = 4 - x^2$ from $x = -1$ to $x = 1$ with $n = 4$ inscribed rectangles.

Draw four rectangles that look like this.

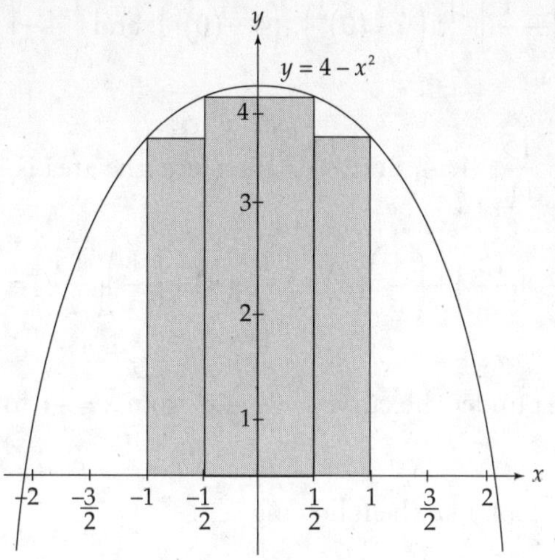

The width of each rectangle is $\frac{1}{2}$. The heights of the rectangles are found by evaluating $y = 4 - x^2$ at the appropriate endpoints.

$$\left(4-(-1)^2\right), \left(4-\left(-\frac{1}{2}\right)^2\right), \left(4-\left(\frac{1}{2}\right)^2\right), \text{ and } \left(4-(1)^2\right)$$

These can be simplified to $3, \frac{15}{4}, \frac{15}{4}$, and 3. Therefore, the area is

$$\left(\frac{1}{2}\right)(3)+\left(\frac{1}{2}\right)\left(\frac{15}{4}\right)+\left(\frac{1}{2}\right)\left(\frac{15}{4}\right)+\left(\frac{1}{2}\right)(3)=\frac{27}{4}$$

Example 9: Find the area under the curve $y = 4 - x^2$ from $x = -1$ to $x = 1$ with $n = 4$ circumscribed rectangles.

Draw four rectangles that look like the following:

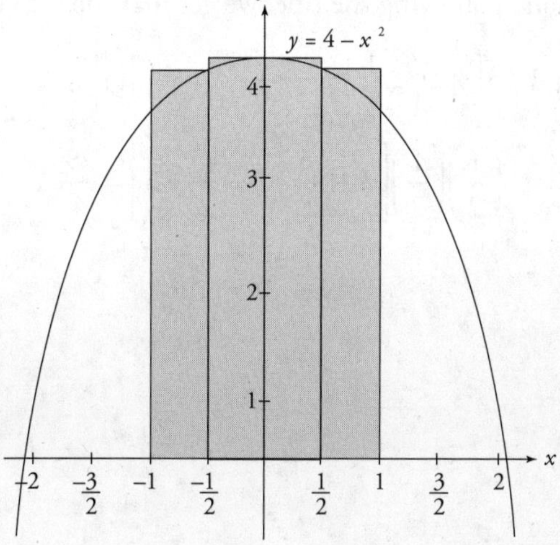

The width of each rectangle is $\frac{1}{2}$. The heights of the rectangles are found by evaluating $y = 4 - x^2$ at the appropriate endpoints.

$$\left(4-\left(-\frac{1}{2}\right)^2\right),\ \left(4-(0)^2\right),\ \left(4-(0)^2\right), \text{ and } \left(4-\left(\frac{1}{2}\right)^2\right)$$

These can be simplified to $\frac{15}{4}$, 4, 4, and $\frac{15}{4}$. Therefore, the area is

$$\left(\frac{1}{2}\right)\left(\frac{15}{4}\right)+\left(\frac{1}{2}\right)(4)+\left(\frac{1}{2}\right)(4)+\left(\frac{1}{2}\right)\left(\frac{15}{4}\right)=\frac{31}{4}$$

Example 10: Find the area under the curve $y = 4 - x^2$ from $x = -1$ to $x = 1$ using the Trapezoid Rule with $n = 4$.

Draw four trapezoids that look like the following:

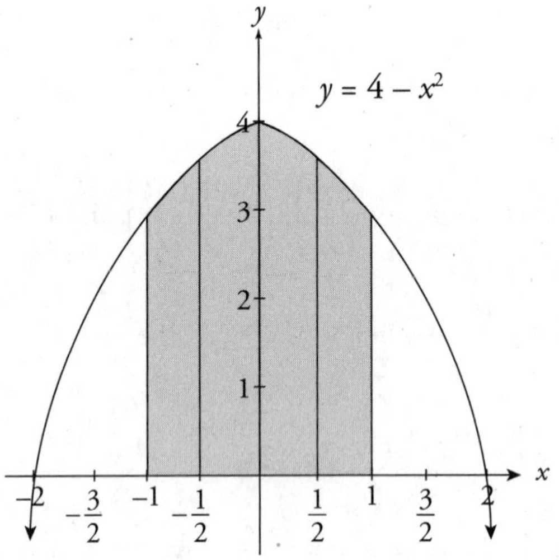

The width of each trapezoid is $\frac{1}{2}$. Evaluate the bases of the trapezoid by calculating $y = 4 - x^2$ at the appropriate endpoints. Following the rule, we get that the area is approximately

$$\left(\frac{1}{2}\right)\left(\frac{1}{2}\right)\left[\left(4-(-1)^2\right)+2\left(4-\left(-\frac{1}{2}\right)^2\right)+2\left(4-(0)^2\right)+2\left(4-\left(\frac{1}{2}\right)^2\right)+\left(4-(1)^2\right)\right]=$$

$$\left(\frac{1}{2}\right)\left(\frac{1}{2}\right)\left[3+\frac{15}{2}+8+\frac{15}{2}+3\right]=\frac{29}{4}$$

Example 11: Find the area under the curve $y = 4 - x^2$ from $x = -1$ to $x = 1$ using the Midpoint Formula with $n = 4$.

Draw four rectangles that look like the following:

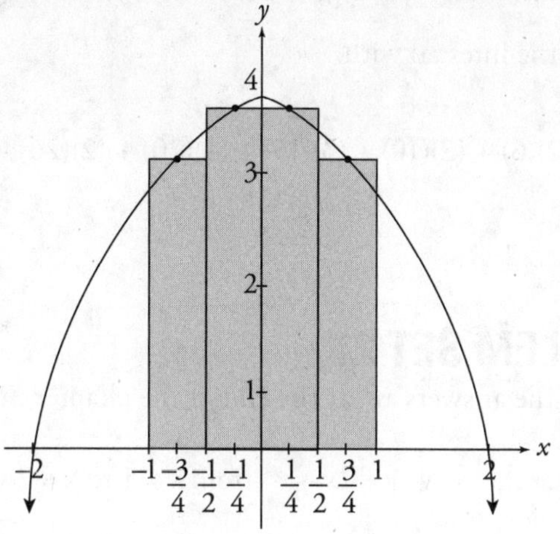

The width of each rectangle is $\frac{1}{2}$. The heights of the rectangles are found by evaluating $y = 4 - x^2$ at the appropriate points.

$$4-\left(-\frac{3}{4}\right)^2=\frac{55}{16};\ 4-\left(-\frac{1}{4}\right)^2=\frac{63}{16};\ 4-\left(\frac{1}{4}\right)^2=\frac{63}{16};\ 4-\left(\frac{3}{4}\right)^2=\frac{55}{16}$$

Multiply each height by the width of $\frac{1}{2}$ and add up the areas.

$$\left(\frac{1}{2}\right)\left(\frac{55}{16}\right)+\left(\frac{1}{2}\right)\left(\frac{63}{16}\right)+\left(\frac{1}{2}\right)\left(\frac{63}{16}\right)+\left(\frac{1}{2}\right)\left(\frac{55}{16}\right)=\frac{59}{8}$$

Example 12: Given the following table of values for t and $f(t)$:

t	0	2	4	7	11	13	14
$f(t)$	5	6	10	15	20	26	30

Use a *right-hand* Riemann sum with 6 subintervals indicated by the data in the table to approximate $\int_0^{14} f(t)\,dt$.

The width of the first rectangle is 2 – 0 = 2; the second width is 4 – 2 = 2; the third is 7 – 4 = 3; the fourth is 11 – 7 = 4; the fifth is 13 – 11 = 2; and the sixth is 14 – 13 = 1. We find the height of each rectangle by evaluating $f(t)$ at the appropriate value of t, the right endpoint of each interval on the t-axis. Here, $y_1 = 6$, $y_2 = 10$, $y_3 = 15$, $y_4 = 20$, $y_5 = 26$, and $y_6 = 30$. Therefore, we can approximate the integral with

$$\int_0^{14} f(t)\,dt = (2)(6) + (2)(10) + (3)(15) + (4)(20) + (2)(26) + (1)(30) = 239$$

PRACTICE PROBLEM SET 21

Now try these problems. The answers are at the end of the chapter, starting on page 414.

1. Find the area under the curve $y = 2x - x^2$ from $x = 1$ to $x = 2$ with $n = 4$ left-endpoint rectangles.

2. Find the area under the curve $y = 2x - x^2$ from $x = 1$ to $x = 2$ with $n = 4$ right-endpoint rectangles.

3. Find the area under the curve $y = 2x - x^2$ from $x = 1$ to $x = 2$ using the Trapezoid Rule with $n = 4$.

4. Find the area under the curve $y = 2x - x^2$ from $x = 1$ to $x = 2$ using the Midpoint Formula with $n = 4$.

5. Suppose we are given the following table of values for x and $g(x)$:

x	0	1	3	5	9	14
$g(x)$	10	8	11	17	20	23

Use a left-hand Riemann sum with 5 subintervals indicated by the data in the table to approximate $\int_0^{14} g(x)dx$.

RIEMANN SUMS, SUMMATION NOTATION, AND DEFINITE INTEGRAL NOTATION

As we saw in the previous section, the formula for a tabular Riemann sum is $\lim\limits_{n\to\infty}\left(\frac{b-a}{n}\right)\left[y_1+y_2+y_3+\ldots+y_n\right]$. Let's use some slightly different notation for the formula. We are going to let $\frac{b-a}{n}=\Delta x$ because it stands for the width of each rectangle along the x-axis. The first y-value is found by starting at a and adding Δx, so $y_1 = f(a + \Delta x)$. To get the next y-value, we add Δx again, so $y_2 = f(a + 2\Delta x)$. To get the next y-value, we add Δx again, so $y_3 = f(a + 3\Delta x)$. And so on, up to $y_n = f(a + n\Delta x)$. Now we can rewrite the formula as $\lim\limits_{n\to\infty}\Delta x\left[f(a+\Delta x)+f(a+2\Delta x)+f(a+3\Delta x)+\ldots+f(a+n\Delta x)\right]$. Finally, using sigma notation, we can write this as $\int_a^b f(x)\,dx=\lim\limits_{n\to\infty}\Delta x\sum_{i=1}^{n}f(a+i\Delta x)$.

This is another topic that rarely appears on the AP Calculus Exam, but which pops up just enough that you should still learn it.

This gives us the Riemann sum formula: $\int_a^b f(x)\,dx=\lim\limits_{n\to\infty}\Delta x\sum_{i=1}^{n}f(a+i\Delta x)$, where $\frac{b-a}{n}=\Delta x$.

Let's redo Examples 2–4 from pages 358–360, but this time, we will use an infinite number of rectangles, each infinitesimally thin, to find the area under the curve. In other words, use the Riemann sum formula.

Example 13: Find the area under the curve $y = x^3$ from $x = 2$ to $x = 3$ using an infinite number of right-hand rectangles. In other words, find $\int_2^3 x^3 dx$.

First, let's find Δx: $\Delta x=\frac{3-2}{n}=\frac{1}{n}$. Next, $f(x) = x^3$. Plugging into the formula, we get $\int_2^3 x^3\,dx=\lim\limits_{n\to\infty}\frac{1}{n}\sum_{i=1}^{n}\left(2+\frac{1}{n}i\right)^3$.

On the AP Exam, you won't be asked to evaluate the Riemann sum. You are only asked to set up the sum. Or, given the Riemann sum, can you figure out what integral it represents? Let's do an example.

Example 14: What integral is represented by $\lim_{n\to\infty}\frac{4}{n}\sum_{i=1}^{n}\left(6+\frac{4}{n}i\right)^2-\left(6+\frac{4}{n}i\right)$?

First, let's look at the term $\frac{4}{n}$. We know that $b - a = 4$, and $a = 6$, so $b = 10$. Next, what is $f(x)$?

See how we have the expression $\left(6+\frac{4}{n}i\right)^2-\left(6+\frac{4}{n}i\right)$. We must have the function $f(x) = x^2 - x$.

Therefore, the limit represents the integral $\lim_{n\to\infty}\frac{4}{n}\sum_{i=1}^{n}\left(6+\frac{4}{n}i\right)^2-\left(6+\frac{4}{n}i\right)=\int_6^{10}\left(x^2-x\right)dx$.

THE FUNDAMENTAL THEOREM OF CALCULUS AND ACCUMULATION FUNCTIONS

As you saw in the last chapter, we've only half-learned the theorem. Here it is in full.

The Fundamental Theorem of Calculus

If $f(x)$ is continuous on $[a, b]$, then the derivative of the function $F(x) = \int_a^x f(t)dt$ is

$$\frac{d}{dx}\int_a^x f(t)dt = f(x)$$

This theorem tells us how to find the derivative of an integral.

Example 15: Find $\frac{d}{dx}\int_1^x \cos t\, dt$.

The Fundamental Theorem says that the derivative of this integral is just $\cos x$.

Example 16: Find $\frac{d}{dx}\int_2^x \left(1-t^3\right) dt$.

Here, the theorem says that the derivative of this integral is just $(1 - x^3)$.

Isn't this easy? Let's add a couple of nuances. First, the constant term in the limits of integration is a "dummy term." Any constant will give the same answer. For example,

$$\frac{d}{dx}\int_{2}^{x}(1-t^3)\,dt = \frac{d}{dx}\int_{-2}^{x}(1-t^3)\,dt = \frac{d}{dx}\int_{\pi}^{x}(1-t^3)\,dt = 1 - x^3$$

In other words, all we're concerned with is the variable term.

Second, if the upper limit is a function of *x*, instead of just plain *x*, we multiply the answer by the derivative of that term. For example,

$$\frac{d}{dx}\int_{2}^{x^2}\left(1-t^3\right)\,dt = \left[1-\left(x^2\right)^3\right](2x) = \left(1-x^6\right)(2x)$$

Example 17: Find $\frac{d}{dx}\int_{0}^{3x^4}\left(t+4t^2\right)dt = \left[3x^4+4(3x^4)^2\right]\left(12x^3\right)$.

Try these solved problems on your own. You know the drill.

Example 18: Find $\frac{d}{dx}\int_{1}^{x}\frac{dt}{1-\sqrt[3]{t}}$.

According to the Fundamental Theorem of Calculus,

$$\frac{d}{dx}\int_{1}^{x}\frac{dt}{1-\sqrt[3]{t}} = \frac{1}{1-\sqrt[3]{x}}$$

Example 19: Find $\frac{d}{dx}\int_{1}^{x^2}\frac{t\,dt}{\sin t}$.

According to the Fundamental Theorem of Calculus,

$$\frac{d}{dx}\int_{1}^{x^2}\frac{t\,dt}{\sin t} = \frac{x^2}{\sin\left(x^2\right)}(2x) = \frac{2x^3}{\sin x^2}$$

INTERPRETING THE BEHAVIOR OF ACCUMULATION FUNCTIONS INVOLVING AREA

As we have seen, the Fundamental Theorem tells us two important things. First, that a definite integral evaluates the antiderivative and finds the difference between the antiderivative at a point and at a second point. Second, it connects the derivative and the integral by showing that the derivative of an integral of a function returns us to that function (sometimes with some adjustments). Now we will use the integral and the Fundamental Theorem of Calculus to find how the area under a function "accumulates" value as the function sweeps out the area from a point to some other point.

Functions of the form $F(x)=\int_0^x f(t)\,dt$ are called **accumulation functions** because the value of the integral is the area under the curve from the constant to the value x, and as x gets bigger, so does the area (it "accumulates"). In these functions, t is a dummy variable that is used as the variable of integration.

Let's do an example.

Example 20: Suppose we have the function $F(x)=\int_0^x t\,dt$. Let's evaluate this at different values of x. First, let's find $F(1)$. Graphically, we are looking for the area under the curve $y = t$ from $t = 0$ to $t = 1$.

It looks like the following:

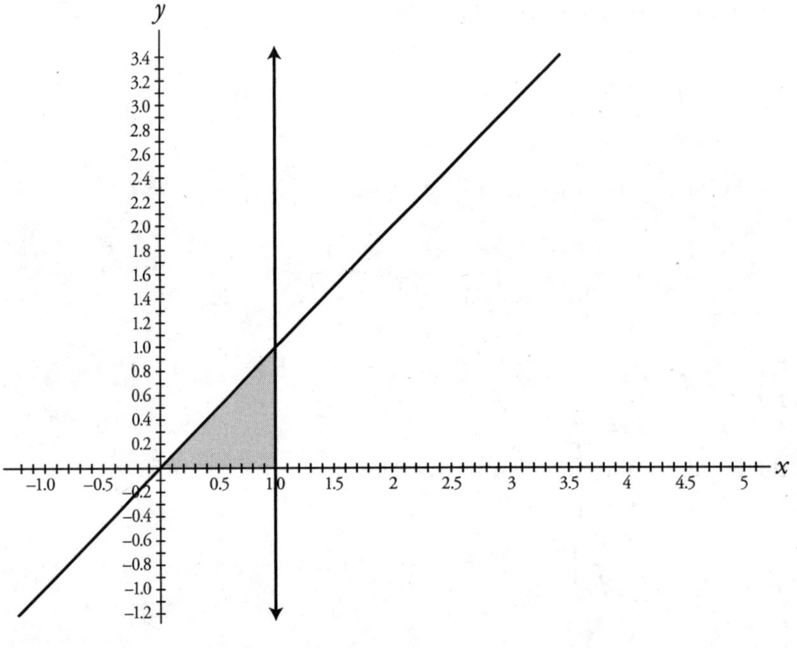

$F(1)$ is just the area of the triangle: $A=\frac{1}{2}(1)(1)=\frac{1}{2}$. If we evaluate the integral, we get $F(1)=\int_0^1 t\,dt=\left(\frac{t^2}{2}\right)\Big|_0^1=\frac{1}{2}-0=\frac{1}{2}$. Let's now find $F(2)$. This is the area under the curve $y = t$ from $t = 0$ to $t = 2$.

It looks like the following:

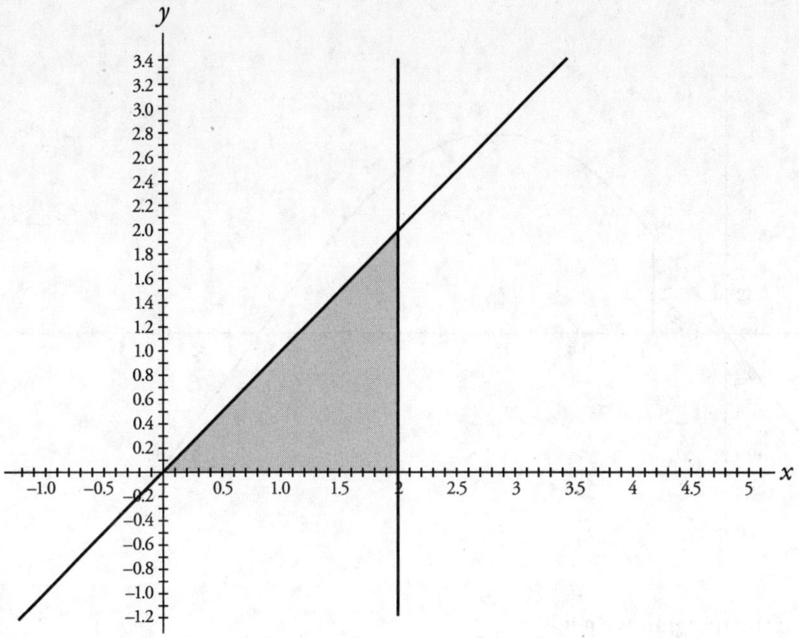

$F(2)$ is just the area of the triangle: $A=\frac{1}{2}(2)(2)=2$. If we evaluate the integral, we get $F(2)=\int_0^2 t\,dt=\left(\frac{t^2}{2}\right)\Big|_0^2=2-0=2$.

We can see that, as x increases, the function increases.

This was a fairly simple example. Let's do another one.

Example 21: Suppose we have the function $F(x)=\int_0^x \sin t\,dt$. Let's evaluate this as x increases from 0 to π. Obviously $F(0) = 0$ because there is no area under the curve. So, first, let's find $F\left(\frac{\pi}{6}\right)$. Graphically, we are looking for the area under the curve $y = \sin t$ from $t = 0$ to $t=\frac{\pi}{6}$.

It looks like the following:

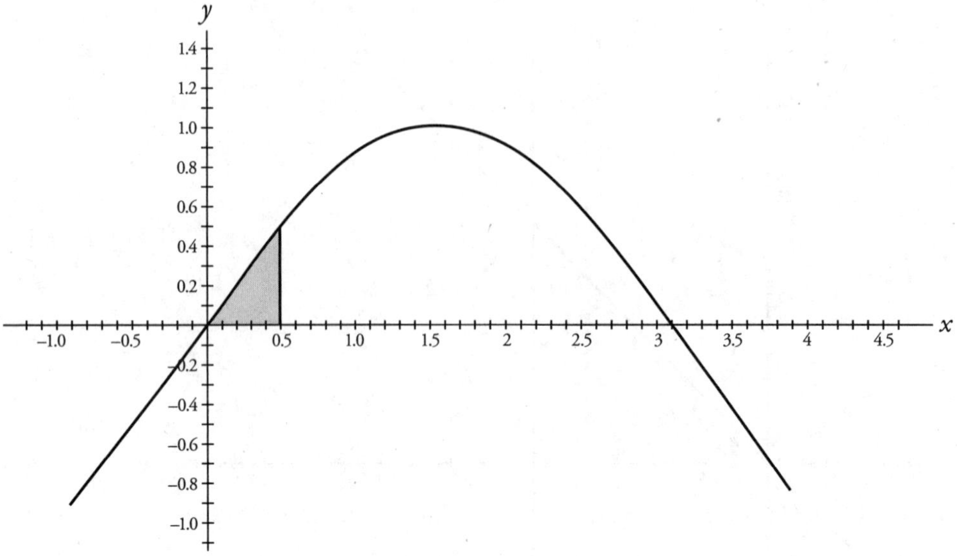

If we evaluate the integral, we get

$$F\left(\frac{\pi}{6}\right) = \int_0^{\frac{\pi}{6}} \sin t\, dt = (-\cos t)\Big|_0^{\frac{\pi}{6}} = -\cos\frac{\pi}{6} + \cos 0 = 1 - \frac{\sqrt{3}}{2} \approx 0.134$$

Now let's find $F\left(\frac{\pi}{4}\right)$. It looks like the following:

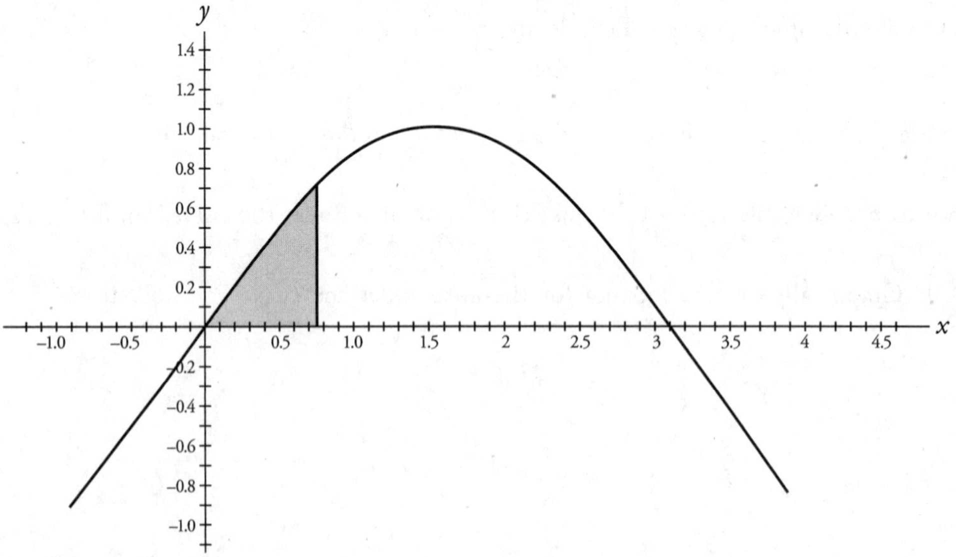

If we evaluate the integral, we get

$$F\left(\frac{\pi}{4}\right) = \int_0^{\frac{\pi}{4}} \sin t\, dt = (-\cos t)\Big|_0^{\frac{\pi}{4}} = -\cos\frac{\pi}{4} + \cos 0 = 1 - \frac{\sqrt{2}}{2} \approx 0.293$$

Let's make a table of values of the accumulation function for different values of x.

x	0	$\frac{\pi}{6}$	$\frac{\pi}{4}$	$\frac{\pi}{3}$	$\frac{\pi}{2}$	π
$F(x)$	0	$1-\frac{\sqrt{3}}{2}$	$1-\frac{\sqrt{2}}{2}$	$\frac{1}{2}$	1	2

We can see that the area is accumulating under the curve as x increases from 0 to π. Naturally, the values will shrink as we move from π to 2π because the values of sin x are negative. But, with accumulation functions, we are usually concerned only with positive areas.

Let's do one more example.

Example 22: Suppose we have the function $F(x)=\int_0^x t^2\,dt$. Let's evaluate this as x increases from 0 to 4. First, let's find $F(1)$. Graphically, we are looking for the area under the curve $y = t^2$ from $t = 0$ to $t = 1$.

It looks like the following:

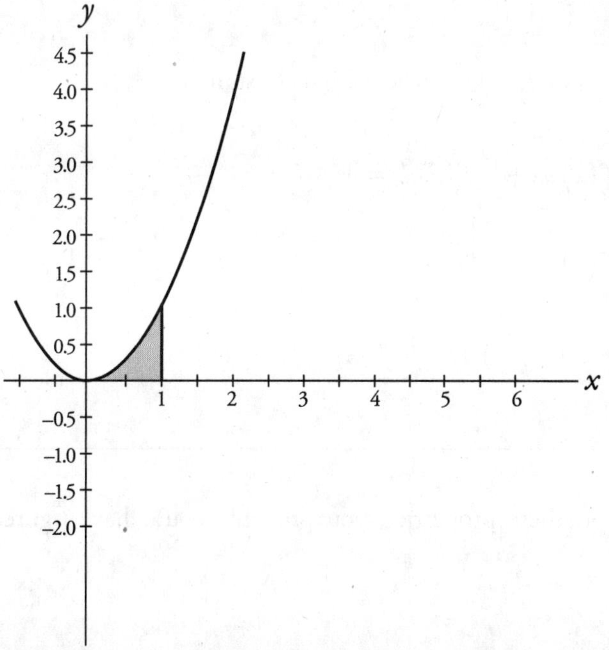

If we evaluate the integral, we get $F(1) = \int_0^1 t^2\, dt = \left(\frac{t^3}{3}\right)\Big|_0^1 = \frac{1}{3}$.

x	1	2	3	4
$F(x)$	$\frac{1}{3}$	$\frac{8}{3}$	9	$\frac{64}{3}$

As in the previous example, we can see that the values of $F(x)$ will increase as x increases.

APPLYING PROPERTIES OF DEFINITE INTEGRALS

As we have said earlier, definite integrals can be used to find the area between a curve and the x-axis. Because of this, definite integrals have some properties.

Suppose we know that $\int_a^b f(x)\, dx = M$ and $\int_b^c f(x)\, dx = N$. Then we have the following properties:

Property #1: $k\int_a^b f(x)\, dx = kM$, where k is a constant

Property #2: $\int_a^b f(x)\, dx + \int_b^c f(x)\, dx = M + N$

Property #3: $\int_b^a f(x)\, dx = -M$

Property #4: $\int_a^b f(x) \pm g(x)\, dx = \int_a^b f(x)\, dx \pm \int_a^b g(x)\, dx$

If we hadn't told you these properties, you probably could have figured them out (except maybe #3)!

Example 23: If $\int_2^5 f(x)\,dx = 12$, $\int_5^{14} f(x)dx = 20$, and $\int_2^{20} f(x)dx = 45$, find

(a) $\int_2^{14} f(x)dx$ and (b) $\int_{14}^{20} f(x)dx$.

First, let's do (a). According to Property #2, we know that $\int_2^5 f(x)\,dx + \int_5^{14} f(x)\,dx = \int_2^{14} f(x)\,dx$. So $12 + 20 = \int_2^{14} f(x)\,dx = 32$.

Now let's do (b). Again, according to Property #2, we know that $\int_2^{14} f(x)\,dx + \int_{14}^{20} f(x)\,dx = \int_2^{20} f(x)\,dx$. In part (a), we found that $\int_2^{14} f(x)\,dx = 32$ and we are given that $\int_2^{20} f(x)dx = 45$, so $32 + \int_{14}^{20} f(x)\,dx = 45$. Therefore, $\int_{14}^{20} f(x)\,dx = 13$.

Example 24: If $\int_2^5 f(x)\,dx = 9$ and $\int_8^5 f(x)dx = 11$, find $\int_2^8 f(x)dx$.

First, use Property #3: $\int_8^5 f(x)dx = 11$, so $\int_5^8 f(x)dx = -11$. Now, use Property #2:

$\int_2^5 f(x)\,dx + \int_5^8 f(x)dx = \int_2^8 f(x)dx$, so $9 - 11 = \int_2^8 f(x)dx = -2$.

THE FUNDAMENTAL THEOREM OF CALCULUS AND DEFINITE INTEGRALS

Before, we said that if you create an infinite number of infinitesimally thin rectangles, you'll get the area under the curve, which is an integral. For the example above, the integral is

$$\int_2^3 x^3 dx$$

There is a rule for evaluating an integral like this. The rule is called the **Fundamental Theorem of Calculus**, and it says

$\int_a^b f(x)dx = F(b) - F(a)$; where $F(x)$ is the antiderivative of $f(x)$.

Using this rule, you can find $\int_2^3 x^3 dx$ by integrating it, and we get $\frac{x^4}{4}$. Now all you do is plug in 3 and 2 and take the difference. We use the following notation to symbolize this

$$\left.\frac{x^4}{4}\right|_2^3$$

Thus, we have

$$\frac{3^4}{4}-\frac{2^4}{4}=\frac{81}{4}-\frac{16}{4}=\frac{65}{4}$$

Because $\frac{65}{4}$ = 16.25, you can see how close we were with our three earlier approximations.

Example 25: Find $\int_1^3 (x^2+2)dx$.

The Fundamental Theorem of Calculus yields the following:

$$\int_1^3 (x^2+2)dx = \left.\left(\frac{x^3}{3}+2x\right)\right|_1^3$$

If we evaluate this, we get the following:

$$\left(\frac{3^3}{3}+2(3)\right)-\left(\frac{1^3}{3}+2(1)\right)=\frac{38}{3}$$

This is the first function for which we found the approximate area by using inscribed rectangles. Our final estimate, where we averaged the inscribed and circumscribed rectangles, was $\frac{51}{4}$, and as you can see, that was very close (off by $\frac{1}{12}$). When we used the midpoints, we were off by $\frac{1}{24}$.

FINDING ANTIDERIVATIVES AND INDEFINITE INTEGRALS: BASIC RULES AND NOTATION

An antiderivative is a derivative in reverse. Therefore, we're going to reverse some of the rules we learned with derivatives and apply them to integrals. For example, we know that the derivative of x^2 is $2x$. If we are given the derivative of a function and have to figure out the original function, we use antidifferentiation. Thus, the antiderivative of $2x$ is x^2. (Actually, the answer is slightly more complicated than that, but we'll get into that in a few moments.)

Now we need to add some info here to make sure that you get this absolutely correct. First, as far as notation goes, it is traditional to write the antiderivative of a function using its uppercase letter, so the antiderivative of $f(x)$ is $F(x)$, the antiderivative of $g(x)$ is $G(x)$, and so on.

The second idea is very important: each function has more than one antiderivative. In fact, there are an infinite number of antiderivatives of a function. Let's go back to our example to help illustrate this.

Remember that the antiderivative of $2x$ is x^2? Well, consider the fact that if you take the derivative of $x^2 + 1$, you get $2x$. The same is true for $x^2 + 2$, $x^2 - 1$, and so on. In fact, if *any* constant is added to x^2, the derivative is still $2x$ because the derivative of a constant is zero.

Because of this, we write the antiderivative of $2x$ as $x^2 + C$, where C stands for any constant.

Finally, whenever you take the integral (or antiderivative) of a function of x, you always add the term dx (or dy if it's a function of y, etc.) to the integrand (the thing inside the integral). You'll learn why later.

For now, just remember that you must always use the dx symbol, and teachers love to take points off for forgetting the dx. Don't ask why, but they do!

Here is the **Power Rule** for antiderivatives.

If $f(x) = x^n$, then $\int f(x)dx = \dfrac{x^{n+1}}{n+1} + C$ (except when $n = -1$).

Example 26: Find $\int x^3 dx$.

Using the Power Rule, we get

$$\int x^3 dx = \frac{x^4}{4} + C$$

Don't forget the constant C, or your teachers will take points off for that too!

Think of a constant as something that is always there, and you won't forget it.

Example 27: Find $\int x^{-3} dx$.

The Power Rule works with negative exponents too.

$$\int x^{-3} dx = \frac{x^{-2}}{-2} + C$$

Not terribly hard, is it? Now it's time for a few more rules that look remarkably similar to the rules for derivatives that we saw in Chapter 4.

$$\int kf(x)dx = k\int f(x)dx$$
$$\int [f(x)+g(x)]dx = \int f(x)dx + \int g(x)dx$$
$$\int k\,dx = kx + C$$

Here are a few more examples to make you an expert.

Example 28: $\int 5\,dx = 5x + C$

Example 29: $\int 7x^3 dx = \frac{7x^4}{4} + C$

Example 30: $\int \left(3x^2 + 2x\right) dx = x^3 + x^2 + C$

Example 31: $\int \sqrt{x}\,dx = \frac{x^{\frac{3}{2}}}{\frac{3}{2}} + C = \frac{2x^{\frac{3}{2}}}{3} + C$

$$\int \sin ax\,dx = -\frac{\cos ax}{a} + C$$

$$\int \cos ax\,dx = \frac{\sin ax}{a} + C$$

$$\int \sec ax \tan ax\,dx = \frac{\sec ax}{a} + C$$

$$\int \sec^2 ax\,dx = \frac{\tan ax}{a} + C$$

$$\int \csc ax \cot ax\,dx = -\frac{\csc ax}{a} + C$$

$$\int \csc^2 ax\,dx = -\frac{\cot ax}{a} + C$$

Integrals of Trig Functions

The integrals of some trigonometric functions follow directly from the derivative formulas in Chapter 4.

We didn't mention the integrals of tangent, cotangent, secant, and cosecant because you need to know some rules about logarithms to figure them out. We'll get to them in a few chapters. Notice also that each of the answers is divided by a constant. This is to account for the Chain Rule. Let's do some examples.

Example 32: Check the integral $\int \sin 5x\,dx = -\frac{\cos 5x}{5} + C$ by differentiating the answer.

$$\frac{d}{dx}\left[-\frac{\cos 5x}{5} + C\right] = -\frac{1}{5}(-\sin 5x)(5) = \sin 5x$$

Notice how the constant is accounted for in the answer?

Example 33: $\int \sec^2 3x\, dx = \frac{\tan 3x}{3} + C$

Example 34: $\int \cos \pi x\, dx = \frac{\sin \pi x}{\pi} + C$

Example 35: $\int \sec\left(\frac{x}{2}\right)\tan\left(\frac{x}{2}\right) dx = 2\sec\left(\frac{x}{2}\right) + C$

If you're not sure if you have the correct answer when you take an integral, you can always check by differentiating the answer and seeing if you get what you started with. Try to get in the habit of doing that at the beginning because it'll help you build confidence in your ability to find integrals properly. You'll see that, although you can differentiate just about any expression that you'll normally encounter, you won't be able to integrate many of the functions you see.

PRACTICE PROBLEM SET 22

Here's a great opportunity to practice finding the area beneath a curve and evaluating integrals. The answers are at the end of the chapter, starting on page 418.

1. Find the area under the curve $y = 2x - x^2$ from $x = 1$ to $x = 2$.

2. Evaluate $\int_{-\frac{\pi}{2}}^{\frac{\pi}{2}} \cos x\, dx$.

3. Evaluate $\int_{1}^{9} 2x\sqrt{x}\, dx$.

4. Evaluate $\int_{-\frac{\pi}{2}}^{\frac{\pi}{2}} \sin x\, dx$.

5. Evaluate $\int \frac{1}{x^4} dx$.

6. Evaluate $\int \frac{5}{\sqrt{x}} dx$.

Addition and Subtraction

By using the rules for addition and subtraction, we can integrate most polynomials.

Example 36: Find $\int\left(x^3+x^2-x\right)\,dx$.

We can break this into separate integrals, which gives us

$$\int x^3dx+\int x^2dx-\int x\,dx$$

Now you can integrate each of these individually.

$$\frac{x^4}{4}+C+\frac{x^3}{3}+C-\frac{x^2}{2}+C$$

You can combine the constants into one constant (it doesn't matter how many C's we use, because their sum is one collective constant whose derivative is zero).

$$\frac{x^4}{4}+\frac{x^3}{3}-\frac{x^2}{2}+C$$

Sometimes you'll be given information about the function you're seeking that will enable you to solve for the constant. Often, this is an "initial value," which is the value of the function when the variable is zero. As we've seen, normally there are an infinite number of solutions for an integral, but when we solve for the constant, there's only one.

Example 37: Find the equation of y where $\frac{dy}{dx}=3x+5$ and $y = 6$ when $x = 0$.

Let's put this in integral form.

$$y=\int(3x+5)\,dx$$

Integrating, we get

$$y=\frac{3x^2}{2}+5x+C$$

Now we can solve for the constant because we know that $y = 6$ when $x = 0$.

$$6=\frac{3(0)^2}{2}+5(0)+C$$

Therefore, $C = 6$ and the equation is

$$y=\frac{3x^2}{2}+5x+6$$

Example 38: Find $f(x)$ if $f'(x) = \sin x - \cos x$ and $f(\pi) = 3$.

Integrate $f'(x)$.

$$f(x) = \int (\sin x - \cos x)\,dx = -\cos x - \sin x + C$$

Now solve for the constant.

$$3 = -\cos(\pi) - \sin(\pi) + C$$

$$C = 2$$

Therefore, the equation becomes

$$f(x) = -\cos x - \sin x + 2$$

Now we've covered the basics of integration. However, integration is a very sophisticated topic, and there are many types of integrals that will cause you trouble. We will need several techniques to learn how to evaluate these integrals. The first and most important is called *u*-substitution, which we will cover in the next section of this chapter.

Example 39: Evaluate $\int x^{\frac{3}{5}}\,dx$.

Here's the Power Rule again.

$$\int x^n\,dx = \frac{x^{n+1}}{n+1} + C$$

Using the rule,

$$\int x^{\frac{3}{5}}\,dx = \frac{x^{\frac{8}{5}}}{\frac{8}{5}} + C$$

You can rewrite it as $\frac{5x^{\frac{8}{5}}}{8} + C$.

Example 40: Evaluate $\int (5x^3 + x^2 - 6x + 4)\,dx$.

We can break this up into several integrals.

$$\int (5x^3 + x^2 - 6x + 4)\,dx = \int 5x^3\,dx + \int x^2\,dx - \int 6x\,dx + \int 4\,dx$$

Each of these can be integrated according to the Power Rule.

$$\frac{5x^4}{4} + C + \frac{x^3}{3} + C - \frac{6x^2}{2} + C + 4x + C$$

This can be rewritten as

$$\frac{5x^4}{4} + \frac{x^3}{3} - 3x^2 + 4x + C$$

Notice that we combine the constant terms into one constant term C.

Example 41: Evaluate $\int (3 - x^2)^2 \, dx$.

First, expand the integrand.

$$\int (9 - 6x^2 + x^4) \, dx$$

Break this up into several integrals.

$$\int 9 \, dx - \int 6x^2 dx + \int x^4 dx$$

And integrate according to the Power Rule.

$$9x - 2x^3 + \frac{x^5}{5} + C$$

Example 42: Evaluate $\int (4\sin x - 3\cos x) \, dx$.

Break this problem into two integrals.

$$4\int \sin x \, dx - 3\int \cos x \, dx$$

Each of these trig integrals can be evaluated according to its rule.

$$-4\cos x - 3\sin x + C$$

Example 43: Evaluate $\int (2\sec^2 x - 5\csc^2 x) \, dx$.

Break the integral in two.

$$2\int \sec^2 x \, dx - 5\int \csc^2 x \, dx$$

Each of these trig integrals can be evaluated according to its rule.

$$2\tan x + 5\cot x + C$$

PRACTICE PROBLEM SET 23

Now evaluate the following integrals. The answers are at the end of the chapter, starting on page 420.

1. $\int_0^1 \left(x^4 - 5x^3 + 3x^2 - 4x - 6\right)dx$

2. $\int \left(5x^4 - 3x^2 + 2x + 6\right)dx$

3. $\int \left(3x^{-3} - 2x^{-2} + x^4 + 16x^7\right)dx$

4. $\int x^{\frac{1}{3}}(2+x)\,dx$

5. $\int \left(x^3 + x\right)^2 dx$

6. $\int \frac{x^6 - 2x^4 + 1}{x^2}\,dx$

7. $\int x(x-1)^3\,dx$

8. $\int \sec x\left(\sec x + \tan x\right)dx$

9. $\int \left(\sec^2 x + x\right)dx$

10. $\int \frac{\cos^3 x + 4}{\cos^2 x}\,dx$

11. $\int \frac{\sin 2x}{\cos x}\,dx$

12. $\int \left(1 + \cos^2 x \sec x\right)dx$

13. $\int \frac{1}{\csc x}\,dx$

14. $\int \left(x - \frac{2}{\cos^2 x}\right)dx$

INTEGRATING USING SUBSTITUTION

When we discussed differentiation, one of the most important techniques we mastered was the Chain Rule. Now, you'll learn the integration corollary of the Chain Rule (called *u*-substitution), which we use when the integrand is a composite function. All you do is replace the function with *u*, and then you can integrate the simpler function using the Power Rule (as shown below).

$$\int u^n du = \frac{u^{n+1}}{n+1} + C$$

Suppose you have to integrate $\int (x-4)^{10}\, dx$. You could expand out this function and integrate each term, but that'll take a while. Instead, you can follow these four steps.

Step 1: Let $u = x - 4$. Then $\frac{du}{dx} = 1$ (rearrange this to get $du = dx$).

Step 2: Substitute $u = x - 4$ and $du = dx$ into the integrand.

$$\int u^{10} du$$

Step 3: Integrate.

$$\int u^{10} du = \frac{u^{11}}{11} + C$$

Step 4: Substitute back for *u*.

$$\frac{(x-4)^{11}}{11} + C$$

That's *u*-substitution. The main difficulty you'll have will be picking the appropriate function to set equal to *u*. The best way to get better is to practice. The objective is to pick a function and replace it with *u*, then take the derivative of *u* to find *du*. If we can't replace *all* of the terms in the integrand, we *can't* do the substitution.

Let's do some examples.

Example 44: $\int 10x\left(5x^2 - 3\right)^6 dx =$

Once again, you could expand this out and integrate each term, but that would be difficult. Use *u*-substitution.

Let $u = 5x^2 - 3$. Then $\frac{du}{dx} = 10x$ and $du = 10x\,dx$. Now you can substitute.

$$\int u^6 du$$

And integrate.

$$\int u^6 du = \frac{u^7}{7} + C$$

Substituting back gives you

$$\frac{\left(5x^2 - 3\right)^7}{7} + C$$

Confirm that this is the integral by differentiating $\frac{\left(5x^2 - 3\right)^7}{7} + C$.

$$\frac{d}{dx}\left[\frac{\left(5x^2 - 3\right)^7}{7} + C\right] = \frac{7\left(5x^2 - 3\right)^6}{7}(10x) = \left(5x^2 - 3\right)^6(10x)$$

Example 45: $\int 2x\sqrt{x^2 - 5}\ dx =$

If $u = x^2 - 5$, then $\frac{du}{dx} = 2x$ and $du = 2x\,dx$. Substitute u into the integrand.

$$\int u^{\frac{1}{2}}\ du$$

Integrate.

$$\int u^{\frac{1}{2}}\ du = \frac{u^{\frac{3}{2}}}{\frac{3}{2}} + C = \frac{2u^{\frac{3}{2}}}{3} + C$$

And substitute back.

$$\frac{2\left(x^2 - 5\right)^{\frac{3}{2}}}{3} + C$$

Note: From now on, we're not going to rearrange $\frac{du}{dx}$; we'll go directly to "$du =$" format. You should be able to do that step on your own.

Example 46: $\int 3\sin(3x-1)dx =$

Let $u = 3x - 1$. Then $du = 3dx$. Substitute the u in the integral.

$$\int \sin u \, du$$

Figure out the integral.

$$\int \sin u \, du = -\cos u + C$$

And throw the xs back in.

$$-\cos(3x - 1) + C$$

So far, this is only the simplest kind of u-substitution; naturally, the process can get worse when the substitution isn't as easy. Usually, you'll have to insert a constant term to put the integrand into a workable form.

Example 47: $\int (5x+7)^{20} dx =$

Let $u = 5x + 7$. Then $du = 5\,dx$. Notice that we can't do the substitution immediately because we need to substitute for dx and we have $5\,dx$. No problem! Because 5 is a constant, just solve for dx.

$$\frac{1}{5}du = dx$$

Now you can substitute.

$$\int (5x+7)^{20} dx = \int u^{20}\left(\frac{1}{5}\right)du$$

Rearrange the integral and solve.

$$\int u^{20}\left(\frac{1}{5}\right)du = \frac{1}{5}\int u^{20} du = \frac{1}{5}\frac{u^{21}}{21} + C = \frac{u^{21}}{105} + C$$

And now it's time to substitute back.

$$\frac{u^{21}}{105} + C = \frac{(5x+7)^{21}}{105} + C$$

Example 48: $\int x\cos\left(3x^2+1\right)dx =$

Let $u = 3x^2 + 1$. Then $du = 6x\,dx$. We need to substitute for $x\,dx$, so we can rearrange the du term.

$$\frac{1}{6}du = x\,dx$$

Now substitute.

$$\int \frac{1}{6}\cos u\,du$$

Evaluate the integral.

$$\int \frac{1}{6}\cos u\,du = \frac{1}{6}\sin u + C$$

And substitute back.

$$\frac{1}{6}\sin u + C = \frac{1}{6}\sin\left(3x^2+1\right) + C$$

Example 49: $\int x\sec^2\left(x^2\right)dx =$

Let $u = x^2$. Then $du = 2x\,dx$ and $\frac{1}{2}du = x\,dx$.

Substitute.

$$\int \frac{1}{2}\sec^2 u\,du$$

Evaluate the integral.

$$\int \frac{1}{2}\sec^2 u\,du = \frac{1}{2}\tan u + C$$

Now the original function goes back in.

$$\frac{1}{2}\tan\left(x^2\right) + C$$

Example 50: Evaluate $\int \sec^2 3x\,dx$.

Let $u = 3x$ and $du = 3dx$. Then $\frac{1}{3}du = dx$.

Substitute and integrate.

$$\frac{1}{3}\int \sec^2 u\, du = \frac{1}{3}\tan u + C$$

Then substitute back.

$$\frac{1}{3}\tan 3x + C$$

Example 51: Evaluate $\int \sqrt{5x-4}\, dx$.

Let $u = 5x - 4$ and $du = 5dx$. Then $\frac{1}{5}du = dx$.

Substitute and integrate.

$$\frac{1}{5}\int u^{\frac{1}{2}}\, du = \frac{2}{15}u^{\frac{3}{2}} + C$$

Then substitute back.

$$\frac{2}{15}(5x-4)^{\frac{3}{2}} + C$$

Example 52: Evaluate $\int x\left(4x^2 - 7\right)^{10} dx$.

Let $u = 4x^2 - 7$ and $du = 8x\, dx$. Then $\frac{1}{8}du = x\, dx$.

Substitute and integrate.

$$\frac{1}{8}\int u^{10}\, du = \frac{1}{88}u^{11} + C$$

Then substitute back.

$$\frac{1}{88}\left(4x^2 - 7\right)^{11} + C$$

Example 53: Evaluate $\int \tan\frac{x}{3}\ \sec^2\frac{x}{3} dx$.

Let $u = \tan\frac{x}{3}$ and $du = \frac{1}{3}\sec^2\frac{x}{3}dx$. Then $3du = \sec^2\frac{x}{3}dx$.

Substituting, we get

$$3\int u\,du = \frac{3}{2}u^2 + C$$

Then substitute back.

$$\frac{3}{2}\tan^2\frac{x}{3} + C$$

PRACTICE PROBLEM SET 24

Now evaluate the following integrals. The answers are at the end of the chapter, starting on page 423.

1. $\int \sin 2x \cos 2x\, dx$

2. $\int \frac{3x\, dx}{\sqrt[3]{10 - x^2}}$

3. $\int x^3 \sqrt{5x^4 + 20}\, dx$

4. $\int \left(x^2 + 1\right)\left(x^3 + 3x\right)^{-5} dx$

5. $\int \frac{1}{\sqrt{x}} \sin\sqrt{x}\, dx$

6. $\int x^2 \sec^2 x^3\, dx$

7. $\int \frac{\cos\left(\frac{3}{x}\right)}{x^2} dx$

8. $\int \sin\left(\sin x\right)\cos x\, dx$

You've learned how to integrate polynomials and some of the trig functions (there are more of them to come), and you have the first technique of integration: u-substitution. Now it's time to learn how to integrate some other functions—namely, exponential and logarithmic functions. The first integral is the natural logarithm.

$$\int \frac{du}{u} = \ln|u| + C$$

Notice the absolute value in the logarithm. This ensures that you aren't taking the logarithm of a negative number. If you know that the term you're taking the log of is positive (for example, $x^2 + 1$), we can dispense with the absolute value marks. Let's do some examples.

Example 54: Find $\int \frac{5dx}{x+3}$.

Whenever an integrand contains a fraction, check to see if the integral is a logarithm. Usually, the process involves u-substitution. Let $u = x + 3$ and $du = dx$. Then,

$$\int \frac{5dx}{x+3} = 5\int \frac{du}{u} = 5\ln|u| + C$$

Substituting back, the final result is

$$5\ln|x + 3| + C$$

Example 55: Find $\int \frac{2x\ dx}{x^2+1}$.

Let $u = x^2 + 1$ and $du = 2x\ dx$, and substitute into the integrand.

$$\int \frac{2x\ dx}{x^2+1} = \int \frac{du}{u} = \ln|u| + C$$

Then substitute back.

$$\ln(x^2 + 1) + C$$

Note that we do not need the absolute value here because $x^2 + 1$ is always positive.

Remember when we started antiderivatives and we didn't do the integral of tangent, cotangent, secant, or cosecant?

Example 56: Find $\int \tan x \, dx$.

First, rewrite this integral as

$$\int \frac{\sin x}{\cos x} dx$$

Now, we let $u = \cos x$ and $du = -\sin x \, dx$ and substitute.

$$\int -\frac{du}{u}$$

Now, integrate and re-substitute.

$$\int -\frac{du}{u} = -\ln|u| = -\ln|\cos x| + C$$

Thus, $\int \tan x \, dx = -\ln|\cos x| + C$.

Example 57: Find $\int \cot x \, dx$.

Just as before, rewrite this integral in terms of sine and cosine.

$$\int \frac{\cos x}{\sin x} dx$$

Now we let $u = \sin x$ and $du = \cos x \, dx$ and substitute.

$$\int \frac{du}{u}$$

Now integrate.

$$\ln|u| + C = \ln|\sin x| + C$$

Therefore, $\int \cot x \, dx = \ln|\sin x| + C$.

Example 58: Find $\int \sec x \, dx$.

You could rewrite this integral as

$$\int \frac{1}{\cos x} \, dx$$

However, if you try *u*-substitution at this point, it won't work. So what should you do? You'll probably never guess, so we'll show you: multiply the $\sec x$ by $\frac{\sec x + \tan x}{\sec x + \tan x}$. This gives you

$$\int \sec x \left(\frac{\sec x + \tan x}{\sec x + \tan x} \right) dx = \int \frac{\sec^2 x + \sec x \tan x}{\sec x + \tan x} \, dx$$

Now you can do *u*-substitution. Let $u = \sec x + \tan x$ $du = (\sec x \tan x + \sec^2 x) \, dx$.

Then rewrite the integral as

$$\int \frac{du}{u}$$

The rest goes according to plan as you integrate.

$$\int \frac{du}{u} = \ln |u| + C = \ln |\sec x + \tan x| + C$$

Therefore, $\int \sec x \, dx = \ln |\sec x + \tan x| + C$.

Example 59: Find $\int \csc x \, dx$.

Multiply $\csc x$ by $\frac{\csc x + \cot x}{\csc x + \cot x} \, dx$. This gives you

$$\int \csc x \left(\frac{\csc x + \cot x}{\csc x + \cot x} \right) dx = \int \frac{\csc^2 x + \csc x \cot x}{\csc x + \cot x} \, dx$$

Let $u = \csc x + \cot x$ and $du = (-\csc x \cot x - \csc^2 x) \, dx$. And, just as in Example 58, you can rewrite the integral as

$$\int -\frac{du}{u}$$

And integrate.

$$\int -\frac{du}{u} = -\ln|u| + C = -\ln|\csc x + \cot x| + C$$

Therefore, $\int \csc x \, dx = -\ln|\csc x + \cot x| + C$.

As we do more integrals, the natural log will turn up over and over. It's important that you get good at recognizing when integrating requires the use of the natural log.

Integrating e^x and a^x

Now let's learn how to find the integral of e^x. Remember that $\frac{d}{dx}e^x = e^x$? Well, you should be able to predict the following formula:

$$\int e^u du = e^u + C$$

As with the natural logarithm, most of these integrals use *u*-substitution.

Example 60: Find $\int e^{7x} dx$.

Let $u = 7x$, $du = 7dx$, and $\frac{1}{7} du = dx$. Then you have

$$\int e^{7x} dx = \frac{1}{7}\int e^u du = \frac{1}{7}e^u + C$$

Substituting back, you get

$$\frac{1}{7}e^{7x} + C$$

In fact, whenever you see $\int e^{kx} dx$, where k is a constant, the integral is

$$\int e^{kx} dx = \frac{1}{k}e^{kx} + C$$

Example 61: Find $\int xe^{3x^2+1}dx$.

Let $u = 3x^2 + 1$, $du = 6x\,dx$, and $\frac{1}{6}du = x\,dx$. The result is

$$\int xe^{3x^2+1}dx = \frac{1}{6}\int e^u\,du = \frac{1}{6}e^u + C$$

Substitute back.

$$\frac{1}{6}e^{3x^2+1} + C$$

Example 62: Find $\int e^{\sin x}\cos x\,dx$.

Let $u = \sin x$ and $du = \cos x\,dx$. The substitution here couldn't be simpler.

$$\int e^u du = e^u + C = e^{\sin x} + C$$

As you can see, these integrals are pretty straightforward. The key is to use *u*-substitution to transform nasty-looking integrals into simple ones.

There's another type of exponential function whose integral you'll have to find occasionally:

$$\int a^u du$$

As you should recall from your rules of logarithms and exponents, the term a^u can be written as $e^{u\ln a}$. Because $\ln a$ is a constant, we can transform $\int a^u du$ into $\int e^{u\ln a}du$. If you integrate this, you'll get

$$\int e^{u\ln a}du = \frac{1}{\ln a}e^{u\ln a} + C$$

Now substituting back a^u for $e^{u\ln a}$,

$$\int a^u du = \frac{1}{\ln a}a^u + C$$

Example 63: Find $\int 5^x\,dx$.

Follow the rule we just derived.

$$\int 5^x\,dx = \frac{1}{\ln 5}5^x + C$$

Because these integrals don't show up too often on the AP Exam, this section is the last you'll see of them in this book. You should, however, be able to integrate them using the rule, or by converting them into a form of $\int e^u\,du$.

Example 64: Evaluate $\int \frac{dx}{3x}$.

Move the constant term outside of the integral, like this.

$$\frac{1}{3}\int \frac{dx}{x}$$

Now you can integrate.

$$\frac{1}{3}\int \frac{dx}{x} = \frac{1}{3}\ln|x| + C$$

Example 65: Evaluate $\int \frac{3x^2}{x^3-1}\,dx$.

Let $u = x^3 - 1$ and $du = 3x^2\,dx$, and substitute.

$$\int \frac{du}{u}$$

Now integrate.

$$\ln|u| + C$$

And substitute back.

$$\ln|x^3 - 1| + C$$

Example 66: Evaluate $\int e^{5x}dx$.

Let $u = 5x$ and $du = 5dx$. Then $\frac{1}{5}du = dx$. Substitute in.

$$\frac{1}{5}\int e^{u}du$$

Integrate.

$$\frac{1}{5}e^{u} + C$$

And substitute back.

$$\frac{1}{5}e^{5x} + C$$

Example 67: Evaluate $\int 2^{3x}dx$.

Let $u = 3x$ and $du = 3dx$. Then $\frac{1}{3}du = dx$. Make the substitution.

$$\frac{1}{3}\int 2^{u}du$$

Integrate according to the rule. Your result should be

$$\frac{1}{3\ln 2}2^{u} + C$$

Now get back to the expression as a function of x.

$$\frac{1}{3\ln 2}2^{3x} + C = \frac{1}{\ln 8}2^{3x} + C$$

PRACTICE PROBLEM SET 25

Evaluate the following integrals. The answers are at the end of the chapter, starting on page 424.

1. $\int \frac{\sec^2 x}{\tan x}\,dx$

2. $\int \frac{\cos x}{1 - \sin x}\,dx$

3. $\int \frac{1}{x \ln x}\,dx$

4. $\int \frac{\sin x - \cos x}{\cos x}\,dx$

5. $\int \frac{dx}{\sqrt{x}(1+2\sqrt{x})}$

6. $\int \frac{e^x\,dx}{1+e^x}$

7. $\int xe^{5x^2-1}\,dx$

8. $\int \frac{e^x + e^{-x}}{e^x - e^{-x}}\,dx$

9. $\int x4^{-x^2}\,dx$

10. $\int 7^{\sin x} \cos x\,dx$

Advanced Integrals of Trig Functions

Earlier, you learned how to find the integrals of some of the trigonometric functions. Now it's time to figure out how to find the integrals of some of the more complicated trig expressions. First, recall some of the basic trigonometric integrals.

$$\int \sin x\,dx = -\cos x + C \qquad \int \cos x\,dx = \sin x + C$$

$$\int \sec^2 x\,dx = \tan x + C \qquad \int \csc^2 x\,dx = -\cot x + C$$

$$\int \sec x \tan x\,dx = \sec x + C \qquad \int \csc x \cot x\,dx = -\csc x + C$$

$$\int \tan x\,dx = -\ln\left|\cos x\right| + C \qquad \int \cot x\,dx = \ln\left|\sin x\right| + C$$

$$\int \sec x\,dx = \ln\left|\sec x + \tan x\right| + C \qquad \int \csc x\,dx = -\ln\left|\csc x + \cot x\right| + C$$

Some of these were derived by reversing differentiation, others by *u*-substitution.

Example 68: Evaluate $\int \sin^2 x\, dx$.

You can start by replacing $\sin^2 x$ with $\dfrac{1 - \cos 2x}{2}$. This comes from taking the Double Angle formula, $\cos 2x = 1 - 2\sin^2 x$, and solving for $\sin^2 x$. After you make that replacement, the integral looks like the following:

$$\int \frac{1 - \cos 2x}{2}\, dx = \frac{1}{2}\int dx - \frac{1}{2}\int \cos 2x\, dx = \frac{x}{2} - \frac{\sin 2x}{4} + C$$

Remember the following substitution:

$$\sin^2 x = \frac{1 - \cos 2x}{2}$$

You could also use the following substitution:

$$\cos^2 x = \frac{1 + \cos 2x}{2}$$

(Note: This comes from taking the Double Angle formula, $\cos 2x = 2\cos^2 x - 1$, and solving for $\cos^2 x$.) You can also use the substitution to find $\int \cos^2 x\, dx$.

Now let's do some variations.

Example 69: Evaluate $\int \sin^2 x \cos x\, dx$.

Here use some simple u-substitution. Let $u = \sin x$ and $du = \cos x\, dx$ and substitute.

$$\int \sin^2 x \cos x\, dx = \int u^2\, du = \frac{u^3}{3} + C$$

When you substitute back, you get

$$\frac{\sin^3 x}{3} + C$$

In fact, you can do any integral of the form $\int \sin^n x \cos x \, dx$ using *u*-substitution, and the result will be

$$\frac{\sin^{n+1} x}{n+1} + C$$

Similarly,

$$\int \cos^n x \sin x \, dx = -\frac{\cos^{n+1}}{n+1} + C$$

How about higher powers of sine and cosine?

Example 70: Evaluate $\int \sin^3 x \, dx$.

First, break the integrand into the following:

$$\int (\sin x)(\sin^2 x) \, dx$$

Next, using trig substitution, you get

$$\int (\sin x)(\sin^2 x) \, dx = \int (\sin x)(1 - \cos^2 x) \, dx$$

You can turn the following into two integrals:

$$\int \sin x \, dx - \int (\sin x \cos^2 x) \, dx$$

You can do both of these integrals.

$$-\cos x + \frac{\cos^3 x}{3} + C$$

Example 71: Evaluate $\int \sin^3 x \cos^2 x \, dx$.

Here, break the integrand into

$$\int (\sin x)(\sin^2 x)(\cos^2 x) \, dx$$

Next, using trig substitution, you get

$$\int(\sin x)(\sin^2 x)(\cos^2 x)\,dx = \int(\sin x)(1 - \cos^2 x)(\cos^2 x)\,dx$$

Now you can use u-substitution. Let $u = \cos x$, $du = -\sin x\,dx$, and substitute.

$$-\int(1 - u^2)u^2\,du = \int(u^4 - u^2)\,du = \frac{u^5}{5} - \frac{u^3}{3} + C$$

Now substitute back, and you're done.

$$\frac{\cos^5 x}{5} - \frac{\cos^3 x}{3} + C$$

As you can see, these integrals all require you to know trig substitutions and u-substitution. The AP Exam doesn't ask too many variations on these integrals, but let's do just a few other types, just in case.

Example 72: Evaluate $\int \tan^2 x\,dx$.

First, use the trigonometric substitution $\tan^2 x = \sec^2 x - 1$.

$$\int \tan^2 x\,dx = \int(\sec^2 x - 1)\,dx$$

Now break this into two integrals that are easy to evaluate.

$$\int \sec^2 x\,dx - \int dx$$

And integrate.

$$\tan x - x + C$$

Example 73: Evaluate $\int \tan^3 x\,dx$.

Recognize what to do here? Right! Break up the integrand!

$$\int(\tan x)(\tan^2 x)\,dx$$

Next, use the trig substitution $\tan^2 x = \sec^2 x - 1$ in the expression.

$$\int(\tan x)(\sec^2 x - 1)\,dx$$

Break this into two integrals.

$$\int \tan x \sec^2 x \, dx - \int \tan x \, dx$$

Tackle the first integral using *u*-substitution. Let $u = \tan x$ and $du = \sec^2 x \, dx$.

$$\int u \, du = \frac{u^2}{2} + C$$

Substituting back gives you

$$\frac{\tan^2 x}{2} + C$$

We've done the second integral before.

$$\int \tan x \, dx = \int \frac{\sin x}{\cos x} dx = \ln |\cos x| + C$$

Thus, the integral is

$$\frac{\tan^2 x}{2} + \ln |\cos x| + C$$

Example 74: $\int \cos^2 x \, dx$

Use the substitution $\frac{1 + \cos 2x}{2}$. Now the integral looks like the following:

$$\int \frac{1 + \cos 2x}{2} dx$$

If you can integrate any power of a trig function up to 4, as well as some of the basic combinations, you'll be able to handle any trigonometric integral that appears on the AP Exam.

Divide this into two integrals.

$$\int \frac{1}{2} dx + \int \frac{\cos 2x}{2} dx$$

When you evaluate each one separately, you get

$$\frac{x}{2} + \frac{\sin 2x}{4} + C$$

Example 75: $\int \cos^3 x \, dx$

Here, split $\cos^3 x$ into $\cos x \cos^2 x$, and replace the second term with $(1 - \sin^2 x)$.

$$\int \cos x \,(1 - \sin^2 x)\, dx$$

This can be rewritten as

$$\int \cos x \, dx - \int \sin^2 x \cos x \, dx$$

The left-hand integral is $\sin x$. Do the right-hand integral using u-substitution. Let $u = \sin x$ and $du = \cos x \, dx$. You get

$$\int u^2 \, du = \frac{u^3}{3}$$

When you substitute back, the second integral becomes

$$\frac{\sin^3 x}{3} + C$$

Thus, the complete answer is

$$\sin x - \frac{\sin^3 x}{3} + C$$

Example 76: $\int \tan^2 x \sec^2 x \, dx$

Use u-substitution, by letting $u = \tan x$ and $du = \sec^2 x \, dx$.

$$\int u^2 \, du = \frac{u^3}{3}$$

The final result when you substitute back is

$$\frac{\tan^3 x}{3} + C$$

PRACTICE PROBLEM SET 26

Evaluate the following integrals. The answers are at the end of the chapter, starting on page 427.

1. $\int \sin^4 x\, dx$

2. $\int \cos^4 x\, dx$

3. $\int \cos^4 x \sin x\, dx$

4. $\int \sin^2 x \cos^2 x\, dx$

5. $\int \tan^3 x \sec^2 x\, dx$

6. $\int \tan^5 x\, dx$

7. $\int \cot^2 x \sec x\, dx$

Integrals that Result in Inverse Trigonometric Functions

Earlier, we learned how to take the derivatives of Inverse Trigonometric Functions. Specifically, we learned that

$$\frac{d}{dx}\left(\sin^{-1} x\right) = \frac{1}{\sqrt{1-x^2}}$$

$$\frac{d}{dx}\left(\tan^{-1} x\right) = \frac{1}{1+x^2}$$

$$\frac{d}{dx}\left(\sec^{-1} x\right) = \frac{1}{x\sqrt{x^2-1}}$$

Now we will go backwards to evaluate integrals that result in inverse trig functions. Note that we are *not* finding the integrals of inverse trig functions, which are not tested on the AP Exam.

We have

$$\int \frac{dx}{\sqrt{1-x^2}} = \sin^{-1} x + C$$

$$\int \frac{dx}{1+x^2} = \tan^{-1} x + C$$

$$\int \frac{dx}{x\sqrt{x^2-1}} = \sec^{-1} x + C$$

By the way, integrals involving inverse secant hardly ever show up, so we will concentrate on the other two inverse trig functions. Let's do some examples.

Example 77: $\int \frac{x\,dx}{\sqrt{1-x^4}} =$

Don't forget about *u*-substitution. Let $u = x^2$ and $du = 2x\,dx$. Thus, $\frac{1}{2}du = x\,dx$. Substituting and integrating, we get

$$\frac{1}{2}\int \frac{du}{\sqrt{1-u^2}} = \frac{1}{2}\sin^{-1} u + C$$

Now substitute back.

$$\frac{1}{2}\sin^{-1}(x^2) + C$$

Example 78: $\int \frac{dx}{\sqrt{4-x^2}} =$

Anytime you see an integral where you have the square root of a constant minus a function, try to turn it into an inverse sine. You can do that here with a little simple algebra.

$$\int \frac{dx}{\sqrt{4-x^2}} = \int \frac{dx}{\sqrt{4\left(1-\frac{x^2}{4}\right)}} = \frac{1}{2}\int \frac{dx}{\sqrt{1-\frac{x^2}{4}}}$$

Now use *u*-substitution: let $u = \frac{x}{2}$ and $du = \frac{1}{2}dx$, so $2du = dx$, and substitute.

$$\frac{1}{2}\int \frac{dx}{\sqrt{1-\frac{x^2}{4}}} = \frac{1}{2}\int \frac{2\,du}{\sqrt{1-u^2}} = \sin^{-1} u + C$$

When you substitute back, you get

$$\sin^{-1}\frac{x}{2} + C$$

Example 79: $\int \frac{e^x\,dx}{1 + e^{2x}} =$

Again, use u-substitution. Let $u = e^x$ and $du = e^x\,dx$, and substitute.

$$\int \frac{du}{1 + u^2} = \tan^{-1} u + C$$

Then substitute back.

$$\tan^{-1}e^x + C$$

Evaluating these integrals involves looking for a particular pattern. If it's there, all you have to do is use algebra and u-substitution to make the integrand conform to the pattern. Once that's accomplished, the rest is easy. These integrals will show up again when we do partial fractions. Otherwise, as far as the AP Exam is concerned, this is all you need to know.

Example 80: Find the derivative of $y = \sin^{-1}\left(\frac{x}{2}\right)$.

The rule is

$$\frac{d}{dx}(\sin^{-1} u) = \frac{1}{\sqrt{1 - u^2}}\frac{du}{dx}$$

The algebra looks like the following:

$$\frac{d}{dx}\left(\sin^{-1}\left(\frac{x}{2}\right)\right) = \frac{1}{\sqrt{1-\left(\frac{x}{2}\right)^2}}\frac{1}{2} = \frac{1}{2}\left(\frac{1}{\sqrt{1-\frac{x^2}{4}}}\right) = \frac{1}{2}\left(\frac{1}{\sqrt{\frac{4-x^2}{4}}}\right) = \frac{1}{2}\left(\frac{2}{\sqrt{4-x^2}}\right) = \frac{1}{\sqrt{4-x^2}}$$

Example 81: Evaluate $\int \frac{dx}{4 + x^2}$.

First, you need to do a little algebra. Factor 4 out of the denominator.

$$\int \frac{dx}{4\left(1 + \frac{x^2}{4}\right)}$$

Next, rewrite the integrand.

$$\frac{1}{4}\int \frac{dx}{\left(1+\left(\frac{x}{2}\right)^2\right)}$$

Now you can use *u*-substitution. Let $u = \frac{x}{2}$ and $du = \frac{1}{2}dx$, so $2du = dx$.

$$\frac{1}{2}\int \frac{du}{1+u^2}$$

Now integrate it.

$$\frac{1}{2}\tan^{-1} u + C$$

And re-substitute.

$$\frac{1}{2}\tan^{-1} \frac{x}{2} + C$$

INTEGRATING FUNCTIONS USING LONG DIVISION AND COMPLETING THE SQUARE

These two topics are rarely tested on the exam, but they're still good to know, and they will give you an excuse to practice intermediate algebra skills like Long Division (or Synthetic Division) and Completing the Square.

Sometimes, we will be given an integral where we will need to rewrite the integral using Long Division (also known as Polynomial Division) in order to evaluate it. How will we know if we need to use Long Division? If the polynomial in the numerator has a degree greater than or equal to the degree of the polynomial in the denominator, we know that Long Division is a good idea.

Let's do an example.

Example 82: Evaluate $\int \frac{4x^2-14x+9}{x-3}dx$.

Based on the denominator, this looks as if it would be some kind of logarithm, but the numerator isn't the derivative of the denominator. Let's use Long Division to simplify the integrand. We get

$$\begin{array}{r} 4x-2+\dfrac{3}{x-3} \\ x-3\overline{\big)\,4x^2-14x+9} \end{array}$$

Now we can rewrite the integrand as $\int \frac{4x^2 - 14x + 9}{x - 3} dx = \int 4x - 2 + \frac{3}{x - 3} dx$. This is easy to integrate. We get $\int \left[4x - 2 + \frac{3}{x - 3} \right] dx = 2x^2 - 2x + 3\ln|x - 3| + C$.

Let's do another one.

Example 83: Evaluate $\int \frac{6x^2 - 7x + 8}{2x + 5} dx$.

Let's use Long Division to simplify the integrand. We get

$$2x+5 \overline{\smash{\big)}\,6x^2 - 7x + 8} \quad = 3x - 11 + \frac{63}{2x + 5}$$

Now we can rewrite the integrand as $\int \frac{6x^2 - 7x + 8}{2x + 5} dx = \int 3x - 11 + \frac{63}{2x + 5} dx$. This is easy to integrate. We get $\int 3x - 11 + \frac{63}{2x + 5} dx = \frac{3x^2}{2} - 11x + \frac{63}{2}\ln|2x + 5| + C$.

As an alternative, or when we can't use Long Division, we can also use Completing the Square to simplify an integrand. When we do, we will almost always get an Inverse Trig integral.

Let's do a couple of examples.

Example 84: Evaluate $\int \frac{4dx}{x^2 - 4x + 5}$.

Note that we can't use Long Division because the numerator is a lower degree than the denominator. And we can't use *u*-substitution because the numerator is not the derivative of the denominator. Let's complete the square in the denominator to rewrite the integrand. We get $\int \frac{4dx}{x^2 - 4x + 5} = \int \frac{4dx}{x^2 - 4x + 4 - 4 + 5} = \int \frac{4dx}{(x - 2)^2 + 1}$. This should look like an Inverse Tangent integral. Let's do *u*-substitution. Let $u = x - 2$ and $du = dx$. Substitute into the integrand: $\int \frac{4dx}{(x - 2)^2 + 1} = \int \frac{4du}{u^2 + 1}$. Integrate: $\int \frac{4du}{u^2 + 1} = 4\tan^{-1} u$. And substitute back: $4\tan^{-1} u = 4\tan^{-1}(x - 2) + C$.

Example 85: Evaluate $\int \frac{8dx}{\sqrt{-8+6x-x^2}}$.

As with the previous example, let's complete the square in the denominator to rewrite the integrand.

We get $\int \frac{8dx}{\sqrt{-8+6x-x^2}} = \int \frac{8dx}{\sqrt{-8-(x^2-6x)}} = \int \frac{8dx}{\sqrt{-8-\left((x-3)^2-9\right)}} = \int \frac{8dx}{\sqrt{1-(x-3)^2}}$.

This should look like an Inverse Sine Integral. Let's do u-substitution. Let $u = x - 3$ and $du = dx$. Substitute into the integrand: $\int \frac{8dx}{\sqrt{1-(x-3)^2}} = \int \frac{8du}{\sqrt{1-u^2}}$. Integrate: $\int \frac{8du}{\sqrt{1-u^2}} = 8\sin^{-1} u$.

And substitute back: $8\sin^{-1} u = 8\sin^{-1}(x - 3) + C$.

Example 86: $\int \frac{dx}{x^2 + 4x + 5} =$

Complete the square in the denominator.

$$x^2 + 4x + 5 = (x + 2)^2 + 1$$

Now rewrite the integral.

$$\int \frac{dx}{1 + (x + 2)^2}$$

Using u-substitution, let $u = x + 2$ and $du = dx$. Then do the substitution.

$$\int \frac{du}{1 + u^2} = \tan^{-1} u + C$$

When you put the function of x back in to the integral, it reads,

$$\tan^{-1}(x + 2) + C$$

Example 87: Evaluate $\int \frac{dx}{\sqrt{-x^2 + 4x - 3}}$.

This time, you need to use some algebra by completing the square of the polynomial under the square root sign.

$$-x^2 + 4x - 3 = -(x^2 - 4x + 3) = -\left[(x-2)^2 - 1\right] = \left[1 - (x-2)^2\right]$$

Now rewrite the integrand.

$$\int \frac{dx}{\sqrt{1-(x-2)^2}}$$

And use u-substitution. Let $u = x - 2$ and $du = dx$.

$$\int \frac{du}{\sqrt{1-u^2}}$$

Now, this looks familiar. Once you integrate, you get

$$\sin^{-1} u + C$$

After you substitute back, it becomes

$$\sin^{-1}(x - 2) + C$$

PRACTICE PROBLEM SET 27

Here is some more practice work on derivatives and integrals of inverse trig functions. The answers are at the end of the chapter, starting on page 431.

1. Find the derivative of $\frac{1}{4}\tan^{-1}\left(\frac{x}{4}\right)$.

2. Find the derivative of $\sin^{-1}\left(\frac{1}{x}\right)$.

3. Find the derivative of $\tan^{-1}(e^x)$.

4. Evaluate $\int \frac{dx}{x\sqrt{x^2 - \pi}}$.

5. Evaluate $\int \frac{dx}{7 + x^2}$.

6. Evaluate $\int \frac{dx}{x(1 + \ln^2 x)}$.

7. Evaluate $\int \frac{\sec^2 x\, dx}{\sqrt{1 - \tan^2 x}}$.

8. Evaluate $\int \frac{e^{3x}\, dx}{1 + e^{6x}}$.

SELECTING TECHNIQUES FOR ANTIDIFFERENTIATION

Now that we have learned a variety of techniques for evaluating integrals, let's practice some where we don't tell you which technique to use. It will be up to you to figure out the best way to evaluate the integral.

Example 88: Evaluate $\int (x-3)\sqrt{4x^2-24x+7}\,dx$.

We should use substitution for this integral. How do we know? Notice that the derivative of what's under the radical is close to the expression in parentheses (except for a factor of 8).

Let $u = 4x^2 - 24x + 7$ and $du = 8x - 24$. Divide both sides of du by 8: $\frac{1}{8}du = (x-3)dx$.

Substitute into the integral: $\int (x-3)\sqrt{4x^2-24x+7}\,dx = \frac{1}{8}\int u^{\frac{1}{2}}du$.

Integrate: $\frac{1}{8}\int u^{\frac{1}{2}}du = \frac{1}{8}\frac{u^{\frac{3}{2}}}{\frac{3}{2}} + C = \frac{1}{12}u^{\frac{3}{2}} + C$.

And substitute back: $\frac{1}{12}u^{\frac{3}{2}} + C = \frac{1}{12}\left(4x^2-24x+7\right)^{\frac{3}{2}} + C$.

Example 89: Evaluate $\int \tan^5\theta\sec^4\theta\,d\theta$.

This is a trigonometric integral (obviously!), so we will look to use trig identities to simplify the integrand. We know that the derivative of $\tan\theta$ is $\sec^2\theta$. Break up the integrand: $\int \tan^5\theta\sec^2\theta\sec^2\theta\,d\theta$. Next, use the identity $1+\tan^2\theta = \sec^2\theta$ to get $\int \tan^5\theta\left(1+\tan^2\theta\right)\sec^2\theta\,d\theta$. Now we can use Substitution. Let $u = \tan\theta$ and $du = \sec^2\theta\,d\theta$. Substitute: $\int \tan^5\theta\left(1+\tan^2\theta\right)\sec^2\theta\,d\theta = \int u^5\left(1+u^2\right)du = \int u^5 + u^7\,du$.

Integrate: $\int u^5 + u^7\,du = \frac{u^6}{6} + \frac{u^8}{8} + C$. And substitute back: $\frac{u^6}{6} + \frac{u^8}{8} + C = \frac{\tan^6\theta}{6} + \frac{\tan^8\theta}{8} + C$.

Chapter 8 Drill

The answers are at the end of the chapter, starting on page 434.

1 Mark for Review

Evaluate $\int \left(3x^4 - 6x^3 + 2x + 7\right) dx$.

Ⓐ $12x^3 - 18x^2 + 2 + C$

Ⓑ $3x^5 - 6x^4 + x^2 + 7x + C$

Ⓒ $\frac{3x^5}{5} - \frac{6x^4}{4} + x^2 + 7x + C$

Ⓓ $\frac{3x^4}{4} - \frac{6x^3}{3} + \frac{x^2}{2} + 7x + C$

2 Mark for Review

Evaluate $\int 3\cos x - \frac{4}{\sqrt{x}} dx$.

Ⓐ $3\sin x - 2\sqrt{x} + C$

Ⓑ $3\sin x - 8\sqrt{x} + C$

Ⓒ $3\sin x + 2\sqrt{x} + C$

Ⓓ $-3\sin x + 8\sqrt{x} + C$

3 Mark for Review

Evaluate $\int \frac{9x^2}{\left(1 - x^3\right)^2} dx$.

Ⓐ $\frac{3}{1 - x^3} + C$

Ⓑ $\frac{-3}{1 - x^3} + C$

Ⓒ $\frac{3}{\left(1 - x^3\right)^3} + C$

Ⓓ $\frac{-3}{\left(1 - x^3\right)^3} + C$

4 Mark for Review

Evaluate $\int \frac{\cos\left(\frac{1}{x}\right)}{x^2} dx$.

Ⓐ $\frac{\sin\left(\frac{1}{x}\right)}{x} + C$

Ⓑ $\frac{-\sin\left(\frac{1}{x}\right)}{x} + C$

Ⓒ $\sin\left(\frac{1}{x}\right) + C$

Ⓓ $-\sin\left(\frac{1}{x}\right) + C$

5 Mark for Review

Evaluate $\int \cos^3(3x)\sin(3x)\, dx$.

Ⓐ $\frac{\cos^4(3x)}{12} + C$

Ⓑ $-\frac{\cos^4(3x)}{12} + C$

Ⓒ $-\frac{\sin^4(3x)}{12} + C$

Ⓓ $\frac{\sin^4(3x)}{12} + C$

6 Mark for Review

Approximate the area under the curve $y = 3 + x^2$ from $x = 1$ to $x = 5$ using $n = 4$ left-endpoint rectangles.

(A) 21

(B) 33

(C) 42

(D) 66

7 Mark for Review

Approximate the area under the curve $y = x^3 + 4$ from $x = 0$ to $x = 8$ using $n = 4$ trapezoids.

(A) 560

(B) 1120

(C) 2240

(D) 4480

8 Mark for Review

Approximate $\int_0^{16} \frac{3}{x}\,dx$ using the Midpoint Formula with $n = 4$.

(A) $\frac{22}{35}$

(B) $\frac{88}{35}$

(C) $\frac{176}{35}$

(D) $\frac{352}{35}$

9 Mark for Review

Find $\frac{d}{dx}\int_2^{x^2} \sin^3 t\,dt$.

(A) $3x^2\sin^2 x\cos x$

(B) $6x\sin^2 x\cos x$

(C) $x^2 \sin^3 x^2$

(D) $2x \sin^3 x^2$

10 Mark for Review

Evaluate $\int \sin^{19} x \cos^3 x\,dx$.

(A) $\frac{\sin^{20} x}{20} - \frac{\cos^4 x}{4} + C$

(B) $\frac{\sin^{20} x}{20} - \frac{\sin^{22} x}{22} + C$

(C) $\frac{\sin^{20} x}{20} + \frac{\sin^{22} x}{22} + C$

(D) $\frac{\sin^{20} x}{20} + \frac{\cos^4 x}{4} + C$

11 Mark for Review

Evaluate $\int \cos^3(3x)\sin(3x)\,dx$.

(A) $\frac{\sin^4(3x)}{12}\frac{\cos^2(3x)}{3} + C$

(B) $-\frac{\sin^4(3x)}{12}\frac{\cos^2(3x)}{3} + C$

(C) $\frac{\cos^4(3x)}{12} + C$

(D) $-\frac{\cos^4(3x)}{12} + C$

12 Mark for Review

Evaluate $\int 16\sin^2(2x)\cos^2(2x)\,dx$.

(A) $2x - \frac{\sin 8x}{4} + C$

(B) $2x + \frac{\sin 8x}{4} + C$

(C) $2x - \frac{\sin 2x}{4} + C$

(D) $2x + \frac{\sin 2x}{4} + C$

13 Mark for Review

Evaluate $\int \sec^4 x\,dx$.

(A) $\frac{\sec^5 x}{5} + C$

(B) $-\frac{\sec^5 x}{5} + C$

(C) $\tan x + \frac{\tan^3 x}{3} + C$

(D) $\tan x - \frac{\tan^3 x}{3} + C$

ANSWERS TO PRACTICE PROBLEM SETS

Practice Problem Set 21

1. $\dfrac{25}{32}$

First, draw a picture.

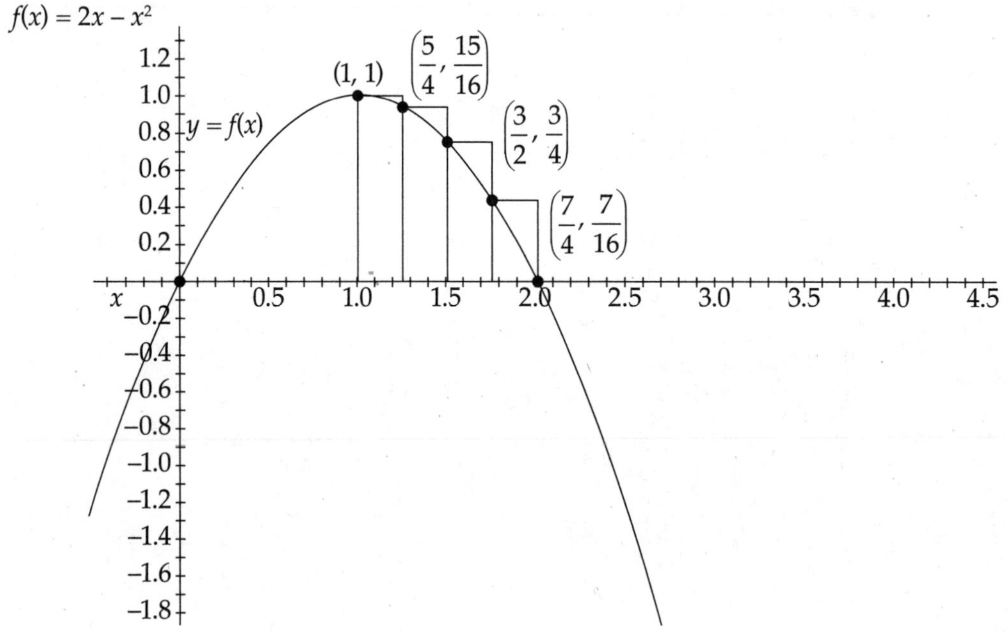

The width of each rectangle is found by taking the difference between the endpoints and dividing by n. Here, the width of each rectangle is $\dfrac{2-1}{4}=\dfrac{1}{4}$.

Find the heights of the rectangles by evaluating $y = 2x - x^2$ at the appropriate endpoints.

$y(1) = 2(1) - (1)^2 = 1$; $y\left(\dfrac{5}{4}\right)=2\left(\dfrac{5}{4}\right)-\left(\dfrac{5}{4}\right)^2=\dfrac{15}{16}$; $y\left(\dfrac{3}{2}\right)=2\left(\dfrac{3}{2}\right)-\left(\dfrac{3}{2}\right)^2=\dfrac{3}{4}$;

and $y\left(\dfrac{7}{4}\right)=2\left(\dfrac{7}{4}\right)-\left(\dfrac{7}{4}\right)^2=\dfrac{7}{16}$.

Therefore, the area is $\left(\dfrac{1}{4}\right)\left(1+\dfrac{15}{16}+\dfrac{3}{4}+\dfrac{7}{16}\right)=\dfrac{25}{32}$.

2. $\frac{17}{32}$

First, draw a picture.

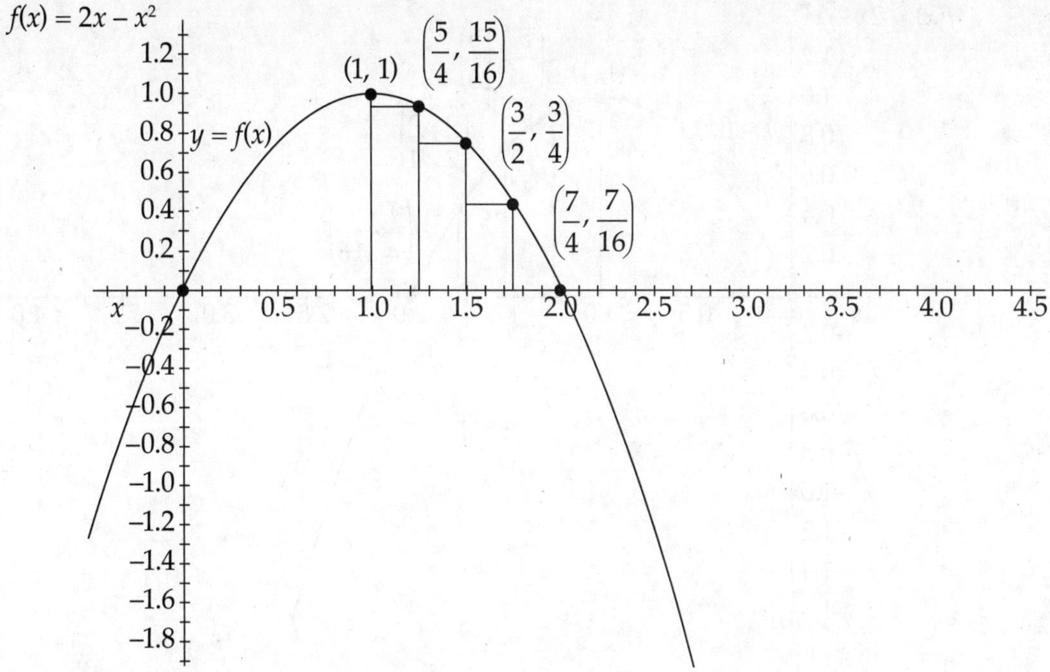

The width of each rectangle is found by taking the difference between the endpoints and dividing by n. Here, the width of each rectangle is $\frac{2-1}{4}=\frac{1}{4}$.

Find the heights of the rectangles by evaluating $y = 2x - x^2$ at the appropriate endpoints.

$y\left(\frac{5}{4}\right)=2\left(\frac{5}{4}\right)-\left(\frac{5}{4}\right)^2=\frac{15}{16}$; $y\left(\frac{3}{2}\right)=2\left(\frac{3}{2}\right)-\left(\frac{3}{2}\right)^2=\frac{3}{4}$; $y\left(\frac{7}{4}\right)=2\left(\frac{7}{4}\right)-\left(\frac{7}{4}\right)^2=\frac{7}{16}$;

and $y(2) = 2(2) - (2)^2 = 0$.

Therefore, the area is $\left(\frac{1}{4}\right)\left(\frac{15}{16}+\frac{3}{4}+\frac{7}{16}+0\right)=\frac{17}{32}$.

3. $\frac{21}{32}$

First, draw a picture.

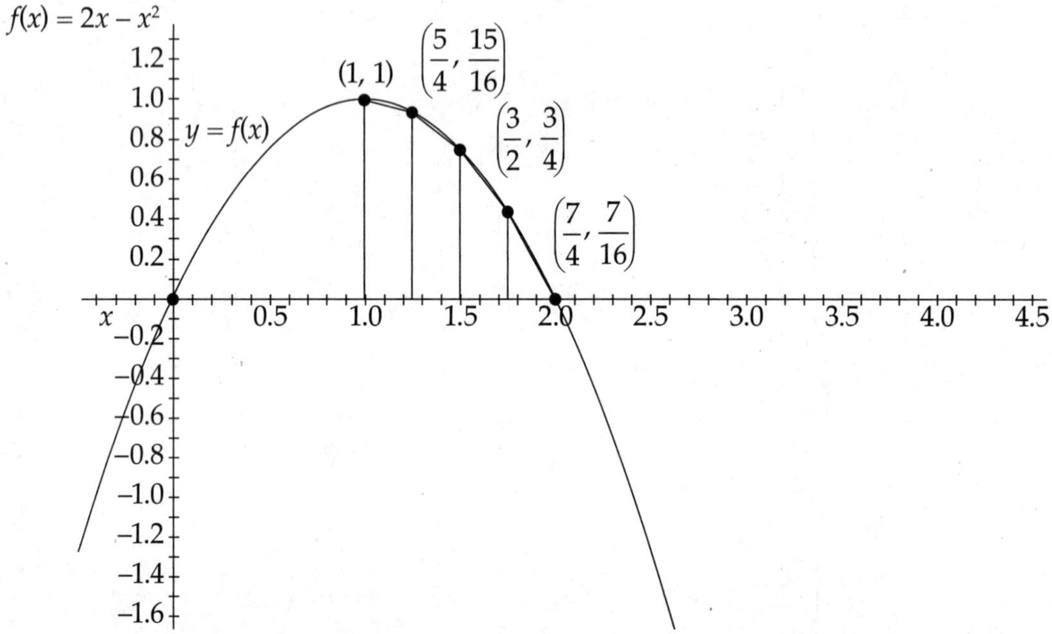

The height of each trapezoid is found by taking the difference between the endpoints and dividing by n. Here, the width of each trapezoid is $\frac{2-1}{4} = \frac{1}{4}$.

Find the bases of the trapezoids by evaluating $y = 2x - x^2$ at the appropriate endpoints.

$y(1) = 2(1) - (1)^2 = 1$; $y\left(\frac{5}{4}\right) = 2\left(\frac{5}{4}\right) - \left(\frac{5}{4}\right)^2 = \frac{15}{16}$; $y\left(\frac{3}{2}\right) = 2\left(\frac{3}{2}\right) - \left(\frac{3}{2}\right)^2 = \frac{3}{4}$; $y\left(\frac{7}{4}\right) = 2\left(\frac{7}{4}\right) - \left(\frac{7}{4}\right)^2 = \frac{7}{16}$; and $y(2) = 2(2) - (2)^2 = 0$.

Therefore, the area is $\left(\frac{1}{2}\right)\left(\frac{1}{4}\right)\left[1 + (2)\left(\frac{15}{16}\right) + (2)\left(\frac{3}{4}\right) + (2)\left(\frac{7}{16}\right) + 0\right] = \frac{21}{32}$.

4. $\frac{43}{64}$

First, draw a picture.

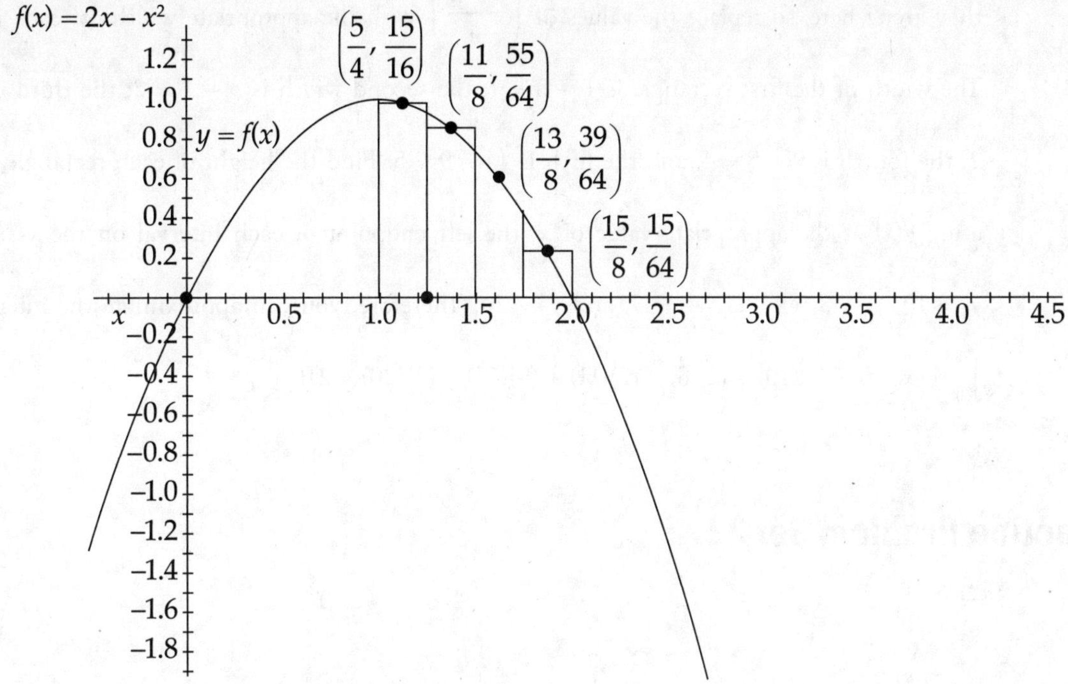

The height of each rectangle is found by taking the difference between the endpoints and dividing by *n*. Here, the width of each rectangle is $\frac{2-1}{4} = \frac{1}{4}$.

Find the bases of the rectangles by evaluating $y = 2x - x^2$ at the appropriate endpoints.

$y\left(\frac{9}{8}\right) = 2\left(\frac{9}{8}\right) - \left(\frac{9}{8}\right)^2 = \frac{63}{64}$; $y\left(\frac{11}{8}\right) = 2\left(\frac{11}{8}\right) - \left(\frac{11}{8}\right)^2 = \frac{55}{64}$; $y\left(\frac{13}{8}\right) = 2\left(\frac{13}{8}\right) - \left(\frac{13}{8}\right)^2 = \frac{39}{64}$;

and $y\left(\frac{15}{8}\right) = 2\left(\frac{15}{8}\right) - \left(\frac{15}{8}\right)^2 = \frac{15}{64}$.

Therefore, the area is $\left(\frac{1}{4}\right)\left(\frac{63}{64} + \frac{55}{64} + \frac{39}{64} + \frac{15}{64}\right) = \frac{43}{64}$.

5. 216

Recall that the formula for finding the area under the curve using the left endpoints is $\left(\frac{b-a}{n}\right)[y_0 + y_1 + y_2 + \ldots + y_{n-1}]$. This formula assumes that the x-values are evenly spaced, but they aren't here, so replace the values of $\left(\frac{b-a}{n}\right)$ with the appropriate widths of each rectangle. The width of the first rectangle is 1 – 0 = 1; the second width is 3 – 1 = 2; the third is 5 – 3 = 2; the fourth is 9 – 5 = 4; and the fifth is 14 – 9 = 5. Find the height of each rectangle by evaluating $g(x)$ at the appropriate value of x, the left endpoint of each interval on the x-axis. Here, $y_0 = 10$, $y_1 = 8$, $y_2 = 11$, $y_3 = 17$, and $y_4 = 20$. Therefore, you can approximate the integral with: $\int_0^{14} g(x)\,dx = (1)(10) + (2)(8) + (2)(11) + (4)(17) + (5)(20) = 216$.

Practice Problem Set 22

1. $\frac{2}{3}$

Find the exact area by evaluating the integral $\int_1^2 \left(2x - x^2\right)dx$. According to the Fundamental Theorem of Calculus: $\int_1^2 \left(2x - x^2\right)dx = \left(\frac{2x^2}{2} - \frac{x^3}{3}\right)\Big|_1^2 = \left(x^2 - \frac{x^3}{3}\right)\Big|_1^2$. Evaluating the integral at the limits, you get $\left[(2)^2 - \left(\frac{2^3}{3}\right)\right] - \left[(1)^2 - \left(\frac{1^3}{3}\right)\right] = \frac{2}{3}$.

2. 2

According to the Fundamental Theorem of Calculus: $\int_{-\frac{\pi}{2}}^{\frac{\pi}{2}} \cos x\,dx = \sin x\Big|_{-\frac{\pi}{2}}^{\frac{\pi}{2}} = \sin\frac{\pi}{2} - \sin\left(-\frac{\pi}{2}\right) = 2$.

3. $\dfrac{968}{5}$

First, rewrite the integrand: $\int_1^9 \left(2x\sqrt{x}\right)dx = \int_1^9 \left(2x^{\frac{3}{2}}\right)dx$. Now, according to the Fundamental Theorem of Calculus: $\int_1^9 \left(2x^{\frac{3}{2}}\right)dx = \left(\dfrac{2x^{\frac{5}{2}}}{\frac{5}{2}}\right)\Bigg|_1^9 = \left(\dfrac{4x^{\frac{5}{2}}}{5}\right)\Bigg|_1^9 = \left(\dfrac{4(9)^{\frac{5}{2}}}{5} - \dfrac{4(1)^{\frac{5}{2}}}{5}\right) = \dfrac{968}{5}$.

4. 0

According to the Fundamental Theorem of Calculus:

$$\int_{-\frac{\pi}{2}}^{\frac{\pi}{2}} \sin x\, dx = -\cos x\Big|_{-\frac{\pi}{2}}^{\frac{\pi}{2}} = -\cos\frac{\pi}{2} + \cos\left(-\frac{\pi}{2}\right) = 0$$

5. $-\dfrac{1}{3x^3} + C$

Use the Power Rule, which says that $\int x^n dx = \dfrac{x^{n+1}}{n+1} + C$. The integral is $\int \dfrac{1}{x^4} dx = \int x^{-4} dx = \dfrac{x^{-3}}{-3} + C = -\dfrac{1}{3x^3} + C$.

6. $10\sqrt{x} + C$

Use the Power Rule, which says that $\int x^n dx = \dfrac{x^{n+1}}{n+1} + C$. The integral is $\int \dfrac{5}{\sqrt{x}} dx = 5\int x^{-\frac{1}{2}} dx = 5\dfrac{x^{\frac{1}{2}}}{\frac{1}{2}} + C = 10\sqrt{x} + C$.

Practice Problem Set 23

1. $-\frac{161}{20}$

According to the Fundamental Theorem of Calculus:

$$\int_0^1 \left(x^4 - 5x^3 + 3x^2 - 4x - 6\right) dx = \left(\frac{x^5}{5} - 5\frac{x^4}{4} + x^3 - 2x^2 - 6x\right)\Bigg|_0^1 =$$

$$\left(\frac{(1)^5}{5} - 5\frac{(1)^4}{4} + (1)^3 - 2(1)^2 - 6(1)\right) - 0 = -\frac{161}{20}$$

2. $x^5 - x^3 + x^2 + 6x + C$

Use the Power Rule, which says that $\int x^n\, dx = \frac{x^{n+1}}{n+1} + C$. The integral is $\int \left(5x^4 - 3x^2 + 2x + 6\right) dx = 5\frac{x^5}{5} - 3\frac{x^3}{3} + 2\frac{x^2}{2} + 6x + C = x^5 - x^3 + x^2 + 6x + C$.

3. $-\frac{3}{2x^2} + \frac{2}{x} + \frac{x^5}{5} + 2x^8 + C$

Use the Power Rule, which says that $\int x^n\, dx = \frac{x^{n+1}}{n+1} + C$. The integral is

$$\int \left(3x^{-3} - 2x^{-2} + x^4 + 16x^7\right) dx = 3\frac{x^{-2}}{-2} - 2\frac{x^{-1}}{-1} + \frac{x^5}{5} + 16\frac{x^8}{8} + C = -\frac{3}{2x^2} + \frac{2}{x} + \frac{x^5}{5} + 2x^8 + C.$$

4. $\frac{3x^{\frac{4}{3}}}{2} + \frac{3x^{\frac{7}{3}}}{7} + C$

Use the Power Rule, which says that $\int x^n\, dx = \frac{x^{n+1}}{n+1} + C$. First, simplify the integrand: $\int x^{\frac{1}{3}}(2 + x)\, dx = \int \left(2x^{\frac{1}{3}} + x^{\frac{4}{3}}\right) dx$. Now, evaluate the integral:

$$\int \left(2x^{\frac{1}{3}} + x^{\frac{4}{3}}\right) dx = 2\frac{x^{\frac{4}{3}}}{\frac{4}{3}} + \frac{x^{\frac{7}{3}}}{\frac{7}{3}} + C = \frac{3x^{\frac{4}{3}}}{2} + \frac{3x^{\frac{7}{3}}}{7} + C.$$

5. $\frac{x^7}{7}+\frac{2x^5}{5}+\frac{x^3}{3}+C$

Use the Power Rule, which says that $\int x^n\,dx = \frac{x^{n+1}}{n+1}+C$. First, simplify the integrand: $\int (x^3 + x)^2\,dx$ $= \int (x^6 + 2x^4 + x^2)\,dx$. Now, evaluate the integral: $\int \left(x^6+2x^4+x^2\right)dx = \frac{x^7}{7}+\frac{2x^5}{5}+\frac{x^3}{3}+C$.

6. $\frac{x^5}{5}-\frac{2x^3}{3}-\frac{1}{x}+C$

Use the Power Rule, which says that $\int x^n\,dx = \frac{x^{n+1}}{n+1}+C$. First, simplify the integrand: $\int \frac{x^6-2x^4+1}{x^2}\,dx = \int\left(\frac{x^6}{x^2}-2\frac{x^4}{x^2}+\frac{1}{x^2}\right)dx = \int\left(x^4-2x^2+x^{-2}\right)dx$.

Now, evaluate the integral: $\int\left(x^4-2x^2+x^{-2}\right)dx = \frac{x^5}{5}-\frac{2x^3}{3}+\frac{x^{-1}}{-1}+C = \frac{x^5}{5}-\frac{2x^3}{3}-\frac{1}{x}+C$.

7. $\frac{x^5}{5}-\frac{3x^4}{4}+x^3-\frac{x^2}{2}+C$

Use the Power Rule, which says that $\int x^n dx = \frac{x^{n+1}}{n+1}+C$.

First, simplify the integrand: $\int x(x-1)^3\,dx = \int x(x^3-3x^2+3x-1)\,dx = \int (x^4-3x^3+3x^2-x)\,dx$.

Now, evaluate the integral:

$$\int\left(x^4-3x^3+3x^2-x\right)dx = \frac{x^5}{5}-\frac{3x^4}{4}+\frac{3x^3}{3}-\frac{x^2}{2}+C = \frac{x^5}{5}-\frac{3x^4}{4}+x^3-\frac{x^2}{2}+C$$

8. $\tan x + \sec x + C$

Use the Rules for the Integrals of Trig Functions, namely, $\int \sec^2 x\,dx = \tan x + C$ and $\int (\sec x \tan x)\,dx =$ $\sec x + C$. Expand the integrand: $\int \sec x\,(\sec x + \tan x)\,dx = \int (\sec^2 x + \sec x \tan x)\,dx$. You get $\int (\sec^2 x + \sec x \tan x)\,dx = \tan x + \sec x + C$.

9. $\tan x + \frac{x^2}{2}+C$

Use the Rules for the Integrals of Trig Functions, namely, $\int \sec^2 x\,dx = \tan x + C$ and the Power Rule, which says that $\int x^n dx = \frac{x^{n+1}}{n+1}+C$. You get $\int\left(\sec^2 x + x\right)dx = \tan x + \frac{x^2}{2}+C$.

10. $\sin x + 4 \tan x + C$

Use the Rules for the Integrals of Trig Functions, namely, $\int \sec^2 x\, dx = \tan x + C$ and $\int \cos x\, dx = \sin x + C$.

First, rewrite the integrand, using trig identities:

$\int \frac{\cos^3 x + 4}{\cos^2 x} dx = \int \left(\frac{\cos^3 x}{\cos^2 x} + \frac{4}{\cos^2 x} \right) dx = \int \left(\cos x + 4\sec^2 x \right) dx$. Now, evaluate the integral: $\int (\cos x + 4\sec^2 x) dx = \sin x + 4\tan x + C$.

11. $-2\cos x + C$

Use the Rules for the Integrals of Trig Functions, namely, $\int \sin x\, dx = -\cos x + C$. First, rewrite the integrand, using trig identities: $\int \frac{\sin 2x}{\cos x} dx = \int \frac{2 \sin x \cos x}{\cos x} dx = \int 2 \sin x\, dx$. Now, evaluate the integral: $\int 2 \sin x\, dx = -2\cos x + C$.

12. $x + \sin x + C$

Use the Rules for the Integrals of Trig Functions, namely, $\int \cos x\, dx = \sin x + C$. First, rewrite the integrand, using trig identities: $\int \left(1 + \cos^2 x \sec x \right) dx = \int \left(1 + \frac{\cos^2 x}{\cos x} \right) dx = \int \left(1 + \cos x \right) dx$. Now, evaluate the integral: $\int (1 + \cos x) dx = x + \sin x + C$.

13. $-\cos x + C$

Use the Rules for the Integrals of Trig Functions, namely, $\int \sin x\, dx = -\cos x + C$. First, rewrite the integrand, using trig identities: $\int \frac{1}{\csc x} dx = \int \sin x\, dx$. Now, evaluate the integral: $\int \sin x\, dx = -\cos x + C$.

14. $\frac{x^2}{2} - 2\tan x + C$

Use the Rules for the Integrals of Trig Functions, namely, $\int \sec^2 x\, dx = \tan x + C$. First, rewrite the integrand, using trig identities: $\int \left(x - \frac{2}{\cos^2 x} \right) dx = \int \left(x - 2\sec^2 x \right) dx$. Now, evaluate the integral: $\int \left(x - 2\sec^2 x \right) dx = \frac{x^2}{2} - 2\tan x + C$.

Practice Problem Set 24

1. $\dfrac{\sin^2 2x}{4}+C$

If you let $u = \sin 2x$, then $du = 2\cos 2x\,dx$. Substitute for $\cos 2x\,dx$, so you can divide the du term by 2: $\dfrac{du}{2} = \cos 2x\,dx$. Next, substitute into the integral: $\int \sin 2x \cos 2x\,dx = \dfrac{1}{2}\int u\,du$. Now integrate: $\dfrac{1}{2}\int u\,du = \dfrac{1}{2}\left(\dfrac{u^2}{2}\right)+C = \dfrac{u^2}{4}+C$. Last, substitute back and get $\dfrac{\sin^2 2x}{4}+C$.

2. $-\dfrac{9}{4}\left(10-x^2\right)^{\frac{2}{3}}+C$

First, pull the constant out of the integrand: $\int \dfrac{3x\,dx}{\sqrt[3]{10-x^2}}dx = 3\int \dfrac{x\,dx}{\sqrt[3]{10-x^2}}dx$. If you let $u = 10 - x^2$, then $du = -2x\,dx$. Substitute for $x\,dx$, so you can divide the du term by -2: $-\dfrac{du}{2} = x\,dx$. Next, substitute into the integral: $3\int \dfrac{x\,dx}{\sqrt[3]{10-x^2}}dx = -\dfrac{3}{2}\int u^{-\frac{1}{3}}du$. Now integrate: $-\dfrac{3}{2}\int u^{-\frac{1}{3}}du = -\dfrac{3}{2}\dfrac{u^{\frac{2}{3}}}{\frac{2}{3}}+C = -\dfrac{9}{4}u^{\frac{2}{3}}+C$. Last, substitute back and get $-\dfrac{9}{4}\left(10-x^2\right)^{\frac{2}{3}}+C$.

3. $\dfrac{1}{30}\left(5x^4+20\right)^{\frac{3}{2}}+C$

If you let $u = 5x^4 + 20$, then $du = 20x^3\,dx$. Substitute for $x^3\,dx$, so you can divide the du term by 20: $\dfrac{du}{20} = x^3\,dx$. Next, $\dfrac{1}{20}\int u^{\frac{1}{2}}du = \dfrac{1}{20}\dfrac{u^{\frac{3}{2}}}{\frac{3}{2}}+C = \dfrac{1}{30}u^{\frac{3}{2}}+C$. Last, substitute back and get $\dfrac{1}{30}\left(5x^4+20\right)^{\frac{3}{2}}+C$.

4. $-\dfrac{1}{12\left(x^3+3x\right)^4}+C$

If you let $u = x^3 + 3x$, then $du = (3x^2 + 3)\,dx$. Substitute for $(x^2 + 1)\,dx$, so you can divide the du term by 3: $\dfrac{du}{3} = \left(x^2+1\right)dx$. Next, substitute into the integral: $\int \left(x^2+1\right)\left(x^3+3x\right)^{-5}dx = \dfrac{1}{3}\int u^{-5}du$. Now integrate: $\dfrac{1}{3}\int u^{-5}du = \dfrac{1}{3}\dfrac{u^{-4}}{-4}+C = -\dfrac{1}{12}\dfrac{1}{u^4}+C$. Last, substitute back and get $-\dfrac{1}{12\left(x^3+3x\right)^4}+C$.

5. $-2\cos\sqrt{x}+C$

If you let $u=\sqrt{x}$, then $du=\frac{1}{2\sqrt{x}}dx$. Substitute for $\frac{1}{\sqrt{x}}dx$, so you can multiply the du term by 2: $2du=\frac{1}{\sqrt{x}}dx$. Next, substitute into the integral: $\int\frac{1}{\sqrt{x}}\sin\sqrt{x}\,dx = 2\int\sin u\,du$. Now integrate: $2\int\sin u\,du = -2\cos u + C$. Last, substitute back and get $-2\cos\sqrt{x}+C$.

6. $\frac{1}{3}\tan\left(x^3\right)+C$

If you let $u = x^3$, then $du = 3x^2\,dx$. Substitute for $x^2\,dx$, so you can divide the du term by 3: $\frac{du}{3}=x^2dx$. Next, substitute into the integral: $\int x^2\sec^2\left(x^3\right)dx=\frac{1}{3}\int\sec^2 u\,du$. Now integrate: $\frac{1}{3}\int\sec^2 u\,du=\frac{1}{3}\tan u+C$. Last, substitute back and get $\frac{1}{3}\tan\left(x^3\right)+C$.

7. $-\frac{1}{3}\sin\left(\frac{3}{x}\right)+C$

If you let $u=\frac{3}{x}$, then $du=-\frac{3}{x^2}dx$. Substitute for $\frac{1}{x^2}dx$, so you can divide the du term by -3: $-\frac{du}{3}=\frac{1}{x^2}dx$. Next, substitute into the integral: $\int\frac{\cos\left(\frac{3}{x}\right)}{x^2}dx=-\frac{1}{3}\int\cos u\,du$. Now integrate: $-\frac{1}{3}\int\cos u\,du=-\frac{1}{3}\sin u+C$. Last, substitute back and get $-\frac{1}{3}\sin\left(\frac{3}{x}\right)+C$.

8. $-\cos(\sin x) + C$

If you let $u = \sin x$, then $du = \cos x\,dx$. Next, substitute into the integral: $\int\sin(\sin x)\cos x\,dx = \int\sin u\,du$. Now integrate: $\int\sin u\,du = -\cos u + C$. Last, substitute back and get $-\cos(\sin x) + C$.

Practice Problem Set 25

1. $\ln|\tan x| + C$

Whenever you have an integral in the form of a quotient, check to see if the solution is a logarithm. A clue is whether the numerator is the derivative of the denominator, as it is here. Use u-substitution. If you let $u = \tan x$, then $du = \sec^2 x\,dx$. If you substitute into the integrand, you get $\int \frac{\sec^2 x}{\tan x}dx = \int \frac{du}{u}$. Recall that $\int \frac{du}{u} = \ln|u| + C$. Substituting back, you get $\int \frac{\sec^2 x}{\tan x}dx = \ln|\tan x| + C$.

2. $-\ln|1 - \sin x| + C$

Whenever you have an integral in the form of a quotient, check to see if the solution is a logarithm. A clue is whether the numerator is the derivative of the denominator, as it is here. Use u-substitution. If you let $u = 1 - \sin x$, then $du = -\cos x\,dx$. If you substitute into the integrand, you get $\int \frac{\cos x}{1-\sin x}dx = -\int \frac{du}{u}$. Recall that $\int \frac{du}{u} = \ln|u| + C$. Substituting back, you get $\int \frac{\cos x}{1-\sin x}dx = -\ln|1-\sin x| + C$.

3. $\ln|\ln x| + C$

Whenever you have an integral in the form of a quotient, check to see if the solution is a logarithm. A clue is whether the numerator is the derivative of the denominator, as it is here. Use u-substitution. If you let $u = \ln x$, then $du = \frac{1}{x}dx$. If you substitute into the integrand, you get $\int \frac{1}{x\ln x}dx = \int \frac{du}{u}$. Recall that $\int \frac{du}{u} = \ln|u| + C$. Substituting back, you get $\int \frac{1}{x\ln x}dx = \ln|\ln x| + C$.

4. $-\ln|\cos x| - x + C$

First, rewrite the integrand.

$$\int \frac{\sin x - \cos x}{\cos x}dx = \int\left(\frac{\sin x}{\cos x} - \frac{\cos x}{\cos x}\right)dx = \int\left(\frac{\sin x}{\cos x} - 1\right)dx = \int\left(\frac{\sin x}{\cos x}\right)dx - \int dx$$

Whenever you have an integral in the form of a quotient, check to see if the solution is a logarithm. A clue is whether the numerator is the derivative of the denominator, as it is in the first integral. Use u-substitution. If you let $u = \cos x$, then $du = -\sin x\,dx$. If you substitute into the integrand, you get $\int\left(\frac{\sin x}{\cos x}\right)dx = -\int \frac{du}{u}$. Recall that $\int \frac{du}{u} = \ln|u| + C$. Substituting back, you get

$\int\left(\frac{\sin x}{\cos x}\right)dx = -\ln|\cos x| + C$. The second integral is simply $\int dx = x + C$. Therefore, the integral is $\int\left(\frac{\sin x}{\cos x}\right)dx - \int dx = -\ln|\cos x| - x + C$.

5. $\ln\left(1+2\sqrt{x}\right)+C$

Whenever you have an integral in the form of a quotient, check to see if the solution is a logarithm. A clue is whether the numerator is the derivative of the denominator, as it is here. Use u-substitution. If you let $u = 1 + 2\sqrt{x}$, then $du = \frac{1}{\sqrt{x}}dx$. If you substitute into the integrand, you get $\int \frac{1}{\sqrt{x}\left(1+2\sqrt{x}\right)}dx = \int \frac{du}{u}$. Recall that $\int \frac{du}{u} = \ln|u| + C$. Substituting back, you get $\int \frac{1}{\sqrt{x}\left(1+2\sqrt{x}\right)}dx = \ln\left(1+2\sqrt{x}\right)+C$. Notice that you don't need the absolute value bars because $1+2\sqrt{x}$ is never negative.

6. $\ln(1 + e^x) + C$

Whenever you have an integral in the form of a quotient, check to see if the solution is a logarithm. A clue is whether the numerator is the derivative of the denominator, as it is here. Use u-substitution. If you let $u = 1 + e^x$, then $du = e^x\,dx$. If you substitute into the integrand, you get $\int \frac{e^x}{1+e^x}dx = \int \frac{du}{u}$. Recall that $\int \frac{du}{u} = \ln|u| + C$. Substituting back, you get $\int \frac{e^x}{1+e^x}dx = \ln\left(1+e^x\right)+C$. Notice that you don't need the absolute value bars because $1 + e^x$ is never negative.

7. $\frac{1}{10}e^{5x^2-1}+C$

Recall that $\int e^u\,du = e^u + C$. Use u-substitution. If you let $u = 5x^2 - 1$, then $du = 10x\,dx$. But you need to substitute for $x\,dx$, so if you divide du by 10, you get $\frac{1}{10}du = x\,dx$. Now, if you substitute into the integrand, you get $\int xe^{5x^2-1}dx = \frac{1}{10}\int e^u\,du = \frac{1}{10}e^u + C$. Substituting back, you get $\int xe^{5x^2-1}dx = \frac{1}{10}e^{5x^2-1}+C$.

8. $\ln|e^x - e^{-x}| + C$

Recall that $\int e^u\, du = e^u + C$. Use u-substitution. If you let $u = e^x - e^{-x}$, then $du = e^x + e^{-x}\, dx$. If you substitute into the integrand, you get $\int \frac{e^x + e^{-x}}{e^x - e^{-x}}\, dx = \int \frac{du}{u}$. Recall that $\int \frac{du}{u} = \ln|u| + C$. Substituting back, you get $\int \frac{e^x + e^{-x}}{e^x - e^{-x}}\, dx = \ln|e^x - e^{-x}| + C$.

9. $-\frac{4^{-x^2}}{\ln 16} + C$

Recall that $\int a^u\, du = \frac{1}{\ln a} a^u + C$. Use u-substitution. If you let $u = -x^2$, then $du = -2x\, dx$. But you need to substitute for $x\, dx$, so if you divide du by -2, you get $-\frac{1}{2}du = x\, dx$. Now, if you substitute into the integrand, you get $\int x4^{-x^2}\, dx = -\frac{1}{2}\int 4^u\, du = -\frac{1}{2}\left(\frac{1}{\ln 4}\right)4^u + C$. Substituting back, you get $\int x4^{-x^2}\, dx = -\frac{1}{2}\left(\frac{1}{\ln 4}\right)4^{-x^2} + C = -\frac{1}{\ln 16}4^{-x^2} + C$.

10. $\frac{7^{\sin x}}{\ln 7} + C$

Recall that $\int a^u\, du = \frac{1}{\ln a} a^u + C$. Use u-substitution. If you let $u = \sin x$, then $du = \cos x\, dx$. If you substitute into the integrand, you get $\int 7^{\sin x} \cos x\, dx = \int 7^u\, du = \frac{1}{\ln 7}7^u + C$. Substituting back, you get $\int 7^{\sin x}\cos x\, dx = \frac{7^{\sin x}}{\ln 7} + C$.

Practice Problem Set 26

1. $\frac{3x}{8} - \frac{\sin 2x}{4} + \frac{\sin 4x}{32} + C$

Use the substitution $\sin^2 x = \frac{1 - \cos 2x}{2}$ into the integrand. You get $\int \sin^4 x\, dx = \int \left(\frac{1 - \cos 2x}{2}\right)^2 dx = \int \left(\frac{1 - 2\cos 2x + \cos^2 2x}{4}\right) dx$. Break this into three integrals: $\int \left(\frac{1 - 2\cos 2x + \cos^2 2x}{4}\right) dx = \frac{1}{4}\int dx - \frac{1}{2}\int \cos 2x\, dx + \frac{1}{4}\int \cos^2 2x\, dx$. The first two integrals are easy: $\frac{1}{4}\int dx - \frac{1}{2}\int \cos 2x\, dx = \frac{x}{4} - \frac{\sin 2x}{4}$. For the third integral, use the substitution

$\cos^2 x = \dfrac{1+\cos 2x}{2}$. You get $\int \cos^2 2x\, dx = \int \dfrac{1+\cos 4x}{2} dx = \dfrac{1}{2}\int dx + \dfrac{1}{2}\int \cos 4x\, dx$. Now integrate these: $\dfrac{1}{2}\int dx + \dfrac{1}{2}\int \cos 4x\, dx = \dfrac{1}{2}x + \dfrac{1}{8}\sin 4x + C$. Now put everything together to get $\dfrac{x}{4} - \dfrac{\sin 2x}{4} + \dfrac{1}{4}\left(\dfrac{1}{2}x + \dfrac{1}{8}\sin 4x\right) + C = \dfrac{3x}{8} - \dfrac{\sin 2x}{4} + \dfrac{\sin 4x}{32} + C.$

2. $\dfrac{3x}{8} + \dfrac{\sin 2x}{4} + \dfrac{\sin 4x}{32} + C$

Use the substitution $\cos^2 x = \dfrac{1+\cos 2x}{2}$ into the integrand. You get $\int \cos^4 x\, dx = \int \left(\dfrac{1+\cos 2x}{2}\right)^2 dx = \int \left(\dfrac{1+2\cos 2x + \cos^2 2x}{4}\right) dx$. Break this into three integrals: $\int \left(\dfrac{1+2\cos 2x + \cos^2 2x}{4}\right) dx = \dfrac{1}{4}\int dx + \dfrac{1}{2}\int \cos 2x\, dx + \dfrac{1}{4}\int \cos^2 2x\, dx$. The first two integrals are easy: $\dfrac{1}{4}\int dx + \dfrac{1}{2}\int \cos 2x\, dx = \dfrac{x}{4} + \dfrac{\sin 2x}{4}$. For the third integral, again use the substitution $\cos^2 x = \dfrac{1+\cos 2x}{2}$. You get $\int \cos^2 2x\, dx = \int \dfrac{1+\cos 4x}{2} dx = \dfrac{1}{2}\int dx + \dfrac{1}{2}\int \cos 4x\, dx$. Now integrate these: $\dfrac{1}{2}\int dx + \dfrac{1}{2}\int \cos 4x\, dx = \dfrac{1}{2}x + \dfrac{1}{8}\sin 4x + C$. Now put everything together to get $\dfrac{x}{4} + \dfrac{\sin 2x}{4} + \dfrac{1}{4}\left(\dfrac{1}{2}x + \dfrac{1}{8}\sin 4x\right) + C = \dfrac{3x}{8} + \dfrac{\sin 2x}{4} + \dfrac{\sin 4x}{32} + C.$

3. $-\dfrac{\cos^5 x}{5} + C$

Use u-substitution. If you let $u = \cos x$, then $du = -\sin x\, dx$. Substituting into the integrand, you get $\int \cos^4 x \sin x\, dx = -\int u^4 du = -\dfrac{u^5}{5} + C$. Now substitute back: $\int \cos^4 x \sin x\, dx = -\dfrac{\cos^5 x}{5} + C$.

4. $\dfrac{x}{8} - \dfrac{\sin 4x}{32} + C$

First, use the trig identity $\sin^2 x = 1 - \cos^2 x$, and substitute into the integrand: $\int \sin^2 x \cos^2 x\, dx = \int (1-\cos^2 x)\cos^2 x\, dx = \int (\cos^2 x - \cos^4 x) dx$. Next, use the substitution

$\cos^2 x = \dfrac{1+\cos 2x}{2}$ into the integrand: $\int \left(\left(\dfrac{1+\cos 2x}{2}\right) - \left(\dfrac{1+\cos 2x}{2}\right)^2\right) dx$. Use a little algebra:

$$\int \left(\left(\frac{1+\cos 2x}{2}\right) - \left(\frac{1+\cos 2x}{2}\right)^2\right) dx =$$

$$\int \left(\left(\frac{1+\cos 2x}{2}\right) - \left(\frac{1+2\cos 2x+\cos^2 2x}{4}\right)\right) dx =$$

$$\int \left(\frac{1}{4} - \frac{\cos^2 2x}{4}\right) dx$$

Now break this into two integrals: $\dfrac{1}{4}\int dx - \int \dfrac{\cos^2 2x}{4} dx$. The first integral is easy: $\dfrac{1}{4}\int dx = \dfrac{x}{4}$. For the second integral, again use the substitution $\cos^2 x = \dfrac{1+\cos 2x}{2}$ into the integrand: $\dfrac{1}{4}\int \dfrac{1+\cos 2x}{2} dx = \dfrac{1}{8}\int dx + \dfrac{1}{8}\int \cos 4x\, dx$. These are both easy to integrate: $\dfrac{1}{8}\int dx + \dfrac{1}{8}\int \cos 4x\, dx = \dfrac{x}{8} + \dfrac{\sin 4x}{32}$. Putting it all together, you get $\dfrac{x}{4} - \left(\dfrac{x}{8} + \dfrac{\sin 4x}{32}\right) = \dfrac{x}{8} - \dfrac{\sin 4x}{32} + C$.

There is another way to do this integral that is easier, but only if you spot it. Recall that $\sin 2x = 2\sin x \cos x$. This means that you could rewrite the integrand as $\int \sin^2 x \cos^2 x\, dx = \int \left(\dfrac{\sin 2x}{2}\right)^2 dx = \dfrac{1}{4}\int \sin^2 2x\, dx$. Substitute in $\sin^2 x = \dfrac{1-\cos 2x}{2}$ to rewrite the integrand: $\dfrac{1}{4}\int \sin^2 2x\, dx = \dfrac{1}{4}\int \dfrac{1-\cos 4x}{2} dx = \dfrac{1}{8}\int (1-\cos 4x) dx$. This is easy to integrate: $\dfrac{1}{8}\int (1-\cos 4x) dx = \dfrac{1}{8}\left(x - \dfrac{\sin 4x}{4}\right) = \dfrac{x}{8} - \dfrac{\sin 4x}{32} + C$. This is why you should know your trigonometric identities really well before you take calculus!

5. $\dfrac{\tan^4 x}{4} + C$

Use u-substitution. If you let $u = \tan x$, then $du = \sec^2 x\, dx$. Substituting into the integrand, you get $\int \tan^3 x \sec^2 x\, dx = \int u^3\, du = \frac{u^4}{4} + C$. Now substitute back: $\int \tan^3 x \sec^2 x\, dx = \frac{\tan^4 x}{4} + C$.

6. $\dfrac{\tan^4 x}{4} - \dfrac{\tan^2 x}{2} - \ln|\cos x| + C$

First, break up the $\tan^5 x$ in the integrand into $\int \tan^5 x\, dx = \int \tan^3 x(\tan^2 x)\, dx$. Next, use the trigonometric identity $\tan^2 x = \sec^2 x - 1$, and rewrite the integrand: $\int (\tan^3 x)(\sec^2 x - 1)\, dx = \int (\tan^3 x \sec^2 x - \tan^3 x)\, dx$. Now, break this into two integrals: $\int (\tan^3 x \sec^2 x)\, dx - \int \tan^3 x\, dx$. Next, break up the integrand of the second integral into $\int \tan^3 x\, dx = \int \tan x(\tan^2 x)\, dx$. Once again, use the trigonometric identity $\tan^2 x = \sec^2 x - 1$ to rewrite the integrand: $\int \tan x(\sec^2 x - 1)\, dx = \int \tan x \sec^2 x\, dx - \int \tan x\, dx$. Putting the integrals together, you get $\int \tan^3 x \sec^2 x\, dx - \int \tan x \sec^2 x\, dx + \int \tan x\, dx$. Use u-substitution to find the first two integrals. If you let $u = \tan x$, then $du = \sec^2 x\, dx$. Substituting into the integrand, you get $\int \tan^3 x \sec^2 x\, dx - \int \tan x \sec^2 x\, dx = \int u^3 du - \int u\, du = \frac{u^4}{4} - \frac{u^2}{2}$. Substituting back, you get $\int \tan^3 x \sec^2 x\, dx - \int \tan x \sec^2 x\, dx = \frac{\tan^4 x}{4} - \frac{\tan^2 x}{2}$. The last integral you have done before (see page 391): $\int \tan x\, dx = -\ln|\cos x| + C$. Therefore, $\int \tan^5 x\, dx = \frac{\tan^4 x}{4} - \frac{\tan^2 x}{2} - \ln|\cos x| + C$.

7. $-\csc x + C$

First, rewrite the integrand in terms of sin x and cos x:

$\int \cot^2 x \sec x\, dx = \int \frac{\cos^2 x}{\sin^2 x}\left(\frac{1}{\cos x}\right)dx = \int \frac{\cos x}{\sin^2 x}dx$. Next, use u-substitution. If you let $u = \sin x$, then $du = \cos x\, dx$. Substituting into the integrand, you get $\int \frac{\cos x}{\sin^2 x}dx = \int u^{-2}du = -\frac{1}{u} + C$.

Substituting back, you get $\int \frac{\cos x}{\sin^2 x}dx = -\frac{1}{\sin x} + C = -\csc x + C$.

Practice Problem Set 27

1. $\frac{1}{16 + x^2}$

Find the derivative of the inverse tangent using the formula $\frac{d}{dx}\left(\tan^{-1} u\right) = \frac{1}{1+u^2}\frac{du}{dx}$. Here you have $u = \frac{x}{4}$, so $\frac{du}{dx} = \frac{1}{4}$. Therefore, $\frac{d}{dx}\left(\frac{1}{4}\tan^{-1}\frac{x}{4}\right) = \frac{1}{4}\frac{1}{1+\left(\frac{x}{4}\right)^2}\left(\frac{1}{4}\right) = \frac{1}{16}\frac{1}{1+\frac{x^2}{16}} = \frac{1}{16+x^2}$.

2. $\frac{-1}{|x|\sqrt{x^2-1}}$

Find the derivative of the inverse sine using the formula $\frac{d}{dx}\left(\sin^{-1} u\right) = \frac{1}{\sqrt{1-u^2}}\frac{du}{dx}$. Here you have $u = \frac{1}{x}$, so $\frac{du}{dx} = -\frac{1}{x^2}$. Therefore, $\frac{d}{dx}\sin^{-1}\left(\frac{1}{x}\right) = \frac{1}{\sqrt{1-\left(\frac{1}{x}\right)^2}}\left(-\frac{1}{x^2}\right) = \frac{-1}{\sqrt{1-\frac{1}{x^2}}}\left(\frac{1}{x^2}\right) = \frac{-1}{|x|\sqrt{x^2-1}}$.

3. $\frac{e^x}{1+e^{2x}}$

Find the derivative of the inverse tangent using the formula $\frac{d}{dx}\left(\tan^{-1}u\right)=\frac{1}{1+u^2}\frac{du}{dx}$. Here you have $u = e^x$, so $\frac{du}{dx}=e^x$. Therefore, $\frac{d}{dx}\left(\tan^{-1}e^x\right)=\frac{1}{1+\left(e^x\right)^2}\left(e^x\right)=\frac{e^x}{1+e^{2x}}$.

4. $\frac{1}{\sqrt{\pi}}\sec^{-1}\frac{x}{\sqrt{\pi}}+C$

Recall that $\int\frac{du}{u\sqrt{u^2-1}}=\sec^{-1}u+C$. Here you have $\int\frac{dx}{x\sqrt{x^2-\pi}}$, and you just need to rearrange the integrand so that it is in the proper form to use the integral formula. If you factor π out of the radicand, you get $\int\frac{dx}{x\sqrt{\pi}\sqrt{\frac{x^2}{\pi}-1}}$. Next, use u-substitution. Let $u=\frac{x}{\sqrt{\pi}}$ and $du=\frac{1}{\sqrt{\pi}}dx$. Multiply both by $\sqrt{\pi}$ so that $u\sqrt{\pi}=x$ and $du\sqrt{\pi}=dx$. Substituting into the integrand, you get $\int\frac{du\sqrt{\pi}}{\pi u\sqrt{u^2-1}}=\frac{1}{\sqrt{\pi}}\int\frac{du}{u\sqrt{u^2-1}}$. Now you get $\frac{1}{\sqrt{\pi}}\int\frac{du}{u\sqrt{u^2-1}}=\frac{1}{\sqrt{\pi}}\sec^{-1}u+C$. Substituting back, you get $\frac{1}{\sqrt{\pi}}\sec^{-1}\frac{x}{\sqrt{\pi}}+C$.

5. $\frac{1}{\sqrt{7}}\tan^{-1}\frac{x}{\sqrt{7}}+C$

Recall that $\int\frac{du}{1+u^2}=\tan^{-1}u+C$. Here you have $\int\frac{dx}{7+x^2}$, and you just need to rearrange the integrand so that it is in the proper form to use the integral formula. If you factor 7 out of the denominator, you get $\int\frac{dx}{7\left(1+\frac{x^2}{7}\right)}=\frac{1}{7}\int\frac{dx}{\left(1+\left(\frac{x}{\sqrt{7}}\right)^2\right)}$. Next, use u-substitution. Let $u=\frac{x}{\sqrt{7}}$

and $du = \frac{1}{\sqrt{7}}dx$. Multiply du by $\sqrt{7}$ so that $du\sqrt{7} = dx$. Substituting into the integrand, you get $\frac{1}{7}\int \frac{du\sqrt{7}}{1+u^2} = \frac{1}{\sqrt{7}}\int \frac{du}{1+u^2}$. Now you get $\frac{1}{\sqrt{7}}\int \frac{du}{1+u^2} = \frac{1}{\sqrt{7}}\tan^{-1} u + C$. Substituting back, you get $\frac{1}{\sqrt{7}}\tan^{-1}\frac{x}{\sqrt{7}} + C$.

6. $\tan^{-1}(\ln x) + C$

Recall that $\int \frac{du}{1+u^2} = \tan^{-1} u + C$. Here you have $\int \frac{dx}{x\left(1+\ln^2 x\right)}$, and you will need to use *u*-substitution. Let $u = \ln x$ and $du = \frac{1}{x}dx$. Substituting into the integrand, you get $\int \frac{du}{1+u^2} = \tan^{-1} u + C$.

Substituting back, you get $\tan^{-1}(\ln x) + C$.

7. $\sin^{-1}(\tan x) + C$

Recall that $\int \frac{du}{\sqrt{1-u^2}} = \sin^{-1} u + C$. Here you have $\int \frac{\sec^2 x\, dx}{\sqrt{1-\tan^2 x}}$, and you will need to use *u*-substitution. Let $u = \tan x$ and $du = \sec^2 x\, dx$. Substituting into the integrand, you get $\int \frac{du}{\sqrt{1-u^2}} = \sin^{-1} u + C$. Substituting back, you get $\sin^{-1}(\tan x) + C$.

8. $\frac{1}{3}\tan^{-1}\left(e^{3x}\right) + C$

Recall that $\int \frac{du}{1+u^2} = \tan^{-1} u + C$. Here you have $\int \frac{e^{3x}dx}{1+e^{6x}}$, and you will need to use *u*-substitution.

Let $u = e^{3x}$ and $du = 3e^{3x}\, dx$. Divide du by 3 so that $\frac{1}{3}du = e^{3x}dx$. Substituting into the integrand, you get $\int \frac{e^{3x}dx}{1+e^{6x}} = \frac{1}{3}\int \frac{du}{1+u^2} = \frac{1}{3}\tan^{-1} u + C$. Substituting back, you get $\frac{1}{3}\tan^{-1}\left(e^{3x}\right) + C$.

ANSWERS TO CHAPTER 8 DRILL

1. **C** Using the Power Rule, you get $\int(3x^4 - 6x^3 + 2x + 7)\,dx = \frac{3x^5}{5} - \frac{6x^4}{4} + x^2 + 7x + C$.

   The answer is (C).

2. **B** $\int 3\cos x - \frac{4}{\sqrt{x}}\,dx = 3\sin x - 8\sqrt{x} + C$

   The answer is (B).

3. **A** Use u-substitution. Let $u = 1 - x^3$ and $du = -3x^2 dx$, so $-3du = 9x^2 dx$.

   Substitute: $\int \frac{9x^2}{(1-x^3)^2}\,dx = -3\int \frac{du}{u^2} = -3\int u^{-2}du = 3u^{-1} + C = \frac{3}{1-x^3} + C$.

   The answer is (A).

4. **D** Use u-substitution. Let $u = \frac{1}{x}$ and $du = -\frac{1}{x^2}dx$.

   Substitute: $\int \frac{\cos\left(\frac{1}{x}\right)}{x^2}\,dx = -\int \cos u\,du = -\sin u = -\sin\left(\frac{1}{x}\right) + C$.

   The answer is (D).

5. **B** Use u-substitution. Let $u = \cos(3x)$ and $du = -3\sin(3x)dx$, so $-\frac{1}{3}du = \sin(3x)dx$. Substitute:

   $\int \cos^3(3x)\sin(3x)\,dx = -\frac{1}{3}\int u^3 du = -\frac{u^4}{12} = -\frac{\cos^4(3x)}{12} + C$.

   The answer is (B).

6. **C** The width of each rectangle is $\frac{5-1}{4} = 1$. Using the formula, the area is approximately

$$(1)[y(1) + y(2) + y(3) + y(4)]$$
$$= [(3 + (1)^2) + (3 + (2)^2) + (3 + (3)^2) + (3 + (4)^2)]$$
$$= [4 + 7 + 12 + 19] = 42$$

   The answer is (C).

7. **B** The width of each trapezoid is $\frac{8-0}{4}=2$. Using the formula, the area is approximately

$$\frac{1}{2}(2)\left[y(0)+2y(2)+2y(4)+2y(6)+y(8)\right]$$

$$=\left[\left(0^3+4\right)+2\left(2^3+4\right)+2\left(4\left(4^3+4\right)\right)+2\left(6^3+4\right)+\left(8^3+4\right)\right]$$

$$=\left[4+2(12)+2(68)+2(220)+(516)\right]=1{,}120$$

The answer is (B).

8. **D** The width of each rectangle is $\frac{16-0}{4}=4$. Using the formula, the area is approximately

$$(4)\left[y(2)+y(6)+y(10)+y(14)\right]$$

$$=(4)\left[\frac{3}{2}+\frac{3}{6}+\frac{3}{10}+\frac{3}{14}\right]=(4)\left(\frac{88}{35}\right)=\frac{352}{35}$$

The answer is (D).

9. **D** $\frac{d}{dx}\int_2^{x^2}\sin^3 t\,dt=\left(\sin^3 x^2\right)(2x)=2x\sin^3 x^2$

The answer is (D).

10. **B** First, break up the integrand into $\int \sin^{19} x\cos^3 x\,dx=\int \sin^{19} x\cos^2 x\cos x\,dx$. Next, use the trig identity $\cos^2 x=1-\sin^2 x$.

Rewrite the integrand as $\int \sin^{19} x\left(1-\sin^2 x\right)\cos x\,dx$.

Now use *u*-substitution. Let $u=\sin x$ and $du=\cos x\,dx$.

Substitute: $\int \sin^{19} x\left(1-\sin^2 x\right)\cos x\,dx=\int u^{19}\left(1-u^2\right)du=\int u^{19}-u^{21}\,du$.

Integrate: $\int u^{19}-u^{21}\,du=\frac{u^{20}}{20}-\frac{u^{22}}{22}=\frac{\sin^{20} x}{20}-\frac{\sin^{22} x}{22}+C$.

The answer is (B).

11. **D** Use *u*-substitution. Let $u=\cos(3x)$ and $du=-3\sin(3x)dx$, so $-\frac{1}{3}du=\sin(3x)dx$. Substitute: $\int \cos^3(3x)\sin(3x)\,dx=-\frac{1}{3}\int u^3du=-\frac{u^4}{12}=-\frac{\cos^4(3x)}{12}+C$.

The answer is (D).

12. **A** Use the substitutions $\sin^2 2x = \frac{1}{2}(1-\cos 4x)$ and $\cos^2 2x = \frac{1}{2}(1+\cos 4x)$. Substitute into the integrand: $\int 16\sin^2(2x)\cos^2(2x)\,dx = 16\int \frac{1}{2}(1-\cos 4x)\frac{1}{2}(1+\cos 4x)\,dx$, which simplifies to $4\int(1-\cos 4x)(1+\cos 4x)\,dx = 4\int\left(1-\cos^2 4x\right)dx$.

Break this into two integrals: $4\int\left(1-\cos^2(4x)\right)dx = 4\int dx - 4\int \cos^2 4x\,dx$.

Now use the substitution $\cos^2(4x) = \frac{1}{2}(1+\cos 8x)$:

$4\int dx - 4\int \cos^2 4x = 4\int dx - 4\int \frac{1}{2}(1+\cos(8x))\,dx$.

This simplifies to $4\int dx - 2\int(1+\cos(8x))\,dx = 4\int dx - 2\int dx - 2\int \cos(8x)\,dx$.

And integrate: $4x - 2x - 2\frac{\sin 8x}{8} = 2x - \frac{\sin 8x}{4} + C$.

The answer is (A).

13. **C** Break up the integrand into $\int \sec^4 x\,dx = \int \sec^2 x \sec^2 x\,dx$. Use the trig identity $1+\tan^2 x = \sec^2 x$ for one of the $\sec^2 x$ terms: $\int\left(1+\tan^2 x\right)\sec^2 x\,dx$.

Now use u-substitution. Let $u = \tan x$ and $du = \sec^2 x\,dx$.

Substitute: $\int 1+u^2\,du = u + \frac{u^3}{3} = \tan x + \frac{\tan^3 x}{3} + C$.

The answer is (C).

Chapter 9
Differential Equations

MODELING SITUATIONS WITH DIFFERENTIAL EQUATIONS

Differential Equations are equations where one of the variables in the equation is a derivative. Why would we want to work with equations like these? When modeling real-world situations, variables often change with time, rather than staying constant. So instead of using, for example, y, in the model, we'll need to use something like $\frac{dy}{dt}$. Suppose we were trying to model the weather. We would need to think about variables like temperature, pressure, and wind, to name a few. These variables are constantly changing. The equations that we would use would reflect that these variables change with time. By the way, modeling the weather is incredibly complicated; no one has come close to a perfect model, and it requires far more mathematics than are covered in AP Calculus! There are, however, situations that can easily be modeled with differential equations. In this chapter, we will learn how to set up and solve a few of these types of differential equations.

VERIFYING SOLUTIONS FOR DIFFERENTIAL EQUATIONS

Derivatives can be used to verify that a function is a solution to a differential equation. Let's look at a couple of examples.

Example 1: Given the differential equation $\frac{dy}{dt} = 6t^2 - 2t + 7$, verify that $y = 2t^3 - t^2 + 7t + 5$ is a solution.

Let's take the derivative of y: $\frac{dy}{dt} = 6t^2 - 2t + 7$. That's it. We have verified the solution.

Note that there are an infinite number of possible solutions to the equation. This is because the derivative of the constant at the end, in this case 5, is zero. In other words, $y = 2t^3 - t^2 + 7t + C$ is a solution for any constant value of C. This is called a **general solution**, whereas $y = 2t^3 - t^2 + 7t + 5$ was a **particular solution**.

Example 2: Given the differential equation $\frac{d^2y}{dx^2} = 4\cos 2x + 6x$, verify that $y = x^3 - \cos 2x$ is a particular solution.

Let's take the first derivative: $\frac{dy}{dx} = 3x^2 + 2\sin 2x$. And the second derivative: $\frac{d^2y}{dx^2} = 6x + 4\cos 2x$.

So if the College Board asks you to "verify a solution," all you do is take the derivative (or derivatives) and plug into the original equation.

Example 3: Verify that $y = e^{-4x}$ is a solution to the differential equation $\frac{dy}{dx} + 4y = 0$.

Let's take the first derivative: $\frac{dy}{dx} = -4e^{-4x}$. Now plug in for the derivative and for y:

$$\frac{dy}{dx} + 4y = -4e^{-4x} + 4e^{-4x} = 0.$$

SKETCHING SLOPE FIELDS

The idea behind **slope fields**, also known as **direction fields**, is to make a graphical representation of the slope of a function at various points in the plane. We are given a differential equation, but not the equation itself. So how do we do this? Well, it's always easiest to start with an example.

Example 4: Given $\frac{dy}{dx} = x$, sketch the slope field of the function.

What does this mean? Look at the equation. It gives us the derivative of the function, which is the slope of the tangent line to the curve at any point x. In other words, the equation tells us that the slope of the curve at any point x is the x-value at that point.

For example, the slope of the curve at $x = 1$ is 1. The slope of the curve at $x = 2$ is 2. The slope of the curve at the origin is 0. The slope of the curve at $x = -1$ is -1. We will now represent these different slopes by drawing small segments of the tangent lines at those points. Let's make a sketch.

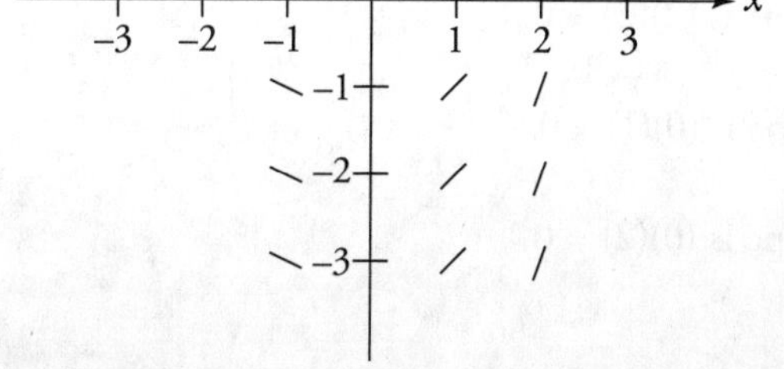

See how all of these slopes are independent of the y-values, so for each value of x, the slope is the same vertically, but it is different horizontally. Compare this slope field to the next example.

Example 5: Given $\frac{dy}{dx} = y$, sketch the slope field of the function.

Here, the slope of the curve at $y = 1$ is 1. The slope of the curve at $y = 2$ is 2. The slope of the curve at the origin is 0. The slope of the curve at $y = -1$ is -1. Let's make a sketch.

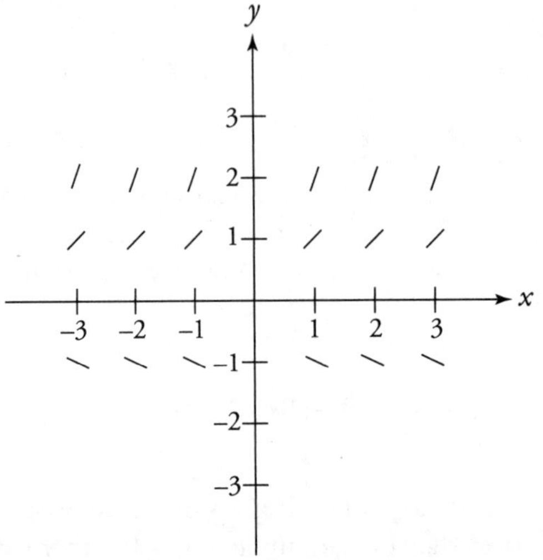

See how all of these slopes are independent of the x-values, so for each value of y, the slope is the same horizontally, but is different vertically.

Now let's do a slightly harder example.

Example 6: Given $\frac{dy}{dx} = xy$, sketch the slope field of the function.

Now, we have to think about both the x- and y-values at each point. Let's calculate a few slopes.

At (0, 0), the slope is (0)(0) = 0.

At (1, 0), the slope is (1)(0) = 0.

At (2, 0), the slope is (2)(0) = 0.

At (0, 1), the slope is (0)(1) = 0.

At (0, 2), the slope is (0)(2) = 0.

So the slope will be zero at any point on the coordinate axes.

At (1, 1), the slope is (1)(1) = 1.

At (1, 2), the slope is (1)(2) = 2.

At (1, –1), the slope is (1)(–1) = –1.

At (1, –2), the slope is (1)(–2) = –2.

So the slope at any point where $x = 1$ will be the y-value. Similarly, you should see that the slope at any point where $y = 1$ will be the x-value. As we move out the coordinate axes, slopes will get steeper—whether positive or negative.

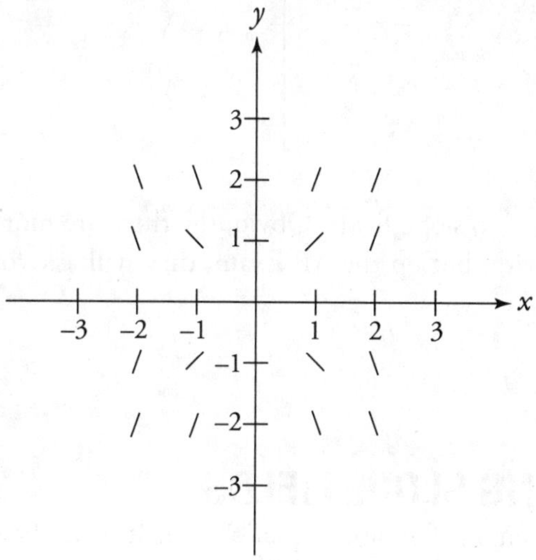

Let's do one more example.

Example 7: Given $\frac{dy}{dx} = y - x$, sketch the slope field of the function.

We have to think about both the x- and y-values at each point. This time, let's make a table of the values of the slope at different points.

	$y = -3$	$y = -2$	$y = -1$	$y = 0$	$y = 1$	$y = 2$	$y = 3$
$x = -3$	0	1	2	3	4	5	6
$x = -2$	–1	0	1	2	3	4	5
$x = -1$	–2	–1	0	1	2	3	4
$x = 0$	–3	–2	–1	0	1	2	3
$x = 1$	–4	–3	–2	–1	0	1	2
$x = 2$	–5	–4	–3	–2	–1	0	1
$x = 3$	–6	–5	–4	–3	–2	–1	0

Now let's make a sketch of the slope field. Notice that the slopes are zero along the line $y = x$ and that the slopes get steeper as we move away from the line in either direction.

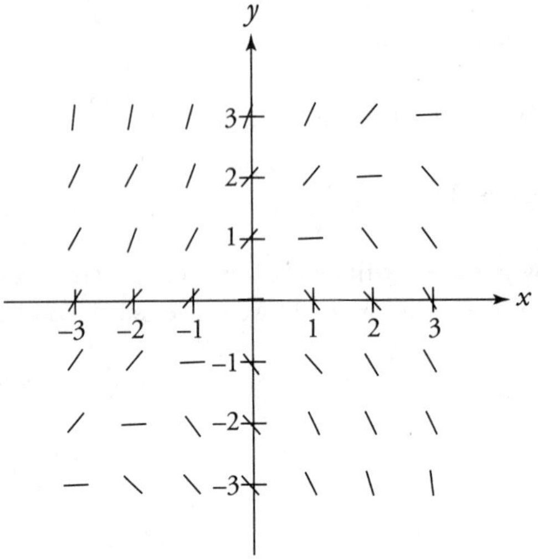

That's really all that there is to slope fields. Obviously, there are more complicated slope fields that one could come up with, but on the AP Exam, they will ask you to sketch only the simplest ones.

REASONING USING SLOPE FIELDS

Sometimes we will be given a differential equation, but it won't be easy to find the solution. Even so, if we know what the slope field looks like, we can make a guess as to the equation. For example, suppose we have the slope field below:

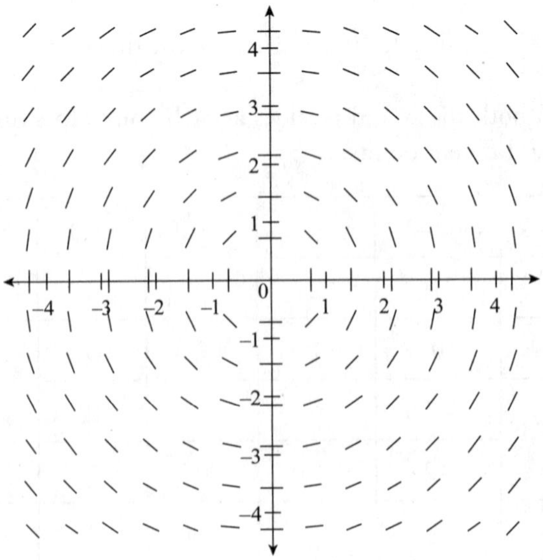

Notice how the segments form a circle. This means that the solution to the differential equation is probably $x^2 + y^2 = C$, where C is a constant.

Or suppose we have this slope field:

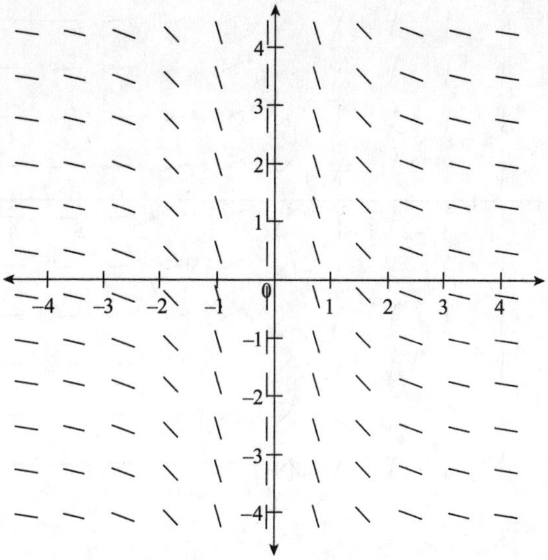

It may be a little harder to see, but this looks like the equation $y = \frac{C}{x}$. In fact, let's superimpose the graph $y = \frac{3}{x}$. The graph would go through (3, 1) and (1, 3):

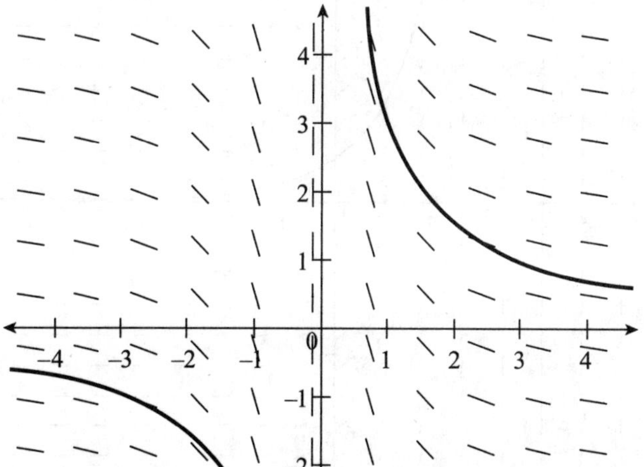

As a matter of fact, if we are given a slope field and a point that the graph of the function goes through, we can make a good guess as to what the function might be.

Let's do one more. Look at the slope field below:

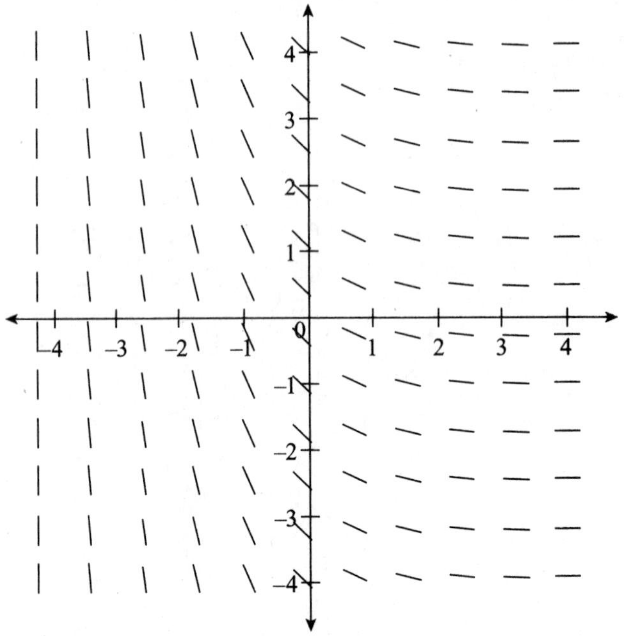

If we are told that the equation goes through the point (0, 1), we can draw a curve through that point following the slope field lines. We get

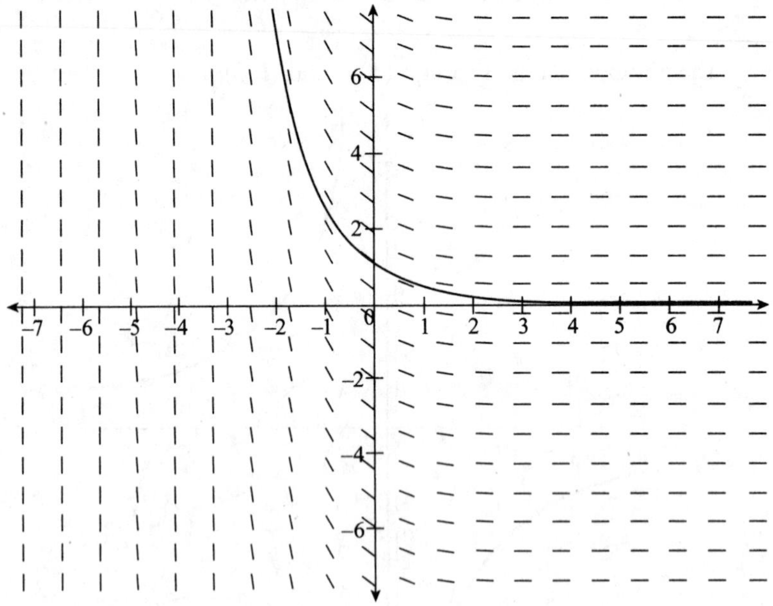

PRACTICE PROBLEM SET 28

Now try these problems. The answers are at the end of the chapter, starting on page 454.

1. Sketch the slope field for $\frac{dy}{dx} = 2x$.

2. Sketch the slope field for $\frac{dy}{dx} = -\frac{x}{y}$.

FINDING GENERAL SOLUTIONS USING SEPARATION OF VARIABLES

Some differential equations can be solved using a method called Separation of Variables. All that we do is take the terms containing one variable and put them on one side of the equals sign, and take the terms containing the other variable and put them on the other side of the equals sign. Then we integrate both sides. That's all there is to it. Let's do an example.

Example 8: Solve for y: $\frac{dy}{dx} = \frac{5x^2}{6y^2}$.

Separate the variables. Put the terms with y on the left and the terms with x on the right. $6y^2dy = 5x^2dx$. Integrate both sides: $\int 6y^2dy = \int 5x^2dx$. Evaluate the integrals: $2y^3 = \frac{5x^3}{3} + C$. Finally, isolate y: $y^3 = \frac{5x^3}{6} + \frac{C}{2}$ so $y = \left(\frac{5x^3}{6} + \frac{C}{2}\right)^{\frac{1}{3}}$. Notice that because C is a constant, $\frac{C}{2}$ is a constant, as well. Because we haven't solved for either constant, we can just write this as $y = \left(\frac{5x^3}{6} + C\right)^{\frac{1}{3}}$. Consolidating the constants to a single constant C is standard practice, and we will do it in subsequent problems.

Example 9: Solve for y: $\frac{dy}{dx} = 6\cos 2x$.

Separate the variables. Put the terms with y on the left and the terms with x on the right. $dy = 6\cos 2x\,dx$. Integrate both sides: $\int dy = \int 6\cos 2x\,dx$. Evaluate the integrals: $y = 3\sin 2x + C$.

How about a slightly harder one?

Example 10: Solve for y: $\frac{dy}{dx} = xy^2$.

Separate the variables. Put the terms with y on the left and the terms with x on the right: $y^{-2}dy = xdx$. Integrate both sides: $\int y^{-2}dy = \int xdx$. Evaluate the integrals: $\frac{-1}{y} = \frac{x^2}{2} + C$. Finally, isolate y: $y = \frac{-1}{\frac{x^2}{2} + C} = \frac{-1}{\frac{x^2 + 2C}{2}} = -\frac{2}{x^2 + 2C}$. Remember that we can just write this as $y = -\frac{2}{x^2 + C}$ because we haven't solved for C yet. We will learn how to do this in the section coming right up!

FINDING PARTICULAR SOLUTIONS USING INITIAL CONDITIONS AND SEPARATION OF VARIABLES

If we are given an equation in which the derivative of a function is equal to some other function, we can determine the original function by integrating both sides of the equation and then solving for the constant term.

Example 11: If $\frac{dy}{dx} = \frac{4x}{y}$ and $y(0) = 5$, find an equation for y in terms of x.

The first step in solving these is to put all of the terms that contain y on the left side of the equals sign and all of the terms that contain x on the right side. We then have $y\,dy = 4x\,dx$. The second step is to integrate both sides.

$$\int y\,dy = \int 4x\,dx$$

And then we integrate.

$$\frac{y^2}{2} = 2x^2 + C$$

The final step is to solve for the constant by plugging in $x = 0$ and $y = 5$.

$$\frac{5^2}{2} = 2(0^2) + C\text{, so } C = \frac{25}{2}$$

The solution is $\frac{y^2}{2} = 2x^2 + \frac{25}{2}$.

That's all there is to it. Separate the variables, integrate both sides, and solve for the constant. Often, the equation will involve a logarithm. Let's do an example.

Example 12: If $\frac{dy}{dx} = 3x^2y$ and $y(0) = 2$, find an equation for y in terms of x.

First, put the y-terms on the left and the x-terms on the right.

$$\frac{dy}{y} = 3x^2\, dx$$

Next, integrate both sides.

$$\int \frac{dy}{y} = \int 3x^2\, dx$$

The result is $\ln y = x^3 + C$. It's customary to solve this equation for y. We can do this by putting both sides into exponential form.

$$y = e^{x^3 + C}$$

This can be rewritten as $y = e^{x^3} \cdot e^c$ and, because e^c is a constant, the equation becomes

$$y = Ce^{x^3}$$

This is the preferred form of the equation. Now, solve for the constant. Plug in $x = 0$ and $y = 2$, and you get $2 = Ce^0$.

Because $e^0 = 1$, $C = 2$. The solution is $y = 2e^{x^3}$.

This is the typical differential equation that you'll see on the AP Exam. Other common problem types involve position, velocity, and acceleration or exponential growth and decay. We did several problems of this type in Chapter 6, before we knew how to use integrals. Now we are given the velocity and acceleration and told to find distance (the reverse of what we did before). We will of do some of these types of problems here and some of them in the next chapter.

Example 13: If the acceleration of a particle is given by $a(t) = -32$ feet per second squared, the velocity of the particle is 64 feet per second, and the height of the particle is 32 feet at time $t = 0$, find: (a) the equation of the particle's velocity at time t; (b) the equation for the particle's height, h, at time t; and (c) the maximum height of the particle.

Part (a): Because acceleration is the rate of change of velocity with respect to time, we can write that $\frac{dv}{dt} = -32$. Now separate the variables and integrate both sides.

$$\int dv = \int -32\, dt$$

Integrating this expression, we get $v = -32t + C$. Now we can solve for the constant by plugging in $t = 0$ and $v = 64$. We get $64 = -32(0) + C$ and $C = 64$. Thus, velocity is $v = -32t + 64$.

Part (b): Because velocity is the rate of change of displacement with respect to time, we know that

$$\frac{dh}{dt} = -32t + 64$$

Separate the variables and integrate both sides.

$$\int dh = \int (-32t + 64)\, dt$$

Integrate the expression: $h = -16t^2 + 64t + C$. Now solve for the constant by plugging in $t = 0$ and $h = 32$.

$$32 = -16(0^2) + 64(0) + C \text{ and } C = 32$$

Thus, the equation for height is $h = -16t^2 + 64t + 32$.

Part (c): In order to find the maximum height, we need to take the derivative of the height with respect to time and set it equal to zero. Notice that the derivative of height with respect to time is the velocity; just set the velocity equal to zero and solve for t.

$$-32t + 64 = 0, \text{ so } t = 2$$

Thus, at time $t = 2$, the height of the particle is a maximum. Now, plug $t = 2$ into the equation for height.

$$h = -16(2)^2 + 64(2) + 32 = 96$$

Therefore, the maximum height of the particle is 96 feet.

Example 14: If $\frac{dy}{dx} = \frac{3x}{2y}$ and $y(0) = 10$, find an equation for y in terms of x.

First, separate the variables.

$$2y\, dy = 3x\, dx$$

Then, we take the integral of both sides.

$$\int 2y\, dy = \int 3x\, dx$$

Next, integrate both sides.

$$y^2 = \frac{3x^2}{2} + C$$

Finally, solve for the constant.

$$10^2 = \frac{3(0)^2}{2} + C, \text{ so } C = 100$$

The solution is $y^2 = \frac{3x^2}{2} + 100$.

Example 15: If $\frac{dy}{dx} = 4xy^2$ and $y(0) = 1$, find an equation for y in terms of x.

First, separate the variables: $\frac{dy}{y^2} = 4x\,dx$. Then, take the integral of both sides.

$$\int \frac{dy}{y^2} = \int 4x\,dx$$

Next, integrate both sides: $-\frac{1}{y} = 2x^2 + C$. We can rewrite this as $y = -\frac{1}{2x^2 + C}$.

Finally, solve for the constant.

$$1 = -\frac{1}{2(0)^2 + C} = \frac{-1}{C}, \text{ so } C = -1$$

The solution is $y = -\frac{1}{2x^2 - 1}$.

Example 16: If $\frac{dy}{dx} = \frac{y^2}{x}$ and $y(1) = \frac{1}{3}$, find an equation for y in terms of x.

This time, separating the variables gives us $\frac{dy}{y^2} = \frac{dx}{x}$.

Then take the integral of both sides: $\int \frac{dy}{y^2} = \int \frac{dx}{x}$.

Next, integrate both sides.

$$-\frac{1}{y} = \ln x + C$$

And rearrange the equation.

$$y = \frac{-1}{\ln x + C}$$

Finally, solve for the constant. $\frac{1}{3} = \frac{-1}{C}$, so $C = -3$. The solution is $y = \frac{-1}{\ln x - 3}$.

EXPONENTIAL MODELS WITH DIFFERENTIAL EQUATIONS

Sometimes a differential equation will yield an exponential equation. This usually occurs when the rate of change of something is proportional to the quantity of that something. We see this a lot with exponential growth and decay. Let's do an example.

Example 17: The rate of growth in the number of bacteria in a petri dish is proportional to the number in the dish at any time. Initially, there are 100 bacteria. Two hours later, there are 160 bacteria. Find an equation for the number of bacteria, B, in the dish at time t, where t is in hours.

"The rate of growth is proportional to the number in the dish" means $\frac{dB}{dt} = kB$, where k is a constant. We also know that $B(0) = 100$ and $B(2) = 160$. Let's solve the differential equation. Separate the variables: $\frac{dB}{B} = kdt$. Integrate both sides: $\int \frac{dB}{B} = \int k\,dt$. We get $\ln B = kt + C$. In order to isolate B, we can exponentiate both sides: $B = e^{kt + C}$. Notice that $e^{kt + C} = e^{kt}e^{C}$, and because e^{C} is a constant, we can rewrite this as $B = Ce^{kt}$. Some people like to use a different constant so as to not confuse C with e^{C} but, as we have pointed out earlier above, until we solve for the constant, it doesn't really matter what we do for the constant.

First, use the initial condition to solve for C: $100 = Ce^{k(0)} = C$. Now the equation is $B = 100e^{kt}$.

Next, solve for k: $160 = 100e^{k(2)}$. Divide both sides by 100: $\frac{8}{5} = e^{2k}$. Take the log of both sides: $\ln\frac{8}{5} = 2k$. Therefore, $k = \frac{1}{2}\ln\frac{8}{5}$. Plugging back into the equation, we get $B = 100e^{\frac{t}{2}\ln\frac{8}{5}}$.

Note that this can be rewritten as $B = 100\left(\frac{8}{5}\right)^{\frac{t}{2}}$. Either is correct.

Example 18: A capacitor is fully charged at 1000 millifarads. After 5 milliseconds, it only has 10 millifarads left. If the amount of charge in the capacitor, C, is proportional to the amount at time t, where t is measured in milliseconds, find (a) an equation for the charge in terms of time and (b) the amount of charge at 8 milliseconds.

"The rate of growth is proportional to the amount of charge" means $\frac{dC}{dt} = kC$, where k is a constant. We also know that $C(0) = 1000$ and $C(5) = 10$. Let's solve the differential equation. Separate the variables: $\frac{dC}{C} = kdt$. Integrate both sides: $\int \frac{dC}{C} = \int k\,dt$. We get $\ln C = kt + A$ (we used a different letter for the constant to avoid confusion). Exponentiate both sides:

$C = e^{kt+A} = Ae^{kt}$. First, use the initial condition to solve for A: $1000 = Ae^{k(0)} = A$. Now the equation is $C = 1000e^{kt}$. Next, solve for k: $10 = 1000e^{k(5)}$. Divide both sides by 1000: $\frac{1}{100} = e^{5k}$. Take the log of both sides: $\ln\frac{1}{100} = 5k$. Therefore, $k = \frac{1}{5}\ln\frac{1}{100} = -\frac{1}{5}\ln 100$. (We use the negative to remind us that the charge is going down, not up.) Plugging back into the equation, we get $C = 1000e^{-\frac{t}{5}\ln 100}$. Note that this can be rewritten as $C = 1000(100)^{-\frac{t}{5}}$. Either is correct.

Now just plug in $t = 8$: $C = 1000(100)^{-\frac{8}{5}} \approx 0.631$ millifarad.

Here are a couple of tips for these types of differential equations. Note that College Board will let you go directly to these solutions on a free-response question without the intermediate steps.

> If the equation is of the form $\frac{dy}{dx} = ky$, then the solution must be $y = Ce^{kt}$.
> If such an equation has the initial condition $y(0) = A$, then the equation becomes $y = Ae^{ky}$.

Example 19: A city had a population of 10,000 in 1980 and 13,000 in 1990. Assuming an exponential growth rate, estimate the city's population in 2000.

The phrase "exponential growth rate" means that $\frac{dy}{dt} = ky$, where k is a constant. Take the integral of both sides.

$$\int \frac{dy}{y} = \int k\, dt$$

Then, integrate both sides ($\ln y = kt + C$) and put them in exponential form.

$$y = e^{kt+C} = Ce^{kt}$$

Next, use the information about the population to solve for the constants. If you treat 1980 as $t = 0$ and 1990 as $t = 10$, then

$$10{,}000 = Ce^{k(0)} \text{ and } 13{,}000 = Ce^{k(10)}$$

So $C = 10{,}000$ and $k = \frac{1}{10}\ln 1.3 \approx 0.0262$.

The equation for population growth is approximately $y = 10{,}000e^{0.0262t}$. We can estimate that the population in 2000 will be

$$y = 10{,}000e^{0.0262(20)} = 16{,}900$$

PRACTICE PROBLEM SET 29

Now try these problems. The answers are at the end of the chapter, starting on page 455.

1. If $\frac{dy}{dx} = \frac{7x^2}{y^3}$ and $y(3) = 2$, find an equation for y in terms of x.

2. If $\frac{dy}{dx} = 5x^2 y$ and $y(0) = 6$, find an equation for y in terms of x.

3. If $\frac{dy}{dx} = \frac{e^x}{y^2}$ and $y(0) = 1$, find an equation for y in terms of x.

4. If $\frac{dy}{dx} = \frac{y^2}{x^3}$ and $y(1) = 2$, find an equation for y in terms of x.

5. If $\frac{dy}{dx} = \frac{\sin x}{\cos y}$ and $y(0) = \frac{3\pi}{2}$, find an equation for y in terms of x.

6. A rock is thrown upward with an initial velocity, $v(t)$, of 18 meters per second from a height, $h(t)$, of 45 meters. If the acceleration of the rock is a constant -9 meters per second squared, find the height of the rock at time $t = 4$.

7. A colony of bacteria grows exponentially and the colony's population is 4000 at time $t = 0$ and 6500 at time $t = 3$. How big is the population at time $t = 10$?

8. A radioactive element decays exponentially in proportion to its mass. One-half of its original amount remains after 5750 years. If 10,000 grams of the element are present initially, how much will be left after 1000 years?

Chapter 9 Drill

The answers are at the end of the chapter, starting on page 461.

1 Mark for Review

If $\frac{dy}{dx} = -4xy^2$ and $y(0) = 1$, find an equation for y in terms of x.

(A) $y = \frac{-1}{2x^2 - 1}$

(B) $y = \frac{1}{2x^2 - 1}$

(C) $y = \frac{1}{2x^2 + 1}$

(D) $y = \frac{-1}{2x^2 + 1}$

2 Mark for Review

If $\sec x \frac{dy}{dx} = 4\tan y$ and $y(0) = \frac{\pi}{6}$, find an equation for y in terms of x.

(A) $\sec^2 y = \frac{\sqrt{3}}{2} e^{4\sin x}$

(B) $\sec^2 y = \frac{1}{2} e^{4\sin x}$

(C) $\sin y = \frac{\sqrt{3}}{2} e^{4\sin x}$

(D) $\sin y = \frac{1}{2} e^{4\sin x}$

3 Mark for Review

Moss is growing on a stone at an exponential rate, where the rate of growth is proportional to the amount at time t. If there are 16 square centimeters of moss on June 1, and 60 square centimeters on July 1 (30 days later), to the nearest square centimeter, how much will there be on September 1 (92 days later)?

(A) 151

(B) 921

(C) 3601

(D) 9601

4 Mark for Review

Polonium-210 has a half-life of 140 days. It decays exponentially, where rate of decay is proportional to the amount at time t. If we start with 100 grams, how much will remain after 10 weeks?

(A) 50

(B) 71

(C) 90

(D) 280

ANSWERS TO PRACTICE PROBLEM SETS

Practice Problem Set 28

1.

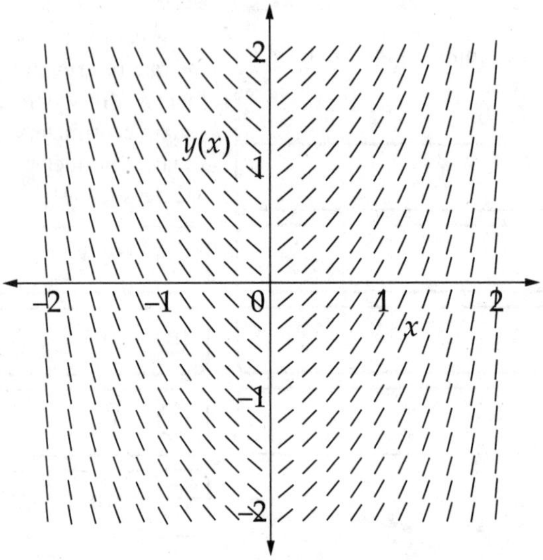

2.

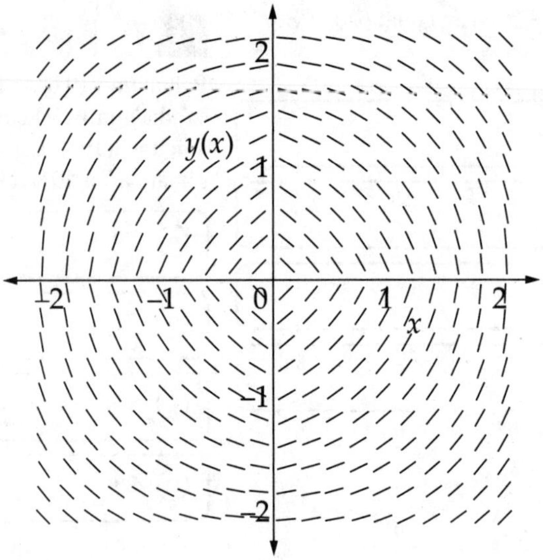

Practice Problem Set 29

1. $y = \sqrt[4]{\dfrac{28x^3}{3} - 236}$

Solve this differential equation by separation of variables. Get all of the y variables on one side of the equals sign and all of the x variables on the other side. Do this easily by cross-multiplying. You get $y^3\,dy = 7x^2\,dx$. Next, integrate both sides.

$$\int y^3\,dy = \int 7x^2\,dx$$

$$\frac{y^4}{4} = 7\frac{x^3}{3} + C$$

$$y^4 = \frac{28x^3}{3} + C$$

Now solve for C. Plug in $x = 3$ and $y = 2$.

$$16 = 252 + C$$

$$C = -236$$

Therefore, $y^4 = \dfrac{28x^3}{3} - 236$.

Now isolate y: $y = \sqrt[4]{\dfrac{28x^3}{3} - 236}$.

2. $y = 6e^{\frac{5x^3}{3}}$

Solve this differential equation by separation of variables. Get all of the y variables on one side of the equals sign and all of the x variables on the other side. Do this easily by dividing both sides by y and multiplying both sides by dx. You get $\dfrac{dy}{y} = 5x^2\,dx$. Next, integrate both sides.

$$\int \frac{dy}{y} = \int 5x^2\,dx$$

$$\ln y = \frac{5x^3}{3} + C_0$$

Now isolate y: $y = e^{\frac{5x^3}{3} + C}$. Rewrite this as $y = e^{\frac{5x^3}{3}}\left(e^{C_0}\right) = Ce^{\frac{5x^3}{3}}$.

Note that we used the letter C in the last equation to distinguish it from the C_0 in the first equation. Now solve for C. Plug in $x = 0$ and $y = 6$: $6 = Ce^0 = C$. Therefore, the equation is $y = 6e^{\frac{5x^3}{3}}$.

3. $y = \sqrt[3]{3e^x - 2}$

Solve this differential equation by separation of variables. Get all of the y variables on one side of the equals sign and all of the x variables on the other side. Do this easily by cross-multiplying. You get $y^2\,dy = e^x\,dx$. Next, integrate both sides.

$$\int y^2\,dy = \int e^x\,dx$$

$$\frac{y^3}{3} = e^x + C$$

Now isolate y.

$$y^3 = 3e^x + C$$

$$y = \sqrt[3]{3e^x + C}$$

Now solve for C. Plug in $x = 0$ and $y = 1$.

$$1 = \sqrt[3]{3e^0 + C}$$

$$C = -2$$

Therefore, the equation is $y = \sqrt[3]{3e^x - 2}$.

4. $y = 2x^2$

Solve this differential equation by separation of variables. Get all of the y variables on one side of the equals sign and all of the x variables on the other side. Do this easily by dividing both sides by y^2 and multiplying both sides by dx. You get $\frac{dy}{y^2} = \frac{dx}{x^3}$.

Next, integrate both sides.

$$\int \frac{dy}{y^2} = \int \frac{dx}{x^3}$$

$$-\frac{1}{y} = -\frac{1}{2x^2} + C$$

Now isolate y.

$$\frac{1}{y} = \frac{1}{2x^2} + C$$

Now solve for C. Plug in $x = 1$ and $y = 2$.

$$\frac{1}{2} = \frac{1}{2(1)^2} + C$$

$$C = 0$$

Therefore, the equation is $\frac{1}{y} = \frac{1}{2x^2} + C$, which can be rewritten as $y = 2x^2$.

5. $y = \sin^{-1}(-\cos x)$

Solve this differential equation by separation of variables. Get all of the y variables on one side of the equals sign and all of the x variables on the other side. Do this easily by cross-multiplying. You get $\cos y\, dy = \sin x\, dx$. Next, integrate both sides.

$$\int \cos y\, dy = \int \sin x\, dx$$

$$\sin y = -\cos x + C$$

Now solve for C. Plug in $x = 0$ and $y = \frac{3\pi}{2}$.

$$\sin \frac{3\pi}{2} = -\cos 0 + C$$

$$-1 = -1 + C$$

$$C = 0$$

Now isolate y to get the equation $y = \sin^{-1}(-\cos x)$.

6. 45 meters

Because acceleration is the derivative of velocity, write $\frac{dv}{dt} = -9$. You are also told that $v = 18$ and $h = 45$ when $t = 0$. Solve this differential equation by separation of variables. Get all of the v variables on one side of the equals sign and all of the t variables on the other side. Do this easily by multiplying both sides by dt: $dv = -9\, dt$. Next, integrate both sides:

$$\int dv = \int -9\, dt$$

$$v = -9 + C_0$$

Now solve for C_0 by plugging in $v = 18$ and $t = 0$: $18 = -9(0) + C_0$, so $C_0 = 18$. Thus, the equation for the velocity of the rock is $v = -9t + 18$. Next, because the height of the rock is the derivative of the velocity, write $\frac{dh}{dt} = -9t + 18$. Again, separate the variables by multiplying both sides by dt: $dh = (-9t + 18)\, dt$.

Integrate both sides:

$$\int dh = \int (-9t + 18)\, dt$$

$$h = -\frac{9t^2}{2} + 18t + C_1$$

Next, plug in $h = 45$ and $t = 0$ to solve for C_1.

$$45 = -\frac{9(0)^2}{2} + 18(0) + C_1 \text{, so } C_1 = 45$$

Therefore, the equation for the height of the rock, h, at time t is $h = -\frac{9t^2}{2} + 18t + 45$.

Finally, solve for the height of the rock at time $t = 4$: $h = -\frac{9t^2}{2} + 18t + 45$, so $h = -\frac{9(4)^2}{2} + 18(4) + 45 = 45$ meters.

7. 20,000 (approximately)

The phrase "exponential growth" means that you can represent the situation with the differential equation $\frac{dy}{dt} = ky$, where k is a constant and y is the population at a time t. You are also told that $y = 4{,}000$ at $t = 0$ and $y = 6{,}500$ at $t = 3$. Solve this differential equation by separation of variables. Get all of the y variables on one side of the equals sign and all of the t variables on the other side. Do this easily by dividing both sides by y and multiplying both sides by dt. You get $\frac{dy}{y} = k\, dt$.

Next, integrate both sides.

$$\int \frac{dy}{y} = \int k\, dt$$

$$\ln y = kt + C_0$$

$$y = Ce^{kt}$$

Next, plug in $y = 4{,}000$ and $t = 0$ to solve for C.

$$4{,}000 = Ce^0$$

$$C = 4{,}000$$

Now, you have $y = 4{,}000e^{kt}$. Next, plug in $y = 6{,}500$ and $t = 3$ to solve for k.

$$6{,}500 = 4{,}000e^{3k}$$

$$k = \frac{1}{3}\ln\frac{13}{8} \approx 0.162$$

Therefore, the equation for the population of bacteria, y, at time t is $y \approx 4{,}000e^{0.162t}$, or to be exact, it is $y = 4{,}000\left(\frac{13}{8}\right)^{\frac{t}{3}}$. Finally, solve for the population at time $t = 10$: $y \approx 4{,}000e^{0.162(10)} \approx 20{,}212$ or $y = 4{,}000\left(\frac{13}{8}\right)^{\frac{10}{3}} \approx 20{,}179$. (Notice how even with an "exact" solution, you still have to round the answer. And, if concerned with significant figures, the population can be written as 20,000.)

8. 8,900 grams (approximately)

Express this situation with the differential equation $\frac{dm}{dt} = -km$, where m is the mass at time t. You are also given that $m = 10{,}000$ when $t = 0$ and $m = 5{,}000$ when $t = 5{,}750$. Solve this differential equation by separation of variables. Get all of the m variables on one side of the equals sign and all of the t variables on the other side. Do this easily by dividing both sides by m and multiplying both sides by dt. You get $\frac{dm}{m} = -k\,dt$. Next, integrate both sides.

$$\int \frac{dm}{m} = \int -k\,dt$$

$$\ln m = -kt + C_0$$

$$m = Ce^{-kt}$$

Now solve for C by plugging in m = 10,000 and t = 0: 10,000 = Ce^0, so C = 10,000. This gives the equation $m = 10{,}000e^{-kt}$. Next, solve for k by plugging in m = 5,000 and t = 5,750.

$$5{,}000 = 10{,}000e^{-5{,}750k}$$

$$\frac{1}{2} = e^{-5{,}750k}$$

$$-\frac{1}{5{,}750}\ln\frac{1}{2} = k \approx 0.000121$$

Therefore, the equation for the mass of the element, m, at time t is $m \approx 10{,}000e^{-.000121t}$, or to be exact, it is $m = 10{,}000\left(\frac{1}{2}\right)^{\frac{t}{5{,}750}}$.

Finally, you can solve for the mass of the element at time t = 1,000: $m \approx 10{,}000e^{-.000121(1{,}000)} \approx 8{,}860$ g or $m = 10{,}000\left(\frac{1}{2}\right)^{\frac{1{,}000}{5{,}750}} \approx 8{,}864$ grams. (And, if concerned with significant figures, the mass can be written as 8,900.)

ANSWERS TO CHAPTER 9 DRILL

1. **C** First, separate the variables: $\frac{dy}{y^2} = -4xdx$. Integrate both sides: $\int \frac{dy}{y^2} = \int -4xdx$, which can be rewritten as $\int y^{-2}dy = -4\int x\,dx$.

$\frac{y^{-1}}{-1} = -4\frac{x^2}{2} + C$, which can be rewritten as $-\frac{1}{y} = -2x^2 + C$.

Isolate y: $\frac{1}{2x^2 + C} = y$. Substitute using the initial condition: $\frac{1}{2(0)^2 + C} = 1$, so $C = 1$. Therefore, the equation is $y = \frac{1}{2x^2 + 1}$.

The answer is (C).

2. **D** First, separate the variables: $\frac{1}{\tan y}dy = \frac{4}{\sec x}dx$. This can be rewritten as $\cot y\,dy = 4\cos x\,dx$.

Integrate both sides: $\int \cot y\,dy = 4\int \cos x\,dx$.

$\ln|\sin y| = 4\sin x + C$. Exponentiate both sides: $\sin y = e^{4\sin x + C} = Ae^{4\sin x}$.

Substitute using the initial condition: $\sin\left(\frac{\pi}{6}\right) = Ae^0 = A$, so $A = \frac{1}{2}$. Therefore, the equation is $\sin y = \frac{1}{2}e^{4\sin x}$.

The answer is (D).

3. **B** The exponential growth means that $\frac{dy}{dt} = ky$, where k is a constant. Separate the variables $\frac{dy}{y} = kdt$. Integrate both sides: $\int \frac{dy}{y} = k\int dt$.

$\ln|y| = kt + C$. Exponentiate both sides: $y = e^{kt+C} = Ae^{kt}$. Next, you are given that initially there are 16 square centimeters, so $16 = Ae^{k(0)}$ and thus $A = 16$. Now the equation is $y = 16e^{kt}$. Next, solve for k:

$$60 = 16e^{k(30)}$$

$$\frac{60}{16} = \frac{15}{4} = e^{k(30)}$$

$\ln\frac{15}{4} = 30k$, so $k = \frac{1}{30}\ln\frac{15}{4}$. If you substitute this for k in the equation, you get $y = 16e^{\left(\frac{1}{30}\ln\frac{15}{4}\right)t}$.

Use the rules of logs to rewrite the exponent: $y = 16e^{\left(\frac{1}{30}\ln\frac{15}{4}\right)t} = 16e^{\left(\ln\frac{15}{4}\right)^{\frac{t}{30}}} = 16\left(\frac{15}{4}\right)^{\frac{t}{30}}$.

Finally, plug in $t = 92$: $y = 16\left(\frac{15}{4}\right)^{\frac{92}{30}} \approx 921$ square centimeters.

The answer is (B).

4. **B** The exponential decay means that $\frac{dy}{dt} = -ky$, where k is a constant. Separate the variables: $\frac{dy}{y} = -kdt$. Integrate both sides: $\int\frac{dy}{y} = -k\int dt$.

$\ln|y| = -kt + C$. Exponentiate both sides: $y = e^{-kt+C} = Ae^{-kt}$. Next, you are given that the half-life is 140 days, so $50 = 100e^{-k(140)}$. Now solve for k: $50 = 100e^{-k(140)}$.

$\frac{1}{2} = e^{-k(140)}$, and $\ln\frac{1}{2} = -140k$, so $k = -\frac{1}{140}\ln\frac{1}{2}$. If you substitute this for k in the equation, you get $y = 100e^{\left(\frac{1}{140}\ln\frac{1}{2}\right)t}$. Use the rules of logs to rewrite the exponent: $y = 100e^{\left(\ln\frac{1}{2}^{\frac{1}{140}}\right)t}$, and $y = 100\frac{1}{2}^{\frac{t}{140}}$.

Finally, plug in $t = 70$ (10 weeks): $y = 100\frac{1}{2}^{\frac{1}{2}} \approx 71$ grams.

The answer is (B).

Chapter 10
Applications of Integration

FINDING THE AVERAGE VALUE OF A FUNCTION ON AN INTERVAL

In Chapter 7, we applied the Mean Value Theorem to derivatives, but now we'll get a chance to use it on integrals. We will refer to this as the "Mean Value Theorem for Integrals" or MVTI. The most important aspect of the MVTI is that it enables us to find the average value of a function. In fact, the AP Exam will often ask you to find the average value of a function, which is just its way of testing your knowledge of the MVTI.

Here's the theorem.

If $f(x)$ is continuous on a closed interval $[a, b]$, then at some point c in the interval $[a, b]$ the following is true:

$$\int_a^b f(x)dx = f(c)(b-a)$$

This tells us that the area under the curve of $f(x)$ on the interval $[a, b]$ is equal to the value of the function at some value c (between a and b) times the length of the interval. If we look at this graphically, we can see that we are finding the area of a rectangle whose base is the interval and whose height is some value of $f(x)$ that creates a rectangle with the same area as the area under the curve.

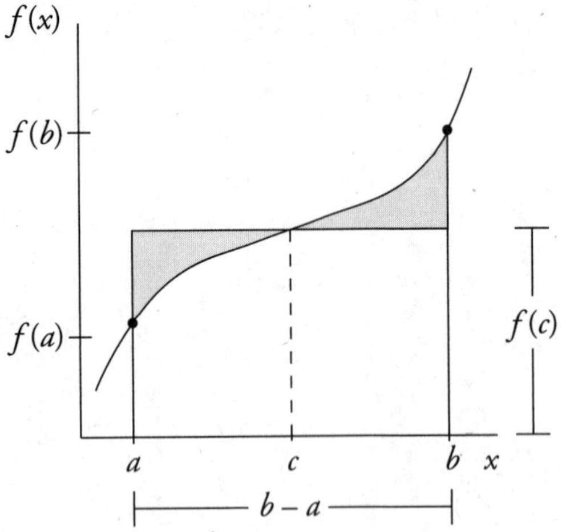

The number $f(c)$ gives us the average value of f on $[a, b]$. Thus, if we rearrange the theorem, we get the formula for finding the average value of $f(x)$ on $[a, b]$.

$$f(c) = \frac{1}{b-a}\int_a^b f(x)dx$$

There's all you need to know about finding average values. Try some examples.

Example 1: Find the average value of $f(x) = x^2$ from $x = 2$ to $x = 4$.

Evaluate the integral $\frac{1}{4-2}\int_2^4 x^2 dx$.

$$\frac{1}{4-2}\int_2^4 x^2 dx = \frac{1}{2}\left[\frac{x^3}{3}\Big|_2^4\right] = \frac{1}{2}\left(\frac{64}{3} - \frac{8}{3}\right) = \frac{28}{3}$$

Example 2: Find the average value of $f(x) = \sin x$ on $[0, \pi]$.

Evaluate $\frac{1}{\pi - 0}\int_0^\pi \sin x \, dx$.

$$\frac{1}{\pi - 0}\int_0^\pi \sin x \, dx = \frac{1}{\pi}(-\cos x)\Big|_0^\pi = \frac{1}{\pi}(-\cos \pi + \cos 0) = \frac{2}{\pi}$$

This is actually the formula for the average value.

Example 3: Find the average value of $f(x) = \frac{1}{x^2}$ on the interval [1, 3].

According to the Mean Value Theorem, the average value is found by evaluating

$$\frac{1}{3-1}\int_1^3 \frac{dx}{x^2}$$

Your result should be

$$\frac{1}{2}\left(-\frac{1}{x}\right)\Big|_1^3 = \frac{1}{2}\left(-\frac{1}{3}+1\right) = \frac{1}{3}$$

Example 4: Find the average value of $f(x) = \sin x$ on the interval $[-\pi, \pi]$.

According to the Mean Value Theorem, the average value is found by evaluating

$$\frac{1}{2\pi}\left(\int_{-\pi}^{\pi} \sin x \, dx\right)$$

Integrating, we get

$$\frac{1}{2\pi}(-\cos x)\Big|_{-\pi}^{\pi} = \frac{1}{2\pi}(-\cos \pi + \cos(-\pi)) = 0$$

PRACTICE PROBLEM SET 30

Now try these problems. The answers are at the end of this chapter, starting on page 489.

1. Find the average value of $f(x) = 4x \cos x^2$ on the interval $\left[0, \sqrt{\frac{\pi}{2}}\right]$.

2. Find the average value of $f(x) = \sqrt{x}$ on the interval [0, 16].

3. Find the average value of $f(x) = 2|x|$ on the interval [–1, 1].

4. Find the average value of $f(x) = 4 - x^2$ on the interval [–2, 2].

5. Find the average value of $f(x) = e^x$ on the interval [–1, 1].

CONNECTING POSITION, VELOCITY, AND ACCELERATION FUNCTIONS USING INTEGRALS

As we saw earlier in Chapter 7, if we are given the position function of an object, we can use the derivative to find its velocity (first derivative) and its acceleration (second derivative). Now we will use integrals to go the other way. In other words, given the acceleration function, we can find the velocity, $v(t)$, and the position, $s(t)$. The formulas are $\int a(t)\,dt + v_0 = v(t)$, where v_0 is the initial velocity, and $\int v(t)\,dt + s_0 = s(t)$, where s_0 is the initial position. Let's do an example.

Example 5: A rock is thrown down from an initial height of 120 meters, a constant acceleration of –10 meters per second squared, and an initial velocity of –8 meters per second. Find the equations of its velocity and position at time t, where t is in seconds.

First, integrate the acceleration to get the velocity: $\int -10\,dt = -10t + C$. The constant is the initial velocity of –8 meters per second, so the velocity equation is $v(t) = -10t - 8$. Now, integrate the velocity to get the position: $\int (-10t - 8)\,dt = -5t^2 - 8t + C$. Here, the constant is the initial position of –120 meters, so the position equation is $s(t) = -5t^2 - 8t + 120$.

Example 6: An object is moving along the x-axis with its acceleration given by $a = 2t - 8$. Initially, it is at the origin, moving at 4 units per second to the right. Find the equation of its position at time t, where t is in seconds.

First, integrate the acceleration to get the velocity: $v(t) = \int (2t - 8)\, dt = t^2 - 8t + C$. Plugging in the initial velocity, we get $v(t) = t^2 - 8t + 4$. Now integrate again to get the position: $x(t) = \int t^2 - 8t + 4\, dt = \frac{t^3}{3} - 4t^2 + 4t + C$. Plugging in the initial position, that is $x = 0$, we get $x(t) = \frac{t^3}{3} - 4t^2 + 4t$.

In the next section, we will explore these a bit more.

USING ACCUMULATION FUNCTIONS AND DEFINITE INTEGRALS IN APPLIED CONTEXTS

Sometimes we will be given a problem where we start with an initial quantity and a function that tells us how much that quantity is increasing or decreasing (an accumulation function). Then we are asked to find the quantity at a later time, using that function. It's very straightforward. Let's do an example.

Example 7: Grain is being stored in a silo. Initially, there are 50 tons of grain in the silo. Grain is being added at a rate given by the function $\frac{dV}{dt} = -t^2 + 14t$ tons per hours. How much grain will there be after 12 hours?

We start with 50 tons and we have the rate at which the grain is accumulating, so we just have to solve $V(t) = 50 + \int_0^{12} (-t^2 + 14t)\, dt$. Evaluate the integral: $\int_0^{12} -t^2 + 14t\, dt = \left(\frac{-t^3}{3} + 7t^2\right)\Bigg|_0^{12} = 432$. Therefore, after 12 hours, the silo contains $50 + 432 = 482$ tons of grain.

How about a slightly harder example?

Example 8: Water is being added to a leaky tank. The rate at which the water is being added is $48t$ gallons per minute. The rate at which the water is leaking is given by $\frac{dV}{dt} = e^{0.03t}$. Initially, there are 120 gallons in the tank. How much water is in the tank at time $t = 30$ minutes?

We start with 120 gallons and we both add water and subtract water to get $V(t) = 120 + 48t - \int_0^{30} e^{0.03t}\,dt$. Evaluate the integral: $\int_0^{30} e^{0.03t}\,dt = \frac{e^{0.03t}}{0.03} \approx 48.653$. Therefore, the amount of water in the tank after 30 minutes is 120 + 48(30) – 48.653 = 1,511.347 gallons.

FINDING THE AREA BETWEEN CURVES EXPRESSED AS FUNCTIONS OF *x*

This topic is almost always tested on the AP Exam. You've already learned that if you want to find the area under a curve, you can integrate the function of the curve by using the endpoints as limits. So far, though, we've talked only about the area between a curve and the x-axis. What if you have to find the area between two curves?

Vertical Slices

Suppose we wanted to find the area between the curve $y = x$ and the curve $y = x^2$ from $x = 2$ to $x = 4$. First, sketch the curves.

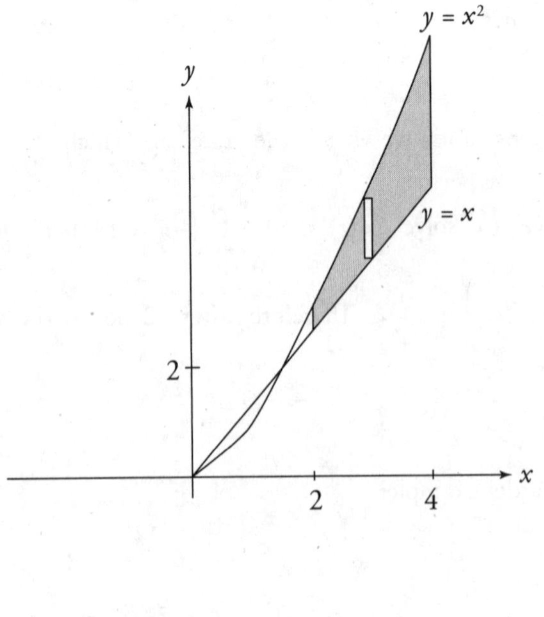

We can find the area by slicing up the region vertically, into a bunch of infinitesimally thin strips, and adding up the areas of all the strips. The height of each strip is $x^2 - x$, and the width of each strip is dx. Add up all the strips by using the integral.

$$\int_2^4 \left(x^2 - x\right)dx$$

Then, evaluate it.

$$\left(\frac{x^3}{3} - \frac{x^2}{2}\right)\Bigg|_2^4 = \left(\frac{64}{3} - \frac{16}{2}\right) - \left(\frac{8}{3} - \frac{4}{2}\right) = \frac{38}{3}$$

That wasn't so hard, was it? Don't worry. The process gets more complicated, but the idea remains the same. Now let's generalize this and come up with a rule.

> If a region is bounded by $f(x)$ above and $g(x)$ below at all points of the interval $[a, b]$, then the area of the region is given by
>
> $$\int_a^b [f(x) - g(x)]\,dx$$

Example 9: Find the area of the region between the parabola $y = 1 - x^2$ and the line $y = 1 - x$.

First, make a sketch of the region.

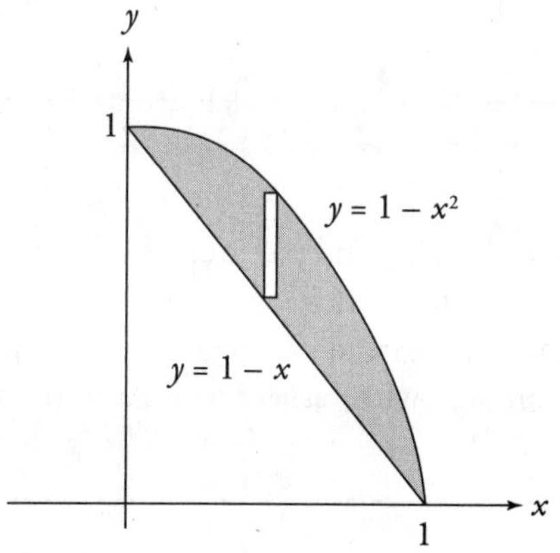

To find the points of intersection of the graphs, set the two equations equal to each other and solve for x.

$$\begin{aligned} 1 - x^2 &= 1 - x \\ x^2 - x &= 0 \\ x(x - 1) &= 0 \\ x &= 0, 1 \end{aligned}$$

The left-hand edge of the region is $x = 0$ and the right-hand edge is $x = 1$, so the limits of integration are from 0 to 1.

Next, note that the top curve is always $y = 1 - x^2$, and the bottom curve is always $y = 1 - x$. (If the region has a place where the top and bottom curves switch, we need to evaluate two integrals, one for each region. Fortunately, that's not the case here.) Thus, we need to evaluate the following:

$$\int_0^1 \left[\left(1 - x^2\right) - \left(1 - x\right)\right] dx$$

$$\int_0^1 \left[\left(1 - x^2\right) - \left(1 - x\right)\right] dx = \int_0^1 \left(-x^2 + x\right) dx = \left(-\frac{x^3}{3} + \frac{x^2}{2}\right)\Bigg|_0^1 = \frac{1}{6}$$

FINDING THE AREA BETWEEN CURVES EXPRESSED AS FUNCTIONS OF *y*

We can slice a region vertically when one function is at the top of our section and a different function is at the bottom. But what if the same function is both the top and the bottom of the slice (what we call a double-valued function)? We have to slice the region horizontally.

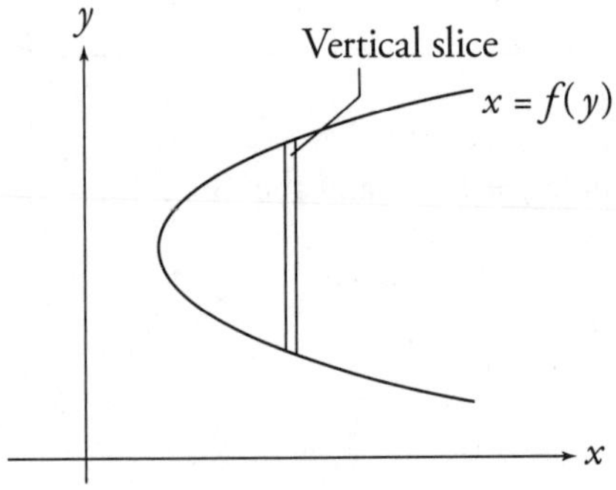

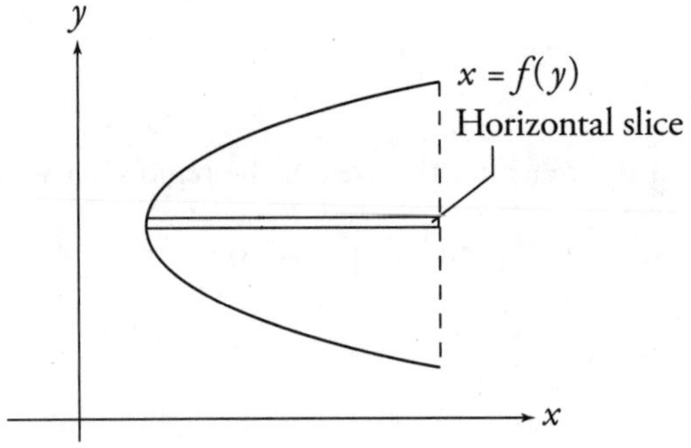

If we were to slice vertically, as in the left-hand picture, we'd have a problem, because we would be subtracting the function from itself. But if we were to slice horizontally, as in the right-hand picture, we don't have a problem. But, instead of integrating the equation $f(x)$ with respect to x, we need to integrate the equation $f(y)$ with respect to y. As a result, our area formula changes a little.

If a region is bounded by $f(y)$ on the right and $g(y)$ on the left at all points of the interval $[c, d]$, then the area of the region is given by

$$\int_c^d \left[f(y) - g(y)\right] dy$$

Example 10: Find the area of the region between the curve $x = y^2$ and the curve $x = y + 6$ from $y = 0$ to $y = 3$.

First, sketch the region.

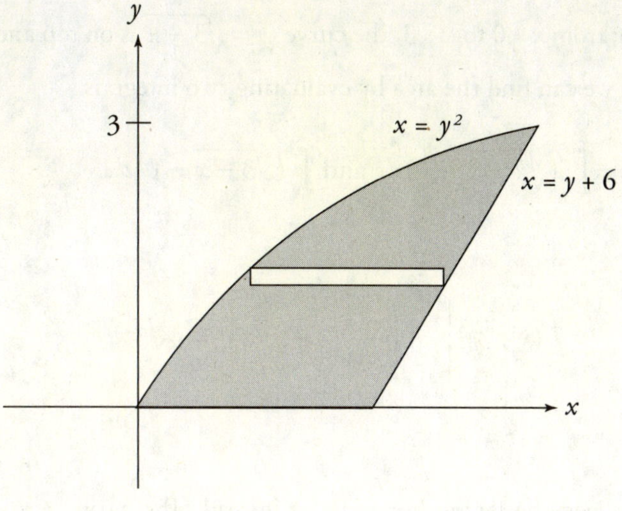

When we slice up the area horizontally, the right end of each section is the curve $x = y + 6$, and the left end of each section is always the curve $x = y^2$. Now set up the integral.

$$\int_0^3 (y + 6 - y^2)\, dy$$

Evaluating this gives us the area.

$$\int_0^3 (y + 6 - y^2)\, dy = \left(\frac{y^2}{2} + 6y - \frac{y^3}{3}\right)\Bigg|_0^3 = \frac{27}{2}$$

Example 11: Find the area between the curve $y = \sqrt{x + 3}$, the curve $y = \sqrt{3 - x}$, and the x-axis from $x = -3$ to $x = 3$.

First, sketch the curves.

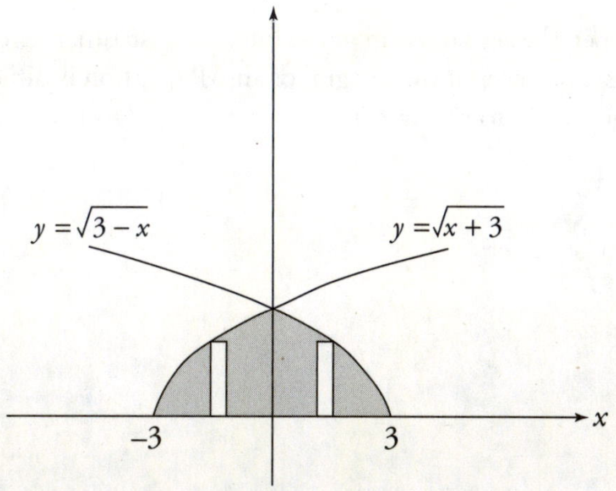

From $x = -3$ to $x = 0$, if we slice the region vertically, the curve $y = \sqrt{x+3}$ is on top, and the x-axis is on the bottom; from $x = 0$ to $x = 3$, the curve $y = \sqrt{3-x}$ is on top and the x-axis is on the bottom. Therefore, we can find the area by evaluating two integrals.

$$\int_{-3}^{0}(\sqrt{x+3} - 0)\,dx \text{ and } \int_{0}^{3}(\sqrt{3-x} - 0)\,dx$$

We get

$$\frac{2}{3}(x+3)^{\frac{3}{2}}\Big|_{-3}^{0} + \left(-\frac{2}{3}(3-x)^{\frac{3}{2}}\right)\Big|_{0}^{3} = 4\sqrt{3}$$

Let's suppose we had sliced the region horizontally instead. The curve $y = \sqrt{x+3}$ is always on the left, and the curve $y = \sqrt{3-x}$ is always on the right. If we solve each equation for x in terms of y, we save some time by using only one integral instead of two.

The two equations are $x = y^2 - 3$ and $x = 3 - y^2$. We also have to change the limits of integration from x-limits to y-limits. The two curves intersect at $y = \sqrt{3}$, so our limits of integration are from $y = 0$ to $y = \sqrt{3}$. The new integral is

$$\int_{0}^{\sqrt{3}}\left[(3-y^2)-(y^2-3)\right]dy = \int_{0}^{\sqrt{3}}(6-2y^2)\,dy = 6y - \frac{2y^3}{3}\Big|_{0}^{\sqrt{3}} = 4\sqrt{3}$$

We get the same answer no matter which way we integrate. The challenge of area problems is determining which way to integrate and then converting the equation to different terms. Unfortunately, there's no simple rule for how to do this. You have to look at the region and figure out its endpoints, as well as where the curves are with respect to each other.

Once you can do that, then the actual set-up of the integral(s) isn't that hard. Sometimes, evaluating the integrals isn't easy; however, if the integral of an AP question is difficult to evaluate, you'll only be required to set it up, not to evaluate it.

PRACTICE PROBLEM SET 31

Find the area of the region between the two curves in each problem, and be sure to sketch each one. (We gave you only endpoints in one of them.) The answers are at the end of the chapter, starting on page 490.

1. The curve $y = x^2 - 2$ and the line $y = 2$

2. The curve $y = x^2$ and the curve $y = 4x - x^2$

3. The curve $y = x^2 - 4x - 5$ and the curve $y = 2x - 5$

4. The curve $y = x^3$ and the x-axis, from $x = -1$ to $x = 2$

5. The curve $x = y^2$ and the line $x = y + 2$

6. The curve $x = y^2$ and the curve $x = 3 - 2y^2$

7. The curve $x = y^2 - 4y + 2$ and the line $x = y - 2$

8. The curve $x = y^{\frac{2}{3}}$ and the curve $x = 2 - y^4$

FINDING THE AREA BETWEEN CURVES THAT INTERSECT AT MORE THAN TWO POINTS

Sometimes we need to find the area of a region that is divided into different smaller regions. We simply integrate each region separately and add the answers.

Example 12: Find the area of the region between the curve $y = \sin x$ and the curve $y = \cos x$ from 0 to $\frac{\pi}{2}$.

First, sketch the region.

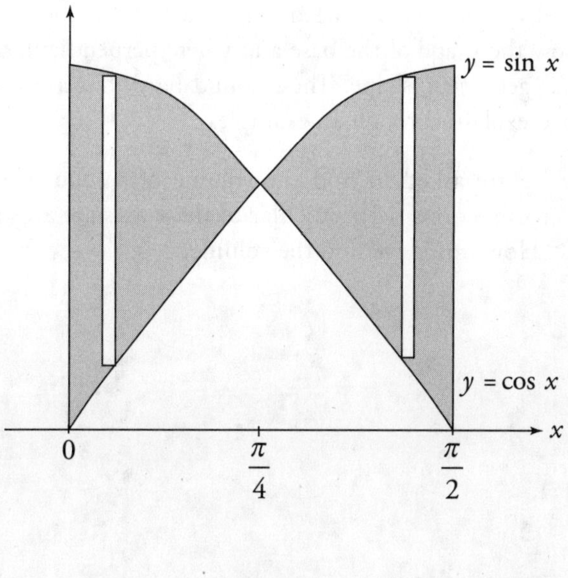

Notice that cos x is on top between 0 and $\frac{\pi}{4}$ and sin x is on top between $\frac{\pi}{4}$ and $\frac{\pi}{2}$. The point where they cross is $\frac{\pi}{4}$, so you have to divide the area into two integrals: one from 0 to $\frac{\pi}{4}$, and the other from $\frac{\pi}{4}$ to $\frac{\pi}{2}$. In the first region, cos x is above sin x, so the integral to evaluate is

$$\int_0^{\frac{\pi}{4}} (\cos x - \sin x)\, dx$$

The integral of the second region is a little different, because sin x is above cos x.

$$\int_{\frac{\pi}{4}}^{\frac{\pi}{2}} (\sin x - \cos x)\, dx$$

If you add the two integrals, you'll get the area of the whole region.

$$\int_0^{\frac{\pi}{4}} (\cos x - \sin x)\, dx = (\sin x + \cos x)\Big|_0^{\frac{\pi}{4}} = \sqrt{2} - 1$$

$$\int_{\frac{\pi}{4}}^{\frac{\pi}{2}} (\sin x - \cos x)\, dx = (-\cos x - \sin x)\Big|_{\frac{\pi}{4}}^{\frac{\pi}{2}} = \sqrt{2} - 1$$

Adding these, we get that the area is $2\sqrt{2} - 2$.

VOLUMES WITH CROSS-SECTIONS: SQUARES AND RECTANGLES

There is one other type of volume that you need to be able to find. Sometimes, we will be given an object where we know the shape of the base and where perpendicular cross-sections are all the same regular, planar geometric shape. These sound hard, but are actually quite straightforward. This is easiest to explain through an example.

Example 13: Suppose we are asked to find the volume of a solid whose base is the circle $x^2 + y^2 = 4$, and where cross-sections perpendicular to the x-axis are all squares whose sides lie on the base of the circle. How would we find the volume?

First, make a drawing of the circle.

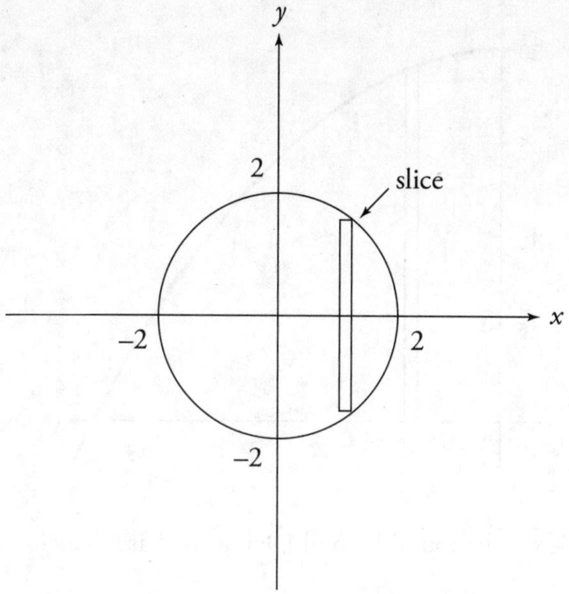

What this problem is telling us is that every time we make a vertical slice, the slice is the length of the base of a square. If we want to find the volume of the solid, all we have to do is integrate the area of the square from one endpoint of the circle to the other.

The side of the square is the vertical slice whose length is $2y$, which we can find by solving the equation of the circle for y and multiplying by 2. We get $y = \sqrt{4 - x^2}$. Then the length of a side of the square is $2\sqrt{4 - x^2}$. Because the area of a square is *side*^{2}, we can find the volume with $\int_{-2}^{2} (16 - 4x^2)\,dx$.

Let's perform the integration, although on some problems, you will be permitted to find the answer with a calculator.

$$\int_{-2}^{2} (16 - 4x^2)\,dx = \left(16x - \frac{4x^3}{3}\right)\Bigg|_{-2}^{2} = \left(32 - \frac{32}{3}\right) - \left(-32 + \frac{32}{3}\right)$$

$$= 64 - \frac{64}{3} = \frac{128}{3}$$

As you can see, the technique is very simple. First, we find the side of the cross-section in terms of y. This will involve a vertical slice. Then, we plug the side into the equation for the area of the cross-section. Then, integrate the area from one endpoint of the base to the other.

Example 14: Find the volume of a solid whose base is the area between the curve $y = 20 - x^2$ in the first quadrant, whose cross-sections perpendicular to the x-axis are rectangles whose bases lie on the region, and whose heights are twice their bases.

First, make a drawing of the region:

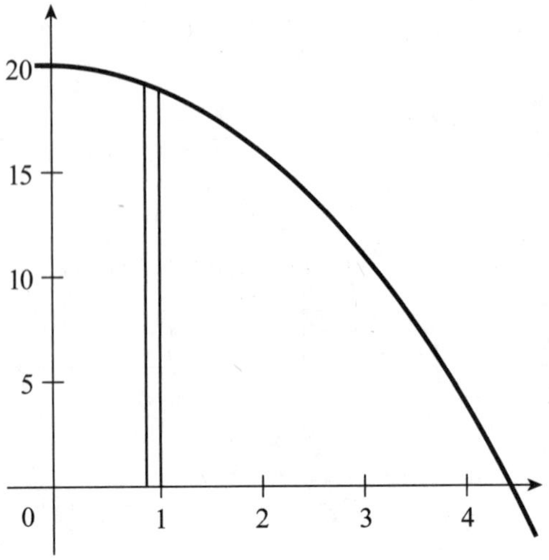

The area of the rectangle can be found by multiplying the base, which is $B = 20 - x^2$, by the height, which is $H = 2(20 - x^2) = 40 - 2x^2$. We get $A = (40 - 2x^2)(20 - x^2) = 800 - 80x^2 + 2x^4$. The curve intersects the x-axis at $x = \sqrt{20}$. Now we simply integrate this area from $x = 0$ to $x = \sqrt{20}$. We get $\int_0^{\sqrt{20}} \left(800 - 80x^2 + 2x^4\right) dx = \left(800x - \frac{80x^3}{3} + \frac{2x^5}{5}\right)\Big|_0^{\sqrt{20}} \approx 1908.111$.

VOLUMES WITH CROSS-SECTIONS: TRIANGLES AND SEMICIRCLES

On the AP Exam, cross-sections will be squares, equilateral triangles, circles, semicircles, or maybe isosceles right triangles. So here are some handy formulas to know.

Given the side of an equilateral triangle, the area is $A = (side)^2 \frac{\sqrt{3}}{4}$.

Given the diameter of a semicircle, the area is $A = (diameter)^2 \frac{\pi}{8}$.

Given the hypotenuse of an isosceles right triangle, the area is $A = \frac{(hypotenuse)^2}{4}$.

Example 15: Use the same base as Example 13, except this time the cross-sections are equilateral triangles. We find the side of the triangle just as we did above. It is $2y$, which is $2\sqrt{4-x^2}$. Now because the area of an equilateral triangle is $(side)^2\,\frac{\sqrt{3}}{4}$, we can find the volume by evaluating the integral: $\frac{\sqrt{3}}{4}\int_{-2}^{2}(4)(4-x^2)\,dx = \sqrt{3}\int_{-2}^{2}(4-x^2)\,dx$.

We get $\sqrt{3}\int_{-2}^{2}(4-x^2)\,dx = \sqrt{3}\left(4x-\frac{x^3}{3}\right)\Bigg|_{-2}^{2} = \frac{32\sqrt{3}}{3}$.

Example 16: Use the same base as Example 13, except this time the cross-sections are semicircles whose diameters lie on the base. We find the side of the semicircle just as we did above. It is $2y$, which is $2\sqrt{4-x^2}$. Now because the area of a semicircle is $(diameter)^2\frac{\pi}{8}$, we can find the volume by evaluating the integral: $\frac{\pi}{8}\int_{-2}^{2}(4)(4-x^2)\,dx = \frac{\pi}{2}\int_{-2}^{2}(4-x^2)\,dx$.

We get $\frac{\pi}{2}\int_{-2}^{2}(4-x^2)\,dx = \frac{\pi}{2}\left(4x-\frac{x^3}{3}\right)\Bigg|_{-2}^{2} = \frac{16\pi}{3}$.

VOLUME WITH DISC METHOD: REVOLVING AROUND THE *x*- OR *y*-AXIS

Let's look at the region between the curve $y=\sqrt{x}$ and the x-axis (the curve $y = 0$), from $x = 0$ to $x = 1$, and revolve it about the x-axis. The picture looks like the following:

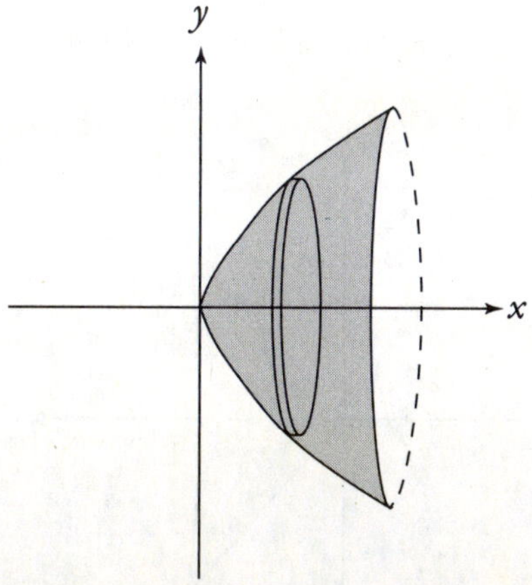

If we slice the resulting solid perpendicular to the x-axis, each cross-section of the solid is a circle, or disc (hence the phrase "disc method"). The radii of the discs vary from one value of

x to the next, but we can find each radius by plugging it into the equation for y: each radius is $\sqrt{x}$. Therefore, the area of each disc is

$$\pi\left(\sqrt{x}\right)^2 = \pi x$$

Each disc is infinitesimally thin, so its thickness is dx; if we add up the volumes of all the discs, we will get the volume. The way to add these up is by using the integral, with the endpoints of the interval as the limits of integration. Therefore, to find the volume, evaluate the integral.

$$\int_0^1 \pi x\, dx = \left.\frac{\pi x^2}{2}\right|_0^1 = \frac{\pi}{2}$$

Now let's generalize this. If we have a region whose area is bounded by the curve $y = f(x)$ and the x-axis on the interval $[a, b]$, each disc has a radius of $f(x)$, and the area of each disc will be

$$\pi\left[f(x)\right]^2$$

To find the volume, evaluate the integral.

$$\pi\int_a^b \left[f(x)\right]^2 dx$$

This is the formula for finding the volume using discs.

Example 17: Find the volume of the solid that results when the region between the curve $y = x$ and the x-axis, from $x = 0$ to $x = 1$, is revolved about the x-axis.

As always, sketch the region to get a better look at the problem.

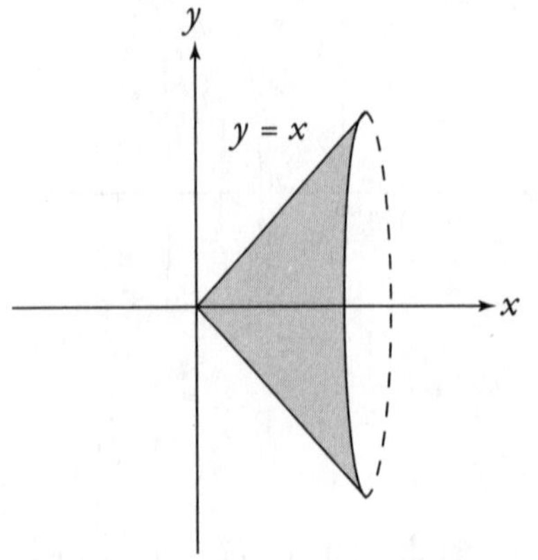

When we slice vertically, the top curve is $y = x$ and the limits of integration are from $x = 0$ to $x = 1$. Using our formula, we evaluate the integral.

$$\pi \int_0^1 x^2 \, dx$$

The result is

$$\pi \int_0^1 x^2 \, dx = \pi \frac{x^3}{3}\Big|_0^1 = \frac{\pi}{3}$$

By the way, did you notice that the solid in the problem is a cone with a height and radius of 1?

The formula for the volume of a cone is $\frac{1}{3}\pi r^2 h$, so you should expect to get $\frac{\pi}{3}$.

Of course, we could revolve around the *y*-axis instead of the *x*-axis. The process is the same, except we are going horizontally instead of vertically. We would have an equation in terms of *y* instead of *x*, from *a* to *b*.

The formula for the volume is

$$V = \pi \int_a^b f(y)^2 \, dy$$

Example 18: The region between the curve $x = 4 - y^2$ in the first quadrant is revolved about the *y*-axis. Find the volume of the solid that results.

First, let's draw a picture.

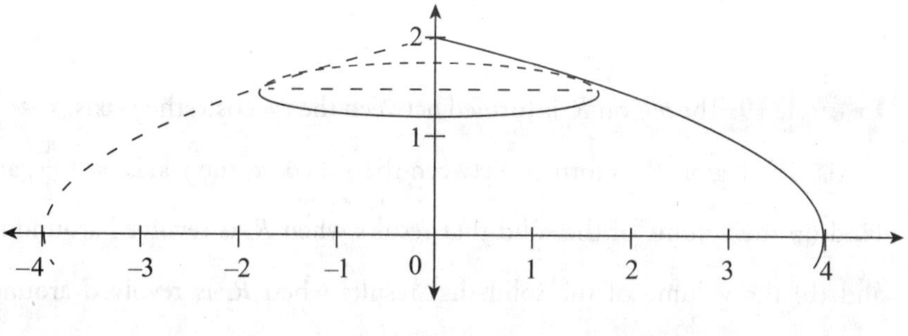

Note that the region goes from $y = 0$ to $y = 2$. Thus, we need to integrate $\pi \int_0^2 (4 - y^2)^2 \, dy$.

We get $\pi \int_0^2 (4 - y^2)^2 \, dy = \pi \int_0^2 (16 - 8y^2 + y^4) dy = \pi \left(16y - \frac{8y^3}{3} + \frac{y^5}{5} \right)\Big|_0^2 = \frac{256\pi}{15}$.

VOLUME WITH DISC METHOD: REVOLVING AROUND OTHER AXES

Sometimes we will revolve the region around an axis other than the x-axis. Suppose we are revolving the region under the graph $y = 2 - x^2$ from $x = 0$ to $x = 2$ around the line $y = -2$. The graphs of $y = 2 - x^2$ and $x = 2$ intersect at $y = -2$.

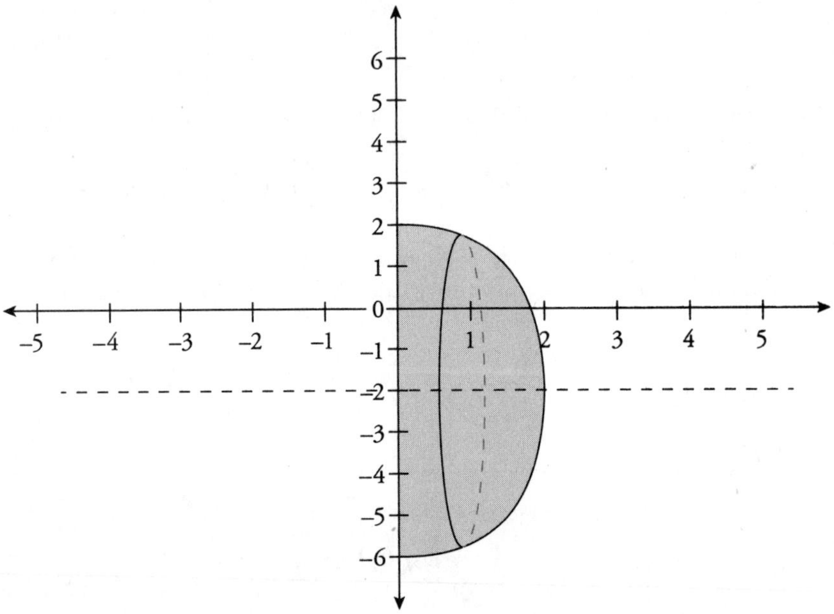

How does this change our formula? Simple. Notice that the radius of the disc is now the distance from $y = 2 - x^2$ to $y = -2$, which is $(2 - x^2) - (-2) = 4 - x^2$. The volume would then be $V = \pi\int_0^2 (4 - x^2)^2\,dx$. Note that this is the same as the volume under $y = 4 - x^2$ about the x-axis from $x = 0$ to $x = 4$. All we have done is move the region down 2 units. We haven't changed its volume.

Usually College Board will only ask you to set up the integrals.

Example 19: The region R_1 is formed between the $y = \cos x$, the y-axis, $x = \frac{\pi}{2}$, and the x-axis. The region R_2 is formed between the $y = \cos x$, the y-axis, $x = \frac{\pi}{2}$, and $y = -1$. Find (a) the volume of the solid that results when R_1 is revolved around the x-axis, and (b) the volume of the solid that results when R_2 is revolved around $y = -1$. Set up but do not evaluate the integrals.

(a) The volume is found by the integral $\pi\int_0^{\frac{\pi}{2}} \cos^2 x\,dx$.

(b) The volume is found by the integral $\pi\int_0^{\frac{\pi}{2}} (\cos x + 1)^2\,dx$.

Of course, we could go around a vertical line as well.

Example 20: The region R_1 is formed between the $y = e^x$, $y = \ln 3$, and the x-axis. The region R_2 is formed between the $y = e^x$, $y = \ln 3$, and $x = -2$. Find (a) the volume of the solid that results when R_1 is revolved around the y-axis, and (b) the volume of the solid that results when R_2 is revolved around $x = -2$. Set up but do not evaluate the integrals.

(a) First, we need to put the equation $y = e^x$ in terms of y: $x = \ln y$. The volume is found by the integral $\pi \int_1^{\ln 3} (\ln y)^2 \, dy$.

(b) The volume is found by the integral $\pi \int_1^{\ln 3} (\ln y + 2)^2 \, dy$.

VOLUME WITH WASHER METHOD: REVOLVING AROUND THE *x*- OR *y*-AXIS

Now let's figure out how to find the volume of the solid that results when we revolve a region that does not touch the x-axis. Consider the region bounded above by the curve $y = x^3$ and below by the curve $y = x^2$, from $x = 2$ to $x = 4$, which is revolved about the x-axis. Sketch the region first.

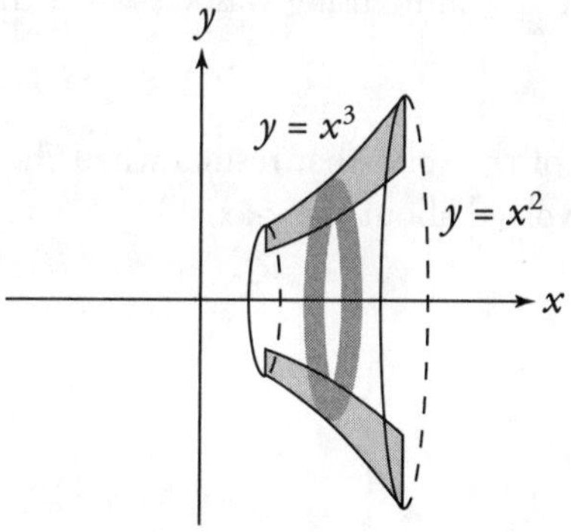

If we slice this region vertically, each cross-section looks like a washer (hence the phrase "washer method").

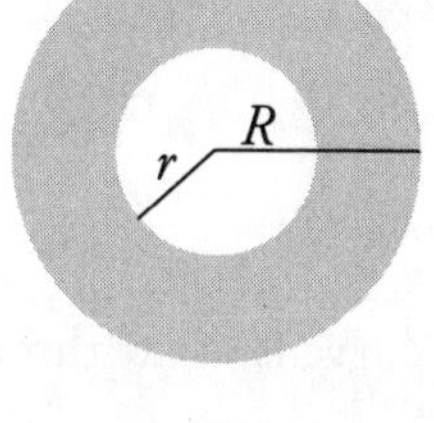

The outer radius is $R = x^3$ and the inner radius is $r = x^2$. To find the area of the region between the two circles, take the area of the outer circle, πR^2, and subtract the area of the inner circle, πr^2.

We can simplify this to

$$\pi R^2 - \pi r^2 = \pi\left(R^2 - r^2\right)$$

Because the outer radius is $R = x^3$ and the inner radius is $r = x^2$, the area of each region is $\pi\left(x^6 - x^4\right)$. You can sum up these regions using the integral.

$$\pi\int_2^4 \left(x^6 - x^4\right) dx = \frac{74{,}336\pi}{35}$$

Here's the general idea: In a region whose area is bounded above by the curve $y = f(x)$ and below by the curve $y = g(x)$, on the interval $[a, b]$, then each washer will have an area of

$$\pi\left[f(x)^2 - g(x)^2\right]$$

To find the volume, evaluate the integral.

$$\pi\int_a^b \left[f(x)^2 - g(x)^2\right]dx$$

This is the formula for finding the volume using washers when the region is rotated around the x-axis.

Example 21: Find the volume of the solid that results when the region bounded by $y = x$ and $y = x^2$, from $x = 0$ to $x = 1$, is revolved about the x-axis.

Sketch it first.

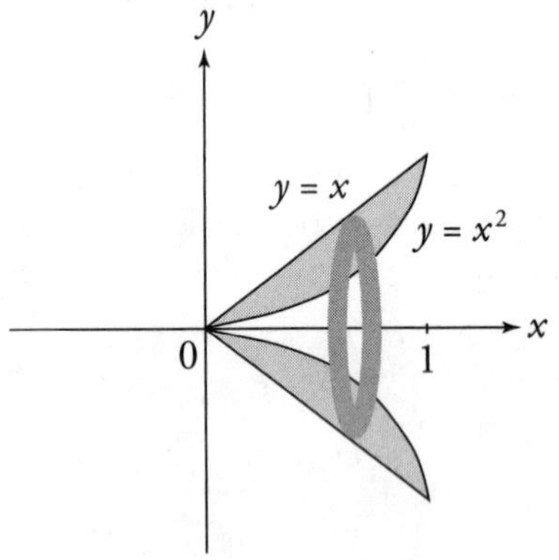

The top curve is $y = x$ and the bottom curve is $y = x^2$ throughout the region. Then our formula tells us that we evaluate the integral.

$$\pi\int_0^1 \left(x^2 - x^4\right)dx$$

The result is

$$\pi\int_0^1 \left(x^2 - x^4\right)dx = \pi\left(\frac{x^3}{3} - \frac{x^5}{5}\right)\Bigg|_0^1 = \frac{2\pi}{15}$$

Suppose the region we're interested in is revolved around the *y*-axis instead of the *x*-axis. Now, to find the volume, you have to slice the region horizontally instead of vertically. We discussed how to do this in the previous chapter on area.

Now, if we have a region whose area is bounded on the right by the curve $x = f(y)$ and on the left by the curve $x = g(y)$, on the interval $[c, d]$, then each washer has an area of

$$\pi\left[f(y)^2 - g(y)^2\right]$$

To find the volume, evaluate the integral.

$$\pi\int_c^d \left[f(y)^2 - g(y)^2\right]dy$$

This is the formula for finding the volume using washers when the region is rotated about the *y*-axis.

Example 22: Find the volume of the solid that results when the region bounded by the curve $x = y^2$ and the curve $x = y^3$, from $y = 0$ to $y = 1$, is revolved about the *y*-axis.

Sketch away.

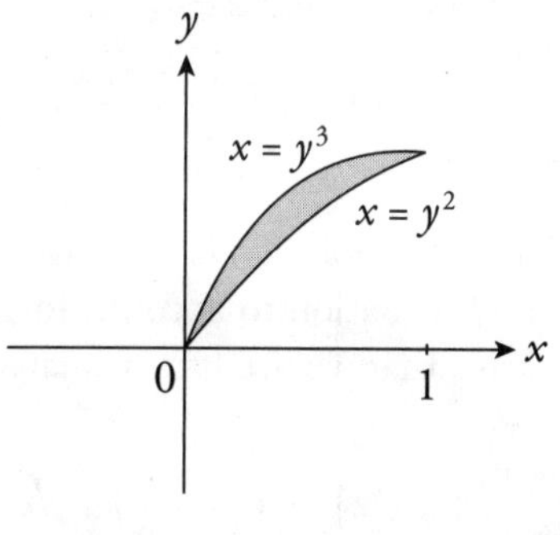

Because $x = y^2$ is always on the outside and $x = y^3$ is always on the inside, we have to evaluate the integral.

$$\pi\int_0^1 \left(y^4 - y^6\right)dy$$

We get the following:

$$\pi\int_0^1 \left(y^4 - y^6\right)dx = \pi\left[\frac{y^5}{5} - \frac{y^7}{7}\right]_0^1 = \frac{2\pi}{35}$$

VOLUME WITH WASHER METHOD: REVOLVING AROUND OTHER AXES

Sometimes we will have to revolve the region about a line instead of one of the axes. If so, this will affect the radii of the washers; we will have to adjust the integral to reflect the shift. Once you draw a picture, it usually isn't too hard to see the difference.

Example 23: Find the volume of the solid that results when the area bounded by the curve $y = x^2$ and the curve $y = 4x$ is revolved about the line $y = -2$. Set up but do not evaluate the integral.

We are not given the limits of integration here, so we need to find where the two curves intersect by setting the equations equal to each other.

$$x^2 = 4x$$
$$x^2 - 4x = 0$$
$$x = 0, 4$$

These will be our limits of integration. Next, sketch the curve.

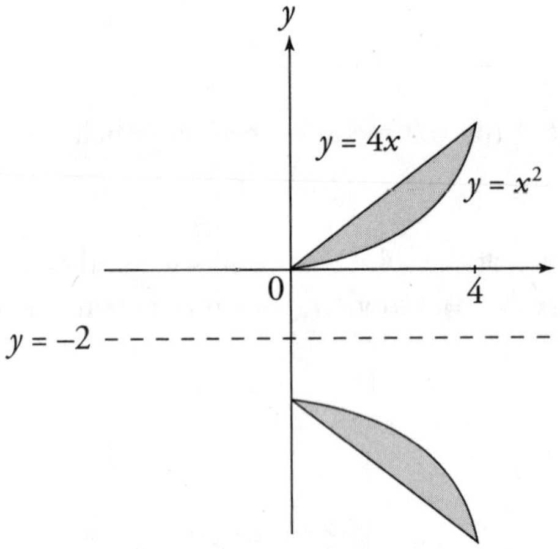

Notice that the distance from the axis of revolution is no longer found by just using each equation. Now, you need to add 2 to each equation to account for the shift in the axis. Thus, the radii are $x^2 + 2$ and $4x + 2$. This means that we need to evaluate the integral.

$$\pi\int_0^4 \left[(4x+2)^2 - (x^2+2)^2\right] dx$$

Suppose instead that the region was revolved about the line $x = -2$. Sketch the region again.

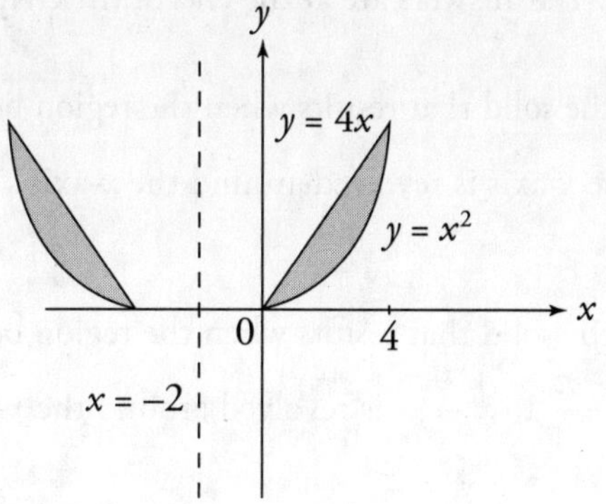

We will have to slice the region horizontally this time; this means we are going to solve each equation for x in terms of y: $x = \sqrt{y}$ and $x = \frac{y}{4}$. We also need to find the y-coordinates of the intersection of the two curves: $y = 0, 16$.

Notice also that, again, each radius is going to be increased by 2 to reflect the shift in the axis of revolution. Thus, we will have to evaluate the integral.

$$\pi \int_0^{16} \left[\left(\sqrt{y} + 2 \right)^2 - \left(\frac{y}{4} + 2 \right)^2 \right] dy$$

Finding the volumes isn't that hard once you've drawn a picture, figured out whether you need to slice vertically or horizontally, and determined whether the axis of revolution has been shifted.

PRACTICE PROBLEM SET 32

Calculate the volumes below. The answers are at the end of the chapter, starting on page 498.

1. Find the volume of the solid that results when the region bounded by $y = \sqrt{9 - x^2}$ and the x-axis is revolved around the x-axis.

2. Find the volume of the solid that results when the region bounded by $y = \sec x$ and the x-axis from $x = -\frac{\pi}{4}$ to $x = \frac{\pi}{4}$ is revolved around the x-axis.

3. Find the volume of the solid that results when the region bounded by $x = 1 - y^2$ and the y-axis is revolved around the y-axis.

4. Find the volume of the solid that results when the region bounded by $x = \sqrt{5}y^2$ and the y-axis from $y = -1$ to $y = 1$ is revolved around the y-axis.

5. Find the volume of the solid whose base is the region between the semi-circle $y = \sqrt{16 - x^2}$ and the x-axis, and whose cross-sections perpendicular to the x-axis are squares with a side on the base.

6. Find the volume of the solid whose base is the region between $y = x^2$ and $y = 4$ and whose perpendicular cross-sections are isosceles right triangles with the hypotenuse on the base.

Chapter 10 Drill

The answers are at the end of the chapter, starting on page 501.

1 Mark for Review

Find the area between the x-axis and $y = \cos x$ from $x = 0$ to $x = \frac{\pi}{2}$.

Ⓐ 0

Ⓑ 1

Ⓒ π

Ⓓ 2π

2 Mark for Review

Find the area between $y = x^2 + 1$ and $y = 5$.

Ⓐ $\frac{16}{3}$

Ⓑ $\frac{32}{3}$

Ⓒ 16

Ⓓ 32

3 Mark for Review

Find the area between $y = 16 - x^2$ and $y = 7$.

Ⓐ 18

Ⓑ 36

Ⓒ 72

Ⓓ 108

4 Mark for Review

Find the area between the y-axis and $x = 25 - y^2$.

Ⓐ $\frac{5}{3}$

Ⓑ $\frac{25}{3}$

Ⓒ $\frac{125}{3}$

Ⓓ $\frac{500}{3}$

5 Mark for Review

Find the area between the x-axis, $x = \sqrt{y}$, and $x = 4$.

Ⓐ $\frac{4}{3}$

Ⓑ $\frac{16}{3}$

Ⓒ $\frac{64}{3}$

Ⓓ $\frac{128}{3}$

6 Mark for Review

Find the volume of the solid that results when the region between $y = \frac{1}{x}$, $x = 1$, $x = 4$, and $y = 0$ is revolved about the x-axis.

(A) $\frac{3}{4}$

(B) $\frac{3\pi}{4}$

(C) $\frac{63}{64}$

(D) $\frac{63\pi}{64}$

7 Mark for Review

Find the volume of the solid that results when the region between $x = \sqrt{1+y}$, $x = 0$, and $y = 3$ is revolved about the y-axis.

(A) 2π

(B) 4π

(C) 8π

(D) 16π

8 Mark for Review

Find the volume of the solid that results when the region between $y = x^2 + 1$ and $y = x + 3$ is revolved about the x-axis.

(A) $\frac{9}{2}$

(B) $\frac{9\pi}{2}$

(C) $\frac{117}{5}$

(D) $\frac{117\pi}{5}$

9 Mark for Review

Find the volume of the solid that results when the region between $x = \csc y$, $y = \frac{\pi}{4}$, $y = \frac{3\pi}{4}$, and $x = 0$ is revolved about the y-axis.

(A) $\frac{\pi}{2}$

(B) π

(C) 2π

(D) 4π

10 Mark for Review

Find the volume of the solid whose base is the region between $y = x^2 + 4$ and $y = 3x + 8$, and whose cross-sections perpendicular to the x-axis are squares with a side on the base.

(A) $\frac{125}{6}$

(B) $\frac{125\pi}{6}$

(C) $\frac{625}{6}$

(D) $\frac{625\pi}{6}$

ANSWERS TO PRACTICE PROBLEM SETS

Practice Problem Set 30

1. $2\sqrt{\frac{2}{\pi}}$

Find the average value of the function, $f(x)$, on the interval $[a, b]$ using the formula $f(c)=\frac{1}{b-a}\int_a^b f(x)\,dx$.

Using the formula, find $\frac{1}{\sqrt{\frac{\pi}{2}}-0}\int_0^{\sqrt{\frac{\pi}{2}}}\left(4x\cos x^2\right)dx=\sqrt{\frac{2}{\pi}}\int_0^{\sqrt{\frac{\pi}{2}}}\left(4x\cos x^2\right)dx$. Use u-substitution to evaluate the integral. Let $u = x^2$ and $du = 2x\ dx$. Substitute for $4x\ dx$, so multiply by 2 to get $2\ du = 4x\ dx$. Now substitute into the integral: $\int(4x\cos x^2)dx = 2\int\cos u\ du = 2\sin u$. If you substitute back, you get $2\sin x^2$. Now evaluate the integral: $\sqrt{\frac{2}{\pi}}\int_0^{\sqrt{\frac{\pi}{2}}}\left(4x\cos x^2\right)dx=\sqrt{\frac{2}{\pi}}\left(2\sin x^2\right)\Big|_0^{\sqrt{\frac{\pi}{2}}}=\sqrt{\frac{2}{\pi}}\left(2\sin\frac{\pi}{2}-2\sin 0\right)=2\sqrt{\frac{2}{\pi}}$.

2. $\frac{8}{3}$

Find the average value of the function, $f(x)$, on the interval $[a, b]$ using the formula $f(c)=\frac{1}{b-a}\int_a^b f(x)\,dx$. Using the formula, find $\frac{1}{16-0}\int_0^{16}\left(\sqrt{x}\right)dx$. You get $\frac{1}{16}\int_0^{16}\left(\sqrt{x}\right)dx=\frac{1}{16}\left(\frac{x^{\frac{3}{2}}}{\frac{3}{2}}\right)\Bigg|_0^{16}=\frac{1}{24}\left(x^{\frac{3}{2}}\right)\Big|_0^{16}=\frac{1}{24}\left((16)^{\frac{3}{2}}-0\right)=\frac{8}{3}$.

3. 1

Find the average value of the function, $f(x)$, on the interval $[a, b]$ using the formula $f(c)=\frac{1}{b-a}\int_a^b f(x)\,dx$. Using the formula, find $\frac{1}{1-(-1)}\int_{-1}^{1}\left(2|x|\right)dx$. Recall that the absolute value function must be rewritten as a piecewise function: $|x| = \begin{cases} x; x \geq 0 \\ -x; x < 0 \end{cases}$. Thus, split the integral into two separate integrals in order to evaluate it: $\frac{1}{1-(-1)}\int_{-1}^{1}(2|x|)\,dx = \int_{-1}^{0}(-x)\,dx + \int_{0}^{1}x\,dx$. You get $\int_{-1}^{0}(-x)\,dx + \int_{0}^{1}x\,dx = \left(-\frac{x^2}{2}\right)\Big|_{-1}^{0} + \left(\frac{x^2}{2}\right)\Big|_{-1}^{0} = \left(0-\left(-\frac{1}{2}\right)\right)+\left(\frac{1}{2}-0\right)=1.$

4. $\frac{8}{3}$

Find the average value of the function $f(x)$ on the interval $[a, b]$ using the formula $f(c)=\frac{1}{b-a}\int_a^b f(x)dx$. Using the formula, find $\frac{1}{2-(-2)}\int_{-2}^{2}4-x^2dx=\frac{1}{4}\int_{-2}^{2}4-x^2dx$. Integrate: $\frac{1}{4}\int_{-2}^{2}4-x^2dx=\frac{1}{4}\left(4x-\frac{x^3}{3}\right)\Big|_{-2}^{2}$. And evaluate: $\frac{1}{4}\left(4x-\frac{x^3}{3}\right)\Big|_{-2}^{2}=\frac{1}{4}\left(8-\frac{8}{3}\right)-\frac{1}{4}\left(-8+\frac{8}{3}\right)=\frac{8}{3}$.

5. $\frac{1}{2}\left(e-\frac{1}{e}\right)\approx 1.175$

Find the average value of the function $f(x)$ on the interval $[a, b]$ using the formula $f(c)=\frac{1}{b-a}\int_a^b f(x)dx$. Using the formula, find $\frac{1}{1-(-1)}\int_{-1}^{1}e^x dx=\frac{1}{2}\int_{-1}^{1}e^x dx$. Integrate: $\frac{1}{2}\int_{-1}^{1}e^x dx=\frac{1}{2}\left(e^x\right)\Big|_{-1}^{1}$. And evaluate: $\frac{1}{2}\left(e^x\right)\Big|_{-1}^{1}=\frac{1}{2}(e)-\frac{1}{2}\left(\frac{1}{e}\right)=\frac{1}{2}\left(e-\frac{1}{e}\right)\approx 1.175.$

Practice Problem Set 31

1. $\frac{32}{3}$

Find the area of a region bounded by $f(x)$ above and $g(x)$ below at all points of the interval $[a, b]$ using the formula $\int_a^b\left[f(x)-g(x)\right]dx$. Here, $f(x) = 2$ and $g(x) = x^2 - 2$.

First, make a sketch of the region.

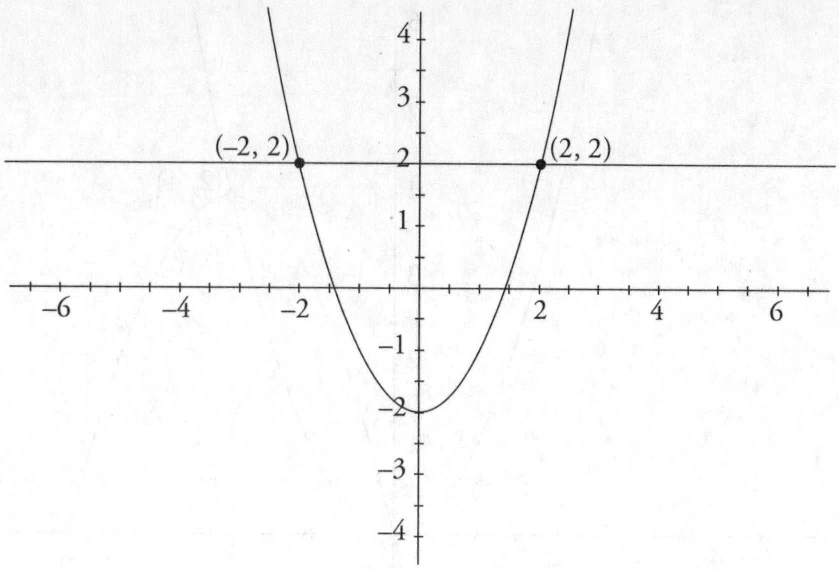

Next, find where the two curves intersect, which will be the endpoints of the region. Do this by setting the two curves equal to each other. You get $x^2 - 2 = 2$. The solutions are (–2, 0) and (2, 0). Therefore, in order to find the area of the region, evaluate the integral $\int_{-2}^{2}\left(2-\left(x^2-2\right)\right)dx = \int_{-2}^{2}\left(4-x^2\right)dx$.

You get $\int_{-2}^{2}\left(4-x^2\right)dx = \left(4x - \frac{x^3}{3}\right)\Big|_{-2}^{2} = \left(4(2)-\frac{(2)^3}{3}\right)-\left(4(-2)-\frac{(-2)^3}{3}\right)=\frac{32}{3}$.

2. $\frac{8}{3}$

Find the area of a region bounded by $f(x)$ above and $g(x)$ below at all points of the interval $[a, b]$ using the formula $\int_a^b\left[f(x)-g(x)\right]dx$. Here, $f(x) = 4x - x^2$ and $g(x) = x^2$.

First, make a sketch of the region.

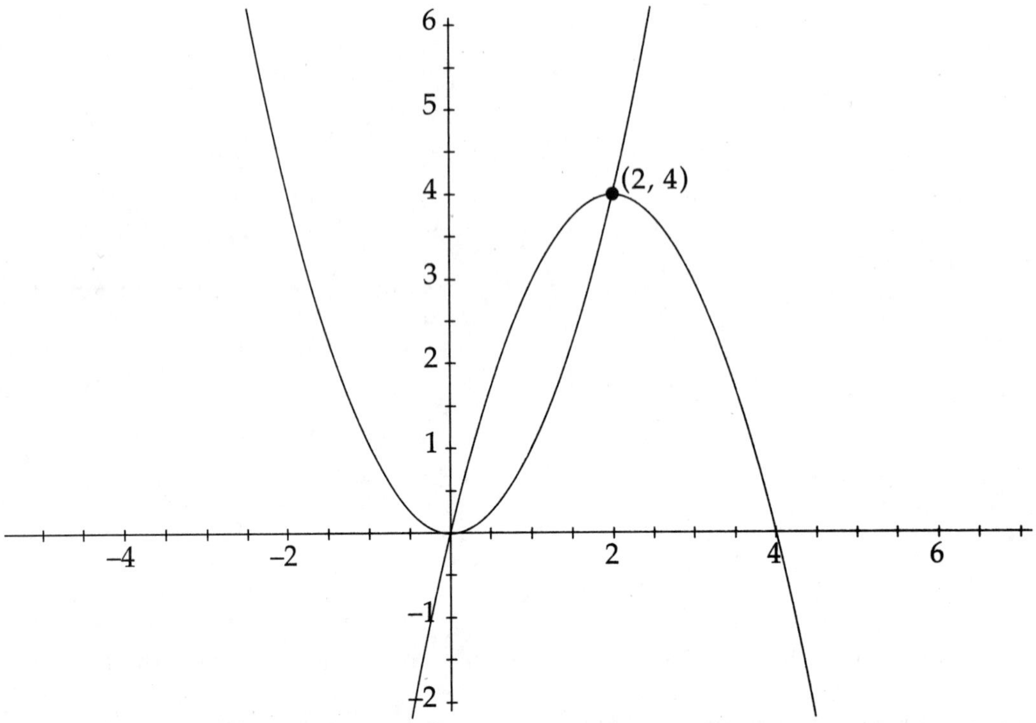

Next, find where the two curves intersect, which will be the endpoints of the region. Do this by setting the two curves equal to each other. You get $4x - x^2 = x^2$. The solutions are (0, 0) and (2, 4). Therefore, in order to find the area of the region, evaluate the integral $\int_0^2 \left(\left(4x - x^2\right) - x^2\right) dx = \int_0^2 \left(4x - 2x^2\right) dx$.

You get $\int_0^2 \left(4x - 2x^2\right) dx = \left(2x^2 - \frac{2x^3}{3}\right)\Bigg|_0^2 = \left(2(2)^2 - \frac{2(2)^3}{3}\right) - 0 = \frac{8}{3}$.

3. 36

Find the area of a region bounded by $f(x)$ above and $g(x)$ below at all points of the interval $[a, b]$ using the formula $\int_a^b \left[f(x) - g(x)\right] dx$. Here, $f(x) = 2x - 5$ and $g(x) = x^2 - 4x - 5$.

First, make a sketch of the region.

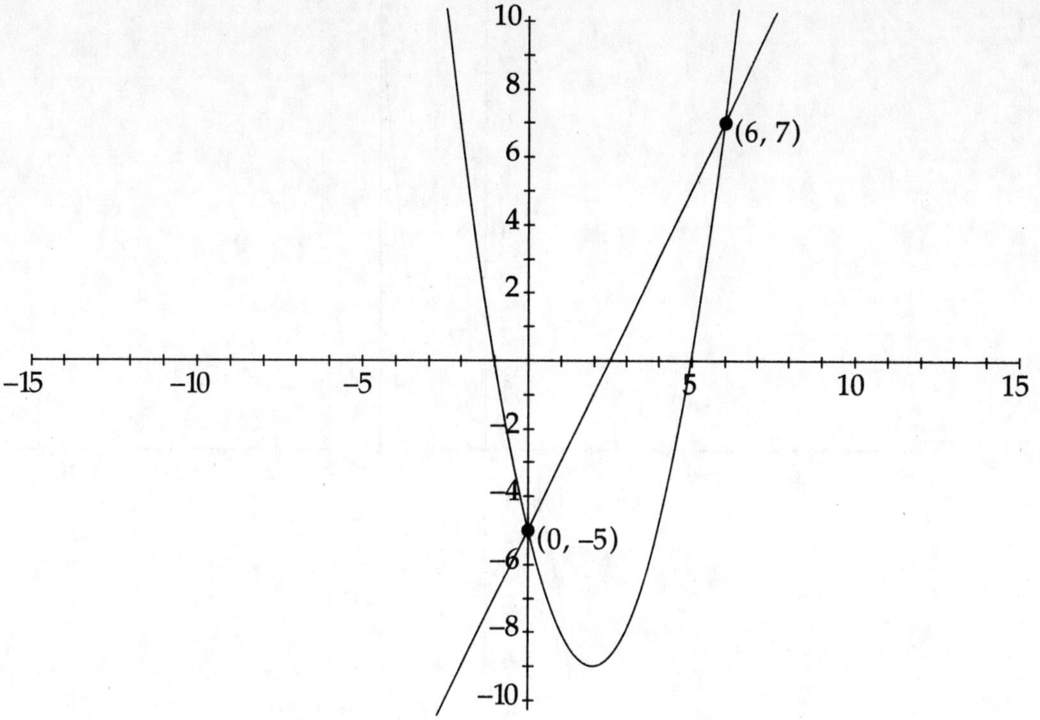

Next, find where the two curves intersect, which will be the endpoints of the region. Do this by setting the two curves equal to each other. You get $x^2 - 4x - 5 = 2x - 5$. The solutions are (0, –5) and (6, 7). Therefore, in order to find the area of the region, evaluate the integral $\int_0^6 \left[(2x-5)-\left(x^2-4x-5\right)\right]dx = \int_0^6 \left(-x^2+6x\right)dx$.

You get $\int_0^6 \left(-x^2+6x\right)dx = \left(-\frac{x^3}{3}+3x^2\right)\Big|_0^6 = \left(-\frac{(6)^3}{3}+3(6)^2\right)-0=36$.

4. $\frac{17}{4}$

Find the area of a region bounded by $f(x)$ above and $g(x)$ below at all points of the interval $[a, b]$ using the formula $\int_a^b \left[f(x)-g(x)\right]dx$.

First, make a sketch of the region.

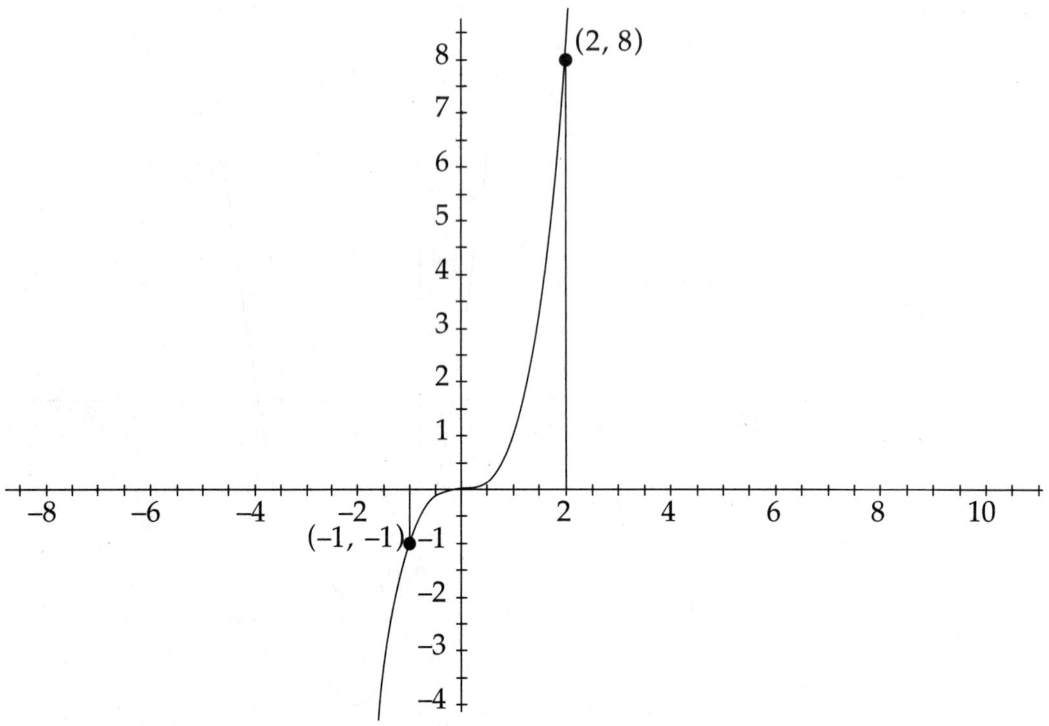

Notice that in the region from $x = -1$ to $x = 0$ the top curve is $f(x) = 0$ (the x-axis) and the bottom curve is $g(x) = x^3$, but from $x = 0$ to $x = 2$ the situation is reversed, so the top curve is $f(x) = x^3$ and the bottom curve is $g(x) = 0$. Thus, split the region into two pieces and find the area by evaluating two integrals and adding the answers: $\int_{-1}^{0}(0 - x^3)\,dx$ and $\int_{0}^{2}(x^3 - 0)\,dx$. You get $\int_{-1}^{0}(0 - x^3)\,dx = \left(-\frac{x^4}{4}\right)\Big|_{-1}^{0} = 0 - \left(-\frac{(-1)^4}{4}\right) = \frac{1}{4}$ and $\int_{0}^{2}(x^3 - 0)\,dx = \left(\frac{x^4}{4}\right)\Big|_{0}^{2} = \frac{(2)^4}{4} - 0 = 4$.

Therefore, the area of the region is $\frac{17}{4}$.

5. $\frac{9}{2}$

Find the area of a region bounded by $f(y)$ on the right and $g(y)$ on the left at all points of the interval $[c, d]$ using the formula $\int_{c}^{d}\left[f(y) - g(y)\right]dy$. Here, $f(y) = y + 2$ and $g(y) = y^2$.

First, make a sketch of the region.

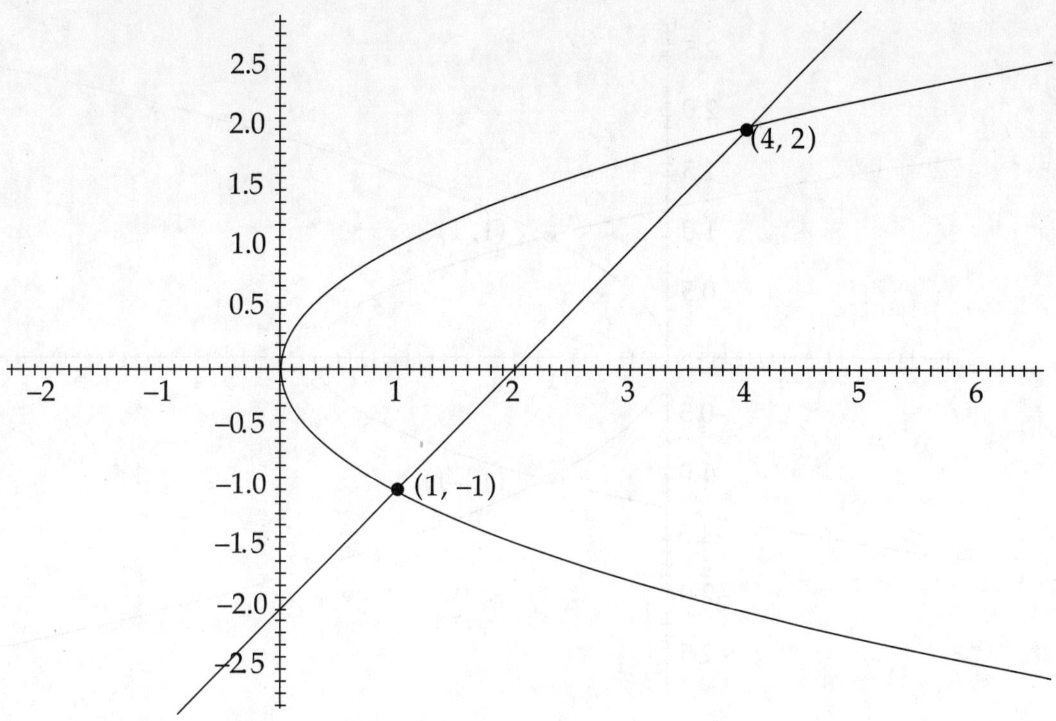

Next, find where the two curves intersect, which will be the endpoints of the region. Do this by setting the two curves equal to each other. You get $y^2 = y + 2$. The solutions are (4, 2) and (1, −1). Therefore, in order to find the area of the region, evaluate the integral $\int_{-1}^{2}\left(y+2-y^2\right)dy$.

You get

$$\int_{-1}^{2}\left(y+2-y^2\right)dy=\left(\frac{y^2}{2}+2y-\frac{y^3}{3}\right)\Bigg|_{-1}^{2}=\left(\frac{(2)^2}{2}+2(2)-\frac{(2)^3}{3}\right)-\left(\frac{(-1)^2}{2}+2(-1)-\frac{(-1)^3}{3}\right)=\frac{9}{2}.$$

6. 4

Find the area of a region bounded by $f(y)$ on the right and $g(y)$ on the left at all points of the interval $[c, d]$ using the formula $\int_{c}^{d}\left[f(y)-g(y)\right]dy$. Here, $f(y) = 3 - 2y^2$ and $g(y) = y^2$.

First, make a sketch of the region.

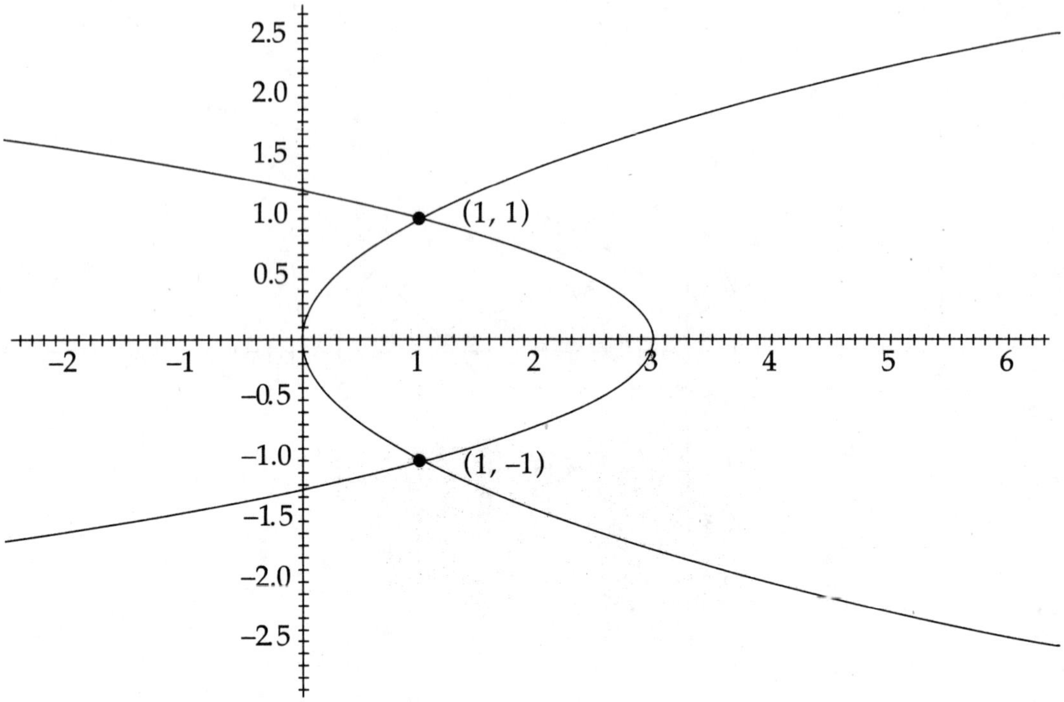

Next, find where the two curves intersect, which will be the endpoints of the region. Do this by setting the two curves equal to each other. You get $y^2 = 3 - 2y^2$. The solutions are (1, 1) and (1, –1). Therefore, in order to find the area of the region, evaluate the integral $\int_{-1}^{1}\left(3-2y^2-y^2\right)dy = \int_{-1}^{1}\left(3-3y^2\right)dy$.

You get $\int_{-1}^{1}\left(3-3y^2\right)dy = \left(3y-y^3\right)\Big|_{-1}^{1} = \left(3(1)-(1)^3\right)-\left(3(-1)-(-1)^3\right) = 4$.

7. $\frac{9}{2}$

Find the area of a region bounded by $f(y)$ on the right and $g(y)$ on the left at all points of the interval $[c, d]$ using the formula $\int_{c}^{d}\left[f(y)-g(y)\right]dy$. Here, $f(y) = y - 2$ and $g(y) = y^2 - 4y + 2$.

First, make a sketch of the region.

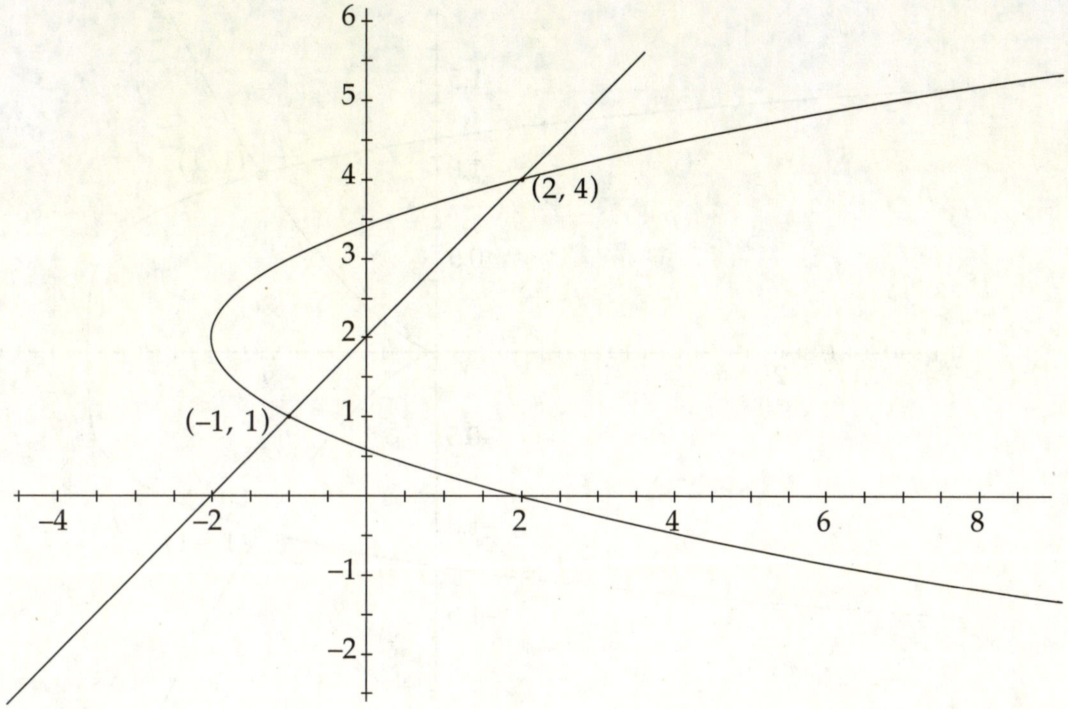

Next, find where the two curves intersect, which will be the endpoints of the region. Do this by setting the two curves equal to each other. You get $y^2 - 4y + 2 = y - 2$. The solutions are (2, 4) and (–1, 1). Therefore, in order to find the area of the region, evaluate the integral $\int_1^4 \left[(y-2)-(y^2-4y+2)\right]dy = \int_1^4 \left[(-y^2+5y-4)\right]dy$. You get $\int_1^4 \left[(-y^2+5y-4)\right]dy =$

$$\left(-\frac{y^3}{3}+\frac{5y^2}{2}-4y\right)\Bigg|_1^4 = \left(-\frac{(4)^3}{3}+\frac{5(4)^2}{2}-4(4)\right)-\left(-\frac{(1)^3}{3}+\frac{5(1)^2}{2}-4(1)\right)=\frac{9}{2}.$$

8. $\frac{12}{5}$

Find the area of a region bounded by $f(y)$ on the right and $g(y)$ on the left at all points of the interval $[c, d]$ using the formula $\int_c^d \left[f(y)-g(y)\right]dy$. Here, $f(y) = 2 - y^4$ and $g(y) = y^{\frac{2}{3}}$.

First, make a sketch of the region.

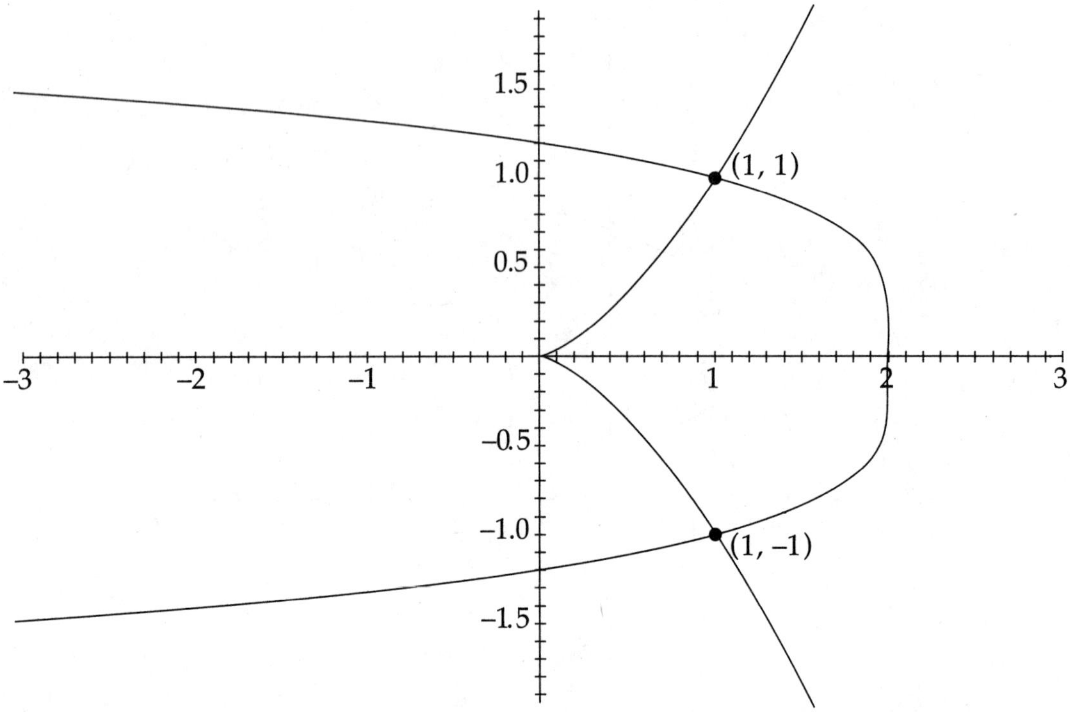

Next, find where the two curves intersect, which will be the endpoints of the region. Do this by setting the two curves equal to each other. You get $2 - y^4 = y^{\frac{2}{3}}$. The solutions are (1, 1) and (1, –1). Therefore, in order to find the area of the region, evaluate the integral $\int_{-1}^{1}\left(2 - y^4 - y^{\frac{2}{3}}\right)dy$.

You get

$$\int_{-1}^{1}\left(2 - y^4 - y^{\frac{2}{3}}\right)dy = \left(2y - \frac{y^5}{5} - \frac{3}{5}y^{\frac{5}{3}}\right)\Bigg|_{-1}^{1} = \left(2(1) - \frac{(1)^5}{5} - \frac{3}{5}(1)^{\frac{5}{3}}\right) - \left(2(-1) - \frac{(-1)^5}{5} - \frac{3}{5}(-1)^{\frac{5}{3}}\right) = \frac{12}{5}.$$

Practice Problem Set 32

1. 36π

When the region you are revolving is defined between a curve $f(x)$ and the x-axis, find the volume using discs. Use the formula $V = \pi\int_a^b \left[f(x)\right]^2 dx$. Here you have a region between $f(x) = \sqrt{9 - x^2}$ and the x-axis. First, find the endpoints of the region. Do this by setting

$f(x) = \sqrt{9 - x^2}$ equal to zero and solving for x. You get $x = -3$ and $x = 3$. Thus, find the volume by evaluating $\pi \int_{-3}^{3} \left(\sqrt{9 - x^2}\right)^2 dx = \pi \int_{-3}^{3} \left(9 - x^2\right) dx$. You get $\pi \int_{-3}^{3} \left(9 - x^2\right) dx = \pi \left(9x - \frac{x^3}{3}\right)\Bigg|_{-3}^{3} = 36\pi$.

2. 2π

When the region you are revolving is defined between a curve $f(x)$ and the x-axis, find the volume using discs. Use the formula $V = \pi \int_a^b \left[f(x)\right]^2 dx$. Here you have a region between $f(x) = \sec x$ and the x-axis. You are given the endpoints of the region: $x = -\frac{\pi}{4}$ and $x = \frac{\pi}{4}$. Thus, find the volume by evaluating $\pi \int_{-\frac{\pi}{4}}^{\frac{\pi}{4}} \sec^2 x\, dx$. You get $\pi \int_{-\frac{\pi}{4}}^{\frac{\pi}{4}} \sec^2 x\, dx = \pi(\tan x)\Big|_{-\frac{\pi}{4}}^{\frac{\pi}{4}} = 2\pi$.

3. $\frac{16\pi}{15}$

When the region you are revolving is defined between a curve $f(y)$ and the y-axis, find the volume using discs. Use the formula $V = \pi \int_a^b \left[f(y)\right]^2 dy$ (see page 479 and note that when $g(y) = 0$, you get discs instead of washers). Here you have a region between $f(y) = 1 - y^2$ and the y-axis. First, find the endpoints of the region. Do this by setting $f(y) = 1 - y^2$ equal to zero and solving for y. You get $y = -1$ and $y = 1$. Thus, find the volume by evaluating $\pi \int_{-1}^{1} \left(1 - y^2\right)^2 dy = \pi \int_{-1}^{1} \left(1 - 2y^2 + y^4\right) dy$. You get $\pi \int_{-1}^{1} \left(1 - 2y^2 + y^4\right) dy = \pi \left(y - \frac{2y^3}{3} + \frac{y^5}{5}\right)\Bigg|_{-1}^{1} = \frac{16\pi}{15}$.

4. 2π

When the region you are revolving is defined between a curve $f(y)$ and the y-axis, find the volume using discs. Use the formula $V = \pi \int_a^b \left[f(y)\right]^2 dy$ (see page 479 and note that when $g(y) = 0$, you get discs instead of washers). Here you have a region between $f(y) = \sqrt{5}y^2$ and the y-axis. You are given the endpoints of the region: $y = -1$ and $y = 1$. Thus, find the volume by evaluating $\pi \int_{-1}^{1} \left(\sqrt{5}y^2\right)^2 dy = \pi \int_{-1}^{1} \left(5y^4\right) dy$. You get $\pi \int_{-1}^{1} \left(5y^4\right) dy = \pi\left(y^5\right)\Big|_{-1}^{1} = 2\pi$.

5. $\frac{256}{3}$

To find the volume of a solid with a cross-section of an isosceles right triangle, integrate the area of the square (*side*²) over the endpoints of the interval. Here, the sides of the squares are found by $f(x)-g(x)=\sqrt{16-x^2}-0$, and the intervals are found by setting $y=\sqrt{16-x^2}$ equal to zero. You get $x = -4$ and $x = 4$. Thus, find the volume by evaluating the integral $\int_{-4}^{4}\left(\sqrt{16-x^2}\right)^2 dx=\int_{-4}^{4}\left(16-x^2\right)dx$. You get $\int_{-4}^{4}\left(16-x^2\right)dx=\left(16x-\frac{x^3}{3}\right)\Big|_{-4}^{4}=\frac{256}{3}$.

6. $\frac{128}{15}$

To find the volume of a solid with a cross-section of a square, integrate the area of the isosceles right triangle, $\frac{hypotenuse^2}{4}$, over the endpoints of the interval. Here, the hypotenuses of the triangles are found by $f(x) - g(x) = 4 - x^2$, and the intervals are found by setting $y = x^2$ equal to $y = 4$. You get $x = -2$ and $x = 2$. Thus, find the volume by evaluating the integral $\int_{-2}^{2}\frac{\left(4-x^2\right)^2}{4}dx=\int_{-2}^{2}\frac{\left(16-8x^2+x^4\right)}{4}dx=\int_{-2}^{2}\left(4-2x^2+\frac{x^4}{4}\right)dx$.
You get $\int_{-2}^{2}\left(4-2x^2+\frac{x^4}{4}\right)dx=\left(4x-\frac{2x^3}{3}+\frac{x^5}{20}\right)\Big|_{-2}^{2}=\frac{128}{15}$.

ANSWERS TO CHAPTER 10 DRILL

1. **B** First, draw a picture.

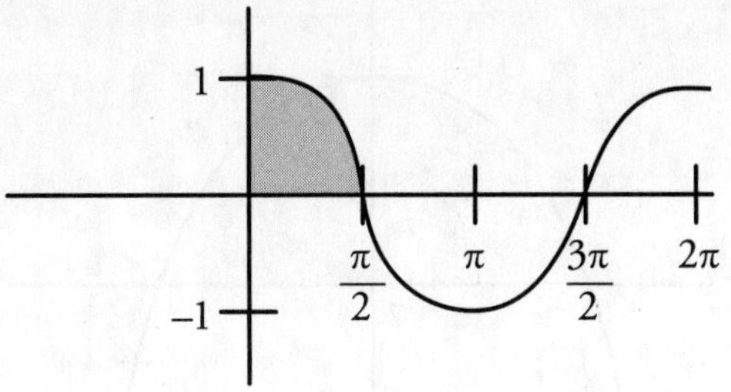

In order to find the area, you need to evaluate the integral $\int_0^{\frac{\pi}{2}} \cos x\, dx$.

You get $\int_0^{\frac{\pi}{2}} \cos x\, dx = \sin x\Big|_0^{\frac{\pi}{2}} = \sin\frac{\pi}{2} - \sin 0 = 1$.

The answer is (B).

2. **B** First, draw a picture.

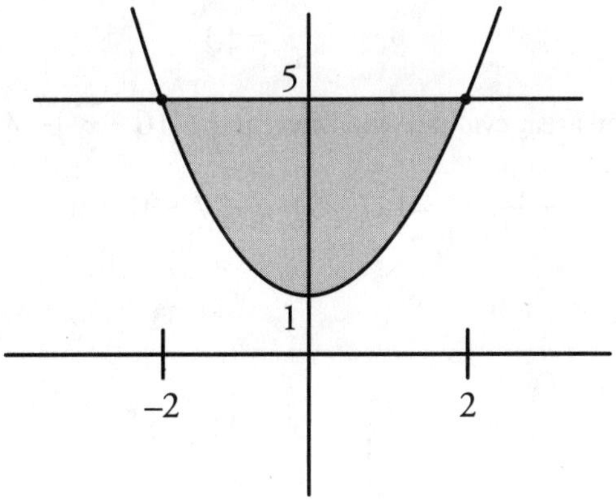

Note that the curve $y = 5$ is on top and the curve $y = x^2 + 1$ is on the bottom. Next, find where the curves intersect. Set them equal to each other and solve for x:

$$x^2 + 1 = 5$$

$$x^2 = 4$$

$$x = \pm 2$$

Thus, in order to find the area, evaluate the integral $\int_{-2}^{2} 5 - \left(x^2 + 1\right) dx = \int_{-2}^{2} 4 - x^2\, dx$.

You get $\int_{-2}^{2} 4 - x^2\, dx = \left(4x - \frac{x^3}{3}\right)\Bigg|_{-2}^{2} = \left(8 - \frac{8}{3}\right) - \left(-8 + \frac{8}{3}\right) = \frac{32}{3}$.

The answer is (B).

3. **B** First, draw a picture.

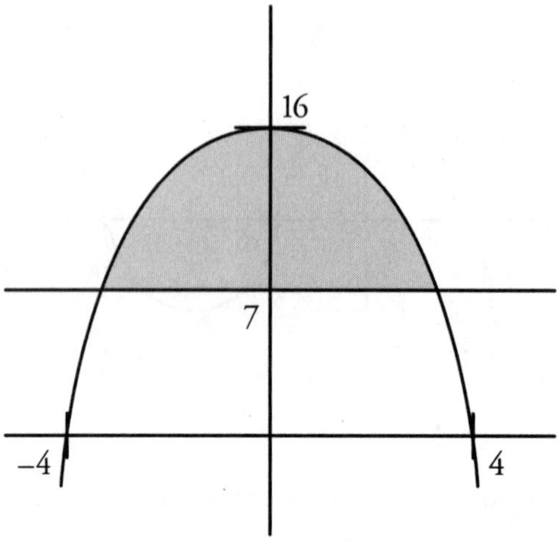

Note that the curve $y = 16 - x^2$ is on top and the curve $y = 7$ is on the bottom. Next, find where the curves intersect. Set them equal to each other and solve for x:

$$16 - x^2 = 7$$
$$x^2 = 9$$
$$x = \pm 3$$

Thus, in order to find the area, evaluate the integral $\int_{-3}^{3} \left(16 - x^2\right) - 7\, dx = \int_{-3}^{3} 9 - x^2\, dx$.

You get $\int_{-3}^{3} 9 - x^2\, dx = \left(9x - \frac{x^3}{3}\right)\Bigg|_{-3}^{3} = (27 - 9) - (-27 + 9) = 36$.

The answer is (B).

4. **D** First, draw a picture.

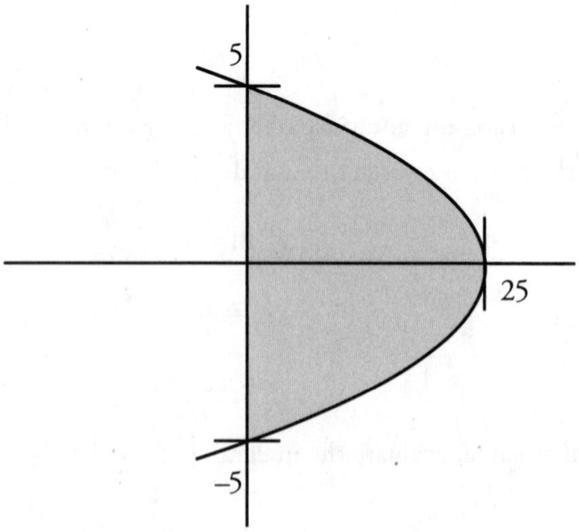

Next, find where the curve $x = 25 - y^2$ intersects the y-axis. Set $25 - y^2 = 0$. You get $y = \pm 5$.

Thus, in order to find the area, evaluate the integral $\int_{-5}^{5} 25 - y^2\, dy$.

You get $\int_{-5}^{5} 25 - y^2\, dy = \left(25y - \frac{y^3}{3}\right)\Big|_{-5}^{5} = \left(125 - \frac{125}{3}\right) - \left(-125 + \frac{125}{3}\right) = \frac{500}{3}$.

The answer is (D).

5. **C** First, draw a picture.

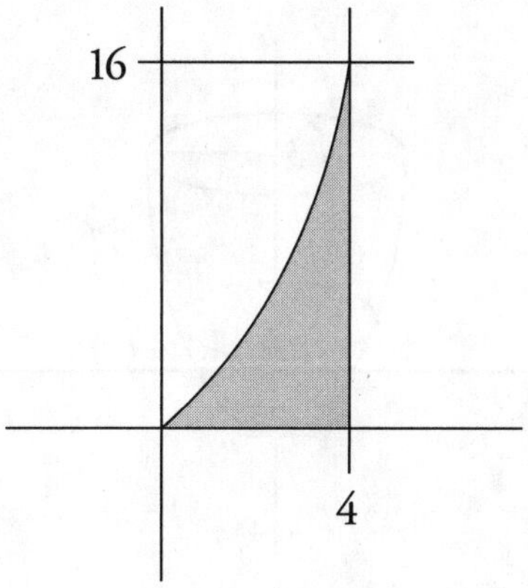

Note that the curve $x = 4$ is farther away from the y-axis than the curve $x = \sqrt{y}$. Next, find where the two curves intersect. Set them equal to each other and solve for y: $\sqrt{y} = 4$, so $y = 16$. Thus, in order to find the area, evaluate the integral $\int_0^{16} 4 - \sqrt{y}\, dy$. You get

$\int_0^{16} 4 - \sqrt{y}\, dy = \left(4y - \frac{2}{3}y^{\frac{3}{2}}\right)\Big|_0^{16} = \left(4y - \frac{2}{3}\sqrt{y^3}\right)\Big|_0^{16} = \left(64 - \frac{128}{3}\right) - 0 = \frac{64}{3}$.

The answer is (C).

6. **B** First, draw a picture.

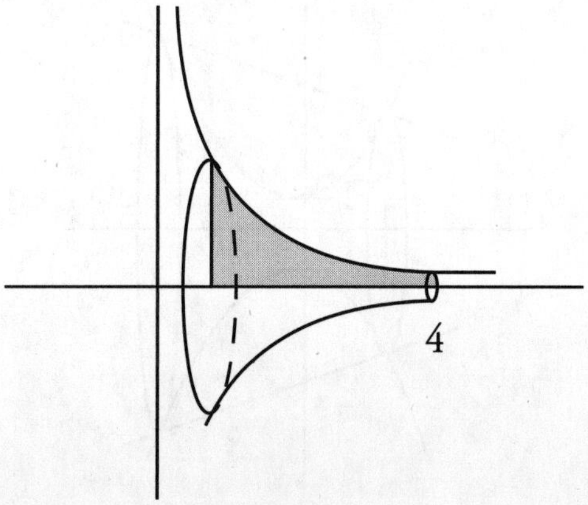

Next, find the boundaries of the region. You are given that they are $x = 1$ and $x = 4$. Thus, in order to find the volume, evaluate the integral $\pi\int_1^4 \left(\frac{1}{x}\right)^2 dx$. You get

$$\pi\int_1^4 \left(\frac{1}{x}\right)^2 dx = \pi\int_1^4 x^{-2}\, dx = \pi\left(\frac{x^{-1}}{-1}\right)\Bigg|_1^4 = \pi\left(-\frac{1}{x}\right)\Bigg|_1^4 = \frac{3\pi}{4}.$$

The answer is (B).

7. **C** First, draw a picture.

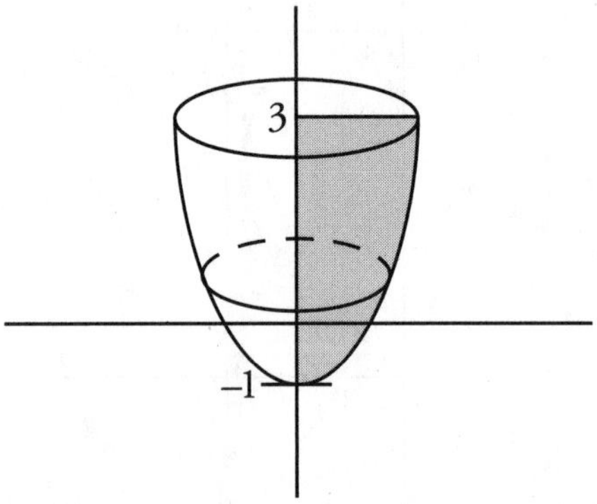

Next, find the boundaries of the region. One of them is $y = 3$ and the other is where the curve intersects the y-axis, which is at $y = -1$. Thus, in order to find the volume, evaluate the integral $\pi\int_{-1}^{3}\left(\sqrt{1+y}\right)^2 dy$. You get

$$\pi\int_{-1}^{3}\left(\sqrt{1+y}\right)^2 dy = \pi\int_{-1}^{3}(1+y)\,dy = \pi\left(y+\frac{y^2}{2}\right)\Bigg|_1^3 = \pi\left(3+\frac{9}{2}\right)-\left(-1+\frac{1}{2}\right) = 8\pi.$$

The answer is (C).

8. **D** First, draw a picture.

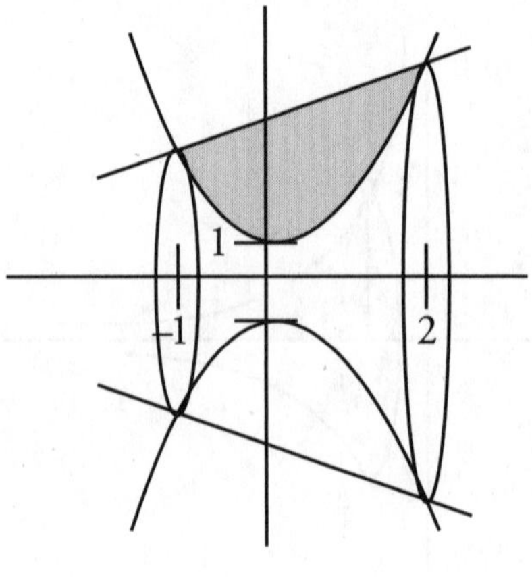

Next, find the boundaries of the region. Set the curves equal to each other and solve for x:

$$x^2 + 1 = x + 3$$

$$x^2 - x - 2 = 0$$

$$(x - 2)(x + 1) = 0$$

$$x = -1, 2$$

Note that the curve $y = x + 3$ is always on top and the curve $y = x^2 + 1$ is always on the bottom. Thus, in order to find the volume, evaluate the integral $\pi\int_{-1}^{2}(x+3)^2 - \left(x^2+1\right)^2 dx$. You get $\pi\int_{-1}^{2}(x+3)^2 - \left(x^2+1\right)^2 dx = \pi\int_{-1}^{2} -x^4 - x^2 + 6x + 8\, dx = \pi\left(-\frac{x^5}{5} - \frac{x^3}{3} + 3x^2 + 8x\right)\Bigg|_{-1}^{2} = \frac{117\pi}{5}$.

The answer is (D).

9. **C** First, draw a picture.

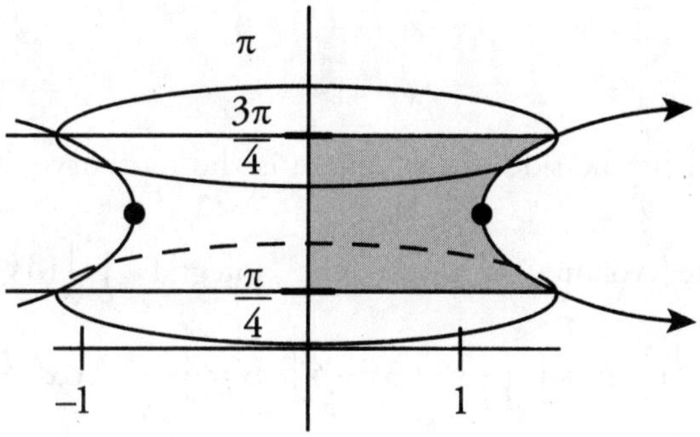

Next, find the boundaries of the region. They are $y = \frac{\pi}{4}$ and $y = \frac{3\pi}{4}$. Thus, in order to find the volume, evaluate the integral $\pi\int_{\frac{\pi}{4}}^{\frac{3\pi}{4}}(\csc y)^2\, dy$. You get $\pi\int_{\frac{\pi}{4}}^{\frac{3\pi}{4}}(\csc y)^2\, dy = -\pi\cot y\Big|_{\frac{\pi}{4}}^{\frac{3\pi}{4}} = 2\pi$.

The answer is (C).

10. **C** First, draw a picture.

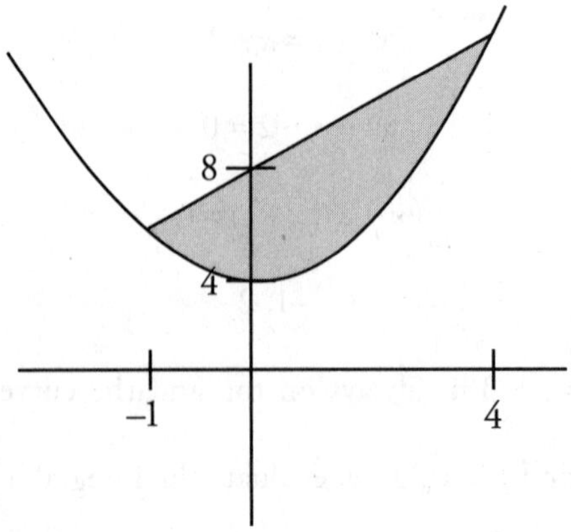

Note that the curve $y = 3x + 8$ is always on top and $y = x^2 + 4$ is always on the bottom of the region. Next, find the boundaries of the region. Set the curves equal to each other and solve for x:

$$x^2 + 4 = 3x + 8$$
$$x^2 - 3x - 4 = 0$$
$$(x - 4)(x + 1) = 0$$
$$x = -1, 4$$

The "slice" that will be the side of the square is the top curve minus the bottom curve, so in order to find the volume, evaluate the integral $\int_{-1}^{4} \left[(3x+8)-(x^2+4)\right]^2 dx$. You get $\int_{-1}^{4} \left[(3x+8)-(x^2+4)\right]^2 dx = \int_{-1}^{4} \left[-x^2+3x+4\right]^2 dx = \int_{-1}^{4} x^4 - 6x^3 + x^2 + 24x + 16\, dx = \frac{625}{6}$.

The answer is (C).

Part VI Additional Practice Tests

- Practice Test 2
- Practice Test 2: Answers and Explanations
- Practice Test 2: Free-Response Questions Grading Rubrics
- How to Score Practice Test 2
- Practice Test 3
- Practice Test 3: Answers and Explanations
- Practice Test 3: Free-Response Questions Grading Rubrics
- How to Score Practice Test 3

Practice Test 2

AP® Calculus AB Exam

SECTION I, PART A: Multiple Choice

At a Glance

Total Time
1 hour
Number of Questions
30
Percent of Total Grade
33 1/3%

DISCLAIMER: The official AP Calculus AB will be administered in a hybrid digital format. Instructions for the digital exam may differ from this practice test.

Instructions

Section I, Part A, of this examination contains 30 multiple-choice questions. Fill in only the ovals for numbers 1 through 30 on your answer sheet.

No calculator is allowed for this part of the exam.

Indicate all of your answers to the multiple-choice questions on the answer sheet. Give only one answer to each question.

Use your time effectively, working as quickly as you can without losing accuracy. Do not spend too much time on any one question. Go on to other questions and come back to the ones you have not answered if you have time. It is not expected that everyone will know the answers to all the multiple-choice questions.

Your total score on the multiple-choice section is based only on the number of questions answered correctly. Points are not deducted for incorrect answers or unanswered questions.

SECTION I, PART B: Multiple Choice

At a Glance

Total Time
45 minutes
Number of Questions
15
Percent of Total Grade
16 2/3%

DISCLAIMER: The official AP Calculus AB will be administered in a hybrid digital format. Instructions for the digital exam may differ from this practice test.

Instructions

Section I, Part B, of this examination contains 15 multiple-choice questions. Fill in only the ovals for numbers 31 through 45 on your answer sheet.

A graphing calculator is required for some questions on this part of the exam. You may use a handheld calculator or a Desmos graphing calculator application. **Make sure your calculator is in radian mode.**

Indicate all of your answers to the multiple-choice questions on the answer sheet. Give only one answer to each question.

Use your time effectively, working as quickly as you can without losing accuracy. Do not spend too much time on any one question. Go on to other questions and come back to the ones you have not answered if you have time. It is not expected that everyone will know the answers to all the multiple-choice questions.

Your total score on the multiple-choice section is based only on the number of questions answered correctly. Points are not deducted for incorrect answers or unanswered questions.

GO ON TO THE NEXT PAGE.

AP CALCULUS AB

SECTION I, Part A

Time—60 Minutes

Number of questions—30

No calculator is allowed for this part of the exam.

Directions: Solve each of the following problems, using the available paper for scratchwork. After examining the choices, select the best of the choices given.

In this test: Unless otherwise specified, the domain of a function f is assumed to be the set of all real numbers x for which $f(x)$ is a real number. The inverse of a trigonometric function f may be indicated using the inverse function notation f^{-1} or with the prefix "arc" (e.g., $\sin^{-1} x = \arcsin x$).

1 Mark for Review

If $g(x) = \frac{1}{32}x^4 - 5x^2$, find $g'(4)$.

(A) -72

(B) -32

(C) 24

(D) 32

---

2 Mark for Review

$\lim\limits_{x \to 0} \frac{8x^2}{\cos x - 1} =$

(A) -16

(B) -1

(C) 6

(D) 8

GO ON TO THE NEXT PAGE.

3 Mark for Review

$\lim\limits_{x \to 5} \frac{x^2 - 25}{x - 5} =$

(A) 0

(B) 5

(C) 10

(D) The limit does not exist.

---

4 Mark for Review

If $f(x) = \frac{x^5 - x + 2}{x^3 + 7}$, find $f'(x)$.

(A) $\frac{(5x^4 - 1)}{(3x^2)}$

(B) $\frac{(x^3 + 7)(5x^4 - 1) - (x^5 - x + 2)(3x^2)}{(x^3 + 7)}$

(C) $\frac{(x^5 - x + 2)(3x^2) - (x^3 + 7)(5x^4 - 1)}{(x^3 + 7)^2}$

(D) $\frac{(x^3 + 7)(5x^4 - 1) - (x^5 - x + 2)(3x^2)}{(x^3 + 7)^2}$

GO ON TO THE NEXT PAGE.

5 Mark for Review

Evaluate $\lim_{h \to 0} \frac{\tan\left(\frac{\pi}{4} + h\right) - 1}{h}$.

(A) 0

(B) 1

(C) 2

(D) The limit does not exist.

---

6 Mark for Review

$\int x\sqrt{3x}\, dx =$

(A) $\frac{2\sqrt{3}}{5} x^{\frac{5}{2}} + C$

(B) $\frac{5\sqrt{3}}{2} x^{\frac{5}{2}} + C$

(C) $\frac{\sqrt{3}}{2} x^{\frac{1}{2}} + C$

(D) $\frac{5\sqrt{3}}{2} x^{\frac{3}{2}} + C$

GO ON TO THE NEXT PAGE.

7 Mark for Review

For what value of k is f continuous at $x = 1$ if

$$f(x) = \begin{cases} x^2 - 3kx + 2; x \le 1 \\ 5x - kx^2; x > 1 \end{cases}?$$

(A) –1

(B) $-\frac{1}{2}$

(C) 2

(D) 8

---

8 Mark for Review

The graph of $f(x)$ is given below. Evaluate $\int_0^7 f(x)\,dx$.

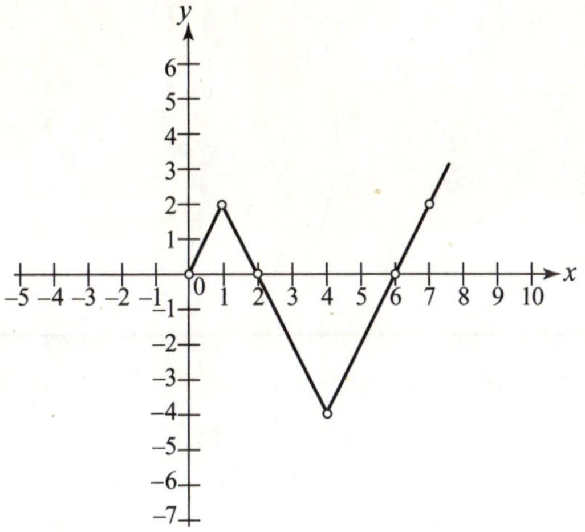

(A) –11

(B) –5

(C) 5

(D) 11

GO ON TO THE NEXT PAGE.

9 Mark for Review

Find $\frac{dy}{dx}$ if $y = \sec(\pi x^2)$.

(A) $\tan(\pi x^2)$

(B) $(2\pi x)\tan(\pi x^2)$

(C) $\sec(\pi x^2)\tan(\pi x^2)$

(D) $\sec(\pi x^2)\tan(\pi x^2)(2\pi x)$

---

10 Mark for Review

Given the curve $y = 5 - (x - 2)^{\frac{2}{3}}$, find $\frac{dy}{dx}$ at $x = 2$.

(A) $-\frac{2}{3}$

(B) $-\frac{2}{3\sqrt[3]{2}}$

(C) 5

(D) The derivative does not exist.

---

11 Mark for Review

$\int_0^{\frac{1}{2}} \frac{2}{\sqrt{1 - x^2}}\,dx =$

(A) $\frac{\pi}{3}$

(B) $-\frac{\pi}{3}$

(C) $\frac{2\pi}{3}$

(D) $-\frac{2\pi}{3}$

GO ON TO THE NEXT PAGE.

12 Mark for Review

Let f be the function $f(x)=\begin{cases} x^2-3bx+2;\ x\leq 2 \\ 2bx^2-8;\ x>2 \end{cases}$. For what value of b is f continuous at $x = 2$?

(A) 0

(B) $\frac{1}{4}$

(C) 1

(D) 4

---

13 Mark for Review

Let f be the function defined by $f(x) = xe^{-x}$. What is the absolute maximum value of f?

(A) 0

(B) $\frac{1}{e}$

(C) 1

(D) e

GO ON TO THE NEXT PAGE.

14 Mark for Review

Find $\frac{dy}{dx}$ at (1, 2) for $y^3 = xy - 2x^2 + 8$.

(A) $-\frac{11}{2}$

(B) $-\frac{2}{11}$

(C) $\frac{2}{11}$

(D) $\frac{11}{2}$

---

15 Mark for Review

$\lim_{x \to 0} \frac{x \cdot 2^x}{2^x - 1} =$

(A) ln 2

(B) 1

(C) 2

(D) $\frac{1}{\ln 2}$

---

16 Mark for Review

$\int x \sec^2(1 + x^2)\,dx =$

(A) $\frac{1}{2}\tan(1 + x^2) + C$

(B) $2\tan(1 + x^2) + C$

(C) $\frac{x}{2}\tan(1 + x^2) + C$

(D) $2x\tan(1 + x^2) + C$

GO ON TO THE NEXT PAGE.

17 Mark for Review

Find the equation of the tangent line to $9x^2 + 16y^2 = 52$ through $(2, -1)$.

(A) $-9x + 8y - 26 = 0$

(B) $9x - 8y - 26 = 0$

(C) $9x - 8y - 106 = 0$

(D) $8x + 9y - 17 = 0$

---

18 Mark for Review

Evaluate $\lim_{x \to \infty} \frac{5x^3 - 4x^2 + 10}{8x - 3x^2 + 9x^3}$.

(A) 0

(B) $\frac{5}{9}$

(C) $\frac{11}{14}$

(D) The limit does not exist.

---

19 Mark for Review

If $f(x) = 3^{\pi x}$, then $f'(x) =$

(A) $\frac{3^{\pi x}}{\ln 3}$

(B) $\frac{3^{\pi x}}{\pi}$

(C) $\pi(3^{\pi x} - 1)$

(D) $\pi \ln 3(3^{\pi x})$

GO ON TO THE NEXT PAGE.

20 Mark for Review

Find the value of k that makes the $f(x)$ continuous for all values of x if $f(x)=\begin{cases}5x^2-7kx+3; & x<3\\ x^3+4x^2+9k; & x\geq 3\end{cases}$.

(A) $-\frac{1}{2}$

(B) $\frac{1}{2}$

(C) $\frac{4}{5}$

(D) $\frac{5}{4}$

---

21 Mark for Review

If $y=\left(x^4+\sin x\right)^6$, then $\frac{dy}{dx}$ =

(A) $6\left(x^4+\sin x\right)^6\left(4x^3-\cos x\right)$

(B) $6\left(x^4+\sin x\right)^5\left(4x^3+\cos x\right)$

(C) $6\left(4x^3+\cos x\right)^5$

(D) $6\left(4x^3-\cos x\right)^5$

GO ON TO THE NEXT PAGE.

22 Mark for Review

Find the slope of the normal line to $y = x + \cos xy$ at (0, 1).

(A) -1

(B) 1

(C) 0

(D) Undefined

---

23 Mark for Review

$\int \frac{\csc^2(\sqrt{x})}{\sqrt{x}}\,dx =$

(A) $2\cot\sqrt{x} + C$

(B) $-2\cot\sqrt{x} + C$

(C) $\frac{\csc^2\sqrt{x}}{3\sqrt{x}} + C$

(D) $\frac{\csc^2\sqrt{x}}{6\sqrt{x}} + C$

GO ON TO THE NEXT PAGE.

24 Mark for Review

$\lim_{x \to 0} \frac{\tan^3(2x)}{x^3} =$

(A) -8

(B) 2

(C) 8

(D) The limit does not exist.

---

25 Mark for Review

Find the value of c that satisfies the Mean Value Theorem if $f(x) = 2x^3 - 4x + 1$ on the interval [1, 2].

(A) 1

(B) $\sqrt{2}$

(C) $\sqrt{\frac{7}{3}}$

(D) There is no value of c in the interval.

---

26 Mark for Review

If $y = \left(\frac{x^3 - 2}{2x^5 - 1}\right)^4$, find $\frac{dy}{dx}$ at $x = 1$.

(A) -52

(B) -28

(C) 13

(D) 52

GO ON TO THE NEXT PAGE.

27 Mark for Review

$\int x\sqrt{5-x}\,dx =$

(A) $-\frac{10}{3}(5-x)^{\frac{3}{2}}$

(B) $\frac{10}{3}\sqrt{\frac{5x^2}{2}-\frac{x^3}{3}}+C$

(C) $10(5-x)^{\frac{1}{2}}+\frac{2}{3}(5-x)^{\frac{3}{2}}+C$

(D) $-\frac{10}{3}(5-x)^{\frac{3}{2}}+\frac{2}{5}(5-x)^{\frac{5}{2}}+C$

---

28 Mark for Review

Given the differential equation $\frac{dy}{dt}=-2y$, where $y(0) = 100$, find $y(2)$.

(A) –200

(B) –4

(C) $\frac{100}{e^{16}}$

(D) $\frac{100}{e^{4}}$

GO ON TO THE NEXT PAGE.

29 Mark for Review

Given $y = x^4 - 6x^3 - 24x^2 + 10$, for which of the following values of x does the graph of y have a point of inflection?

(A) 1

(B) 2

(C) 3

(D) 4

---

30 Mark for Review

$\int_0^1 \tan x \, dx =$

(A) 0

(B) ln(cos(1))

(C) ln(sec(1))

(D) ln(sec(1)) – 1

END OF PART A, SECTION I

IF YOU FINISH BEFORE TIME IS CALLED, YOU MAY CHECK YOUR WORK ON PART A ONLY.

DO NOT GO ON TO PART B UNTIL YOU ARE TOLD TO DO SO.

AP CALCULUS AB

SECTION I, Part B

Time—45 Minutes

Number of questions—15

A graphing calculator is required for some questions on this part of the exam. You may use a handheld calculator or a Desmos graphing calculator application. **Make sure your calculator is in radian mode.**

Directions: Solve each of the following problems, using the available paper for scratchwork. After examining the choices, select the best of the choices given.

In this test:

1. The **exact** numerical value of the correct answer does not always appear among the choices given. When this happens, select from among the choices the number that best approximates the exact numerical value.

2. Unless otherwise specified, the domain of a function f is assumed to be the set of all real numbers x for which $f(x)$ is a real number. The inverse of a trigonometric function f may be indicated using the inverse function notation f^{-1} or with the prefix "arc" (e.g., $\sin^{-1} x = \arcsin x$).

---

31 Mark for Review

The graph of $y=\frac{1}{2}+\cos x$ has a zero on the interval $[0, \pi]$.

What is the slope of the tangent line to the graph at that point?

Ⓐ $-\frac{\sqrt{3}}{2}$

Ⓑ $-\frac{\sqrt{2}}{2}$

Ⓒ $\frac{\sqrt{2}}{2}$

Ⓓ $\frac{\sqrt{3}}{2}$

GO ON TO THE NEXT PAGE.

32 Mark for Review

$\frac{d}{dx}\int_0^{x^2} \sin^2 t \, dt =$

Ⓐ $x^2 \sin^2(x^2)$

Ⓑ $2x \sin^2(x^2)$

Ⓒ $\sin^2(x^2)$

Ⓓ $x^2 \cos^2(x^2)$

---

33 Mark for Review

Given $y = x^{\cos 4x}$, find $\frac{dy}{dx}$.

Ⓐ $\frac{dy}{dx} = x^{\cos 4x}\left[-\left(\frac{1}{x}\right)(4\sin 4x)\right]$

Ⓑ $\frac{dy}{dx} = x^{\cos 4x}\left[(\cos 4x)\left(\frac{1}{x}\right) - \ln x(4\sin 4x)\right]$

Ⓒ $\frac{dy}{dx} = (\cos 4x)x^{(\cos 4x)-1}$

Ⓓ $\frac{dy}{dx} = \left[(\cos 4x)x^{(\cos 4x)-1}\right](-4\sin 4x)$

GO ON TO THE NEXT PAGE.

34 Mark for Review

If f is defined by $f(x)=x+e^{-x^2}$ on the interval [0, 10], then f has a point of inflection at which of the following values of x?

(A) 0.379

(B) 0.5

(C) 0.707

(D) 0.947

---

35 Mark for Review

Estimate $\int_0^2 3e^x + 1\,dx$ using a Riemann sum with $n = 4$ right-hand rectangles.

(A) 26.357

(B) 33.546

(C) 52.713

(D) 56.713

---

36 Mark for Review

The volume generated by revolving about the x-axis the region above the curve $y = x^3$, below the line $y = 1$, and between $x = 0$ and $x = 1$ is

(A) $\frac{\pi}{42}$

(B) 0.143π

(C) 0.643π

(D) $\frac{6\pi}{7}$

GO ON TO THE NEXT PAGE.

37 Mark for Review

Given two numbers x and y, such that $x^2 + y = 48$, what is maximum value of the product of the two numbers, P?

(A) 4

(B) 32

(C) 128

(D) 512

---

38 Mark for Review

The function f is given by $f(x) = 2x^3 - 5$ on the interval [1, 5]. Which of the following is a possible value of c guaranteed by the Mean Value Theorem on the interval (1, 5)?

(A) 3.136

(B) 3.215

(C) 3.225

(D) 4.160

---

39 Mark for Review

Find two non-negative numbers x and y whose sum is 100 and for which x^2y is a maximum.

(A) $x = 50$ and $y = 50$

(B) $x = 33.333$ and $y = 66.667$

(C) $x = 100$ and $y = 0$

(D) $x = 66.667$ and $y = 33.333$

GO ON TO THE NEXT PAGE.

40 Mark for Review

An object is moving along a line with its velocity given by $v(t) = t^2 \sin t$, for time $t \geq 0$. If the object's position at time $t = 0$ is 4, what is its position at time $t = 2$?

(A) 0.469

(B) 2.469

(C) 4.469

(D) 6.469

---

41 Mark for Review

$\int \sin^4(\pi x)\cos(\pi x)\,dx =$

(A) $\dfrac{\sin^5(\pi x)}{5\pi} + C$

(B) $\dfrac{\sin^5(\pi x)}{2\pi} + C$

(C) $-\dfrac{\cos^5(\pi x)}{5\pi} + C$

(D) $-\dfrac{\cos^5(\pi x)}{2\pi} + C$

GO ON TO THE NEXT PAGE.

42 Mark for Review

A balloon is inflating at the rate $\frac{dV}{dt} = 300 - t\ln t$ cubic inches per second, where t is the number of seconds that the balloon has been inflating. If the initial volume of the balloon is 100 cubic inches, what is the volume of the balloon after it has been inflating for 8 seconds, to the nearest 10 cubic inches?

(A) 150 cubic inches

(B) 280 cubic inches

(C) 320 cubic inches

(D) 2450 cubic inches

---

43 Mark for Review

If $x^2 + 3x^2y + y^3 = 13$, find $\frac{dy}{dx}$ at (2, 1).

(A) $-\frac{16}{15}$

(B) $-\frac{4}{15}$

(C) $\frac{16}{15}$

(D) $\frac{19}{15}$

GO ON TO THE NEXT PAGE.

44 Mark for Review

Find the equation of the line tangent to $y = x\tan x$ at $x = 1$.

Ⓐ $y = 4.983x + 3.426$

Ⓑ $y = 4.983x - 3.426$

Ⓒ $y = 4.983x + 6.540$

Ⓓ $y = 4.983x - 6.540$

---

45 Mark for Review

If $f(x)$ is continuous and differentiable and

$$f(x) = \begin{cases} ax^4 + 5x; & x \le 2 \\ bx^2 - 3x; & x > 2 \end{cases}, \text{ then } b =$$

Ⓐ 0

Ⓑ 2

Ⓒ 6

Ⓓ There is no value of b.

STOP

END OF PART B, SECTION I

IF YOU FINISH BEFORE TIME IS CALLED, YOU MAY CHECK YOUR WORK ON PART B ONLY.

DO NOT GO ON TO SECTION II UNTIL YOU ARE TOLD TO DO SO.

AP CALCULUS AB

SECTION II

GENERAL INSTRUCTIONS

You may wish to look over the problems before starting to work on them, since it is not expected that everyone will be able to complete all parts of all problems. All problems are given equal weight, but the parts of a particular problem are not necessarily given equal weight.

A GRAPHING CALCULATOR IS REQUIRED FOR SOME PROBLEMS OR PARTS OF PROBLEMS ON THIS SECTION OF THE EXAMINATION.

- You may use the available paper for scratch work, but you must write your answers in the free-response booklet. In the free-response booklet, write your solution for each part of each question in the space provided for that part. For questions that have sub-parts, be sure to label those clearly in your solution. Use a pencil or pen with black or dark blue ink.
- You are permitted to use your calculator to solve an equation, find the derivative of a function at a point, or calculate the value of a definite integral. However, you must clearly indicate the setup of your problem, namely the equation, function, or integral you are using. If you use other built-in features or programs, you must show the mathematical steps necessary to produce your results.
- Show all of your work, even though a question may not explicitly remind you to do so. Clearly label any functions, graphs, tables, or other objects that you use. Justifications require you to give mathematical reasons and that you verify the needed conditions under which relevant theorems, properties, definitions, or tests applied. Your work will be scored on the correctness and completeness of your methods as well as your answers. Answers without supporting work will usually not receive any credit.
- Your work must be expressed in standard mathematical notation rather than calculator syntax. For example, $\int_1^5 x^2\,dx$ may not be written as fnInt (X^2, X, 1, 5).
- Unless otherwise specified, answers (numeric or algebraic) need not be simplified. If you use decimal approximations in your calculations, your work will be scored on accuracy. Unless otherwise specified, your final answers should be accurate to three places after the decimal point.
- Unless otherwise specified, the domain of a function f is assumed to be the set of all real numbers x for which $f(x)$ is a real number.

GO ON TO THE NEXT PAGE.

AP CALCULUS AB

SECTION II, PART A

Time—30 minutes

Number of problems—2

A graphing calculator is required for some questions on this part of the exam. You may use a handheld calculator or a Desmos graphing calculator application. **Make sure your calculator is in radian mode.**

During the timed portion for Part A, you may work only on the problems in Part A.

On Part A, you are permitted to use your calculator to solve an equation, find the derivative of a function at a point, or calculate the value of a definite integral. However, you must clearly indicate the setup of your problem, namely the equation, function, or integral you are using. If you use other built-in features or programs, you must show the mathematical steps necessary to produce your results.

1 Mark for Review

The temperature on New Year's Day in Hinterland was given by $T(H) = -A - B\cos\left(\frac{\pi H}{12}\right)$, where T is the temperature in degrees Fahrenheit and H is the number of hours from midnight ($0 \le H < 24$).

A. The initial temperature at midnight was $-15°$F, and the temperature at noon of New Year's Day was $5°$F. Find A and B.

B. Find the average temperature for the first 10 hours.

C. Use the Trapezoid Rule with four equal subdivisions to estimate $\int_6^8 T(H)\, dH$.

D. Find an expression for the rate that the temperature is changing with respect to H.

GO ON TO THE NEXT PAGE.

2 Mark for Review

Sea grass grows on a lake. The rate of growth of the grass is $\frac{dG}{dt} = kG$, where k is a constant.

A. Find an expression for G, the amount of grass in the lake (in tons), in terms of t, the number of years, if the amount of grass is 100 tons initially and 120 tons after one year.

B. In how many years will the amount of grass available be 300 tons?

C. If fish are now introduced into the lake and consume a consistent 80 tons of sea grass per year, how long will it take for the lake to be completely free of sea grass?

END OF PART A, SECTION II

IF YOU FINISH BEFORE TIME IS CALLED, YOU MAY CHECK YOUR WORK ON PART A ONLY.

DO NOT GO ON TO PART B UNTIL YOU ARE TOLD TO DO SO.

AP CALCULUS AB

SECTION II, PART B

Time—1 hour

Number of problems—4

A calculator is not allowed for this part of the exam.

During the timed portion for Part B, you may continue to work on the problems in Part A without the use of any calculator.

3 Mark for Review

The functions f and g are twice-differentiable and have the following table of values:

Table 1

x	$f(x)$	$f'(x)$	$g(x)$	$g'(x)$
1	3	4	–2	–4
2	2	3	4	–2
3	5	2	–1	3
4	–1	–6	–8	0

A. Let $h(x) = f(g(x))$. Find the equation of the tangent line to h at $x = 2$.

B. Let $j(x) = f(x)g(x)$. Find $j'(3)$.

C. Evaluate $\int_{1}^{4} 3f''(x)\,dx$.

GO ON TO THE NEXT PAGE.

4 Mark for Review

Water is being poured into a hemispherical bowl of radius 6 inches at the rate of 4 cubic inches per second.

Figure 1

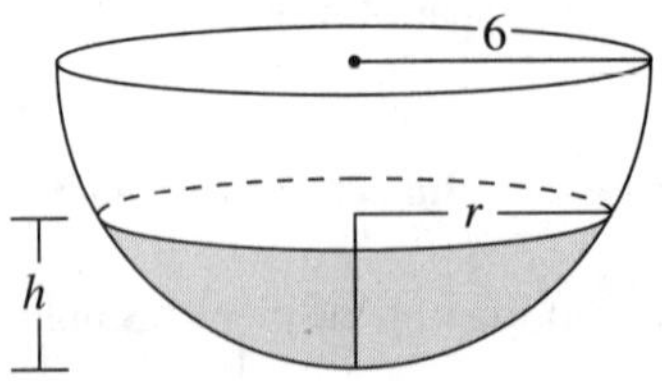

A. Given that the volume of the water in the spherical segment shown in Figure 1 is $V = \pi h^2\left(R - \frac{h}{3}\right)$, where R is the radius of the *sphere,* find the rate that the water level is rising when the water is 2 inches deep.

B. Find an expression for r, the radius of the *surface of the spherical segment* of water, in terms of h.

C. How fast is the circular area of the surface of the spherical segment of water growing (in square inches per second) when the water is 2 inches deep?

GO ON TO THE NEXT PAGE.

5 Mark for Review

Let R be the region in the first quadrant bounded by $y^2 = x$ and $x^2 = y$.

A. Find the area of region R.

B. Find the volume of the solid generated when R is revolved about the x-axis.

C. The section of a certain solid cut by any plane perpendicular to the x-axis is a circle with the endpoints of its diameter lying on the parabolas $y^2 = x$ and $x^2 = y$. Find the volume of the solid.

GO ON TO THE NEXT PAGE.

6 Mark for Review

For time $t \geq 0$, a particle moves along the x-axis. The velocity of the particle at time t is given by $v(t) = 1 - 2\cos\left(\frac{\pi}{3}t\right)$. The particle's position at time $t = 0$ is $x(0) = 8$.

A. Is the particle speeding up or slowing down at time $t = \frac{1}{2}$? Justify your answer.

B. When does the particle change direction in the interval $0 \leq t \leq 2$? Justify your answer.

C. What is the particle's position at time $t = 2$?

D. What is the total distance traveled from time $t = 0$ to time $t = 2$?

STOP

END OF EXAM

Practice Test 2: Answers and Explanations

ANSWER KEY

Section I

1. B	11. A	21. B	31. A	41. A
2. A	12. C	22. A	32. B	42. D
3. C	13. B	23. B	33. B	43. A
4. D	14. B	24. C	34. C	44. B
5. C	15. D	25. C	35. A	45. C
6. A	16. A	26. A	36. D	
7. A	17. B	27. D	37. C	
8. B	18. B	28. D	38. B	
9. D	19. D	29. D	39. D	
10. D	20. A	30. C	40. D	

ANSWERS AND EXPLANATIONS TO SECTION I

1. **B** First, take the derivative.

$$g'(x) = \frac{1}{32}\left(4x^3\right) - 5(2x) = \frac{x^3}{8} - 10x$$

Now, plug in 4 for x.

$$\frac{(4)^3}{8} - 10(4) = 8 - 40 = -32$$

2. **A** If you take the limit as x goes to 0, you get an indeterminate form $\frac{0}{0}$, so use L'Hospital's Rule. Take the derivative of the numerator and the denominator to get $\lim_{x\to 0}\frac{8x^2}{\cos x - 1} = \lim_{x\to 0}\frac{16x}{\left(-\sin x\right)}$. When you take the limit, you again get an indeterminate form $\frac{0}{0}$, so use L'Hospital's Rule a second time. Take the derivative of the numerator and the denominator to get $\lim_{x\to 0}\frac{16x}{\left(-\sin x\right)} = \lim_{x\to 0}\frac{16}{\left(-\cos x\right)}$. Now, when you take the limit, you get $\lim_{x\to 0}\frac{16}{\left(-\cos x\right)} = -16$.

3. **C** Notice that if you plug 5 into the expressions in the numerator and the denominator, you get $\frac{0}{0}$, which is undefined. Before you give up, you need to see if you can simplify the limit so that it can be evaluated. If you factor the expression in the numerator, you get $\frac{(x+5)(x-5)}{(x-5)}$, which can be simplified to $x + 5$. Now, if you take the limit (by plugging in 5 for x), you get 10.

4. **D** You need to use the Quotient Rule, which is

Given $f(x) = \frac{g(x)}{h(x)}$, then $f'(x) = \frac{h(x)g'(x) - g(x)h'(x)}{\left[h(x)\right]^2}$.

Here you have

$$f'(x) = \frac{(x^3+7)(5x^4-1) - (x^5 - x + 2)(3x^2)}{(x^3+7)^2}$$

5. **C** There are two ways you could evaluate this limit. First, you could recognize that this limit is in the form of the definition of the derivative: $\lim_{h\to 0}\frac{f(x+h)-f(x)}{h}$. Then the limit is the derivative of $f(x) = \tan x$ at $x = \frac{\pi}{4}$. The derivative of $f(x) = \tan x$ is $f'(x) = \sec^2 x$, and at $x = \frac{\pi}{4}$, you get $f'\left(\frac{\pi}{4}\right) = \sec^2\left(\frac{\pi}{4}\right) = 2$. Second, you have a limit of the indeterminate form $\frac{0}{0}$, so you can use L'Hospital's Rule to find the limit. Take the derivative of the numerator and the denominator: $\lim_{h\to 0}\frac{\tan\left(\frac{\pi}{4}+h\right)-1}{h} = \lim_{h\to 0}\frac{\sec^2\left(\frac{\pi}{4}+h\right)}{1}$. Now you take the limit to get $\sec^2\left(\frac{\pi}{4}\right) = 2$.

6. **A** First, rewrite the integral as $\int x\left(\sqrt{3}\right)x^{\frac{1}{2}}\,dx$.

   Now, you can simplify the integral to $\sqrt{3}\int x^{\frac{3}{2}}\,dx$.

   Next, use the Power Rule for integrals, which is $\int x^n\,dx = \frac{x^{n+1}}{n+1} + C$.

   Then, you get $\sqrt{3}\int x^{\frac{3}{2}}\,dx = \sqrt{3}\,\frac{x^{\frac{5}{2}}}{\frac{5}{2}} + C = \frac{2\sqrt{3}}{5}x^{\frac{5}{2}} + C$.

7. **A** The simplest thing to do here is to find $\lim_{x\to 1}$ of both pieces of the function and set them equal to each other. You can do this by plugging $x = 1$ into both pieces: $1 - 3k + 2 = 5 - k$. If you solve for k, you get $k = -1$.

8. **B** You have to find the areas of the regions between the curve and the x-axis on the interval from $x = 0$ to $x = 7$. The regions are all triangles, so you can easily find the areas and add them up. Note that the region from $x = 2$ to $x = 6$ is negative. You get $\frac{1}{2}(2)(2)-\frac{1}{2}(4)(4)+\frac{1}{2}(1)(2) = -5$.

9. **D** Use the Chain Rule: $\frac{dy}{dx} = \sec(\pi x^2)\tan(\pi x^2)(2\pi x)$.

10. **D** First, take the derivative: $\frac{dy}{dx} = -\frac{2}{3}(x-2)^{-\frac{1}{3}}$, which can be rewritten as $\frac{dy}{dx} = -\frac{2}{3\sqrt[3]{x-2}}$. If you plug in $x = 2$, you get zero in the denominator, so the derivative does not exist at $x = 2$.

11. **A** This integral is of the form $\int \frac{dx}{\sqrt{a^2 - x^2}} = \sin^{-1}\left(\frac{x}{a}\right) + C$, where $a = 1$.

Thus, you get

$$\int_0^{\frac{1}{2}} \frac{2dx}{\sqrt{1 - x^2}} = 2\sin^{-1}(x)\Big|_0^{\frac{1}{2}} = 2\left[\sin^{-1}\left(\frac{1}{2}\right) - \sin^{-1}(0)\right] = 2\left(\frac{\pi}{6} - 0\right) = \frac{\pi}{3}$$

12. **C** In order for f to be continuous, three conditions must be met. First, $f(2)$ must exist: $f(2) = 4 - 6b + 2 = 6 - 6b$. Next, $\lim_{x\to 2} f(x)$ must exist. The limit from the left is $\lim_{x\to 2^-} f(x) = 6 - 6b$, and the limit from the right is $\lim_{x\to 2^+} f(x) = 8b - 8$. You set the two limits equal to each other and solve for b: $6 - 6b = 8b - 8$. So $b = 1$. The final condition is that $\lim_{x\to 2} f(x) = f(2)$. At $b = 1$, $\lim_{x\to 2} f(x) = f(2) = 0$. Therefore, b equals 1, and the answer is (C), 1.

13. **B** First, take the derivative of f using the Product Rule: $f'(x) = e^{-x} - xe^{-x}$. Next, set it equal to zero and solve: $e^{-x} - xe^{-x} = 0$. Factor: $e^{-x}(1 - x) = 0$. Because e^{-x} is always positive, the only solution is $x = 1$. Next, you need to verify that $x = 1$ is a maximum. Plug in values less than and greater than 1 to make sure that the derivative changes sign there: $f'(0) = e^0 - (0)e^0 > 0$ and $f'(2) = e^{-2} - (2)e^{-2} < 0$. So, $x = 1$ is the x-coordinate of the maximum. The maximum value occurs at the y-coordinate, which you find by plugging $x = 1$ into the original equation: $f(1) = (1)e^{-1} = \frac{1}{e}$.

14. **B** You can use implicit differentiation to find $\frac{dy}{dx}$. First, differentiate with respect to x: $3y^2\frac{dy}{dx} = x\frac{dy}{dx} + y - 4x$. Next, plug in (1, 2) for x and y: $3(2)^2\frac{dy}{dx} = (1)\frac{dy}{dx} + (2) - 4(1)$. Simplify: $12\frac{dy}{dx} = \frac{dy}{dx} - 2$. Solve for $\frac{dy}{dx}$ on the left and the terms without $\frac{dy}{dx}$ on the right: $\frac{dy}{dx} = -\frac{2}{11}$.

15. **D** If you take the limit as x goes to 0, you get an indeterminate form $\frac{0}{0}$, so use L'Hospital's Rule. Take the derivative of the numerator and the denominator to get $\lim_{x\to 0}\frac{x\cdot 2^x}{2^x - 1} = \lim_{x\to 0}\frac{x\cdot 2^x \ln 2 + 2^x}{2^x \ln 2}$.

Now, when you take the limit, you get $\lim_{x\to 0}\frac{x\cdot 2^x \ln 2 + 2^x}{2^x \ln 2} = \frac{1}{\ln 2}$.

16. **A** You can evaluate this integral using u-substitution. Let $u = 1 + x^2$ and $du = 2x\,dx$, so $\frac{1}{2}du = x\,dx$. Substitute into the integrand: $\int x\sec^2(1 + x^2)dx = \frac{1}{2}\int \sec^2 u\, du$. Integrate: $\frac{1}{2}\int \sec^2 u\, du = \frac{1}{2}\tan u + C$ and substitute back: $\frac{1}{2}\tan u + C = \frac{1}{2}\tan(1 + x^2) + C$.

17. **B** First, you need to find $\frac{dy}{dx}$. It's simplest to find it implicitly.

$$18x + 32y\frac{dy}{dx} = 0$$

Now, solve for $\frac{dy}{dx}$.

$$\frac{dy}{dx} = -\frac{18x}{32y} = -\frac{9x}{16y}$$

Next, plug in $x = 2$ and $y = -1$ to get the slope of the tangent line at the point.

$$\frac{dy}{dx} = \frac{-18}{-16} = \frac{9}{8}$$

Now, use the point-slope formula to find the equation of the tangent line.

$$(y + 1) = \frac{9}{8}(x - 2)$$

If you multiply through by 8, you get $8y + 8 = 9x - 18$ or $9x - 8y - 26 = 0$.

18. **B** To evaluate the limit to infinity of a polynomial divided by another polynomial, look at the term with the highest power in the numerator (call it ax^m) and the term with the highest power in the denominator (call it bx^n). This gives you $\frac{ax^m}{bx^n}$. The rule is very simple. If $m < n$, then the limit is 0. If $m > n$, then the limit is $\pm\infty$. If $m = n$, then the limit is $\frac{a}{b}$. Here, you have $\lim_{x\to\infty}\frac{5x^3 - 4x^2 + 10}{8x - 3x^2 + 9x^3}$. Looking at the highest power terms, you get $\lim_{x\to\infty}\frac{5x^3}{9x^3}$. The highest power of the numerator and the denominator is the same (it's 3), so the limit is $\frac{5}{9}$. The answer is (B).

19. **D** The derivative of an expression of the form a^u, where u is a function of x, is

$$\frac{d}{dx} a^u = a^u (\ln a) \frac{du}{dx}$$

Here you get

$$\frac{d}{dx} 3^{\pi x} = 3^{\pi x} (\ln 3)\pi$$

20. **A** Note that the two pieces of the function are both polynomials, which are continuous for all values of x in their domains, so the only potential problem is where the two pieces of the function meet. In order for $f(x)$ to be continuous, the limit as x approaches 3 from the left has to equal the limit from the right. All you have to do then is plug in 3 for both pieces of the function and set them equal to each other. You get $f(x)=\begin{cases}5(3)^2-7(3)k+3; & x<3\\(3)^3+4(3)^2+9k; & x\geq 3\end{cases}$ and $45 - 21k + 3 = 27 + 36 + 9k$. This reduces to $30k = -15$, so $k=-\frac{1}{2}$. The answer is (A).

21. **B** Use the Chain Rule to find the derivative: $\frac{dy}{dx}=6\left(x^4+\sin x\right)^5\left(4x^3+\cos x\right)$.

22. **A** First, find $\frac{dy}{dx}$ using implicit differentiation.

$$\frac{dy}{dx} = 1 - \left(x\frac{dy}{dx} + y\right)\sin xy$$

Rather than simplifying this, simply plug in (0, 1) to find $\frac{dy}{dx}$.

You get $\frac{dy}{dx} = 1$.

This means that the slope of the tangent line at (0, 1) is 1, so the slope of the normal line at (0, 1) is the negative reciprocal, which is –1.

23. **B** You can evaluate this integral using u-substitution. Let $u=\sqrt{x}$ and $du=\frac{1}{2\sqrt{x}}dx$, so $2du=\frac{1}{\sqrt{x}}dx$. Substituting into the integrand, you get $2\int\csc^2 u\,du=-2\cot u+C$. Now you substitute back: $-2\cot u+C=-2\cot\sqrt{x}+C$.

24. **C** You will need to use the fact that $\lim_{x \to 0} \frac{\sin x}{x} = 1$ to find the limit.

First, rewrite the limit as

$$\lim_{x \to 0} \frac{\sin^3(2x)}{x^3 \cos^3(2x)}$$

Next, break the fraction into

$$\lim_{x \to 0} \left(\frac{\sin^3(2x)}{x^3} \quad \frac{1}{\cos^3(2x)} \right)$$

Now, if you multiply the top and bottom of the first fraction by 8, you get

$$\lim_{x \to 0} \frac{8\sin^3(2x)}{(2x)^3} \quad \frac{1}{\cos^3(2x)}$$

Now, you can take the limit, which gives you 8(1)(1) = 8.

25. **C** The Mean Value Theorem states that there is a value c in the interval (1, 2) such that $\frac{f(2)-f(1)}{2-1} = f'(c)$. First, let's evaluate $\frac{f(2)-f(1)}{2-1}$: $\frac{f(2)-f(1)}{2-1} = \frac{9-(-1)}{1} = 10$. Next, find $f'(c)$. You get $f'(x) = 6x^2 - 4$, so $f'(c) = 6c^2 - 4$. Setting them equal to each other, you get $6c^2 - 4 = 10$, so $c = \pm\sqrt{\frac{7}{3}}$. The negative answer is not in the interval (1, 2). Therefore, $c = \sqrt{\frac{7}{3}}$. The answer is (C).

26. **A** Use the Chain Rule and the Quotient Rule.

$$\frac{dy}{dx} = 4\left(\frac{x^3 - 2}{2x^5 - 1}\right)^3 \left[\frac{(2x^5 - 1)(3x^2) - (x^3 - 2)(10x^4)}{(2x^5 - 1)^2}\right]$$

If you plug in 1 for x, you get

$$\frac{dy}{dx} = 4(-1)^3\left[\frac{3 + 10}{1^2}\right] = -52$$

27. **D** You can evaluate this integral using u-substitution. Let $u = 5 - x$ and $5 - u = x$. Then $-du = dx$.

Substituting, you get

$$-\int (5-u)u^{\frac{1}{2}}\,du$$

The integral can be rewritten as

$$-\int \left(5u^{\frac{1}{2}} - u^{\frac{3}{2}}\right) du$$

Evaluating the integral, you get

$$-5\frac{u^{\frac{3}{2}}}{\frac{3}{2}} + \frac{u^{\frac{5}{2}}}{\frac{5}{2}} + C$$

This can be simplified to

$$-\frac{10}{3}u^{\frac{3}{2}} + \frac{2}{5}u^{\frac{5}{2}} + C$$

Finally, substituting back, you get

$$-\frac{10}{3}(5-x)^{\frac{3}{2}} + \frac{2}{5}(5-x)^{\frac{5}{2}} + C$$

28. **D** First, separate the variables: $\frac{dy}{y} = -2dt$. Integrate both sides: $\int \frac{dy}{y} = -2\int dt$, and $\ln y = -2t + C$. You can isolate y by exponentiating both sides: $y = e^{-2t+C} = Ae^{-2t}$. Now, plug in the initial condition to solve for the constant: $100 = Ae^{0} = A$. You get $y = 100e^{-2t}$. Therefore, $y(2) = 100e^{-2(2)} = 100e^{-4} = \frac{100}{e^{4}}$.

29. **D** In order to find possible points of inflection, you first need to find the second derivative:

$$\frac{dy}{dx} = 4x^3 - 18x^2 - 48x$$

$$\frac{d^2y}{dx^2} = 12x^2 - 36x - 48$$

Next, set the second derivative equal to zero and solve: $12x^2 - 36x - 48 = 0$. Divide through by 12: $x^2 - 3x - 4 = 0$, which factors to $(x-4)(x+1) = 0$. You get $x = 4$ and $x = -1$. In order to be sure that $x = 4$ is a point of inflection, you need to check that the second derivative changes sign there. Plug in values less than and greater than 4 and check the sign of the second derivative: $12(3)^2 - 36(3) - 48 < 0$ and $12(5)^2 - 36(5) - 48 > 0$, so $x = 4$ is a point of inflection.

30. **C** First, rewrite the integral as $\int_0^1 \frac{\sin x}{\cos x}\,dx$.

Now, you can use *u*-substitution to evaluate the integral. Let $u = \cos x$. Then $du = -\sin x$. You can also change the limits of integration. The lower limit becomes $\cos 0 = 1$ and the upper limit becomes $\cos 1$, which you leave alone. Now you perform the substitution, and you get

$$-\int_1^{\cos 1} \frac{du}{u}$$

Evaluating the integral, you get $-\ln u\Big|_1^{\cos 1} = -\ln(\cos 1) + \ln 1 = -\ln(\cos 1)$. This log is also equal to $\ln(\sec 1)$.

31. **A** You can find the slope of the tangent line by taking the derivative: $\frac{dy}{dx} = -\sin x$. Next, you need to find where the graph has a zero: $\frac{1}{2} + \cos x = 0$. Therefore, $\cos x = -\frac{1}{2}$ in the interval at $x = \frac{2\pi}{3}$.

Plug this into the derivative: $\frac{dy}{dx} = -\sin\left(\frac{2\pi}{3}\right) = -\frac{\sqrt{3}}{2}$.

32. **B** The Second Fundamental Theorem of Calculus tells you how to find the derivative of an integral. It says that $\frac{d}{dx}\int_c^u f(t)\,dt = f(u)\frac{du}{dx}$, where c is a constant and u is a function of x.

Here you can use the theorem to get

$$\frac{d}{dx}\int_0^{x^2} \sin^2 t\,dt = (\sin^2(x^2))(2x), \text{ or } 2x\sin^2(x^2)$$

33. **B** Here you have a function raised to another function. The best way to find the derivative is with logarithmic differentiation. First, take the log of both sides: $\ln y = \ln x^{\cos 4x}$. Next, use the log laws to rewrite the right-hand side: $\ln y = (\cos 4x)\ln x$. Now take the derivative of both sides: $\frac{1}{y}\frac{dy}{dx} = (\cos 4x)\left(\frac{1}{x}\right) - \ln x(4\sin 4x)$. Multiply both sides by y: $\frac{dy}{dx} = y\left[(\cos 4x)\left(\frac{1}{x}\right) - \ln x(4\sin 4x)\right]$. Remember that $y = x^{\cos 4x}$, so the derivative is $\frac{dy}{dx} = x^{\cos 4x}\left[(\cos 4x)\left(\frac{1}{x}\right) - \ln x(4\sin 4x)\right]$.

34. **C** You need to take the second derivative of f: $f'(x) = 1 - 2xe^{-x^2}$ and $f''(x) = -2x\left(-2xe^{-x^2}\right) - 2e^{-x^2} = 4x^2e^{-x^2} - 2e^{-x^2} = e^{-x^2}\left(4x^2 - 2\right)$. Now you set the second derivative equal to zero. Note that e^{-x^2} is always positive, so you get $4x^2 - 2 = 0$ and $x = \pm 0.707$. Only 0.707 is a choice, so there is no need to go further.

35. **A** The width of each rectangle of the Riemann sum will be $\frac{2-0}{4} = \frac{1}{2}$, so the sum is $\frac{1}{2}\left[f\left(\frac{1}{2}\right) + f(1) + f\left(\frac{3}{2}\right) + f(2)\right]$. Evaluate this with a calculator and you get 26.357.

36. **D** First, make a quick sketch of the region.

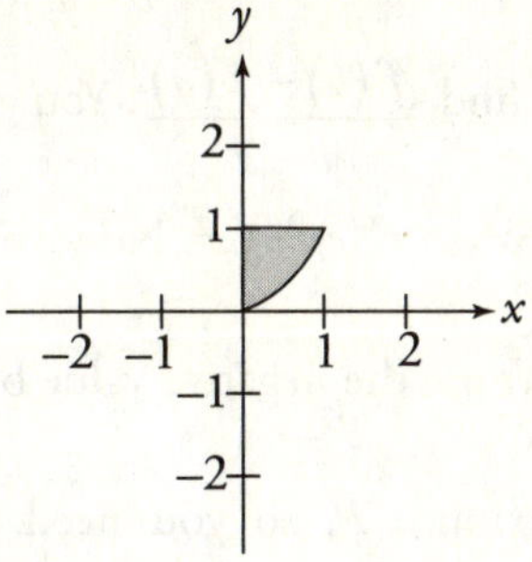

You can find the volume by taking a vertical slice of the region. The formula for the volume of a solid of revolution around the x-axis, using a vertical slice bounded from above by the curve $f(x)$ and from below by $g(x)$, on the interval $[a, b]$, is

$$\pi\int_a^b \left[f(x)^2 - g(x)^2\right] dx$$

Here you get

$$\pi\int_0^1 [(1)^2 - (x^3)^2]\, dx$$

Now you have to evaluate the integral. First, expand the integrand to get

$$\pi\int_0^1 (1 - x^6)\, dx$$

Next integrate to get

$$\pi\left(x - \frac{x^7}{7}\right)\Bigg|_0^1 = \pi\left(1 - \frac{1}{7}\right) = \frac{6\pi}{7}$$

37. **C** To maximize $P = xy$, subject to the constraint $x^2 + y = 48$, isolate y: $y = 48 - x^2$. Now plug this into P for y: $P = xy = x(48 - x^2) = 48x - x^3$. Differentiate: $\frac{dP}{dx} = 48 - 3x^2$. Set the derivative equal to zero and solve for x to get $48 - 3x^2 = 0$ and $x = \pm 4$. Plug each value of x back into the equation $y = 48 - x^2$. If $x = 4$, then $y = 32$. If $x = -4$, then $y = 32$. This gives you either $P = 128$ or $P = -128$. Therefore, the maximum product is $P = 128$. The answer is (C).

38. **B** The Mean Value Theorem states that if a function is continuous on the interval $[a, b]$, then there exists a value c in the interval (a, b) such that $\frac{f(b) - f(a)}{b - a} = f'(c)$. Here, f is a polynomial, so it is continuous everywhere. First, find $\frac{f(b) - f(a)}{b - a}$. You get $\frac{f(5) - f(1)}{5 - 1} = \frac{245 - (-3)}{4} = 62$. Next, you need $f'(c)$. You get $f'(x) = 6x^2$ so $f'(c) = 6c^2$. Now you set them equal to each other and solve: $6c^2 = 62$, so $c = \pm 3.215$. You don't use the negative value because it is not in the interval.

39. **D** Set $P = x^2y$. You want to maximize P, so you need to eliminate one of the variables. You are also given that $x + y = 100$, so you can solve this for y and substitute: $y = 100 - x$, so $P = x^2(100 - x) = 100x^2 - x^3$.

Now you can take the derivative.

$$\frac{dP}{dx} = 200x - 3x^2$$

Set the derivative equal to zero and solve for x.

$$200x - 3x^2 = 0$$
$$x(200 - 3x) = 0$$
$$x = 0 \text{ or } x = \frac{200}{3} \approx 66.667$$

Now you can use the second derivative to find the maximum: $\frac{d^2P}{dx^2} = 200 - 6x$. If you plug in $x = 66.667$, the second derivative is negative, so P is a maximum at $x = 66.667$. Solving for y, you get $y \approx 33.333$.

40. **D** First, you need to see if the velocity changes sign in the interval from $t = 0$ to $t = 2$. If you graph the velocity equation on your calculator, you will see that all of the y-values are positive, so the object will always be moving to the right. Next, see how far the object travels in those 2 seconds. You can find this by evaluating $\int_0^2 t^2 \sin t \, dt$. If you plug this into your calculator, you get 2.469. (If you are unsure how to evaluate the integral on your calculator, check the online Appendix for some tips on using the Desmos calculator.) Because the object's initial position is 4, the position at time $t = 2$ will be 6.469.

41. **A** You can evaluate this integral using u-substitution. Let $u = \sin(\pi x)$ and $du = \pi\cos(\pi x)dx$, so $\frac{1}{\pi}du = \cos(\pi x)dx$. Substituting into the integrand, you get $\int \sin^4(\pi x)\cos(\pi x)\,dx = \frac{1}{\pi}\int u^4 du$. Evaluate: $\frac{1}{\pi}\int u^4 du = \frac{1}{\pi}\frac{u^5}{5} + C = \frac{u^5}{5\pi} + C$. Now substitute back: $\frac{u^5}{5\pi} + C = \frac{\sin^5(\pi x)}{5\pi} + C$.

42. **D** If you want to find the amount that the balloon has inflated in the first 8 seconds, you need to evaluate $\int_0^8 300 - t\ln t \, dt$. Use your calculator to get 2,349.458. (If you are unsure how to evaluate the integral on your calculator, check the online Appendix for some tips on using a Desmos calculator.) Add the original 100 cubic inches to get 2,449.458, which rounds to 2,450 cubic inches.

43. **A** You can find $\frac{dy}{dx}$ using implicit differentiation. You get $2x + 3x^2\frac{dy}{dx} + 6xy + 3y^2\frac{dy}{dx} = 0$. Plug in the point (2, 1) and solve for $\frac{dy}{dx}$:

$$2(2) + 3(2)^2\frac{dy}{dx} + 6(2)(1) + 3(1)^2\frac{dy}{dx} = 0$$

$$4 + 12\frac{dy}{dx} + 12 + 3\frac{dy}{dx} = 0$$

$$\frac{dy}{dx} = -\frac{16}{15}$$

44. **B** First, you will need to find the y-coordinate that corresponds to $x = 1$: $y = \tan 1 \approx 1.557$. Next, you need to find the derivative: $\frac{dy}{dx} = \tan x + x\sec^2 x$. You plug in $x = 1$ to get the slope of the tangent line: $\frac{dy}{dx} = \tan 1 + \sec^2 1 \approx 4.983$. Now you can plug this into the equation of a line: $y - 1.557 = 4.983(x - 1)$ or $y = 4.983x - 3.426$.

45. **C** In order to solve for b, you need $f(x)$ to be continuous at $x = 2$. If you plug $x = 2$ into both pieces of this piecewise function, you get

$$f(x) = \begin{cases} 16a + 10; & x \le 2 \\ 4b - 6; & x > 2 \end{cases}$$

So, you need $16a + 10 = 4b - 6$. Now, if you take the derivative of both pieces of this function and plug in $x = 2$, you get

$$f'(x) = \begin{cases} 32a + 5; & x \le 2 \\ 4b - 3; & x > 2 \end{cases}, \text{ so you need } 32a + 5 = 4b - 3$$

Solving the simultaneous equations, you get $a = \frac{1}{2}$ and $b = 6$.

ANSWERS AND EXPLANATIONS TO SECTION II

1. The temperature on New Year's Day in Hinterland was given by $T(H) = -A - B\cos\left(\frac{\pi H}{12}\right)$, where T is the temperature in degrees Fahrenheit and H is the number of hours from midnight $(0 \le H < 24)$.

A. The initial temperature at midnight was –15°F, and the temperature at noon of New Year's Day was 5°F. Find A and B.

Simply plug in the temperature, –15, for T and the time, midnight ($H = 0$), for H into the equation. You get $-15 = -A - B\cos 0$, which simplifies to $-15 = -A - B$.

Now plug the temperature, 5, for T and the time, noon ($H = 12$), for H into the equation. You get $5 = -A - B\cos(\pi)$, which simplifies to $5 = -A + B$.

Now you can solve the pair of simultaneous equations for A and B, and you get $A = 5$°F and $B = 10$°F.

B. Find the average temperature for the first 10 hours.

In order to find the average value, you use the Mean Value Theorem for Integrals, which says that the average value of $f(x)$ on the interval $[a, b]$ is

$$\frac{1}{b-a}\int_a^b f(x)\,dx$$

Here, you have $\frac{1}{10-0}\int_0^{10}\left(-5-10\cos\left(\frac{\pi H}{12}\right)\right)dH$.

Evaluating the integral, you get

$$\frac{1}{10}\left[\left(-5H-\frac{120}{\pi}\sin\left(\frac{\pi H}{12}\right)\right)\right]_0^{10}=\frac{1}{10}\left[\left(-50-\frac{120}{\pi}\sin\left(\frac{5\pi}{6}\right)\right)\right]=\frac{1}{10}\left[\left(-50-\frac{60}{\pi}\right)\right]\approx -6.910°\text{F}$$

C. Use the Trapezoid Rule with four equal subdivisions to estimate $\int_6^8 T(H)\,dH$.

The Trapezoid Rule enables you to approximate the area under a curve with a fair degree of accuracy. The rule says that the area between the x-axis and the curve $y = f(x)$ on the interval $[a, b]$, with n trapezoids, is

$$\frac{1}{2}\frac{b-a}{n}\left[y_0+2y_1+2y_2+2y_3+\ldots+2y_{n-1}+y_n\right]$$

Using the rule here, with $n = 4$, $a = 6$, and $b = 8$, you get

$$\frac{1}{2}\cdot\frac{1}{2}\left[\left(-5-10\cos\frac{6\pi}{12}\right)+2\left(-5-10\cos\frac{13\pi}{24}\right)+2\left(-5-10\cos\frac{7\pi}{12}\right)+2\left(-5-10\cos\frac{15\pi}{24}\right)+\left(-5-10\cos\frac{8\pi}{12}\right)\right]$$

This is approximately –4.890 degrees Fahrenheit-hours.

D. Find an expression for the rate that the temperature is changing with respect to H.

You simply take the derivative with respect to H.

$$\frac{dT}{dH}=-10\left(\frac{\pi}{12}\right)\left(-\sin\frac{\pi H}{12}\right)=\frac{5\pi}{6}\sin\frac{\pi H}{12}$$

2. Sea grass grows on a lake. The rate of growth of the grass is $\frac{dG}{dt} = kG$, where k is a constant.

A. Find an expression for G, the amount of grass in the lake (in tons), in terms of t, the number of years, if the amount of grass is 100 tons initially and 120 tons after one year.

You solve this differential equation using separation of variables.

First, move the G to the left side and the dt to the right side to get $\frac{dG}{G} = k\,dt$.

Now, integrate both sides.

$$\int \frac{dG}{G} = k\int dt$$

$$\ln G = kt + C$$

Next, solve for G by exponentiating both sides to the base e. You get $G = e^{kt+C}$.

Using the rules of exponents, you can rewrite this as $G = e^{kt}e^{C}$. Finally, because e^{C} is a constant, you can rewrite the equation as $G = Ce^{kt}$.

Now you can use the initial condition that $G = 100$ at time $t = 0$ to solve for C.

$$100 = Ce^{0} = C(1) = C$$

This gives you $G = 100e^{kt}$.

Next, use the condition that $G = 120$ at time $t = 1$ to solve for k.

$$120 = 100e^{k}$$

$$1.2 = e^{k}$$

$$\ln 1.2 = k \approx 0.1823$$

This gives you $G = 100e^{0.1823t}$.

B. In how many years will the amount of grass available be 300 tons?

All you need to do is set G equal to 300 and solve for t.

$$300 = 100e^{0.1823t}$$

$$3 = e^{0.1823t}$$

$$\ln 3 = 0.1823t$$

$$t \approx 6.026 \text{ years}$$

C. If fish are now introduced into the lake and consume a consistent 80 tons of sea grass per year, how long will it take for the lake to be completely free of sea grass?

Now you have to account for the fish's consumption of the sea grass. So you have to evaluate the differential equation $\frac{dG}{dt} = kG - 80$.

First, separate the variables to get

$$\frac{dG}{kG - 80} = dt$$

Now, integrate both sides.

$$\int \frac{dG}{kG - 80} = \int dt \text{ or } \int \frac{dG}{G - \frac{80}{k}} = k\int dt$$

$$\ln\left(G - \frac{80}{k}\right) = kt + C$$

Next, exponentiate both sides to the base *e*. You get

$$G - \frac{80}{k} = Ce^{kt}$$

Solving for *G*, you get

$$G = \left(G_0 - \frac{80}{k}\right)e^{kt} + \frac{80}{k}$$

Set $G = 0$ to get

$$0 = \left(G_0 - \frac{80}{k}\right)e^{kt} + \frac{80}{k}$$

Now, set $G_0 = 300$, and solve for e^{kt}.

$$e^{kt} = \frac{\frac{-80}{k}}{300 - \frac{80}{k}} = \frac{80}{80 - 300k}$$

Take the log of both sides.

$$kt = \ln\left(\frac{80}{80 - 300k}\right)$$

$$\text{and } t = \frac{1}{k}\ln\left(\frac{80}{80 - 300k}\right)$$

Plug in the value for *k* that you got in part A above to get $t \approx 6.313$ years.

3. The functions f and g are twice-differentiable and have the following table of values:

x	$f(x)$	$f'(x)$	$g(x)$	$g'(x)$
1	3	4	–2	–4
2	2	3	4	–2
3	5	2	–1	3
4	–1	–6	–8	0

A. Let $h(x) = f(g(x))$. Find the equation of the tangent line to h at $x = 2$.

You can find the derivative of $h(x)$ using the Chain Rule: $h'(x) = f'(g(x))g'(x)$. At $x = 2$, you get $h'(2) = f'(g(2))g'(2) = f'(4)g'(2) = (-6)(-2) = 12$.

And, at $x = 2$, $h(2) = f(g(2)) = f(4) = -1$.

Therefore, the equation of the tangent line is $y + 1 = 12(x - 2)$.

B. Let $j(x) = f(x)g(x)$. Find $j'(3)$.

You can find the derivative of $j(x)$ using the Product Rule: $j'(x) = f(x)g'(x) + f'(x)g(x)$. At $x = 3$, you get $j'(3) = f(3)g'(3) + f'(3)g(3) = (5)(3) + (2)(-1) = 13$.

C. Evaluate $\int_1^4 3f''(x)\,dx$.

You simply evaluate $\int_1^4 3f''(x)\,dx = 3f'(x)\Big|_1^4 = 3f'(4) - 3f'(1) = 3(-6) - 3(4) = -30$.

4. Water is being poured into a hemispherical bowl of radius 6 inches at the rate of 4 cubic inches per second.

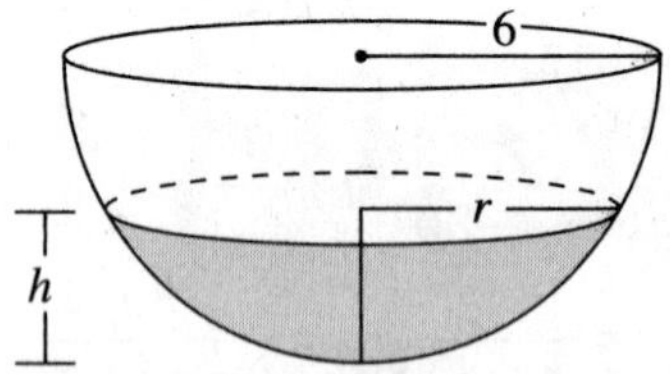

A. Given that the volume of the water in the spherical segment shown above is $V = \pi h^2\left(R - \frac{h}{3}\right)$, where R is the radius of the *sphere,* find the rate that the water level is rising when the water is 2 inches deep.

First, rewrite the equation as

$$V = \pi R h^2 - \frac{\pi}{3}h^3$$

Take the derivative of the equation with respect to t.

$$\frac{dV}{dt} = 2\pi Rh\frac{dh}{dt} - \pi h^2\frac{dh}{dt}$$

If you plug in $\frac{dV}{dt} = 4$, $R = 6$, and $h = 2$, you get

$$4 = 20\pi\frac{dh}{dt} \text{ or } \frac{dh}{dt} = \frac{4}{20\pi} = \frac{1}{5\pi}$$

B. Find an expression for r, the radius of the *surface of the spherical segment* of water, in terms of h.

Notice that you can construct a right triangle using the radius of the sphere and the radius of the surface of the water.

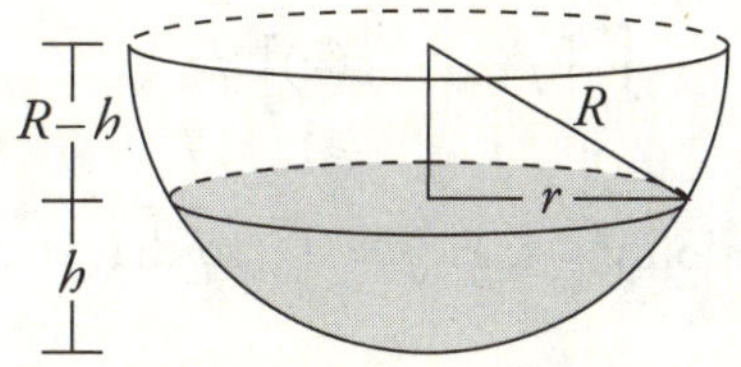

Notice that the distance from the center of the sphere to the surface of the water is $R - h$. Now, you can use the Pythagorean Theorem to find r.

$$R^2 = (R - h)^2 + r^2$$

You can rearrange this to get

$$r = \sqrt{R^2 - (R-h)^2} = \sqrt{2Rh - h^2}$$

Because $R = 6$, you get

$$r = \sqrt{12h - h^2}$$

C. How fast is the circular area of the surface of the spherical segment of water growing (in square inches per second) when the water is 2 inches deep?

The area of the surface of the water is $A = \pi r^2$, where $r = \sqrt{12h - h^2}$. Thus, $A = \pi(12h - h^2)$.

Taking the derivative of the equation with respect to t, you get

$$\frac{dA}{dt} = \pi\left(12\frac{dh}{dt} - 2h\frac{dh}{dt}\right)$$

You found in part A above that $\frac{dh}{dt} = \frac{1}{5\pi}$, so $\frac{dA}{dt} = \pi\left(\frac{12}{5\pi} - \frac{4}{5\pi}\right) = \frac{8}{5}$ square inches per second.

5. Let R be the region in the first quadrant bounded by $y^2 = x$ and $x^2 = y$.

A. Find the area of region R.

First, sketch the region.

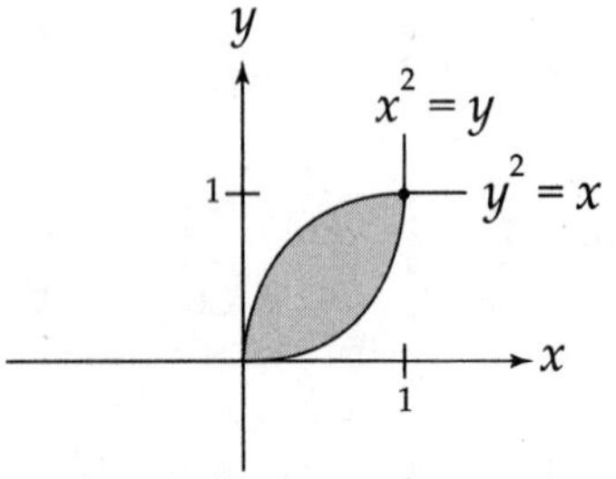

In order to find the area, "slice" the region vertically and add up all of the slices. Now, use the formula for the area of the region between $y = f(x)$ and $y = g(x)$, from $x = a$ to $x = b$.

$$\int_a^b \left[f(x) - g(x)\right] dx$$

You need to rewrite the equation $y^2 = x$ as $y = \sqrt{x}$ so that you can integrate with respect to x. The integral for the area is

$$\int_0^1 \left(\sqrt{x} - x^2\right) dx$$

Evaluating the integral, you get

$$\left(\frac{2x^{\frac{3}{2}}}{3} - \frac{x^3}{3}\right)\Bigg|_0^1 = \frac{2}{3} - \frac{1}{3} = \frac{1}{3}$$

B. Find the volume of the solid generated when R is revolved about the x-axis.

In order to find the volume of a region between $y = f(x)$ and $y = g(x)$, from $x = a$ to $x = b$, when it is revolved around the x-axis, you use the following formula:

$$\pi\int_a^b \left[f(x)^2 - g(x)^2\right] dx$$

Here, the integral for the area is

$$\pi\int_0^1 \left(x - x^4\right) dx$$

Evaluating the integral, you get

$$\pi\left(\frac{x^2}{2} - \frac{x^5}{5}\right)\Bigg|_0^1 = \pi\left(\frac{1}{2} - \frac{1}{5}\right) = \frac{3\pi}{10}$$

C. The section of a certain solid cut by any plane perpendicular to the x-axis is a circle with the endpoints of its diameter lying on the parabolas $y^2 = x$ and $x^2 = y$. Find the volume of the solid.

Whenever you want to find the volume of a solid formed by the region between $y = f(x)$ and $y = g(x)$ with a known cross-section, from $x = a$ to $x = b$, when it is revolved around the x-axis, you use the following formula:

$$\int_a^b A(x)\,dx$$

(Note: $A(x)$ is the area of the cross-section.) Find the area of the cross-section by using the vertical slice formed by $f(x) - g(x)$ and then plugging it into the appropriate area formula. In the case of a circle, $f(x) - g(x)$ gives you the length of the diameter and you use the following formula:

$$A(x) = \frac{\pi(diameter)^2}{4}$$

This gives you the integral,

$$\int_0^1 \frac{\pi}{4}\left(\sqrt{x} - x^2\right)^2 dx$$

Expand the integrand.

$$\int_0^1 \frac{\pi}{4}\left(\sqrt{x} - x^2\right)^2 dx = \frac{\pi}{4}\int_0^1 \left(x - 2x^{\frac{5}{2}} + x^4\right) dx$$

Evaluate the integral.

$$\frac{\pi}{4}\int_0^1 \left(x - 2x^{\frac{5}{2}} + x^4\right) dx = \frac{\pi}{4}\left(\frac{x^2}{2} - \frac{4x^{\frac{7}{2}}}{7} + \frac{x^5}{5}\right)\Bigg|_0^1 = \frac{\pi}{4}\left(\frac{1}{2} - \frac{4}{7} + \frac{1}{5}\right) = \frac{9\pi}{280}$$

6. For time $t \geq 0$, a particle moves along the x-axis. The velocity of the particle at time t is given by $v(t) = 1 - 2\cos\left(\frac{\pi}{3}t\right)$. The particle's position at time $t = 0$ is $x(0) = 8$.

A. Is the particle speeding up or slowing down at time $t = \frac{1}{2}$? Justify your answer.

An object is speeding up if its velocity and acceleration have the same sign at a given time, and it is slowing down if they have opposite signs at a given time. You can find the acceleration by taking the derivative of the velocity: $a(t) = \frac{2\pi}{3}\sin\left(\frac{\pi}{3}t\right)$. Now you can check the signs of the velocity and acceleration at time $t = \frac{1}{2}$: $v = 1 - 2\cos\frac{\pi}{6} = 1 - 2\left(\frac{\sqrt{3}}{2}\right) = 1 - \sqrt{3}$ and $a = \frac{2\pi}{3}\sin\frac{\pi}{6} = \frac{\pi}{3}$. The acceleration and velocity have opposite signs, so the particle is slowing down.

B. When does the particle change direction in the interval $0 \le t \le 2$? Justify your answer.

The particle changes direction at times when the velocity is zero and changes sign. Find when the velocity is zero.

$$1 - 2\cos\left(\frac{\pi}{3}t\right) = 0$$

$$\cos\left(\frac{\pi}{3}t\right) = \frac{1}{2}$$

$$\frac{\pi}{3}t = \frac{\pi}{3}$$

$$t = 1$$

The next solution ($t = 5$) is outside the interval. Now pick a value less than 1, such as $t = \frac{1}{2}$ and note that the velocity is negative (see the explanation in part A). Pick another value greater than 1, such as $t = 2$. The velocity is $1 - 2\cos\frac{2\pi}{3} = 2$, which is positive. Therefore, the particle is changing direction at time $t = 1$.

C. What is the particle's position at time $t = 2$?

$$x(t) = \int v(t)\,dt = \int \left[1 - 2\cos\left(\frac{\pi t}{3}\right)\right]dt = t - \frac{6}{\pi}\sin\left(\frac{\pi t}{3}\right) + C$$

Use the initial conditions given.

$$x(0) = 0 - 0 + C = 8$$

Therefore, $C = 8$, and $x(t) = t - \frac{6}{\pi}\sin\frac{\pi t}{3} + 8$. The position at $t = 2$ is

$$x(2) = 2 - \frac{6}{\pi}\sin\left(\frac{2\pi}{3}\right) + 8 = 10 - \frac{6}{\pi}\frac{\sqrt{3}}{2} = 10 - \frac{3\sqrt{3}}{\pi}$$

D. What is the total distance traveled from time $t = 0$ to $t = 2$?

From part B, you know that there's a change in direction at $t = 1$. Thus, you calculate the integral in two parts.

$$\int_0^1 \left[1 - 2\cos\left(\frac{\pi t}{3}\right)\right]dx = t - \frac{6}{\pi}\sin\left(\frac{\pi t}{3}\right)\Big|_0^1 = 1 - \frac{6}{\pi}\frac{\sqrt{3}}{2} = 1 - \frac{3\sqrt{3}}{\pi}$$

$$\int_1^2 \left[1 - 2\cos\left(\frac{\pi t}{3}\right)\right]dx = t - \frac{6}{\pi}\sin\left(\frac{\pi t}{3}\right)\Big|_1^2 = \left(2 - \frac{6}{\pi}\frac{\sqrt{3}}{2}\right) - \left(1 - \frac{6}{\pi}\frac{\sqrt{3}}{2}\right) = 1$$

Now, add the absolute values.

$$\left|1 - \frac{3\sqrt{3}}{\pi}\right| + |1| = \frac{3\sqrt{3}}{\pi} - 1 + 1 = \frac{3\sqrt{3}}{\pi}$$

PRACTICE TEST 2: FREE-RESPONSE QUESTIONS GRADING RUBRICS

Scoring Guidelines for Question 1

A.	Correct equations	**2 points**
	Answer	**1 point**
B.	Correct integral	**1 point**
	Answer	**1 point**
C.	Correct setup of trapezoid rule	**1 point**
	Answer	**1 point**
D.	Correct derivative	**2 points**
Total		**9 points**

Scoring Guidelines for Question 2

A.	Correct integral	**1 point**
	Correct equation	**1 point**
	Correct constant of integration	**1 point**
	Answer	**1 point**
B.	Answer	**2 points**
C.	Correct equation	**1 point**
	Correct integration	**1 point**
	Answer	**1 point**
Total		**9 points**

Scoring Guidelines for Question 3

A.	Correct derivative	**1 point**
	Finds slope correctly	**1 point**
	Correct equation	**1 point**
B.	Correct derivative	**1 point**
	Plugs in correctly	**1 point**
	Answer	**1 point**
C.	Correct integration	**2 points**
	Answer	**1 point**
Total		**9 points**

Scoring Guidelines for Question 4

A.	Correctly rewrites equation in terms of h	**1 point**
	Correct derivative	**1 point**
	Answer	**1 point**
B.	Correct use of Pythagorean Theorem	**1 point**
	Answer	**2 points**
C.	Correct area	**1 point**
	Correct derivative	**1 point**
	Answer	**1 point**
Total		**9 points**

Scoring Guidelines for Question 5

A.	Correct integral	**1 point**
	Correct integration	**1 point**
	Answer	**1 point**
B.	Correct equation	**1 point**
	Correct integration	**1 point**
	Answer	**1 point**
C.	Correct equation	**1 point**
	Correct integration	**1 point**
	Answer	**1 point**
Total		**9 points**

Scoring Guidelines for Question 6

A.	Correctly finds acceleration	**1 point**
	Answer	**1 point**
	Justification	**1 point**
B.	Answer	**1 point**
	Justification	**1 point**
C.	Correct integral	**1 point**
	Answer	**1 point**
D.	Correct integral	**1 point**
	Answer	**1 point**
Total		**9 points**

HOW TO SCORE PRACTICE TEST 2

Section I: Multiple-Choice

_______________ × 1.6667 = _______________

Number of Correct (out of 45) — Weighted Section I Score (Do not round)

Exact scoring can vary from administration to administration. Therefore, this scoring should only be used as an estimate.

Section II: Free-Response

(See if you can find a teacher or classmate to score your Free-Response questions.)

Question 1	_______________ (out of 9)	× 1.3889	=	_______________ (Do not round)
Question 2	_______________ (out of 9)	× 1.3889	=	_______________ (Do not round)
Question 3	_______________ (out of 9)	× 1.3889	=	_______________ (Do not round)
Question 4	_______________ (out of 9)	× 1.3889	=	_______________ (Do not round)
Question 5	_______________ (out of 9)	× 1.3889	=	_______________ (Do not round)
Question 6	_______________ (out of 9)	× 1.3889	=	_______________ (Do not round)

Sum = _______________

Weighted Section II Score (Do not round)

AP Score Conversion Chart Calculus AB

Composite Score Range	AP Score
112–150	5
98–111	4
80–97	3
55–79	2
0–54	1

Composite Score

_______________ + _______________ = _______________

Weighted Section I Score + Weighted Section II Score = Composite Score (Round to nearest whole number)

Practice Test 3

AP® Calculus AB Exam

SECTION I, PART A: Multiple Choice

At a Glance

Total Time
1 hour
Number of Questions
30
Percent of Total Grade
33 1/3%

DISCLAIMER: The official AP Calculus AB will be administered in a hybrid digital format. Instructions for the digital exam may differ from this practice test.

Instructions

Section I, Part A, of this examination contains 30 multiple-choice questions. Fill in only the ovals for numbers 1 through 30 on your answer sheet.

No calculator is allowed for this part of the exam.

Indicate all of your answers to the multiple-choice questions on the answer sheet. Give only one answer to each question.

Use your time effectively, working as quickly as you can without losing accuracy. Do not spend too much time on any one question. Go on to other questions and come back to the ones you have not answered if you have time. It is not expected that everyone will know the answers to all the multiple-choice questions.

Your total score on the multiple-choice section is based only on the number of questions answered correctly. Points are not deducted for incorrect answers or unanswered questions.

SECTION I, PART B: Multiple Choice

At a Glance

Total Time
45 minutes
Number of Questions
15
Percent of Total Grade
16 2/3%

DISCLAIMER: The official AP Calculus AB will be administered in a hybrid digital format. Instructions for the digital exam may differ from this practice test.

Instructions

Section I, Part B, of this examination contains 15 multiple-choice questions. Fill in only the ovals for numbers 31 through 45 on your answer sheet.

A graphing calculator is required for some questions on this part of the exam. You may use a handheld calculator or a Desmos graphing calculator application. **Make sure your calculator is in radian mode.**

Indicate all of your answers to the multiple-choice questions on the answer sheet. Give only one answer to each question.

Use your time effectively, working as quickly as you can without losing accuracy. Do not spend too much time on any one question. Go on to other questions and come back to the ones you have not answered if you have time. It is not expected that everyone will know the answers to all the multiple-choice questions.

Your total score on the multiple-choice section is based only on the number of questions answered correctly. Points are not deducted for incorrect answers or unanswered questions.

AP CALCULUS AB

SECTION I, Part A

Time—60 Minutes

Number of questions—30

No calculator is allowed for this part of the exam.

Directions: Solve each of the following problems, using the available paper for scratchwork. After examining the choices, select the best of the choices given.

In this test: Unless otherwise specified, the domain of a function f is assumed to be the set of all real numbers x for which $f(x)$ is a real number. The inverse of a trigonometric function f may be indicated using the inverse function notation f^{-1} or with the prefix "arc" (e.g., $\sin^{-1} x = \arcsin x$).

1 Mark for Review

$\lim_{\theta \to 0} \frac{\sin 3\theta}{\theta} =$

(A) 0

(B) 1

(C) 3

(D) The limit does not exist.

GO ON TO THE NEXT PAGE.

2 Mark for Review

The region R is bounded from above by $y = x + 9$ and below by $y = x^2 + 3$. Find the area of R.

(A) $\frac{95}{6}$

(B) $\frac{125}{6}$

(C) $\frac{13}{6}$

(D) $\frac{1}{6}$

---

3 Mark for Review

Evaluate $\lim_{x \to 10^-} \frac{\pi}{x - 10}$.

(A) $-\infty$

(B) 0

(C) π

(D) $+\infty$

---

4 Mark for Review

Find $\frac{dy}{dx}$ if $y = \sin^3(x^2)$.

(A) $6x\sin^2(x^2)\cos(x^2)$

(B) $3\cos^2(x^3)$

(C) $\cos^3(2x)$

(D) $6x\sin(x^2)\cos(x^2)$

GO ON TO THE NEXT PAGE.

5 Mark for Review

The Intermediate Value Theorem guarantees a value c, such that $f(c) = 0$ for $f(x) = 2x^2 + x - 4$ on which of the following intervals?

(A) $(-1, 0)$

(B) $(0, 1)$

(C) $(1, 2)$

(D) $(2, 3)$

---

6 Mark for Review

Find $y(2)$ if $\frac{dy}{dx} = 8y$ and $y(0) = 10$.

(A) $10e^8$

(B) $10e^{16}$

(C) 26

(D) 16

---

7 Mark for Review

Find $\frac{dy}{dx}$ if $y = \frac{x^6 - 4x^4 + 3x^3}{x^2}$.

(A) $4x^3 - 8x + 3$

(B) $\frac{6x^5 - 16x^3 + 9x^2}{2x}$

(C) $3x^2 - 8x + \frac{9}{2}x^{\frac{1}{2}}$

(D) $4x^3 - 16x^3 + 9x^2$

GO ON TO THE NEXT PAGE.

8 Mark for Review

$\lim_{x \to 5} \frac{x^2 - 3x - 10}{x^2 - x - 20} =$

Ⓐ 0

Ⓑ $\frac{7}{9}$

Ⓒ 1

Ⓓ The limit does not exist.

---

9 Mark for Review

An object's height above the ground is given by the equation $y(t) = -5t^2 + 20t + 8$, $t \geq 0$, and its horizontal location is given by the equation $x(t) = 12t + 10$. What is its horizontal location when the object reaches its maximum height?

Ⓐ –14

Ⓑ 10

Ⓒ 24

Ⓓ 34

---

10 Mark for Review

Find $\frac{dy}{dx}$ if $y = \frac{\sec^2(x) - \tan^2(x)}{\csc(x)}$.

Ⓐ $-\csc x \cot x$

Ⓑ $\cos x$

Ⓒ $2\sin x \cos x$

Ⓓ $\sin x$

GO ON TO THE NEXT PAGE.

11 Mark for Review

What value of k makes $f(x)$ continuous for all values of x?

$$f(x) = \begin{cases} 3x^2 + 5kx - 44; x \geq 2 \\ 2kx - x^3; x < 2 \end{cases}$$

(A) -4

(B) $\frac{12}{7}$

(C) 4

(D) $\frac{20}{3}$

---

12 Mark for Review

The velocity of a particle at certain times is given in the table below. Approximate the total distance traveled on the interval [0, 3].

Time	Velocity
0 seconds	18 miles per hour
1	12
2	10
3	16

(A) 14

(B) 30

(C) 39

(D) 80

GO ON TO THE NEXT PAGE.

13 Mark for Review

Which of the following is a solution to the differential equation $\frac{d^2y}{dx^2} = -y$?

(A) $y = \sin^2 x$

(B) $y = \cos^2 x$

(C) $y = \sin x + \cos x$

(D) $y = \sin^2 x + \cos^2 x$

---

14 Mark for Review

Find $\left[f^{-1}(4)\right]'$ if $f(x) = \frac{x^3 + 7}{2}$.

(A) $\frac{2}{71}$

(B) $\frac{2}{3}$

(C) $\frac{3}{2}$

(D) $\frac{71}{2}$

---

15 Mark for Review

Find $\frac{d^2y}{dx^2}$ if $y = xe^{-x}$.

(A) $-e^{-x}$

(B) $e^{-x}(x + 2)$

(C) $e^{-x}(x - 2)$

(D) e^{-x}

GO ON TO THE NEXT PAGE.

16 Mark for Review

If $f(x)=\frac{3x}{2\sin x}$, find $f'(x)$.

Ⓐ $\frac{3}{2\cos x}$

Ⓑ $-\frac{3}{2\sin x}$

Ⓒ $\frac{6\sin x+6x\cos x}{2\sin^2 x}$

Ⓓ $\frac{6\sin x-6x\cos x}{4\sin^2 x}$

---

17 Mark for Review

If a spherical balloon's volume is increasing at 72 cubic inches per second, how fast is the radius of the balloon increasing when the radius is 6 inches?

Ⓐ $\frac{1}{2\pi}$ inch per second

Ⓑ $\frac{2}{\pi}$ inch per second

Ⓒ $\frac{\pi}{2}$ inches per second

Ⓓ 2π inches per second

GO ON TO THE NEXT PAGE.

18 Mark for Review

Find $\frac{dy}{dx}$ if $y = \frac{\sin(e^x)}{1+\cos(e^x)}$.

(A) $\frac{e^x}{1+\cos(e^x)}$

(B) $\frac{e^x}{1-\cos(e^x)}$

(C) $\frac{e^x}{(1+\cos(e^x))^2}$

(D) $\frac{e^x}{(1-\cos(e^x))^2}$

---

19 Mark for Review

Evaluate $\lim_{x \to 2} \frac{4e^{x-2} - \cos\left(\frac{\pi}{x}\right) - x^2}{x-2}$.

(A) 0

(B) $-\frac{\pi}{4}$

(C) $-\frac{\pi}{2}$

(D) The limit does not exist.

GO ON TO THE NEXT PAGE.

20 Mark for Review

Given $f(x)$ below, for what values of a and b is $f(x)$ differentiable for all values of x?

$$f(x) = \begin{cases} 3ax^2 + 2bx + 1; \; x < 1 \\ -4bx^2 + 7x; \; x \geq 1 \end{cases}$$

(A) $a = \frac{1}{2}$ and $b = -3$

(B) $a = -\frac{1}{2}$ and $b = 3$

(C) $a = 3$ and $b = -\frac{5}{2}$

(D) $a = -3$ and $b = \frac{5}{2}$

---

21 Mark for Review

Find $\frac{dy}{dx}$ if $y = \arcsin\left(\sqrt{x}\right)$.

(A) $\frac{1}{\sqrt{x\left(1-x^2\right)}}$

(B) $\frac{1}{2\sqrt{x}\left(1-x^2\right)}$

(C) $\frac{1}{\sqrt{x}\left(1-x^2\right)}$

(D) $\frac{1}{2\sqrt{x-x^2}}$

GO ON TO THE NEXT PAGE.

22 Mark for Review

If $f(x)=(2x^3+33)(\sqrt[5]{x}-2x)$, then $f'(x)=$

(A) $(2x^3+33)\left(\frac{1}{5\sqrt[5]{x^4}}-2\right)+6x^2(\sqrt[5]{x}-2x)$

(B) $(2x^3+33)\left(\frac{1}{5\sqrt[5]{x^4}}-2\right)+6x^3(\sqrt[5]{x}-2x)$

(C) $(2x^3+33)\left(\frac{1}{5}\sqrt[5]{x^4}-2\right)+6x^2(\sqrt[5]{x}-2x)$

(D) $(2x^3+33)\left(\frac{1}{5\sqrt[5]{x^4}}-2\right)+66x^2(\sqrt[5]{x}-2x)$

---

23 Mark for Review

Evaluate $\int x\sqrt{5-x}\,dx$.

(A) $-\frac{2}{3}(5-x)^{\frac{3}{2}}+C$

(B) $\frac{2}{3}(5-x)^{\frac{3}{2}}+C$

(C) $\frac{2}{5}(x-5)^{\frac{5}{2}}-\frac{10}{3}(x-5)^{\frac{3}{2}}+C$

(D) $\frac{2}{5}(5-x)^{\frac{5}{2}}-\frac{10}{3}(5-x)^{\frac{3}{2}}+C$

GO ON TO THE NEXT PAGE.

24 Mark for Review

On what interval(s) is the graph of $y = xe^x$ concave up?

(A) $(-\infty, 2)$

(B) $(2, \infty)$

(C) $(-\infty, 2) \cup (2, \infty)$

(D) $(-2, \infty)$

---

25 Mark for Review

A corporation's revenue can be found by the function $R(x) = -x^3 + x^2 + 5$, where x is the number of units sold (in thousands). Its cost can be found by the function $C(x) = 9 + x - x^2$. The maximum profit for the corporation will be when it sells how many units?

(A) 100

(B) 300

(C) $\frac{1000}{3}$

(D) 1000

GO ON TO THE NEXT PAGE.

26 Mark for Review

Evaluate $\int \frac{x\,dx}{14-x^2}$.

Ⓐ $\frac{1}{2}\frac{x^2}{(14-x^2)^2}$

Ⓑ $-\frac{1}{2}\frac{x^2}{(14-x^2)^2}$

Ⓒ $-\frac{1}{2}\ln|14-x^2|+C$

Ⓓ $\frac{1}{2}\ln|14-x^2|+C$

---

27 Mark for Review

If the position of a particle is given by the function $s(t) = 2t^3 - 24t^2 + 72t + 4$, $t \geq 0$, for what value(s) of t is the particle changing direction?

Ⓐ $t = 2$

Ⓑ $t = 6$

Ⓒ $t = 2, 6$

Ⓓ $t = 6, 12$

GO ON TO THE NEXT PAGE.

28 Mark for Review

Find $\frac{dy}{dx}$ at the point (3, 1) if $x^2 - 2xy + 4y^3 = 7$.

(A) $-\frac{3}{4}$

(B) $-\frac{2}{3}$

(C) $\frac{3}{5}$

(D) $\frac{2}{3}$

---

29 Mark for Review

Use linear approximation to estimate $(5.1)^3$.

(A) 125.1

(B) 125.5

(C) 127.5

(D) 132.5

GO ON TO THE NEXT PAGE.

30 Mark for Review

Which of the following is the slope field for $\frac{dy}{dx} = (x-y)^2$?

(A)

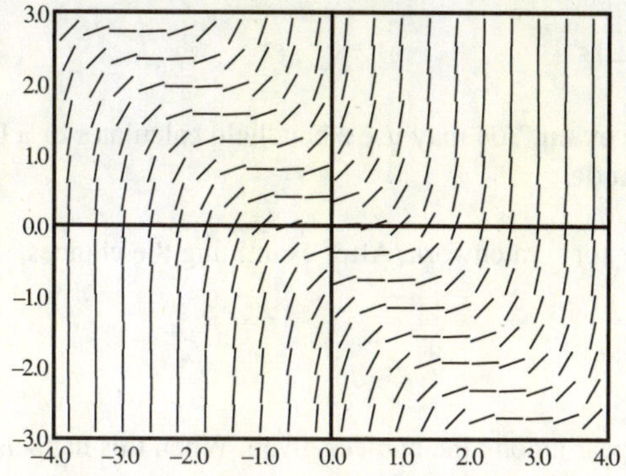

(C)

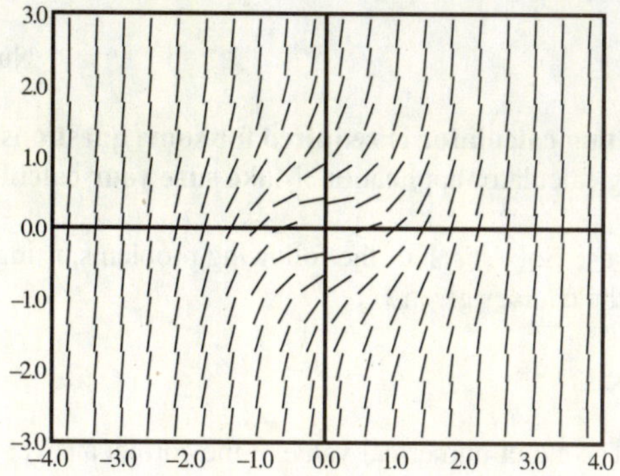

(B)

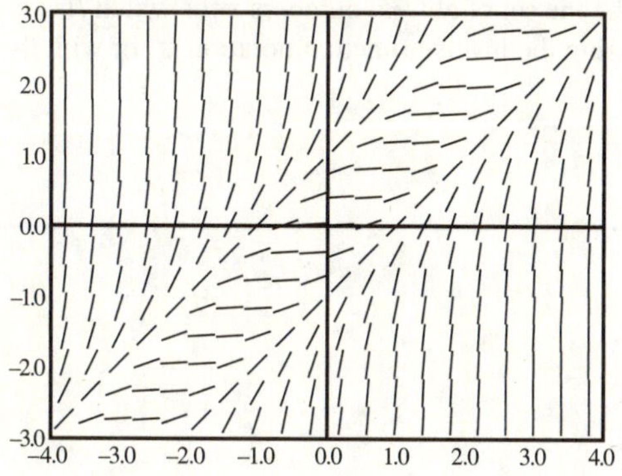

(D)

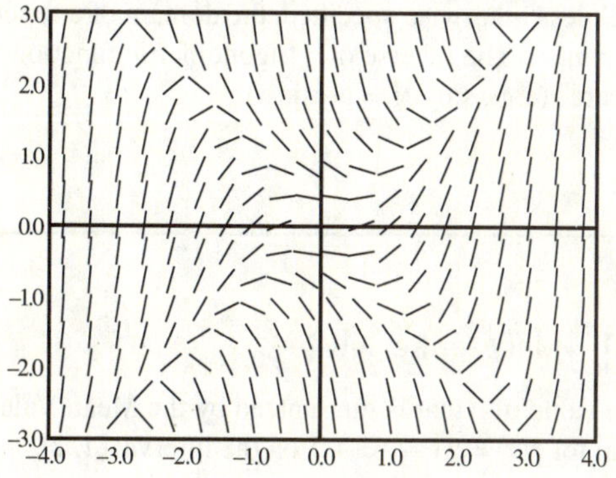

END OF PART A, SECTION I

IF YOU FINISH BEFORE TIME IS CALLED, YOU MAY CHECK YOUR WORK ON PART A ONLY.

DO NOT GO ON TO PART B UNTIL YOU ARE TOLD TO DO SO.

AP CALCULUS AB

SECTION I, Part B

Time—45 Minutes

Number of questions—15

A graphing calculator is required for some questions on this part of the exam. You may use a handheld calculator or a Desmos graphing calculator application. **Make sure your calculator is in radian mode.**

Directions: Solve each of the following problems, using the available paper for scratchwork. After examining the choices, select the best of the choices given.

In this test:

1. The **exact** numerical value of the correct answer does not always appear among the choices given. When this happens, select from among the choices the number that best approximates the exact numerical value.

2. Unless otherwise specified, the domain of a function f is assumed to be the set of all real numbers x for which $f(x)$ is a real number. The inverse of a trigonometric function f may be indicated using the inverse function notation f^{-1} or with the prefix "arc" (e.g., $\sin^{-1} x = \arcsin x$).

---

31 Mark for Review

Find the value of c that is guaranteed by the Mean Value Theorem for $f(x) = 2x^3 - 6x + 1$ on the interval [1, 2].

(A) 1.500

(B) 1.528

(C) 1.725

(D) There is no value of c.

GO ON TO THE NEXT PAGE.

32 Mark for Review

On what interval(s) is $f(x) = x\sin x$ increasing on $[0, \pi]$?

(A) (0, 1.571)

(B) (0, 2.029)

(C) (1.571, 3.141)

(D) (2.029, 3.141)

---

33 Mark for Review

Find $\frac{d}{dx}\int_{10}^{x^3} \cos\left(t^3\right)dt$.

(A) $3x^2\cos x^3$

(B) $3x^2\cos x^9$

(C) $\cos x^9$

(D) $\cos x^3$

GO ON TO THE NEXT PAGE.

34 Mark for Review

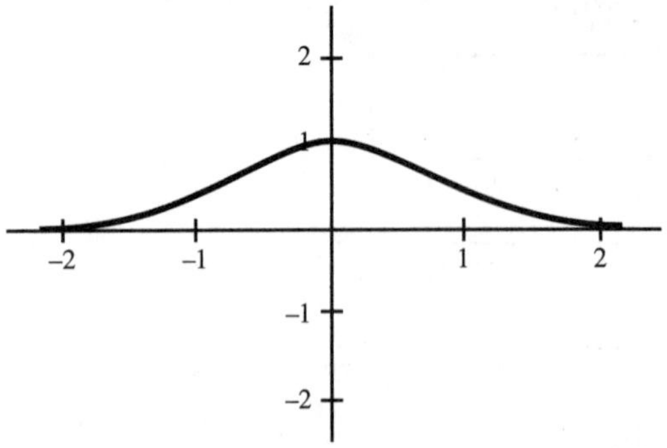

The graph of $f(x)$ is shown above. Which of the following could be the graph of $f'(x)$?

(A)

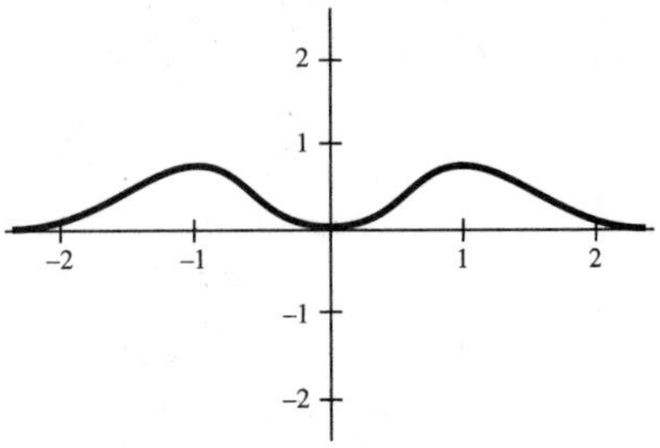

(B)

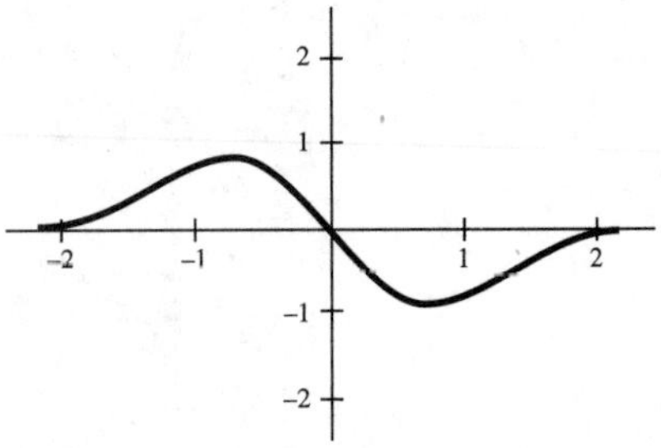

(C)

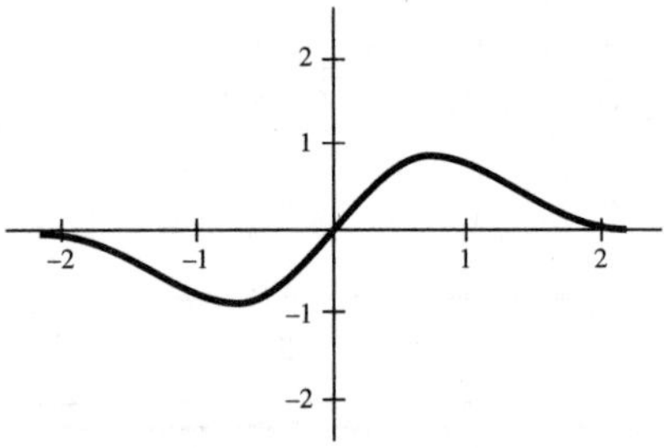

(D)

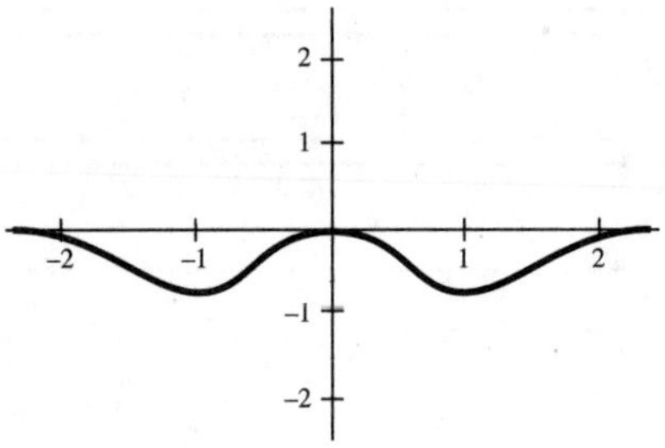

GO ON TO THE NEXT PAGE.

35 Mark for Review

Approximate the area between the parabola $y = 12x - x^2$ and the x-axis using four left-hand rectangles on the interval [0, 12].

Ⓐ 135

Ⓑ 270

Ⓒ 273

Ⓓ 288

---

36 Mark for Review

If $\int_2^5 f(x)\,dx = 11, \int_5^8 f(x)\,dx = 23$, and $\int_{12}^8 f(x)\,dx = 14$, what is $\int_2^{12} f(x)\,dx$?

Ⓐ –48

Ⓑ –20

Ⓒ 20

Ⓓ 48

GO ON TO THE NEXT PAGE.

37 Mark for Review

Evaluate $\int \frac{\cos x}{4 + \sin x} dx$.

(A) $\ln |4 + \sin x| + C$

(B) $\ln |4 - \sin x| + C$

(C) $\frac{(4+\sin x)^2}{2} + C$

(D) $-\frac{(4+\sin x)^2}{2} + C$

---

38 Mark for Review

Find y if $\frac{dy}{dx} = \frac{3}{x+1}$ and $y(0) = 4$.

(A) $y = 3\ln|x + 1| + 4$

(B) $y = 3\ln|x + 1| + \ln 4$

(C) $3\frac{(x+1)^2}{2} + \frac{5}{2}$

(D) $3\frac{(x+1)^2}{2} + 4$

GO ON TO THE NEXT PAGE.

39 Mark for Review

Evaluate $\lim_{h \to 0} \frac{\sqrt{16+h}-4}{h}$.

Ⓐ 0

Ⓑ $\frac{1}{8}$

Ⓒ 4

Ⓓ The limit does not exist.

---

40 Mark for Review

If $g(x) = \int_0^x (t^2 - 2t + 5)dt$, find $g(3)$.

Ⓐ 0

Ⓑ 8

Ⓒ 15

Ⓓ 33

---

41 Mark for Review

If a particle's acceleration is given by $a(t) = 2t + 5$, $t \geq 0$, with its initial velocity $v(0) = 8$ and initial position $s(0) = 100$, find the position equation $s(t)$.

Ⓐ $s(t) = t^2 + 95$

Ⓑ $s(t) = t^2 + 3$

Ⓒ $s(t) = t^3 + 5t^2 + 8t + 100$

Ⓓ $s(t) = \frac{t^3}{3} + \frac{5t^2}{2} + 8t + 100$

GO ON TO THE NEXT PAGE.

42 Mark for Review

Find the area between $y = \cos x$ and $y = \sin x$ on the interval $\left[0, \frac{\pi}{2}\right]$.

(A) -1.414

(B) -0.828

(C) 0.828

(D) 1.414

---

43 Mark for Review

The region R is bounded by $y = \sqrt{\cos x}$, $x = 0$, $y = 0$, and $x = \frac{\pi}{2}$. Find the volume of the solid that results when R is revolved about the x-axis.

(A) 0

(B) $\frac{\pi}{2}$

(C) π

(D) 2π

GO ON TO THE NEXT PAGE.

44 Mark for Review

If the velocity of a particle (in meters per second) is given by $v(t) = t^2 - 10t + 21$, $t \geq 0$, find the distance that the particle travels in the time interval [0, 10].

(A) 10.667

(B) 43.333

(C) 64.667

(D) 108

---

45 Mark for Review

The table below gives values for $f(x)$ and $g(x)$, and their derivatives, for certain values of x. If $h(x) = \dfrac{x^2 + f(x)}{g(x)}$, find $h'(3)$.

x	$f(x)$	$g(x)$	$f'(x)$	$g'(x)$
–6	5	1	–2	0
–3	3	0	2	1
3	2	4	–1	3
6	4	2	5	–1
9	–1	3	0	–2

(A) $-\dfrac{13}{16}$

(B) $\dfrac{13}{16}$

(C) $\dfrac{11}{4}$

(D) 8

STOP

END OF PART B, SECTION I

IF YOU FINISH BEFORE TIME IS CALLED, YOU MAY CHECK YOUR WORK ON PART B ONLY.

DO NOT GO ON TO SECTION II UNTIL YOU ARE TOLD TO DO SO.

AP CALCULUS AB

SECTION II

GENERAL INSTRUCTIONS

You may wish to look over the problems before starting to work on them, since it is not expected that everyone will be able to complete all parts of all problems. All problems are given equal weight, but the parts of a particular problem are not necessarily given equal weight.

A GRAPHING CALCULATOR IS REQUIRED FOR SOME PROBLEMS OR PARTS OF PROBLEMS ON THIS SECTION OF THE EXAMINATION.

- You may use the available paper for scratch work, but you must write your answers in the free-response booklet. In the free-response booklet, write your solution for each part of each question in the space provided for that part. For questions that have sub-parts, be sure to label those clearly in your solution. Use a pencil or pen with black or dark blue ink.
- You are permitted to use your calculator to solve an equation, find the derivative of a function at a point, or calculate the value of a definite integral. However, you must clearly indicate the setup of your problem, namely the equation, function, or integral you are using. If you use other built-in features or programs, you must show the mathematical steps necessary to produce your results.
- Show all of your work, even though a question may not explicitly remind you to do so. Clearly label any functions, graphs, tables, or other objects that you use. Justifications require you to give mathematical reasons and that you verify the needed conditions under which relevant theorems, properties, definitions, or tests applied. Your work will be scored on the correctness and completeness of your methods as well as your answers. Answers without supporting work will usually not receive any credit.
- Your work must be expressed in standard mathematical notation rather than calculator syntax. For example, $\int_1^5 x^2\,dx$ may not be written as fnInt (X^2, X, 1, 5).
- Unless otherwise specified, answers (numeric or algebraic) need not be simplified. If you use decimal approximations in your calculations, your work will be scored on accuracy. Unless otherwise specified, your final answers should be accurate to three places after the decimal point.
- Unless otherwise specified, the domain of a function f is assumed to be the set of all real numbers x for which $f(x)$ is a real number.

GO ON TO THE NEXT PAGE.

AP CALCULUS AB

SECTION II, PART A

Time—30 minutes

Number of problems—2

A graphing calculator is required for some questions on this part of the exam. You may use a handheld calculator or a Desmos graphing calculator application. **Make sure your calculator is in radian mode.**

During the timed portion for Part A, you may work only on the problems in Part A.

On Part A, you are permitted to use your calculator to solve an equation, find the derivative of a function at a point, or calculate the value of a definite integral. However, you must clearly indicate the setup of your problem, namely the equation, function, or integral you are using. If you use other built-in features or programs, you must show the mathematical steps necessary to produce your results.

1 Mark for Review

An object is moving along the x-axis with its velocity given by $v(t)=\frac{2\sin(1.2t^3)}{1+t^2}$, where t is time measured in seconds and $0 \le t \le 5$. The object's initial position is at $x = 8$.

A. Find the acceleration of the object at the time $t = 4$.

B. Find the position of the object at $t = 4$.

C. What is the distance that the object travels in the interval $0 \le t \le 5$?

D. What is the displacement of the object in the interval $0 \le t \le 5$?

E. If a different object moves along the x-axis with its position given by $x_2 = t^3 - t^2$, at what time t are the two objects traveling with the same velocity?

GO ON TO THE NEXT PAGE.

2 Mark for Review

People board a train at a rate modeled by the function B given by $B(t)=\begin{cases}1800\left(\frac{t}{10}\right)^2\left(1-\frac{t}{10}\right)^3; & 0\le t\le 10\\ 0; & t>10\end{cases}$, where $B(t)$ is measured in people per minute and t is measured in minutes. As people board the train, they exit at a constant rate of 3.2 people per minute.

There are initially 30 people on the train.

A. How many people board the train during the time interval $0 \le t \le 10$?

B. How many people are on the train at time $t = 10$?

C. No one boards the train after time $t = 10$, so at what time will there be no people on the train?

D. At what time t is the number of people on the train a maximum?

END OF PART A, SECTION II

IF YOU FINISH BEFORE TIME IS CALLED, YOU MAY CHECK YOUR WORK ON PART A ONLY.

DO NOT GO ON TO PART B UNTIL YOU ARE TOLD TO DO SO.

AP CALCULUS AB

SECTION II, PART B

Time—1 hour

Number of problems—4

A calculator is not allowed for this part of the exam.

During the timed portion for Part B, you may continue to work on the problems in Part A without the use of any calculator.

3 Mark for Review

Grain is filling a container that is in the shape of a right circular cylinder with a diameter of 4 feet. The rate of change of the height of the grain in the cylinder is given by $\frac{dh}{dt} = 12\sqrt{4 + h}$, where h is measured in feet and t is time measured in minutes.

A. Find the rate of change of the volume of grain in the container with respect to time when $h = 12$ feet.

B. When the height of the grain is 12 feet, how fast is the rate of change of the height decreasing?

C. Initially, the container of grain is empty. Find an expression for h in terms of t.

GO ON TO THE NEXT PAGE.

4 Mark for Review

Figure 1

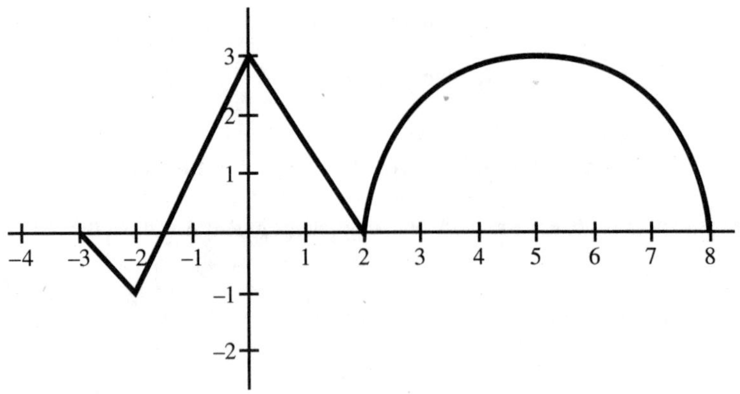

The continuous function f is defined on the interval [−3, 10]. The graph on the interval [2, 8] is a semicircle. Figure 1 shows a portion of the graph of f. The graph contains the point $\left(3, \sqrt{5}\right)$ and crosses the x-axis at the point $\left(-\frac{3}{2}, 0\right)$.

A. If $\int_0^{10} f(x)\,dx = 20$, what is the value of $\int_8^{10} f(x)\,dx$?

B. Evaluate $\int_2^3 4f'(x)\,dx$.

C. If $g(x) = \int_{-3}^{x} f(x)\,dx$, find the absolute maximum of g on the interval [−3, 8]. Justify your answer.

D. Evaluate $\lim\limits_{x \to -1} \frac{3f(x) - 3}{2(x+1)}$.

GO ON TO THE NEXT PAGE.

5 Mark for Review

Figure 1

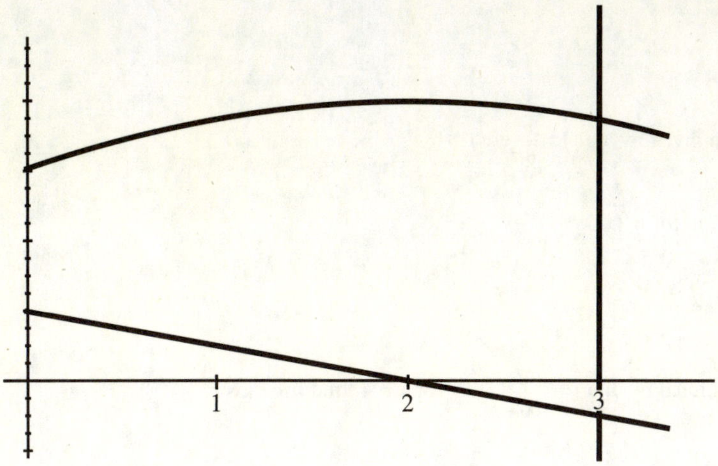

Let R be the region enclosed by the graphs of $f(x) = 16 - (x - 2)^2$, $g(x) = 4 - 2x$, the y-axis, and the line $x = 3$, as shown in Figure 1.

A. Find the area of R.

B. If the region R is the base of a solid and at each x, the perpendicular cross-section to the x-axis has an area $A(x) = e^x$, find the volume of the solid.

C. *Set up but do not evaluate* the integral that gives the volume of the solid that is generated when R is rotated about the x-axis.

GO ON TO THE NEXT PAGE.

6 Mark for Review

Functions f, g, and h are differentiable functions and $f(4) = g(4) = 10$. The line $y = 6 - 3(x - 1)$ is tangent to the graphs of f and g at $x = 4$.

A. Find $f'(4)$.

B. Let w be the function given by $w(x)=\frac{x^4}{64}g(x)$.

i. Find an expression for $w'(x)$.

ii. Evaluate $w'(4)$.

C. If the function h is defined by $h(x)=\frac{16-x^2}{100-g(x)^2}$ for $x \neq 4$, find $\lim_{x\to 4} h(x)$.

STOP

END OF EXAM

Practice Test 3: Answers and Explanations

ANSWER KEY

Section I

1. C	11. C	21. D	31. B	41. D
2. B	12. C	22. A	32. B	42. C
3. A	13. C	23. D	33. B	43. C
4. A	14. B	24. D	34. B	44. C
5. C	15. C	25. D	35. B	45. A
6. B	16. D	26. C	36. C	
7. A	17. A	27. C	37. A	
8. B	18. A	28. B	38. A	
9. D	19. B	29. D	39. B	
10. B	20. D	30. B	40. C	

ANSWERS AND EXPLANATIONS TO SECTION I

1. **C** Multiply the numerator and denominator by 3: $\lim_{\theta \to 0} \frac{3\sin 3\theta}{3\theta}$. Now evaluate the limit: $\lim_{\theta \to 0} \frac{3\sin 3\theta}{3\theta} = 3\lim_{\theta \to 0} \frac{\sin 3\theta}{3\theta} = 3(1) = 3$.

   Here's a shortcut: Any limit of the form $\lim_{\theta \to 0} \frac{\sin a\theta}{\theta} = a$.

2. **B** Sketch the region:

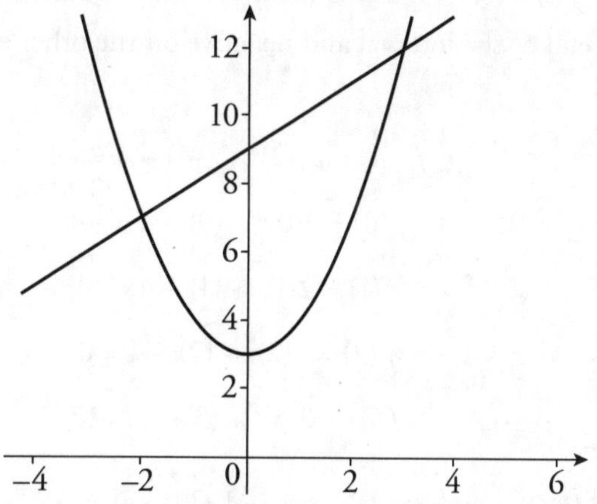

You can find the area of the region bounded from above by $f(x)$ and below by $g(x)$ on the interval $[a, b]$ by evaluating the integral $\int_a^b [f(x) - g(x)]\, dx$. Here, you first need to find the boundaries of the region. Set the two equations equal to each other and solve for x:

$$x + 9 = x^2 + 3$$

$$x^2 - x - 6 = 0$$

$$(x - 3)(x + 2) = 0$$

$$x = -2, 3$$

Thus, you need to evaluate the integral $\int_{-2}^{3} [(x+9) - (x^2+3)]\, dx$. This simplifies to $\int_{-2}^{3} (-x^2 + x + 6)\, dx$. Integrate: $\int_{-2}^{3} (-x^2 + x + 6)\, dx = \left(-\frac{x^3}{3} + \frac{x^2}{2} + 6x\right)\Big|_{-2}^{3}$. Therefore, the area is $\left(-\frac{x^3}{3} + \frac{x^2}{2} + 6x\right)\Big|_{-2}^{3} = \left(-9 + \frac{9}{2} + 18\right) - \left(\frac{8}{3} + 2 - 12\right) = \frac{27}{2} + \frac{22}{3} = \frac{125}{6}$. The answer is (B).

3. **A** Notice that if you plug in 10 for x, you get zero in the denominator of the fraction. But because you are taking the limit, you are not actually plugging in 10 but a number infinitesimally close to 10, but less. If you do that, the fraction will be negative (because the denominator is less than zero) and very large. This means that the limit will approach negative infinity. The answer is (A).

4. **A** Use the Chain Rule: $\frac{dy}{dx} = 3\sin^2\left(x^2\right)\cos\left(x^2\right)(2x) = 6x\sin^2\left(x^2\right)\cos\left(x^2\right)$.

5. **C** The Intermediate Value Theorem guarantees that if a function $f(x)$ is continuous on the interval $[a, b]$ and if $f(a)$ and $f(b)$ have opposite signs, then there exists at least one value c in the interval (a, b) such that $f(c) = 0$. What this means is that you are looking for the interval where $f(x)$ is positive on one end of the interval and negative on the other end. Find the values of $f(x)$ at each of the endpoints:

$$f(-1) = 2(-1)^2 + (-1) - 4 = -3$$

$$f(0) = 2(0)^2 + (0) - 4 = -4$$

$$f(1) = 2(1)^2 + (1) - 4 = -1$$

$$f(2) = 2(2)^2 + (2) - 4 = 6$$

$$f(3) = 2(3)^2 + (3) - 4 = 17$$

On the interval (1, 2), you have $f(1) < 0$ and $f(2) > 0$.

6. **B** Use Separation of Variables to solve the differential equation: $\frac{dy}{y} = 8dt$. Integrate both sides:

$$\int \frac{dy}{y} = \int 8dt$$

$$\ln|y| = 8t + C$$

$$y = e^{8t + C} = Ae^{8t}$$

Now substitute the initial condition to solve for the constant. You get $10 = Ae^{8(0)} = A$, so $y = 10e^{8t}$. Therefore, $y(2) = 10e^{8(2)} = 10e^{16}$.

7. **A** Before you reflexively use the Quotient Rule, notice that you can factor x^2 out of every term in the numerator. You get $y = \frac{x^6 - 4x^4 + 3x^3}{x^2} = x^4 - 4x^2 + 3x$. Now just use the Power Rule to find the derivative: $\frac{dy}{dx} = 4x^3 - 8x + 3$.

8. **B** If you plug in 5 for x, you get $\lim_{x\to 5}\frac{x^2-3x-10}{x^2-x-20}=\frac{(5)^2-3(5)-10}{(5)^2-(5)-20}=\frac{0}{0}$. Before you decide that the limit does not exist, factor the numerator and the denominator:

$\lim_{x\to 5}\frac{x^2-3x-10}{x^2-x-20}=\lim_{x\to 5}\frac{(x-5)(x+2)}{(x-5)(x+4)}$. Now you can cancel $(x - 5)$ from the numerator and denominator: $\lim_{x\to 5}\frac{(x-5)(x+2)}{(x-5)(x+4)}=\lim_{x\to 5}\frac{(x+2)}{(x+4)}$. Now if you take the limit, you get $\lim_{x\to 5}\frac{(x+2)}{(x+4)}=\frac{7}{9}$.

Anytime you have the limit of a polynomial divided by another polynomial, where the limit to a number, k, gives $\frac{0}{0}$, you will be able to factor out $(x - k)$ from the numerator and denominator and cancel.

9. **D** First, find the derivative of y: $\frac{dy}{dt}=-10t+20$. Set the derivative equal to zero and solve for t. You get $t = 2$. This will be the time when the object reaches its maximum height. Next, plug $t = 2$ into the equation for the horizontal location: $x(2) = 12(2) + 10 = 34$.

10. **B** Before you use the Quotient Rule, see if you can simplify the expression. Using the trigonometric identity $1+\tan^2(x)=\sec^2(x)$, you can rewrite the numerator as $y=\frac{\sec^2(x)-\tan^2(x)}{\csc(x)}=\frac{1+\tan^2(x)-\tan^2(x)}{\csc(x)}=\frac{1}{\csc(x)}$. And, because $\csc x=\frac{1}{\sin x}$, you can simplify this more to $\frac{1}{\csc(x)}=\sin x$. The derivative then is $\cos x$.

Remember that the Quotient Rule tends to give "messy" derivatives, so always look for a way to simplify the quotient first.

11. **C** You simply plug 2 into the top and bottom expressions and set them equal to each other. You get $f(x)=\begin{cases}3(2)^2+5k(2)-44=10k-32\\ 2k(2)-(2)^3=4k-8\end{cases}$. You have to solve $10k - 32 = 4k - 8$. You get $k = 4$.

12. **C** The approximate total distance is the sum of the width of each time interval times the velocity during that interval. The width of each time interval is 1. If you use the left end of each interval, you get that the total distance traveled is (1)[18 + 12 + 10] = 40. If you use the right end of each interval, you get that the total distance traveled is (1)[12 + 10 + 16] = 38. If you average the two values, you get that the total distance traveled is 39.

13. **C** The simplest thing to do is to take the second derivative of each of the answer choices to see which one satisfies the differential equation.

Choice (A): $\frac{dy}{dx} = 2\sin x\cos x$; $\frac{d^2y}{dx^2} = 2\sin x(-\sin x) + 2\cos x\cos x = 2(\cos^2 x - \sin^2 x)$, which is not $-\sin^2 x$.

Choice (B): $\frac{dy}{dx} = -2\sin x\cos x$; $\frac{d^2y}{dx^2} = -2\sin x(-\sin x) - 2\cos x\cos x = 2(\sin^2 x - \cos^2 x)$, which is not $-\cos^2 x$.

Choice (C): $\frac{dy}{dx} = \cos x - \sin x$; $\frac{d^2y}{dx^2} = -\sin x - \cos x = -(\sin x + \cos x)$, which works.

14. **B** First, you need to find the x-value when $f(x) = 4$. Solve $4 = \frac{x^3+7}{2}$ to get $x = 1$. Next, find $f'(x)$: $f'(x) = \frac{1}{2}(3x^2) = \frac{3x^2}{2}$. Finally, find $\frac{1}{f'(1)}$: $\frac{1}{\frac{3(1)^2}{2}} = \frac{2}{3}$.

15. **C** Use the Product Rule to find $\frac{dy}{dx}$: $\frac{dy}{dx} = x(-e^{-x}) + e^{-x}(1) = (1-x)e^{-x}$. Now use the Product Rule again to find $\frac{d^2y}{dx^2}$: $\frac{d^2y}{dx^2} = (1-x)(-e^{-x}) + (-1)(e^{-x}) = e^{-x}(x-2)$.

16. **D** Use the Quotient Rule: $f'(x) = \frac{(2\sin x)(3) - (3x)(2\cos x)}{(2\sin x)^2}$, which simplifies to $\frac{6\sin x - 6x\cos x}{4\sin^2 x}$. The answer is (D).

17. **A** You need to find a relationship between the radius of a sphere and its volume. The volume of a sphere is $V = \frac{4}{3}\pi r^3$. You have $\frac{dV}{dt} = 72$ and you want to find $\frac{dr}{dt}$ when $r = 6$. Take the derivative with respect to t: $\frac{dV}{dt} = \frac{4}{3}\pi(3r^2)\frac{dr}{dt} = 4\pi r^2\frac{dr}{dt}$. Now plug in and solve for $\frac{dr}{dt}$:

$$72 = 4\pi(6)^2\frac{dr}{dt}$$

$$\frac{dr}{dt} = \frac{1}{2\pi}$$

18. **A** Use the Quotient Rule: $\frac{dy}{dx} = \frac{(1+\cos(e^x))(e^x\cos(e^x)) - (\sin(e^x))(-e^x\sin(e^x))}{(1+\cos(e^x))^2}$.

This simplifies to

$$\frac{e^x\cos(e^x) + e^x\cos^2(e^x) + e^x\sin^2(e^x)}{(1+\cos(e^x))^2} = \frac{e^x(\cos(e^x) + \cos^2(e^x) + \sin^2(e^x))}{(1+\cos(e^x))^2} = \frac{e^x(\cos(e^x)+1)}{(1+\cos(e^x))^2} = \frac{e^x}{1+\cos(e^x)}$$

19. **B** First, check that you can use L'Hospital's Rule. Plug in $x = 2$: $\dfrac{4e^0 - \cos\left(\dfrac{\pi}{2}\right) - 2^2}{2-2} = \dfrac{0}{0}$. So yes, you can use L'Hospital's Rule. Take the derivative of the numerator and the denominator:

$$\lim_{x\to2}\frac{4e^{x-2} - \cos\left(\frac{\pi}{x}\right) - x^2}{x-2} = \lim_{x\to2}\frac{4e^{x-2} + \sin\left(\frac{\pi}{x}\right)\left(-\frac{\pi}{x^2}\right) - 2x}{1} = \lim_{x\to2}\left(4e^{x-2} + \sin\left(\frac{\pi}{x}\right)\left(-\frac{\pi}{x^2}\right) - 2x\right).$$

Now plug in $x = 2$ to evaluate the limit: $\lim_{x\to2}\left(4e^{2-2} + \sin\left(\frac{\pi}{2}\right)\left(-\frac{\pi}{2^2}\right) - 2(2)\right) = -\frac{\pi}{4}$.

20. **D** First, you need to show that $f(x)$ is continuous. Plug 1 into the top and bottom expressions and set them equal to each other:

$$f(x) = \begin{cases} 3a(1)^2 + 2b(1) + 1 = 3a + 2b + 1 \\ -4b(1)^2 + 7(1) = -4b + 7 \end{cases}$$

You get $3a + 2b + 1 = -4b + 7$, which simplifies to $3a + 6b = 6$ or $a + 2b = 2$. Next, take the derivative of the top and bottom expressions: $f'(x) = \begin{cases} 6ax + 2b;\ x < 1 \\ -8bx + 7;\ x > 1 \end{cases}$. Now plug 1 into the top and bottom expressions and set them equal to each other: $f'(x) = \begin{cases} 6a(1) + 2b \\ -8b(1) + 7 \end{cases}$. You get $6a + 2b = -8b + 7$, which simplifies to $6a + 10b = 7$. Now you solve the two equations. You get $a = -3$ and $b = \frac{5}{2}$.

21. **D** Remember, the derivative of $\arcsin(u) = \frac{1}{\sqrt{1-u^2}}\frac{du}{dx}$. Here you get

$$\frac{dy}{dx} = \frac{1}{\sqrt{1-\left(\sqrt{x}\right)^2}}\frac{1}{2\sqrt{x}} = \frac{1}{\sqrt{1-x}}\frac{1}{2\sqrt{x}} = \frac{1}{2\sqrt{x-x^2}}$$

22. **A** Use the Product Rule: $u\frac{dv}{dx} + v\frac{du}{dx}$. Let $u = 2x^3 + 33$ and $v = \sqrt[5]{x} - 2x = x^{\frac{1}{5}} - 2x$. Then, $\frac{du}{dx} = 6x^2$ and $\frac{dv}{dx} = x^{-\frac{4}{5}}\left(\frac{1}{5}\right) - 2 = \frac{1}{5\sqrt[5]{x^4}} - 2$. Therefore, $f'(x) = \left(2x^3 + 33\right)\left(\frac{1}{5\sqrt[5]{x^4}} - 2\right) + 6x^2\left(\sqrt[5]{x} - 2x\right)$.

23. **D** You can use u-substitution. Let $u = 5 - x$. Then $x = 5 - u$ and $du = -dx$. Substituting into the integrand, you get $\int x\sqrt{5-x}\,dx = \int (u-5)u^{\frac{1}{2}}du = \int\left(u^{\frac{3}{2}} - 5u^{\frac{1}{2}}\right)du$.

Integrate: $\int\left(u^{\frac{3}{2}} - 5u^{\frac{1}{2}}\right)du = \dfrac{u^{\frac{5}{2}}}{\frac{5}{2}} - 5\dfrac{u^{\frac{3}{2}}}{\frac{3}{2}} + C = \frac{2}{5}u^{\frac{5}{2}} - \frac{10}{3}u^{\frac{3}{2}} + C$. And substitute back:

$\frac{2}{5}(5-x)^{\frac{5}{2}} - \frac{10}{3}(5-x)^{\frac{3}{2}} + C$.

24. **D** First, you will need to take the first and second derivatives. Use the Product Rule: $\frac{dy}{dx} = xe^x + e^x = e^x(x+1)$. And use the Product Rule again: $\frac{d^2y}{dx^2} = e^x(x+1) + e^x = e^x(x+2)$. In order to find where the graph of y is concave up, you need to know where the second derivative is positive. Note that e^x is always positive, so you just need to know where $x + 2$ is positive, which is where $x > -2$. So the interval is $(-2, \infty)$.

25. **D** The profit is found by revenue minus cost, namely $P = (-x^3 + x^2 + 5) - (9 + x - x^2) = -x^3 + 2x^2 - x - 4$.

Take the derivative: $\frac{dP}{dx} = -3x^2 + 4x - 1$. Set it equal to zero and solve for x:

$$-3x^2 + 4x - 1 = 0$$

$$-(3x - 1)(x - 1) = 0$$

$$x = \frac{1}{3} \text{ or } x = 1$$

Make a number line and place $x = \frac{1}{3}, 1$ on the line. Choose a value of x less than $\frac{1}{3}$ (say, 0), between $\frac{1}{3}$ and 1 (say, $\frac{2}{3}$), and greater than 1 (say, 2) and plug each value into the derivative to check the sign of the derivative:

$$\text{At } x = 0, \ \frac{dP}{dx} = -1$$

$$\text{At } x = \frac{2}{3}, \ \frac{dP}{dx} = \frac{1}{3}$$

$$\text{At } x = 2, \ \frac{dP}{dx} = -5$$

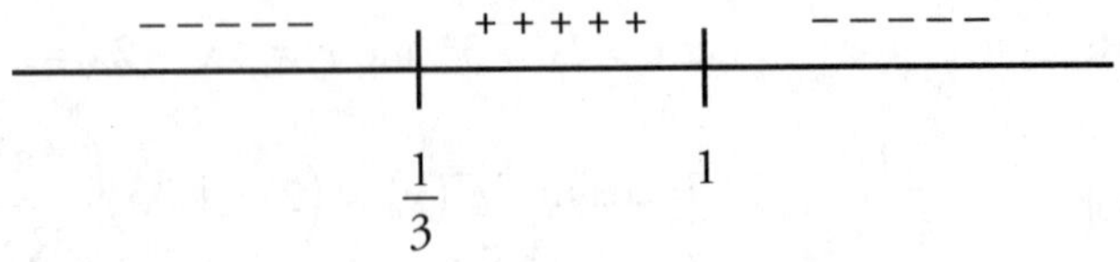

Thus, the derivative goes from positive to negative at $x = 1$, or 1,000 units.

26. **C** Evaluate this integral using u-substitution. Let $u = 14 - x^2$ and $du = -2x\,dx$. Then $-\frac{1}{2}du = x\,dx$.

Substitute into the integrand: $\int \frac{x\,dx}{14 - x^2} = -\frac{1}{2}\int \frac{du}{u}$. Integrate: $-\frac{1}{2}\int \frac{du}{u} = -\frac{1}{2}\ln|u| + C$. And substitute back: $-\frac{1}{2}\ln|u| + C = -\frac{1}{2}\ln|14 - x^2| + C$. The answer is (C).

27. **C** The particle is changing direction at value(s) of t where the velocity changes sign. First, find the velocity by taking the derivative: $v(t) = 6t^2 - 48t + 72$. Next, set the velocity equal to zero and solve for t:

$$6t^2 - 48t + 72 = 0$$

$$t^2 - 8t + 12 = 0$$

$$(t - 2)(t - 6) = 0$$

$$t = 2, 6$$

You need to make sure that the velocity is changing sign at each value of t. Make a number line and place $t = 2, 6$ on the line. Choose a value of t less than 2 (say, 0), between 2 and 6 (say, 3), and greater than 6 (say, 7) and plug each value into the derivative to check the sign of the velocity:

$$\text{At } t = 0,\ v(0) = 72$$

$$\text{At } t = 3,\ v(3) = -18$$

$$\text{At } t = 7,\ v(7) = 30$$

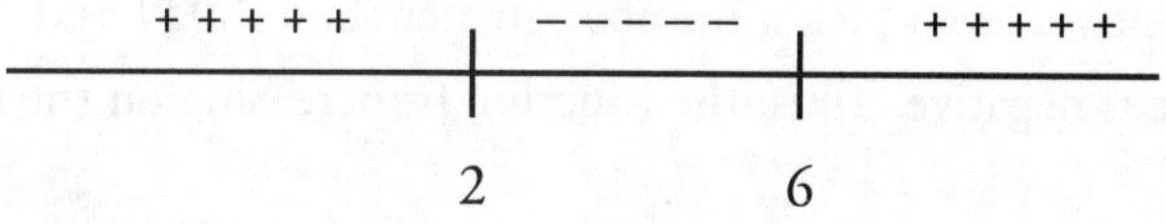

The velocity is thus changing sign at $t = 2, 6$, so those are the values of t where the particle is changing direction.

28. **B** You need to use Implicit Differentiation to find $\frac{dy}{dx}$. Take the derivative of every term: $2x - 2\left(x\frac{dy}{dx} + y\right) + 12y^2\frac{dy}{dx} = 0$. Next, plug in (3, 1): $2(3) - 2\left((3)\frac{dy}{dx} + 1\right) + 12(1)^2\frac{dy}{dx} = 0$. Now you can solve for $\frac{dy}{dx}$:

$$6 - 6\frac{dy}{dx} - 2 + 12\frac{dy}{dx} = 0$$

$$4 + 6\frac{dy}{dx} = 0$$

$$\frac{dy}{dx} = -\frac{2}{3}$$

29. **D** You can use the approximation $f(x + \Delta x) \approx f(x) + f'(x)\Delta x$. Here $f(x) = x^3$, with $x = 5$ and Δx, so you need to find $f(5 + 0.1) \approx f(5) + f'(5)(0.1)$. Take the derivative: $f'(x) = 3x^2$. Plug into the approximation: $f(5 + 0.1) \approx 5^3 + 3(5^2)(0.1) \approx 125 + 7.5 \approx 132.5$.

30. **B** The simplest thing is just to try some points to eliminate answer choices. Note that at (0, 0) the slope is zero. Unfortunately, all four choices have a slope of zero at the origin. Next, let's try (1, 1). The slope should still be $\frac{dy}{dx} = (1-1)^2 = 0$. Only (B) has a flat slope at (1, 1), so (B) is the correct answer.

31. **B** The Mean Value Theorem says that if $f(x)$ is continuous on the interval $[a, b]$ and differentiable on the interval (a, b), then there exists at least one value c on the interval (a, b) such that $f'(c) = \frac{f(b)-f(a)}{b-a}$. Here, you need to find c such that $f'(c) = \frac{f(2)-f(1)}{2-1} = \frac{5-(-3)}{1} = 8$. The derivative is $f'(x) = 6x^2 - 6$, so you need $6c^2 - 6 = 8$, and $c = \pm 1.528$. You can throw out the negative answer because it isn't in the interval.

32. **B** First, take the derivative using the Product Rule: $f'(x) = x\cos x + \sin x$. Next, set the derivative equal to zero and solve. Use your calculator to get $x = 2.029$. Now, put the value of x on the number line and sign test it. Pick a number less than 2.029 and plug it into the derivative. For example, at $x = 1$, the derivative is positive. Then pick a number greater than 2.029 and sign test it. For example, at $x = 3$, the derivative is negative. Thus, the function is increasing on the interval (0, 2.029).

33. **B** The Second Fundamental Theorem of Calculus says $\frac{d}{dx}\int_c^{f(x)} g(t)\,dt = g(f(x))f'(x)$. Here, you get $\frac{d}{dx}\int_{10}^{x^3} \cos(t^3)\,dt = \left((x^3)^3\right)(3x^2) = 3x^2 \cos x^9$.

34. **B** Starting from the left end of the graph ($x = -2$), the slope starts at zero, is positive, and then gets back to zero at $x = 0$. Then the slope is negative and gets back to zero at $x = 2$. The maximum slope is at approximately $x = -1$ and the minimum is at approximately $x = 1$. Put them all together, and you get a graph like this:

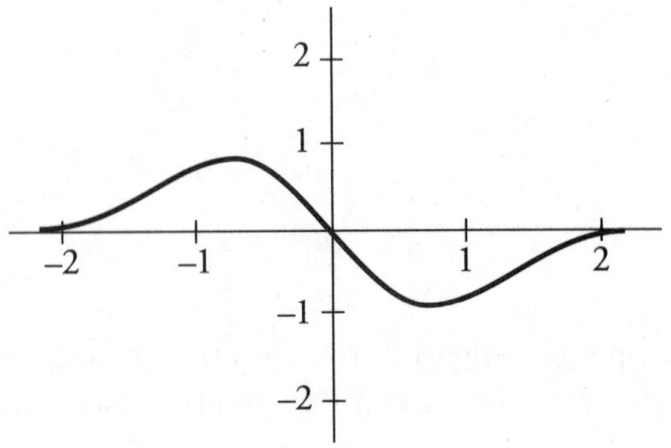

35. **B** First, draw a figure:

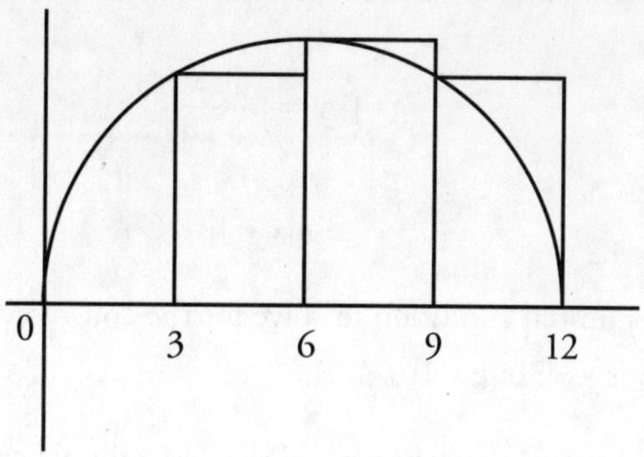

You divide the interval from $x = 0$ to $x = 12$ into four equal intervals. The width of each rectangle is thus $\frac{12-0}{4} = 3$. You can find the height of each rectangle by plugging in the value of x on the left end of each interval. Then the area of each rectangle is the width times the height. You get

$$y(0) = 0$$

$$y(3) = 12(3) - (3)^2 = 27$$

$$y(6) = 12(6) - (6)^2 = 36$$

$$y(9) = 12(9) - (9)^2 = 27$$

Thus, the areas of the rectangles are

$$(3)(0) = 0$$

$$(3)(27) = 81$$

$$(3)(36) = 108$$

$$(3)(27) = 81$$

The area under the parabola is approximately $0 + 81 + 108 + 81 = 270$.

36. **C** According to the First Fundamental Theorem of Calculus, $\int_a^b f(x)dx + \int_b^c f(x)dx = \int_a^c f(x)dx$. This means that $\int_2^5 f(x)dx + \int_5^8 f(x)dx + \int_8^{12} f(x)dx = \int_2^{12} f(x)dx$. Also, note that if $\int_{12}^8 f(x)dx = 14$, then $\int_8^{12} f(x)dx = -14$. Putting all of these together, you get $\int_2^{12} f(x)dx = 11 + 23 - 14 = 20$.

37. **A** You can use u-substitution. Let $u = 4 + \sin x$. Then $du = \cos x\, dx$. Substituting into the integrand, you get $\int \frac{\cos x}{4+\sin x}dx = \int \frac{du}{u}$. Integrate: $\int \frac{du}{u} = \ln|u| + C$. And substitute back: $\ln|4 + \sin x| + C$.

38. **A** Use Separation of Variables to solve the differential equation: $dy = \frac{3dx}{x+1}$. Integrate both sides:

$$\int dy = \int \frac{3dx}{x+1}$$

$$y = 3\ln|x + 1| + C$$

Now substitute the initial condition to solve for the constant. You get $4 = 3\ln|0 + 1| + C$, so $C = 4$. Thus, the solution is $y = 3\ln|x + 1| + 4$.

39. **B** This limit is in the form of the definition of the derivative where $f(x) = \sqrt{x}$, and you are evaluating the derivative at $x = 16$. In other words, $\lim_{h\to 0} \frac{f(16+h)-f(16)}{h} = \lim_{h\to 0} \frac{\sqrt{16+h}-4}{h}$. The derivative of $f(x) = \sqrt{x}$ is $f'(x) = \frac{1}{2\sqrt{x}}$ $f'(x) = \frac{1}{2\sqrt{x}}$. So you just need to find $f'(16) = \frac{1}{2\sqrt{16}} = \frac{1}{8}$.

40. **C** To find $g(3)$, all you need to do is evaluate the integral $g(3) = \int_0^3 (t^2 - 2t + 5)\,dt$. You get $\int_0^3 (t^2 - 2t + 5)\,dt = \left.\frac{t^3}{3} - t^2 + 5t\right|_0^3 = 15$.

41. **D** First, you can find the particle's velocity by integrating the acceleration equation with respect to t: $v(t) = \int (2t+5)\,dt = t^2 + 5t + C$. Next, use the initial condition to solve for the constant: $8 = (0)^2 + 5(0) + C$, so $C = 8$. The velocity equation is thus $v(t) = t^2 + 5t + 8$. Next, you can find the particle's position by integrating the velocity equation with respect to t: $s(t) = \int (t^2 + 5t + 8)\,dt = \frac{t^3}{3} + \frac{5t^2}{2} + 8t + C_1$. Next, use the initial condition to solve for the constant: $100 = \frac{(0)^3}{3} + \frac{5(0)^2}{2} + 8(0) + C_1$, so $C_1 = 100$. Therefore, the position equation is $s(t) = \frac{t^3}{3} + \frac{5t^2}{2} + 8t + 100$.

42. **C** You can find the area by finding the integral of the difference between the two curves over the interval.

First, draw the graph:

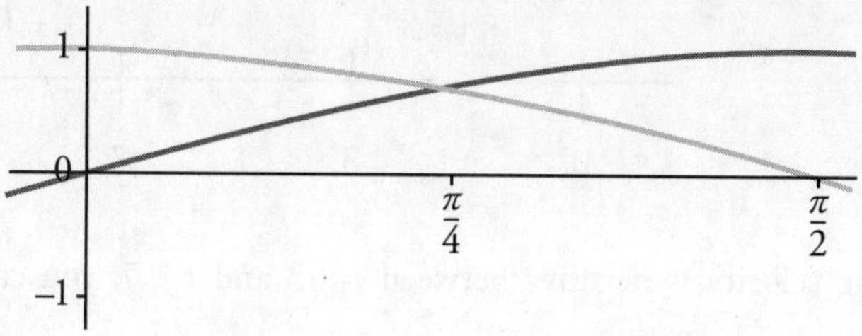

Note that between $x = 0$ and $x = \frac{\pi}{4}$, the curve $y = \cos x$ is above the curve $y = \sin x$ and between $x = \frac{\pi}{4}$ and $x = \frac{\pi}{2}$, the curve $y = \sin x$ is above the curve $y = \cos x$. This means that you will need to evaluate the sum of two integrals: $\int_0^{\frac{\pi}{4}} (\cos x - \sin x)\, dx$ and $\int_{\frac{\pi}{4}}^{\frac{\pi}{2}} (\sin x - \cos x)\, dx$. Use your calculator to get $\int_0^{\frac{\pi}{4}} (\cos x - \sin x)\, dx \approx 0.414$ and $\int_{\frac{\pi}{4}}^{\frac{\pi}{2}} (\sin x - \cos x)\, dx \approx 0.414$. The sum of the two integrals is approximately 0.828.

43. **C** Sketch the region:

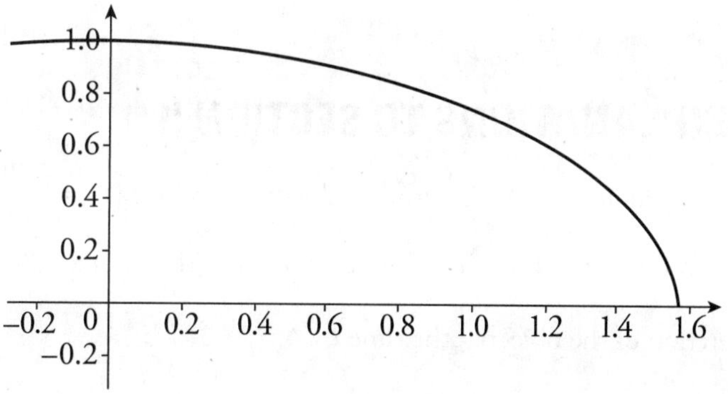

You can find the volume of the region bounded from above by $f(x)$ and below by $g(x)$ on the interval $[a, b]$ by evaluating the integral $\pi\int_a^b \left[f(x)\right]^2 dx$. Here, you get $\pi\int_0^{\frac{\pi}{2}}\left[\left(\sqrt{\cos x}\right)^2\right]dx = \int_0^{\frac{\pi}{2}} \cos x\, dx$.

Integrate: $\pi\int_0^{\frac{\pi}{2}} \cos x\, dx = \pi(\sin x)\Big|_0^{\frac{\pi}{2}}$. Thus, the volume is $\pi(\sin x)\Big|_0^{\frac{\pi}{2}} = \pi\left(\sin\frac{\pi}{2} - \sin 0\right) = \pi$. The answer is (C).

44. **C** First, check the sign of the velocity over the interval. Factor the equation for the velocity to get $v(t) = (t - 3)(t - 7)$. Next use a number line and sign test the velocity.

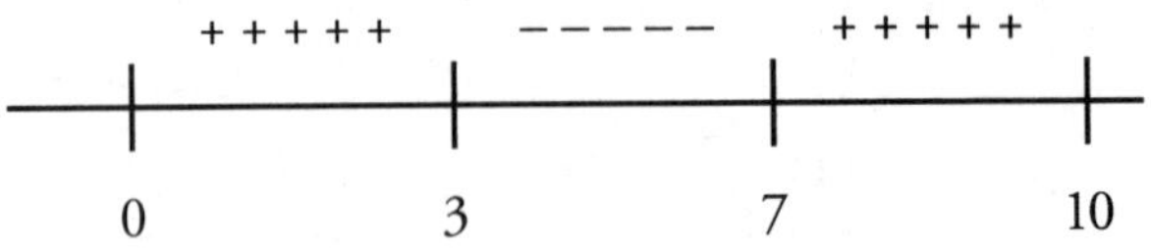

Because the velocity is negative between $t = 3$ and $t = 7$, you can find the distance traveled by evaluating $\int_0^3 (t^2 - 10t + 21)\,dt - \int_3^7 (t^2 - 10t + 21)\,dt + \int_7^{10} (t^2 - 10t + 21)\,dt$. Use your calculator to get $27 + \frac{32}{3} + 27 \approx 64.667$.

45. **A** You can find $h'(x)$ using the Quotient Rule. You get

$$h'(x) = \frac{g(x)(2x + f'(x)) - (x^2 + f(x))g'(x)}{[g(x)]^2}$$

Then $h'(3) = \frac{g(3)(2(3) + f'(3)) - ((3)^2 + f(3))g'(3)}{[g(3)]^2}$. Using the table, you get

$$\frac{(4)(6 + (-1)) - (9 + (2))(3)}{[(4)]^2} = \frac{20 - 33}{16} = -\frac{13}{16}$$

ANSWERS AND EXPLANATIONS TO SECTION II

1. An object is moving along the x-axis with its velocity given by $v(t) = \frac{2\sin(1.2t^3)}{1 + t^2}$, where t is time measured in seconds and $0 \le t \le 5$. The object's initial position is at $x = 8$.

A. Find the acceleration of the object at the time $t = 4$.

You can find the acceleration by taking the derivative of the velocity with respect to time:

$$a(t) = \frac{(1 + t^2)(2\cos(1.2t^3)(3.6t^2)) - (2\sin(1.2t^3))(2t)}{(1 + t^2)^2}$$. Plug in $t = 4$ to get $a(4) = 1.085$.

B. Find the position of the object at $t = 4$.

You need to integrate the velocity over the interval $0 \le t \le 4$, which will tell you how far the object has traveled from its initial position at $x = 8$. Integrate: $\int_0^4 \frac{2\sin(1.2t^3)}{1 + t^2}\,dt = 0.495$. Therefore, the object is at $x = 8.495$.

C. What is the distance that the object travels in the interval $0 \le t \le 5$?

You find the distance traveled by evaluating $\int_0^5 \left| \frac{2\sin\left(1.2t^3\right)}{1+t^2} \right| dt = 1.130$.

D. What is the displacement of the object in the interval $0 \le t \le 5$?

Evaluate $\int_0^5 \frac{2\sin\left(1.2t^3\right)}{1+t^2} dt = 0.495$.

E. If a different object moves along the x-axis with its position given by $x_2 = t^3 - t^2$, at what time t are the two objects traveling with the same velocity?

You can find the velocity of the second object by taking the derivative with respect to time: $v_2 = 3t^2 - 2t$. Set the two equal to each other and use the calculator to find when they are the same. You get time $t = 0.980$ seconds.

2. People board a train at a rate modeled by the function B given by

$$B(t) = \begin{cases} 1800\left(\frac{t}{10}\right)^2\left(1-\frac{t}{10}\right)^3 ; 0 \le t \le 10 \\ 0; t > 10 \end{cases}$$

where $B(t)$ is measured in people per minute and t is measured in minutes. As people board the train, they exit at a constant rate of 3.2 people per minute. There are initially 30 people on the train.

A. How many people board the train during the time interval $0 \le t \le 10$?

You simply integrate $B(t)$ over the time interval: $\int_0^{10} B(t)dt = 300$ people.

B. How many people are on the train at time $t = 10$?

The number of people on the train at time t can be found by $B(t) - 3.2$, so you need to evaluate $30 + \int_0^{10} \left(B(t) - 3.2\right) dt = 298$ people.

C. No one boards the train after time $t = 10$, so at what time will there be no people on the train?

The number of people on the train at time $t = 10$ is 298 (see part B). Therefore, the first time that there are no people on the train is $10 + \frac{298}{3.2} = 103.125$ minutes.

D. At what time t is the number of people on the train a maximum?

The total number of people on the train is found by $30+\int_0^t (B(t)-3.2)\,dt$. Take the derivative, $B(t) - 3.2$, and set it equal to zero. You get $B(t) = 3.2$. Using the calculator, you get $t = 0.452$ minutes and $t = 8.667$ minutes. Now evaluate the function for the total number of people at these two times and the endpoints:

t	Total number of people
0	30
0.452	29.053
8.667	301.131
10	298

Therefore, the maximum number of people on the train is at time $t = 8.667$ and there are 301 people on the train at that time.

3. Grain is filling a container that is in the shape of a right circular cylinder with a diameter of 4 feet. The rate of change of the height of the grain in the cylinder is given by $\frac{dh}{dt} = 12\sqrt{4+h}$, where h is measured in feet and t is time measured in minutes.

A. Find the rate of change of the volume of grain in the container with respect to time when $h = 12$ feet.

The volume of a cylinder with a radius r and a height h is given by $V = \pi r^2 h$. Here the radius of the container is 2 feet and is fixed, so the volume of the grain can be found by $V = 4\pi h$. Take the derivative of both sides with respect to time: $\frac{dV}{dt} = 4\pi\frac{dh}{dt}$. Next, find $\frac{dh}{dt}$ when $h = 12$:

$\frac{dh}{dt} = 12\sqrt{4+12} = 48$. Therefore, $\frac{dV}{dt} = 4\pi(48) = 4\pi(48) = 192\pi$ cubic feet per minute.

B. When the height of the grain is 12 feet, how fast is the rate of change of the height decreasing?

Take the second derivative of h: $\frac{d^2h}{dt^2} = 12\left(\frac{1}{2}\right)(4+h)^{-\frac{1}{2}}(1)\frac{dh}{dt} = \frac{6}{\sqrt{4+h}}\frac{dh}{dt}$. And plug in

$h = 12$: $\frac{d^2h}{dt^2} = \frac{6}{\sqrt{4+12}}(48) = 72$ feet per minute squared.

C. Initially, the container of grain is empty. Find an expression for h in terms of t.

Solve the differential equation using Separation of Variables:

$$\frac{dh}{dt} = 12\sqrt{4+h}$$

$$\frac{dh}{\sqrt{4+h}} = 12dt$$

$$\int \frac{dh}{\sqrt{4+h}} = \int 12dt$$

$$\frac{(4+h)^{\frac{1}{2}}}{\frac{1}{2}} = 12t + C$$

$$2\sqrt{4+h} = 12t + C$$

Plug in the initial condition to solve for C: $2\sqrt{4+0} = 12(0) + C$, so $C = 4$. Therefore, $2\sqrt{4+h} = 12t + 4$.

Finally, you need to isolate h:

$$\sqrt{4+h} = 6t + 2$$

$$4 + h = (6t + 2)^2$$

$$h = (6t + 2)^2 - 4$$

4. The continuous function f is defined on the interval [–3, 10]. The graph on the interval [2, 8] is a semicircle. The figure below shows a portion of the graph of f. The graph contains the point $\left(3, \sqrt{5}\right)$ and crosses the x-axis at the point $\left(-\frac{3}{2}, 0\right)$.

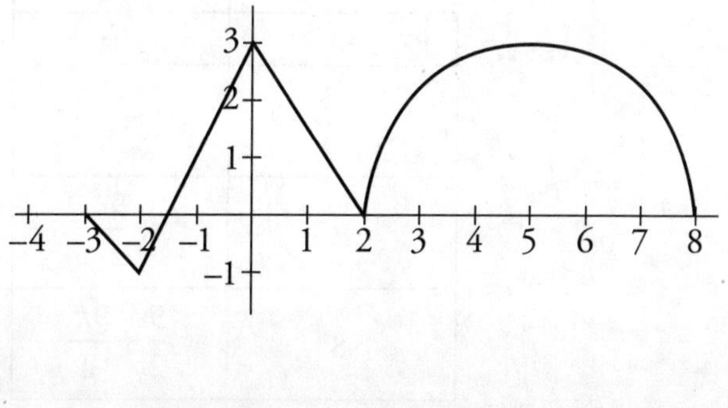

A. If $\int_0^{10} f(x)dx = 20$, what is the value of $\int_8^{10} f(x)dx$?

The Fundamental Theorem of Calculus states that $\int_0^8 f(x)dx + \int_8^{10} f(x)dx = \int_0^{10} f(x)dx$. In order to evaluate $\int_0^8 f(x)dx$, add up the area of the triangle and the semicircle. The integral $\int_0^2 f(x)dx = \frac{1}{2}(2)(3) = 3$ and $\int_2^8 f(x)dx = \frac{1}{2}\pi(3)^2 = \frac{9\pi}{2}$. Thus, $\int_0^8 f(x)dx = 3 + \frac{9\pi}{2}$.

Now you have $\left(3 + \frac{9\pi}{2}\right) + \int_8^{10} f(x)dx = 20$. Therefore, $\int_8^{10} f(x)dx = 17 - \frac{9\pi}{2}$.

B. Evaluate $\int_2^3 4f'(x)dx$.

According to the Fundamental Theorem of Calculus, $\int_a^b f'(x)dx = f(b) - f(a)$. Here you get $\int_2^3 4f'(x)dx = 4\int_2^3 f'(x)dx = 4[f(3) - f(2)]$. Now you use the graph to find $f(2) = 0$, and you know that $f(3) = \sqrt{5}$, so $4[f(3) - f(2)] = 4\sqrt{5}$.

C. If $g(x) = \int_{-3}^{x} f(x)dx$, find the absolute maximum of g on the interval [–3, 8].

Justify your answer.

From the Second Fundamental Theorem of Calculus, you know that $g'(x) = f(x)$. Therefore, the critical values are $x = -3$, $x = -\frac{3}{2}$, $x = 2$, and $x = 8$. Evaluate g at those values. The largest value will be the absolute maximum:

x	$g(x)$
-3	0
$-\frac{3}{2}$	$-\frac{3}{4}$
2	$\frac{9}{2}$
8	$\frac{9}{2} + \frac{9\pi}{2}$

Therefore, the absolute maximum of g on the interval [–3, 8] is $\frac{9}{2} + \frac{9\pi}{2}$.

D. Evaluate $\lim_{x\to -1} \frac{3f(x)-3}{2(x+1)}$.

Note that $f(-1) = 1$, so you have a limit of the indeterminate form $\frac{0}{0}$. You can use L'Hospital's Rule to evaluate the limit. Take the derivative of the numerator and the denominator: $\lim_{x\to -1} \frac{3f(x)-3}{2(x+1)} = \lim_{x\to -1} \frac{3f'(x)}{2}$. The slope of the line segment at $x = -1$ is 2, so $\lim_{x\to -1} \frac{3f'(x)}{2} = \frac{3(2)}{2} = 3$.

5.

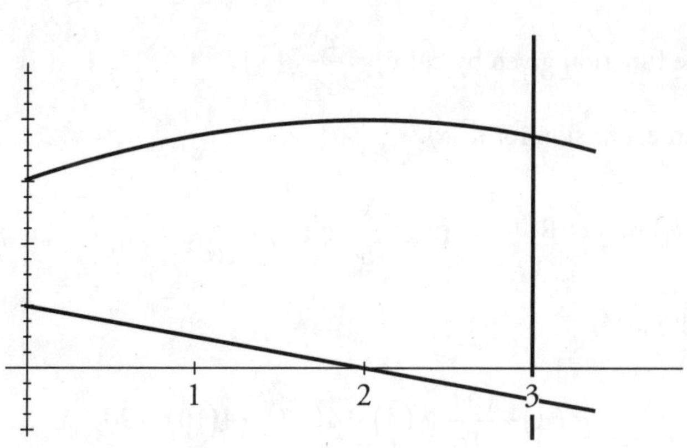

Let R be the region enclosed by the graphs of $f(x) = 16 - (x-2)^2$, $g(x) = 4 - 2x$, the y-axis, and the line $x = 3$, as shown above.

A. Find the area of R.

You can find the area by evaluating the integral $\int_0^3 (f(x) - g(x))\,dx$.

$$\int_0^3 (f(x)-g(x))\,dx = \int_0^3 \left[\left[16-(x-2)^2\right]-(4-2x)\right]dx$$

$$= \int_0^3 \left[\left(16 - x^2 + 4x - 4\right) - (4-2x)\right]dx$$

$$= \int_0^3 \left(8 - x^2 + 6x\right)dx$$

$$= 8x - \frac{x^3}{3} + 3x^2 \Big|_0^3 = 42$$

B. If the region R is the base of a solid and at each x, the perpendicular cross-section to the x-axis has an area $A(x) = e^x$, find the volume of the solid.

You need to evaluate $\int_0^3 A(x)\,dx$: $\int_0^3 A(x)\,dx = \int_0^3 e^x\,dx = e^3 - 1$.

C. *Set up but do not evaluate* the integral that gives the volume of the solid that is generated when R is rotated about the x-axis.

You need to evaluate $\pi\int_0^3\left[\left(16-(x-2)^2\right)^2-(4-2x)^2\right]dx$.

6. Functions f, g, and h are differentiable functions and $f(4) = g(4) = 10$. The line $y = 6 - 3(x - 1)$ is tangent to the graphs of f and g at $x = 4$.

A. Find $f'(4)$.

The derivative of f at 4 will be the same as the slope of the tangent line there. Thus, $f'(4) = -3$.

B. Let w be the function given by $w(x)=\frac{x^4}{64}g(x)$.

i. Find an expression for $w'(x)$.

Use the Product Rule: $w'(x)=\frac{x^4}{64}g'(x)+\frac{x^3}{16}g(x)$.

ii. Evaluate $w'(4)$.

$$w'(4)=\frac{(4)^4}{64}g'(4)+\frac{(4)^3}{16}g(4)=4(-3)+4(10)=28$$

C. If the function h is defined by $h(x)=\frac{16-x^2}{100-g(x)^2}$ for $x \neq 4$, find $\lim_{x\to4}h(x)$.

If you plug $x = 4$ into the limit, you get $\lim_{x\to4}\frac{16-x^2}{100-g(x)^2}=\frac{0}{0}$, which is an indeterminate form.

You can use L'Hospital's Rule. Take the derivative of the numerator and the denominator:

$\lim_{x\to4}\frac{16-x^2}{100-g(x)^2}=\lim_{x\to4}\frac{-2x}{-2g(x)g'(x)}$, which simplifies to $\lim_{x\to4}\frac{x}{g(x)g'(x)}$. Now evaluate the limit:

$$\lim_{x\to4}\frac{x}{g(x)g'(x)}=\frac{4}{(10)(-3)}=-\frac{2}{15}.$$

PRACTICE TEST 3: FREE-RESPONSE QUESTIONS GRADING RUBRICS

Scoring Guidelines for Question 1

A.	Correct derivative	**1 point**
	Answer	**1 point**
B.	Correct integration	**1 point**
	Answer	**1 point**
C.	Correct integration	**1 point**
	Answer	**1 point**
D.	Answer	**1 point**
E.	Correct derivative	**1 point**
	Answer	**1 point**
Total		**9 points**

Scoring Guidelines for Question 2

A.	Correct setup	**1 point**
	Answer	**1 point**
B.	Correct setup	**1 point**
	Answer	**1 point**
C.	Correct setup	**1 point**
	Answer	**1 point**
D.	Correct derivative	**1 point**
	Answer	**2 points**
Total		**9 points**

Scoring Guidelines for Question 3

A.	Correct equation for volume in terms of h	**1 point**
	Correct derivative	**1 point**
	Answer	**1 point**
B.	Correct second derivative	**2 points**
	Answer	**1 point**
C.	Correct integration	**1 point**
	Correct constant of integration	**1 point**
	Answer	**1 point**
Total		**9 points**

Scoring Guidelines for Question 4

A.	Correct integration	**2 points**
	Answer	**1 point**
B.	Correct setup	**1 point**
	Answer	**1 point**
C.	Correct critical values	**1 point**
	Answer	**1 point**
D.	Correct use of L'Hospital's Rule	**1 point**
	Answer	**1 point**
Total		**9 points**

Scoring Guidelines for Question 5

A.	Correct integral	**1 point**
	Correct integration	**1 point**
	Answer	**1 point**
B.	Correct integral	**2 points**
	Answer	**1 point**
C.	Correct integral	**3 points**
Total		**9 points**

Scoring Guidelines for Question 6

A.	Answer	**3 points**
B.	Correct use of Product Rule	**1 point**
	Correct plug in	**1 point**
	Answer	**1 point**
C.	Correct use of L'Hospital's Rule	**2 points**
	Answer	**1 point**
Total		**9 points**

HOW TO SCORE PRACTICE TEST 3

Section I: Multiple-Choice

_______ × 1.6667 = _______

Number of Correct (out of 45) — Weighted Section I Score (Do not round)

Exact scoring can vary from administration to administration. Therefore, this scoring should only be used as an estimate.

Section II: Free-Response

(See if you can find a teacher or classmate to score your Free-Response questions.)

Question 1	_______ (out of 9)	× 1.3889	=	_______ (Do not round)
Question 2	_______ (out of 9)	× 1.3889	=	_______ (Do not round)
Question 3	_______ (out of 9)	× 1.3889	=	_______ (Do not round)
Question 4	_______ (out of 9)	× 1.3889	=	_______ (Do not round)
Question 5	_______ (out of 9)	× 1.3889	=	_______ (Do not round)
Question 6	_______ (out of 9)	× 1.3889	=	_______ (Do not round)

Sum = _______

Weighted Section II Score (Do not round)

AP Score Conversion Chart Calculus AB

Composite Score Range	AP Score
112–150	5
98–111	4
80–97	3
55–79	2
0–54	1

Composite Score

_______ + _______ = _______

Weighted Section I Score — Weighted Section II Score — Composite Score (Round to nearest whole number)

About the Author

David S. Kahn studied applied mathematics and physics at the University of Wisconsin and has taught courses in calculus, precalculus, algebra, trigonometry, and geometry at the college and high school levels. He has worked as an educational consultant for many years and tutored more students in mathematics than he can count! He has worked for The Princeton Review since 1989, and, in addition to AP Calculus, he has taught math and verbal courses for the SAT, SAT Subject Tests, LSAT, GMAT, and the GRE, trained other teachers, and written several other math books.

The Princeton Review®

Completely darken bubbles with a No. 2 pencil. If you make a mistake, be sure to erase mark completely. Erase all stray marks.

1.

YOUR NAME: ______________________________ (Print) Last First M.I.

SIGNATURE: ______________________________ DATE: ___/___/___

HOME ADDRESS: ______________________________ (Print) Number and Street

______________________________ City State Zip Code

PHONE NO.: ______________________________ (Print)

IMPORTANT: Please fill in these boxes exactly as shown on the back cover of your test book.

2. TEST FORM

____

3. TEST CODE

0	A	J	0	0
1	B	K	1	1
2	C	L	2	2
3	D	M	3	3
4	E	N	4	4
5	F	O	5	5
6	G	P	6	6
7	H	Q	7	7
8	I	R	8	8
9			9	9

4. REGISTRATION NUMBER

0	0	0	0	0	0
1	1	1	1	1	1
2	2	2	2	2	2
3	3	3	3	3	3
4	4	4	4	4	4
5	5	5	5	5	5
6	6	6	6	6	6
7	7	7	7	7	7
8	8	8	8	8	8
9	9	9	9	9	9

5. YOUR NAME

First 4 letters of last name					FIRST INIT	MID INIT
A	A	A	A		A	A
B	B	B	B		B	B
C	C	C	C		C	C
D	D	D	D		D	D
E	E	E	E		E	E
F	F	F	F		F	F
G	G	G	G		G	G
H	H	H	H		H	H
I	I	I	I		I	I
J	J	J	J		J	J
K	K	K	K		K	K
L	L	L	L		L	L
M	M	M	M		M	M
N	N	N	N		N	N
O	O	O	O		O	O
P	P	P	P		P	P
Q	Q	Q	Q		Q	Q
R	R	R	R		R	R
S	S	S	S		S	S
T	T	T	T		T	T
U	U	U	U		U	U
V	V	V	V		V	V
W	W	W	W		W	W
X	X	X	X		X	X
Y	Y	Y	Y		Y	Y
Z	Z	Z	Z		Z	Z

6. DATE OF BIRTH

Month	Day		Year	
JAN				
FEB	0	0	0	0
MAR	1	1	1	1
APR	2	2	2	2
MAY	3	3	3	3
JUN		4	4	4
JUL		5	5	5
AUG		6	6	6
SEP		7	7	7
OCT		8	8	8
NOV		9	9	9
DEC				

The Princeton Review®

1. A B C D
2. A B C D
3. A B C D
4. A B C D
5. A B C D
6. A B C D
7. A B C D
8. A B C D
9. A B C D
10. A B C D
11. A B C D
12. A B C D
13. A B C D
14. A B C D
15. A B C D
16. A B C D
17. A B C D
18. A B C D
19. A B C D
20. A B C D
21. A B C D
22. A B C D
23. A B C D
24. A B C D
25. A B C D
26. A B C D
27. A B C D
28. A B C D
29. A B C D
30. A B C D
31. A B C D
32. A B C D
33. A B C D
34. A B C D
35. A B C D
36. A B C D
37. A B C D
38. A B C D
39. A B C D
40. A B C D
41. A B C D
42. A B C D
43. A B C D
44. A B C D
45. A B C D

The Princeton Review®

Completely darken bubbles with a No. 2 pencil. If you make a mistake, be sure to erase mark completely. Erase all stray marks.

1.
YOUR NAME: ______________________________
(Print) Last First M.I.

SIGNATURE: ______________________________ DATE: ___/___/___

HOME ADDRESS: ______________________________
(Print) Number and Street

____
City State Zip Code

PHONE NO.: ______________________________
(Print)

IMPORTANT: Please fill in these boxes exactly as shown on the back cover of your test book.

2. TEST FORM

3. TEST CODE

0	A	J	0	0
1	B	K	1	1
2	C	L	2	2
3	D	M	3	3
4	E	N	4	4
5	F	O	5	5
6	G	P	6	6
7	H	Q	7	7
8	I	R	8	8
9			9	9

4. REGISTRATION NUMBER

0	0	0	0	0	0
1	1	1	1	1	1
2	2	2	2	2	2
3	3	3	3	3	3
4	4	4	4	4	4
5	5	5	5	5	5
6	6	6	6	6	6
7	7	7	7	7	7
8	8	8	8	8	8
9	9	9	9	9	9

5. YOUR NAME

First 4 letters of last name				FIRST INIT	MID INIT
A	A	A	A	A	A
B	B	B	B	B	B
C	C	C	C	C	C
D	D	D	D	D	D
E	E	E	E	E	E
F	F	F	F	F	F
G	G	G	G	G	G
H	H	H	H	H	H
I	I	I	I	I	I
J	J	J	J	J	J
K	K	K	K	K	K
L	L	L	L	L	L
M	M	M	M	M	M
N	N	N	N	N	N
O	O	O	O	O	O
P	P	P	P	P	P
Q	Q	Q	Q	Q	Q
R	R	R	R	R	R
S	S	S	S	S	S
T	T	T	T	T	T
U	U	U	U	U	U
V	V	V	V	V	V
W	W	W	W	W	W
X	X	X	X	X	X
Y	Y	Y	Y	Y	Y
Z	Z	Z	Z	Z	Z

6. DATE OF BIRTH

Month	Day		Year	
JAN				
FEB	0	0	0	0
MAR	1	1	1	1
APR	2	2	2	2
MAY	3	3	3	3
JUN		4	4	4
JUL		5	5	5
AUG		6	6	6
SEP		7	7	7
OCT		8	8	8
NOV		9	9	9
DEC				

The Princeton Review®

1. A B C D
2. A B C D
3. A B C D
4. A B C D
5. A B C D
6. A B C D
7. A B C D
8. A B C D
9. A B C D
10. A B C D
11. A B C D
12. A B C D
13. A B C D
14. A B C D
15. A B C D
16. A B C D
17. A B C D
18. A B C D
19. A B C D
20. A B C D
21. A B C D
22. A B C D
23. A B C D
24. A B C D
25. A B C D
26. A B C D
27. A B C D
28. A B C D
29. A B C D
30. A B C D
31. A B C D
32. A B C D
33. A B C D
34. A B C D
35. A B C D
36. A B C D
37. A B C D
38. A B C D
39. A B C D
40. A B C D
41. A B C D
42. A B C D
43. A B C D
44. A B C D
45. A B C D

The Princeton Review®

Completely darken bubbles with a No. 2 pencil. If you make a mistake, be sure to erase mark completely. Erase all stray marks.

1.

YOUR NAME: ______________________________ (Print) Last / First / M.I.

SIGNATURE: ______________________________ DATE: ___/___/___

HOME ADDRESS: ______________________________ (Print) Number and Street

______________________________ City / State / Zip Code

PHONE NO.: ______________________________ (Print)

IMPORTANT: Please fill in these boxes exactly as shown on the back cover of your test book.

2. TEST FORM

____

3. TEST CODE

0	A	J	0	0
1	B	K	1	1
2	C	L	2	2
3	D	M	3	3
4	E	N	4	4
5	F	O	5	5
6	G	P	6	6
7	H	Q	7	7
8	I	R	8	8
9			9	9

4. REGISTRATION NUMBER

0	0	0	0	0	0
1	1	1	1	1	1
2	2	2	2	2	2
3	3	3	3	3	3
4	4	4	4	4	4
5	5	5	5	5	5
6	6	6	6	6	6
7	7	7	7	7	7
8	8	8	8	8	8
9	9	9	9	9	9

5. YOUR NAME

First 4 letters of last name				FIRST INIT	MID INIT
A	A	A	A	A	A
B	B	B	B	B	B
C	C	C	C	C	C
D	D	D	D	D	D
E	E	E	E	E	E
F	F	F	F	F	F
G	G	G	G	G	G
H	H	H	H	H	H
I	I	I	I	I	I
J	J	J	J	J	J
K	K	K	K	K	K
L	L	L	L	L	L
M	M	M	M	M	M
N	N	N	N	N	N
O	O	O	O	O	O
P	P	P	P	P	P
Q	Q	Q	Q	Q	Q
R	R	R	R	R	R
S	S	S	S	S	S
T	T	T	T	T	T
U	U	U	U	U	U
V	V	V	V	V	V
W	W	W	W	W	W
X	X	X	X	X	X
Y	Y	Y	Y	Y	Y
Z	Z	Z	Z	Z	Z

6. DATE OF BIRTH

Month	Day		Year	
JAN				
FEB	0	0	0	0
MAR	1	1	1	1
APR	2	2	2	2
MAY	3	3	3	3
JUN		4	4	4
JUL		5	5	5
AUG		6	6	6
SEP		7	7	7
OCT		8	8	8
NOV		9	9	9
DEC				

The Princeton Review®

1. A B C D
2. A B C D
3. A B C D
4. A B C D
5. A B C D
6. A B C D
7. A B C D
8. A B C D
9. A B C D
10. A B C D
11. A B C D
12. A B C D
13. A B C D
14. A B C D
15. A B C D
16. A B C D
17. A B C D
18. A B C D
19. A B C D
20. A B C D
21. A B C D
22. A B C D
23. A B C D
24. A B C D
25. A B C D
26. A B C D
27. A B C D
28. A B C D
29. A B C D
30. A B C D
31. A B C D
32. A B C D
33. A B C D
34. A B C D
35. A B C D
36. A B C D
37. A B C D
38. A B C D
39. A B C D
40. A B C D
41. A B C D
42. A B C D
43. A B C D
44. A B C D
45. A B C D

NOTES